Holt Science & Technology Short Courses

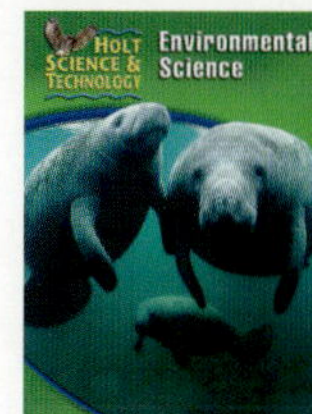

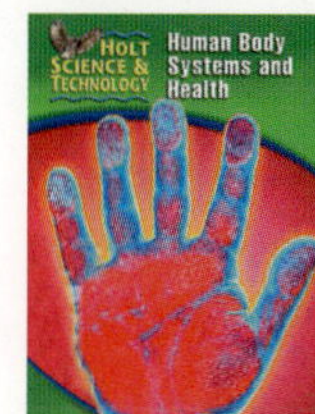

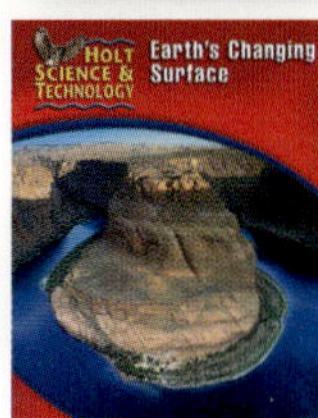

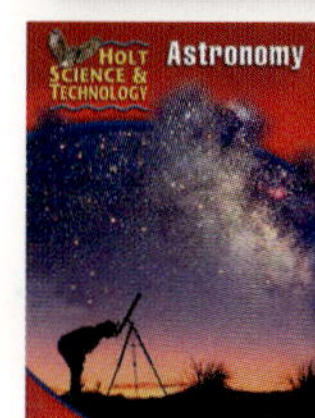

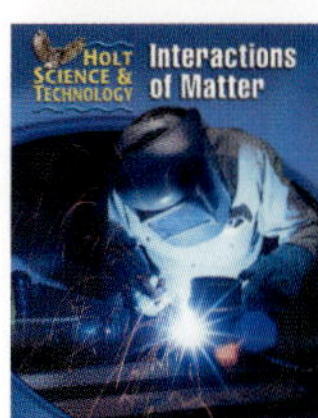

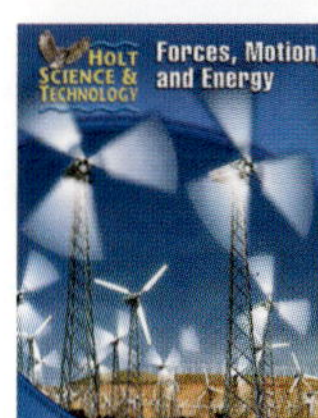

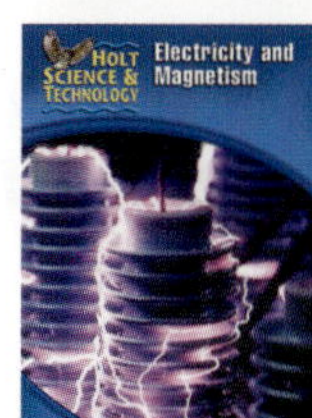

Teacher Edition WALK-THROUGH

Student Edition CONTENTS IN BRIEF

HOLT, RINEHART AND WINSTON
A Harcourt Education Company
Orlando • **Austin** • New York • San Diego • Toronto • London

Designed to meet the needs of all students

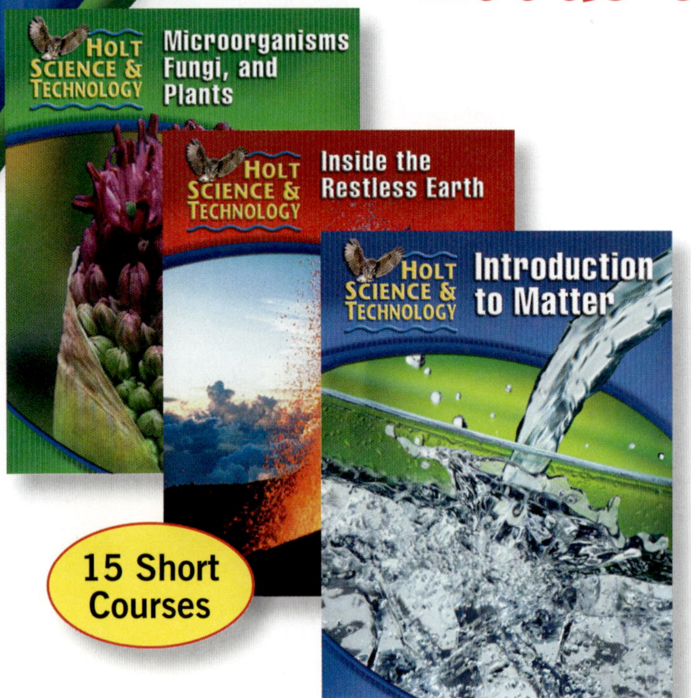

Holt Science & Technology: Short Course Series allows you to match your curriculum by choosing from 15 books covering life, earth, and physical sciences. The program reflects current curriculum developments and includes the strongest skills-development strand of any middle school science series. Students of all abilities will develop skills that they can use both in science as well as in other courses.

STUDENTS OF ALL ABILITIES RECEIVE THE READING HELP AND TAILORED INSTRUCTION THEY NEED.

- The *Student Edition* is accessible with a clean, easy-to-follow design and highlighted vocabulary words.
- Inclusion strategies and different learning styles help support all learners.
- Comprehensive **Section** and **Chapter Reviews** and **Standardized Test Preparation** allow students to practice their test-taking skills.
- **Reading Comprehension Guide** and **Guided Reading Audio CDs** help students better understand the content.

CROSS-DISCIPLINARY CONNECTIONS LET STUDENTS SEE HOW SCIENCE RELATES TO OTHER DISCIPLINES.

- **Mathematics, reading,** and **writing skills** are integrated throughout the program.
- Cross-discipline **Connection To** features show students how science relates to language arts, social studies, and other sciences.

A FLEXIBLE LABORATORY PROGRAM HELPS STUDENTS BUILD IMPORTANT INQUIRY AND CRITICAL-THINKING SKILLS.

- The laboratory program includes labs in each chapter, labs in the **LabBook** at the end of the text, six different lab books, and **Video Labs.**
- All labs are teacher-tested and rated by difficulty in the *Teacher Edition,* so you can be sure the labs will be appropriate for your students.
- A variety of labs, from **Inquiry Labs** to **Skills Practice Labs,** helps you meet the needs of your curriculum and work within the time constraints of your teaching schedule.

INTEGRATED TECHNOLOGY AND ONLINE RESOURCES EXPAND LEARNING BEYOND CLASSROOM WALLS.

- An **Enhanced Online Edition** or **CD-ROM Version** of the student text lightens your students' load.
- **SciLinks,** a Web service developed and maintained by the National Science Teachers Association (NSTA), contains current prescreened links directly related to the textbook.
- **Brain Food Video Quizzes** on videotape and DVD are game-show style quizzes that assess students' progress and motivate them to study.
- The **One-stop Planner® CD-ROM** with **Exam View® Test Generator** contains all of the resources you need including an *Interactive Teacher Edition,* worksheets, customizable lesson plans, **Holt Calendar Planner,** a powerful test generator, **Lab Materials QuickList Software,** and more.
- Spanish Resources include **Guided Reading Audio CD** in Spanish.

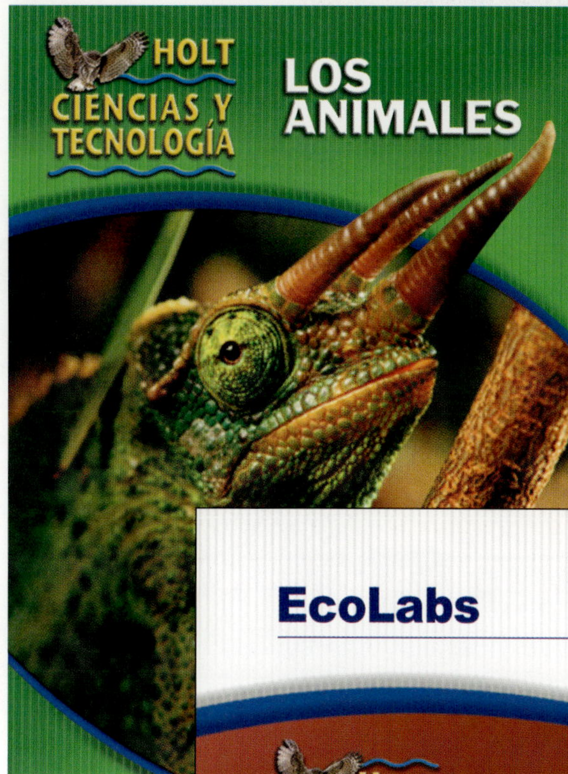

Life Science

	A MICROORGANISMS, FUNGI, AND PLANTS	B ANIMALS
CHAPTER 1	**It's Alive!! Or, Is It?** • Characteristics of living things • Homeostasis • Heredity and DNA • Producers, consumers, and decomposers • Biomolecules	**Animals and Behavior** • Characteristics of animals • Classification of animals • Animal behavior • Hibernation and estivation • The biological clock • Animal communication • Living in groups
CHAPTER 2	**Bacteria and Viruses** • Binary fission • Characteristics of bacteria • Nitrogen-fixing bacteria • Antibiotics • Pathogenic bacteria • Characteristics of viruses • Lytic cycle	**Invertebrates** • General characteristics of invertebrates • Types of symmetry • Characteristics of sponges, cnidarians, arthropods, and echinoderms • Flatworms versus roundworms • Types of circulatory systems
CHAPTER 3	**Protists and Fungi** • Characteristics of protists • Types of algae • Types of protozoa • Protist reproduction • Characteristics of fungi and lichens	**Fishes, Amphibians, and Reptiles** • Characteristics of vertebrates • Structure and kinds of fishes • Development of lungs • Structure and kinds of amphibians and reptiles • Function of the amniotic egg
CHAPTER 4	**Introduction to Plants** • Characteristics of plants and seeds • Reproduction and classification • Angiosperms versus gymnosperms • Monocots versus dicots • Structure and functions of roots, stems, leaves, and flowers	**Birds and Mammals** • Structure and kinds of birds • Types of feathers • Adaptations for flight • Structure and kinds of mammals • Function of the placenta
CHAPTER 5	**Plant Processes** • Pollination and fertilization • Dormancy • Photosynthesis • Plant tropisms • Seasonal responses of plants	
CHAPTER 6		
CHAPTER 7		

PROGRAM SCOPE AND SEQUENCE

Selecting the right books for your course is easy. Just review the topics presented in each book to determine the best match to your district curriculum.

C CELLS, HEREDITY, & CLASSIFICATION

Cells: The Basic Units of Life
- Cells, tissues, and organs
- Populations, communities, and ecosystems
- Cell theory
- Surface-to-volume ratio
- Prokaryotic versus eukaryotic cells
- Cell organelles

The Cell in Action
- Diffusion and osmosis
- Passive versus active transport
- Endocytosis versus exocytosis
- Photosynthesis
- Cellular respiration and fermentation
- Cell cycle

Heredity
- Dominant versus recessive traits
- Genes and alleles
- Genotype, phenotype, the Punnett square and probability
- Meiosis
- Determination of sex

Genes and Gene Technology
- Structure of DNA
- Protein synthesis
- Mutations
- Heredity disorders and genetic counseling

The Evolution of Living Things
- Adaptations and species
- Evidence for evolution
- Darwin's work and natural selection
- Formation of new species

The History of Life on Earth
- Geologic time scale and extinctions
- Plate tectonics
- Human evolution

Classification
- Levels of classification
- Cladistic diagrams
- Dichotomous keys
- Characteristics of the six kingdoms

D HUMAN BODY SYSTEMS & HEALTH

Body Organization and Structure
- Homeostasis
- Types of tissue
- Organ systems
- Structure and function of the skeletal system, muscular system, and integumentary system

Circulation and Respiration
- Structure and function of the cardiovascular system, lymphatic system, and respiratory system
- Respiratory disorders

The Digestive and Urinary Systems
- Structure and function of the digestive system
- Structure and function of the urinary system

Communication and Control
- Structure and function of the nervous system and endocrine system
- The senses
- Structure and function of the eye and ear

Reproduction and Development
- Asexual versus sexual reproduction
- Internal versus external fertilization
- Structure and function of the human male and female reproductive systems
- Fertilization, placental development, and embryo growth
- Stages of human life

Body Defenses and Disease
- Types of diseases
- Vaccines and immunity
- Structure and function of the immune system
- Autoimmune diseases, cancer, and AIDS

Staying Healthy
- Nutrition and reading food labels
- Alcohol and drug effects on the body
- Hygiene, exercise, and first aid

E ENVIRONMENTAL SCIENCE

Interactions of Living Things
- Biotic versus abiotic parts of the environment
- Producers, consumers, and decomposers
- Food chains and food webs
- Factors limiting population growth
- Predator-prey relationships
- Symbiosis and coevolution

Cycles in Nature
- Water cycle
- Carbon cycle
- Nitrogen cycle
- Ecological succession

The Earth's Ecosystems
- Kinds of land and water biomes
- Marine ecosystems
- Freshwater ecosystems

Environmental Problems and Solutions
- Types of pollutants
- Types of resources
- Conservation practices
- Species protection

Energy Resources
- Types of resources
- Energy resources and pollution
- Alternative energy resources

Earth Science

	F INSIDE THE RESTLESS EARTH	G EARTH'S CHANGING SURFACE
CHAPTER 1	**Minerals of the Earth's Crust** • Mineral composition and structure • Types of minerals • Mineral identification • Mineral formation and mining	**Maps as Models of the Earth** • Structure of a map • Cardinal directions • Latitude, longitude, and the equator • Magnetic declination and true north • Types of projections • Aerial photographs • Remote sensing • Topographic maps
CHAPTER 2	**Rocks: Mineral Mixtures** • Rock cycle and types of rocks • Rock classification • Characteristics of igneous, sedimentary, and metamorphic rocks	**Weathering and Soil Formation** • Types of weathering • Factors affecting the rate of weathering • Composition of soil • Soil conservation and erosion prevention
CHAPTER 3	**The Rock and Fossil Record** • Uniformitarianism versus catastrophism • Superposition • The geologic column and unconformities • Absolute dating and radiometric dating • Characteristics and types of fossils • Geologic time scale	**Agents of Erosion and Deposition** • Shoreline erosion and deposition • Wind erosion and deposition • Erosion and deposition by ice • Gravity's effect on erosion and deposition
CHAPTER 4	**Plate Tectonics** • Structure of the Earth • Continental drifts and sea floor spreading • Plate tectonics theory • Types of boundaries • Types of crust deformities	
CHAPTER 5	**Earthquakes** • Seismology • Features of earthquakes • P and S waves • Gap hypothesis • Earthquake safety	
CHAPTER 6	**Volcanoes** • Types of volcanoes and eruptions • Types of lava and pyroclastic material • Craters versus calderas • Sites and conditions for volcano formation • Predicting eruptions	

H WATER ON EARTH

The Flow of Fresh Water
- Water cycle
- River systems
- Stream erosion
- Life cycle of rivers
- Deposition
- Aquifers, springs, and wells
- Ground water
- Water treatment and pollution

Exploring the Oceans
- Properties and characteristics of the oceans
- Features of the ocean floor
- Ocean ecology
- Ocean resources and pollution

The Movement of Ocean Water
- Types of currents
- Characteristics of waves
- Types of ocean waves
- Tides

I WEATHER AND CLIMATE

The Atmosphere
- Structure of the atmosphere
- Air pressure
- Radiation, convection, and conduction
- Greenhouse effect and global warming
- Characteristics of winds
- Types of winds
- Air pollution

Understanding Weather
- Water cycle
- Humidity
- Types of clouds
- Types of precipitation
- Air masses and fronts
- Storms, tornadoes, and hurricanes
- Weather forecasting
- Weather maps

Climate
- Weather versus climate
- Seasons and latitude
- Prevailing winds
- Earth's biomes
- Earth's climate zones
- Ice ages
- Global warming
- Greenhouse effect

J ASTRONOMY

Studying Space
- Astronomy
- Keeping time
- Types of telescope
- Radioastronomy
- Mapping the stars
- Scales of the universe

Stars, Galaxies, and the Universe
- Composition of stars
- Classification of stars
- Star brightness, distance, and motions
- H-R diagram
- Life cycle of stars
- Types of galaxies
- Theories on the formation of the universe

Formation of the Solar System
- Birth of the solar system
- Structure of the sun
- Fusion
- Earth's structure and atmosphere
- Planetary motion
- Newton's Law of Universal Gravitation

A Family of Planets
- Properties and characteristics of the planets
- Properties and characteristics of moons
- Comets, asteroids, and meteoroids

Exploring Space
- Rocketry and artificial satellites
- Types of Earth orbit
- Space probes and space exploration

Physical Science

	K INTRODUCTION TO MATTER	L INTERACTIONS OF MATTER
CHAPTER 1	**The Properties of Matter** • Definition of matter • Mass and weight • Physical and chemical properties • Physical and chemical change • Density	**Chemical Bonding** • Types of chemical bonds • Valence electrons • Ions versus molecules • Crystal lattice
CHAPTER 2	**States of Matter** • States of matter and their properties • Boyle's and Charles's laws • Changes of state	**Chemical Reactions** • Writing chemical formulas and equations • Law of conservation of mass • Types of reactions • Endothermic versus exothermic reactions • Law of conservation of energy • Activation energy • Catalysts and inhibitors
CHAPTER 3	**Elements, Compounds, and Mixtures** • Elements and compounds • Metals, nonmetals, and metalloids (semiconductors) • Properties of mixtures • Properties of solutions, suspensions, and colloids	**Chemical Compounds** • Ionic versus covalent compounds • Acids, bases, and salts • pH • Organic compounds • Biomolecules
CHAPTER 4	**Introduction to Atoms** • Atomic theory • Atomic model and structure • Isotopes • Atomic mass and mass number	**Atomic Energy** • Properties of radioactive substances • Types of decay • Half-life • Fission, fusion, and chain reactions
CHAPTER 5	**The Periodic Table** • Structure of the periodic table • Periodic law • Properties of alkali metals, alkaline-earth metals, halogens, and noble gases	
CHAPTER 6		

M FORCES, MOTION, AND ENERGY

Matter in Motion
- Speed, velocity, and acceleration
- Measuring force
- Friction
- Mass versus weight

Forces in Motion
- Terminal velocity and free fall
- Projectile motion
- Inertia
- Momentum

Forces in Fluids
- Properties in fluids
- Atmospheric pressure
- Density
- Pascal's principle
- Buoyant force
- Archimedes' principle
- Bernoulli's principle

Work and Machines
- Measuring work
- Measuring power
- Types of machines
- Mechanical advantage
- Mechanical efficiency

Energy and Energy Resources
- Forms of energy
- Energy conversions
- Law of conservation of energy
- Energy resources

Heat and Heat Technology
- Heat versus temperature
- Thermal expansion
- Absolute zero
- Conduction, convection, radiation
- Conductors versus insulators
- Specific heat capacity
- Changes of state
- Heat engines
- Thermal pollution

N ELECTRICITY AND MAGNETISM

Introduction to Electricity
- Law of electric charges
- Conduction versus induction
- Static electricity
- Potential difference
- Cells, batteries, and photocells
- Thermocouples
- Voltage, current, and resistance
- Electric power
- Types of circuits

Electromagnetism
- Properties of magnets
- Magnetic force
- Electromagnetism
- Solenoids and electric motors
- Electromagnetic induction
- Generators and transformers

Electronic Technology
- Properties of semiconductors
- Integrated circuits
- Diodes and transistors
- Analog versus digital signals
- Microprocessors
- Features of computers

O SOUND AND LIGHT

The Energy of Waves
- Properties of waves
- Types of waves
- Reflection and refraction
- Diffraction and interference
- Standing waves and resonance

The Nature of Sound
- Properties of sound waves
- Structure of the human ear
- Pitch and the Doppler effect
- Infrasonic versus ultrasonic sound
- Sound reflection and echolocation
- Sound barrier
- Interference, resonance, diffraction, and standing waves
- Sound quality of instruments

The Nature of Light
- Electromagnetic waves
- Electromagnetic spectrum
- Law of reflection
- Absorption and scattering
- Reflection and refraction
- Diffraction and interference

Light and Our World
- Luminosity
- Types of lighting
- Types of mirrors and lenses
- Focal point
- Structure of the human eye
- Lasers and holograms

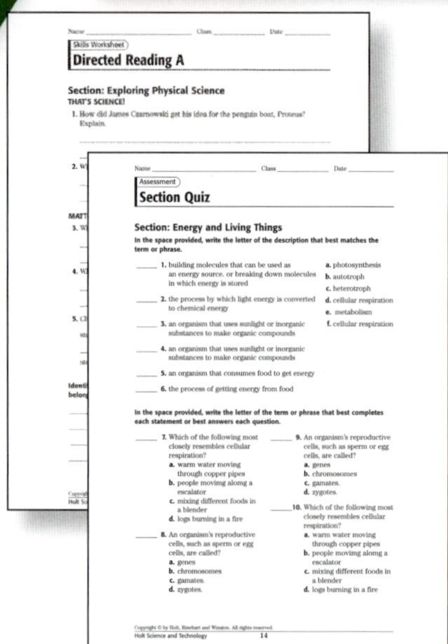

Program resources make teaching and learning easier.

CHAPTER RESOURCES

A ***Chapter Resources book*** accompanies each of the 15 *Short Courses.* Here you'll find everything you need to make sure your students are getting the most out of learning science—all in one book.

Skills Worksheets

- Directed Reading A: Basic
- Directed Reading B: Special Needs
- Vocabulary and Chapter Summary
- Section Reviews
- Chapter Reviews
- Reinforcement
- Critical Thinking

Labs & Activities

- Datasheets for Chapter Labs
- Datasheets for Quick Labs
- Datasheets for LabBook
- Vocabulary Activity
- SciLinks® Activity

Assessments

- Section Quizzes
- Chapter Tests A: General
- Chapter Tests B: Advanced
- Chapter Tests C: Special Needs
- Performance-Based Assessments
- Standardized Test Preparation

Teacher Resources

- Lab Notes and Answers
- Teacher Notes for Performance-Based Assessment
- Answer Keys
- Lesson Plans
- Test Item Listing for ExamView® Test Generator
- Full-color **Teaching Transparencies,** plus section **Bellringers, Concept Mapping,** and **Chapter Starter Transparencies.**

SPANISH RESOURCES

Spanish materials are available for each *Short Course:*

- *Student Edition*
- ***Spanish Resources*** booklet contains worksheets and assessments translated into Spanish with an English **Answer Key.**
- **Guided Reading Audio CD Program**

ONLINE RESOURCES

- *Enhanced Online Editions* engage students and assist teachers with a host of interactive features that are available anytime and anywhere you can connect to the Internet.
- **CNNStudentNews.com** provides award-winning news and information for both teachers and students.
- **SciLinks**—a Web service developed and maintained by the National Science Teachers Association—links you and your students to up-to-date online resources directly related to chapter topics.
- **go.hrw.com** links you and your students to online chapter activities and resources.
- **Current Science** articles relate to students' lives.

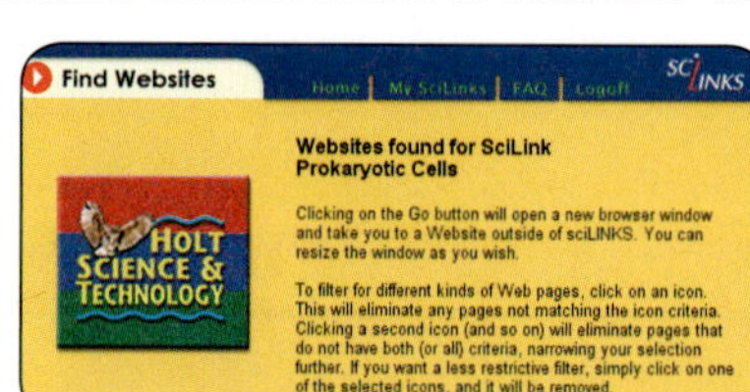

ADDITIONAL LAB AND SKILLS RESOURCES

- ***Calculator-Based Labs*** incorporates scientific instruments, offering students insight into modern scientific investigation.
- ***EcoLabs & Field Activities*** develops awareness of the natural world.
- ***Holt Science Skills Workshop: Reading in the Content Area*** contains exercises that target reading skills key.
- ***Inquiry Labs*** taps students' natural curiosity and creativity with a focus on the process of discovery.
- ***Labs You Can Eat*** safely incorporates edible items into the classroom.
- ***Long-Term Projects & Research Ideas*** extends and enriches lessons.
- ***Math Skills for Science*** provides additional explanations, examples, and math problems so students can develop their skills.
- ***Science Skills Worksheets*** helps your students hone important learning skills.
- ***Whiz-Bang Demonstrations*** gets your students' attention at the beginning of a lesson.

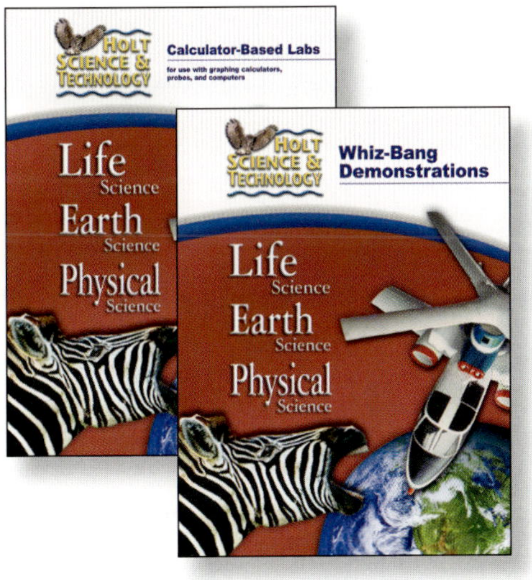

ADDITIONAL RESOURCES

- ***Assessment Checklists & Rubrics*** gives you guidelines for evaluating students' progress.
- ***Holt Anthology of Science Fiction*** sparks your students' imaginations with thought-provoking stories.
- ***Holt Science Posters*** visually reinforces scientific concepts and themes with seven colorful posters including **The Periodic Table of the Elements.**

- ***Professional Reference for Teachers*** contains professional articles that discuss a variety of topics, such as classroom management.
- ***Program Introduction Resource File*** explains the program and its features and provides several additional references, including lab safety, scoring rubrics, and more.
- ***Science Fair Guide*** gives teachers, students, and parents tips for planning and assisting in a science fair.
- ***Science Puzzlers, Twisters & Teasers*** activities challenge students to think about science concepts in different ways.

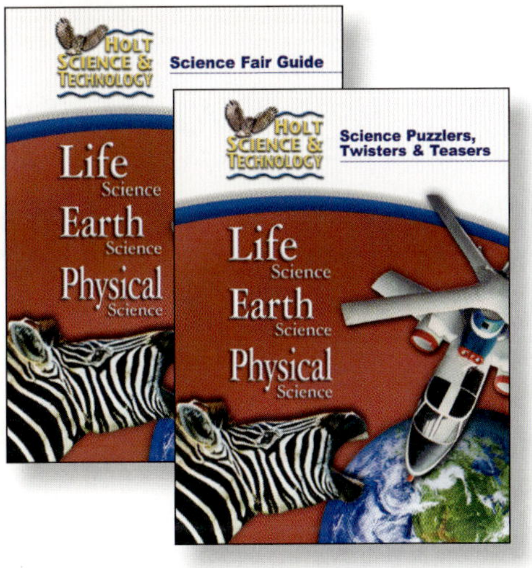

TECHNOLOGY RESOURCES

- ***CNN Presents Science in the News: Video Library*** helps students see the impact of science on their everyday lives with actual news video clips.
  - Multicultural Connections
  - Science, Technology & Society
  - Scientists in Action
  - Eye on the Environment
- ***Guided Reading Audio CD Program***, available in English and Spanish, provides students with a direct read of each section.
- ***HRW Earth Science Videotape*** takes your students on a geology "field trip" with full-motion video.
- ***Interactive Explorations CD-ROM Program*** develops students' inquiry and decision-making skills as they investigate science phenomena in a virtual lab setting.

- ***One-Stop Planner CD-ROM®*** organizes everything you need on one disc, including printable worksheets, customizable lesson plans, a powerful test generator, **PowerPoint® LectureNotes, Lab Materials QuickList Software, Holt Calendar Planner, Interactive Teacher Edition,** and more.
- ***Science Tutor CD-ROMs*** help students practice what they learn with immediate feedback.
- ***Lab Videos*** make it easier to integrate more experiments into your lessons without the preparation time and costs. Available on DVD and VHS.
- **Brain Food Video Quizzes** are game-show style quizzes that assess students' progress. Available on DVD and VHS.
- ***Visual Concepts CD-ROMs*** include graphics, animations, and movie clips that demonstrate key chapter concepts.

Science and Math Worksheets

The ***Holt Science & Technology*** program helps you meet the needs of a wide variety of students, regardless of their skill level. The following pages provide examples of the worksheets available to improve your students' science and math skills whether they already have a strong science and math background or are weak in these areas. Samples of assessment checklists and rubrics are also provided.

In addition to the skills worksheets represented here, ***Holt Science & Technology*** provides a variety of worksheets that are correlated directly with each chapter of the program. Representations of these worksheets are found at the beginning of each chapter in this *Teacher Edition*.

Many worksheets are also available on the Holt Web site. The address is **go.hrw.com.**

Science Skills Worksheets: Thinking Skills

BEING FLEXIBLE

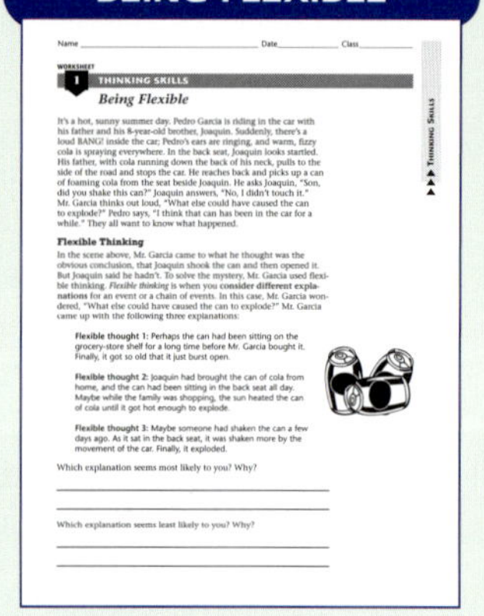
Worksheet 1 — Thinking Skills — *Being Flexible*

USING YOUR SENSES

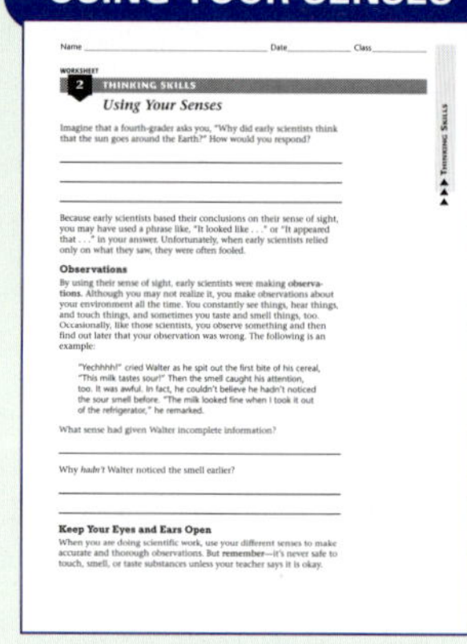
Worksheet 2 — Thinking Skills — *Using Your Senses*

THINKING OBJECTIVELY

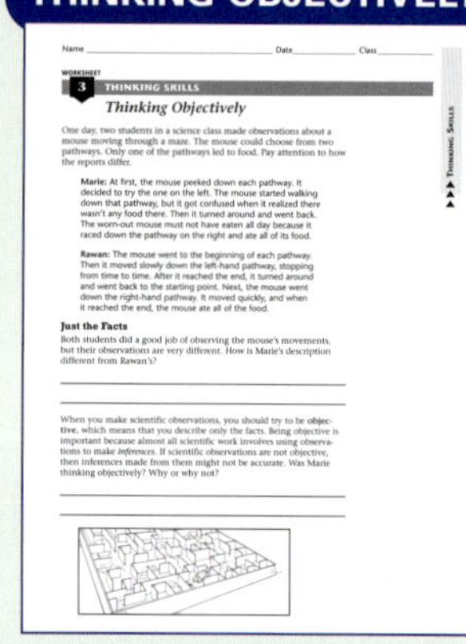
Worksheet 3 — Thinking Skills — *Thinking Objectively*

UNDERSTANDING BIAS

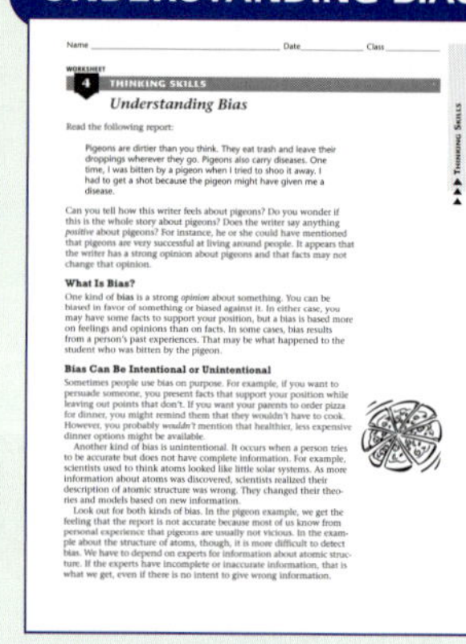
Worksheet 4 — Thinking Skills — *Understanding Bias*

USING LOGIC

Worksheet 5 — Thinking Skills — *Using Logic*

BOOSTING YOUR MEMORY

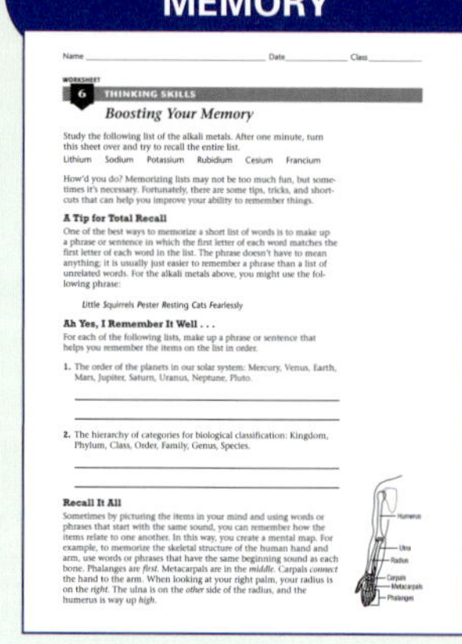
Worksheet 6 — Thinking Skills — *Boosting Your Memory*

IMPROVING YOUR STUDY HABITS

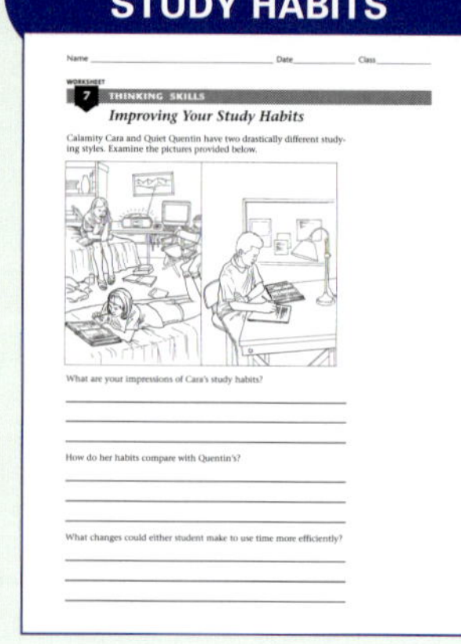
Worksheet 7 — Thinking Skills — *Improving Your Study Habits*

READING A SCIENCE TEXTBOOK

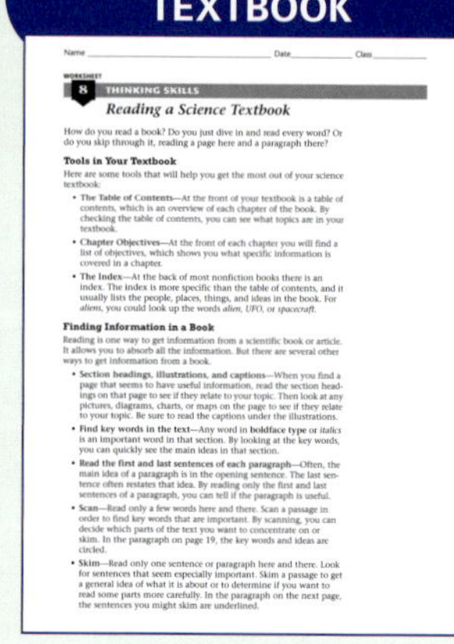
Worksheet 8 — Thinking Skills — *Reading a Science Textbook*

Science Skills Worksheets: Experimenting Skills

SAFETY RULES!

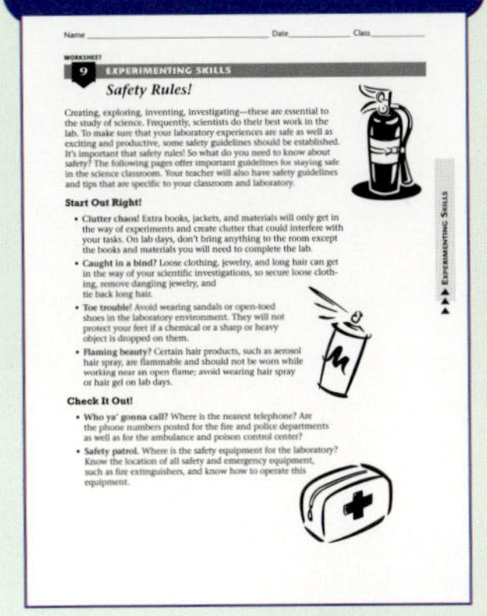

DOING A LAB WRITE-UP

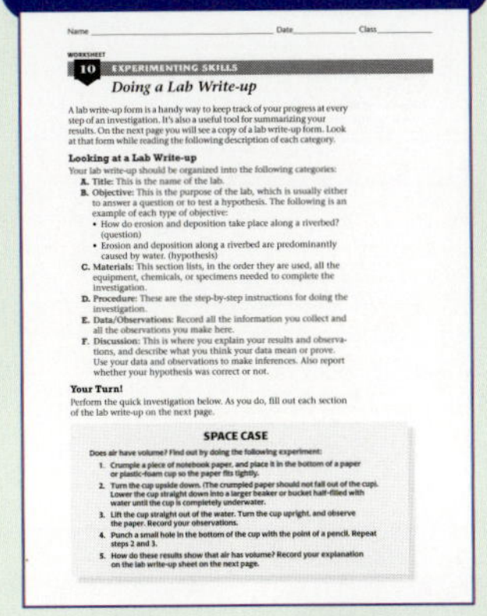

UNDERSTANDING VARIABLES

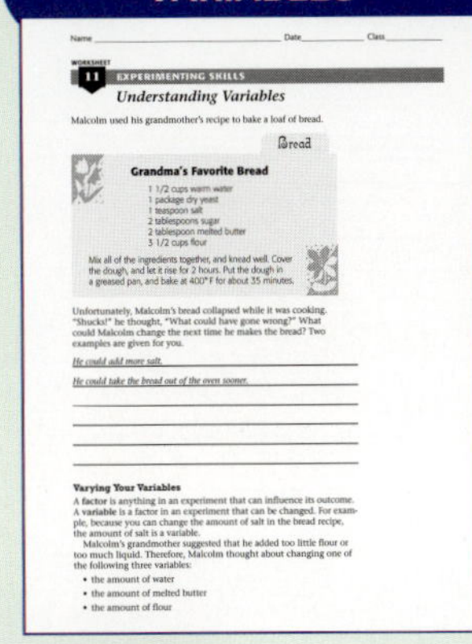

WORKING WITH HYPOTHESES

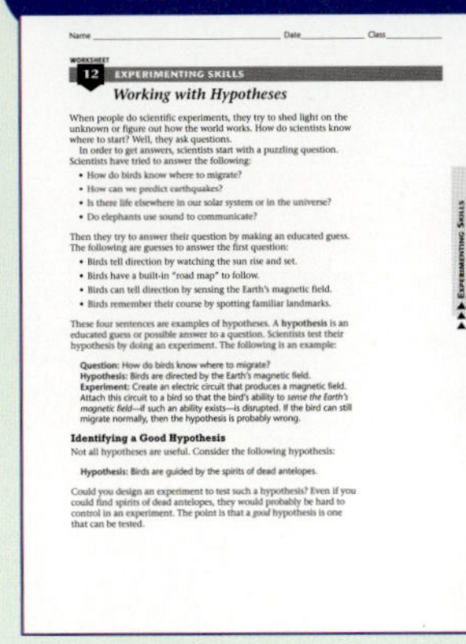

DESIGNING AN EXPERIMENT

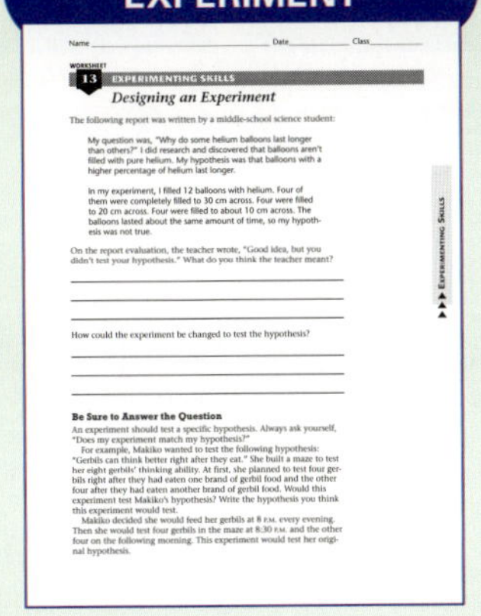

USING THE INTERNATIONAL SYSTEM OF UNITS (SI)

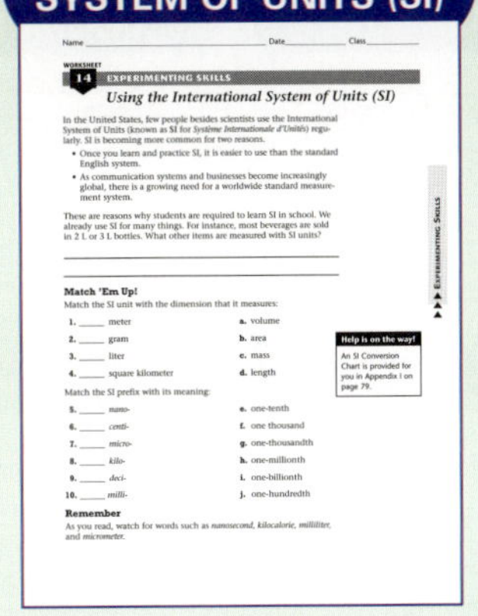

MEASURING

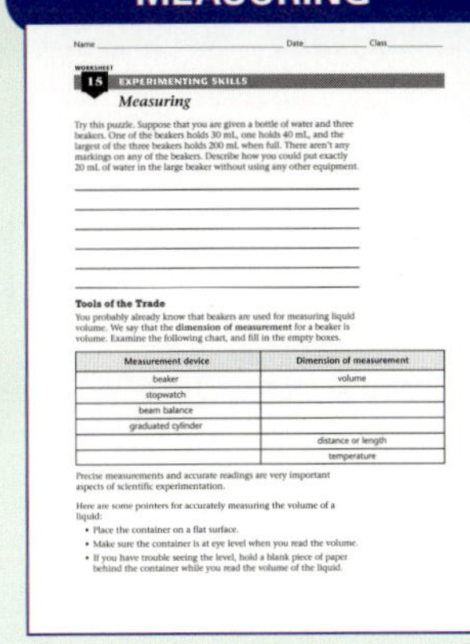

Science Skills Worksheets: Researching Skills

CHOOSING YOUR TOPIC

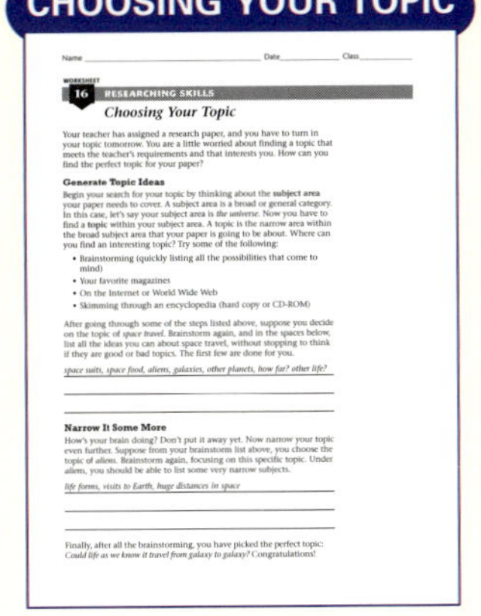

ORGANIZING YOUR RESEARCH

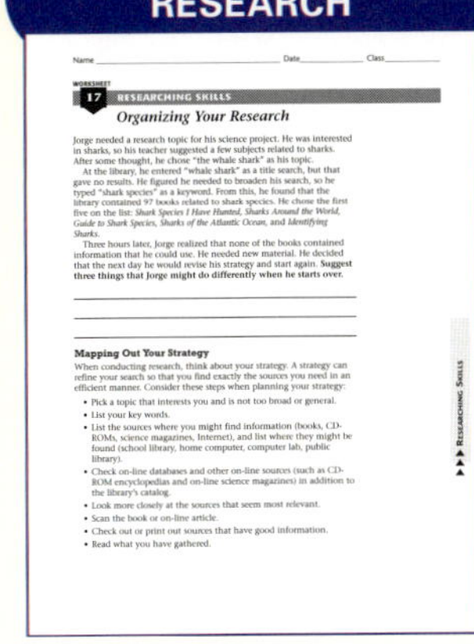

FINDING USEFUL SOURCES

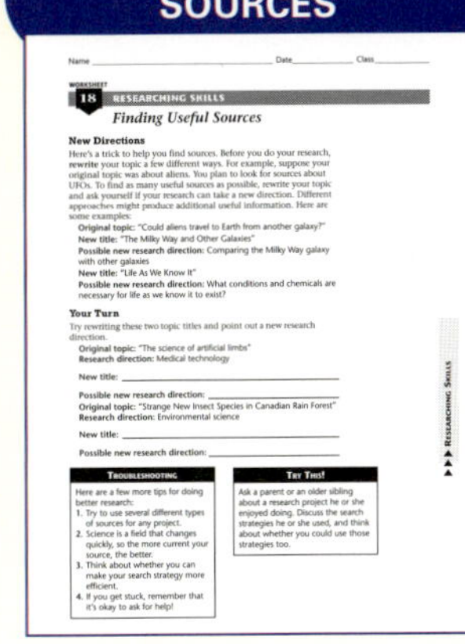

RESEARCHING ON THE WEB

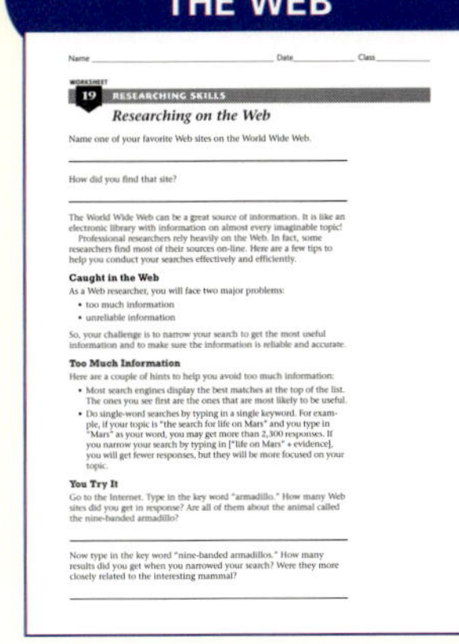

Science Skills Worksheets: Researching Skills (continued)

IDENTIFYING BIAS

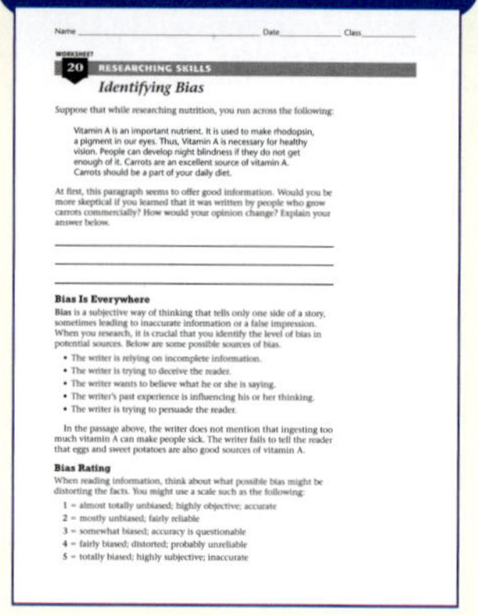

TAKING NOTES

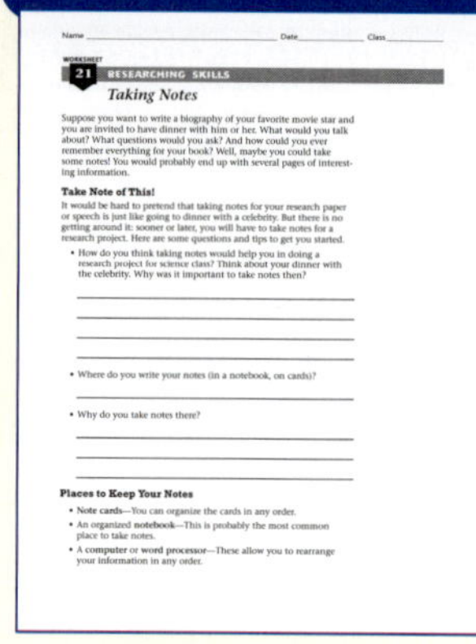

Science Skills Worksheets: Communicating Skills

SCIENCE WRITING

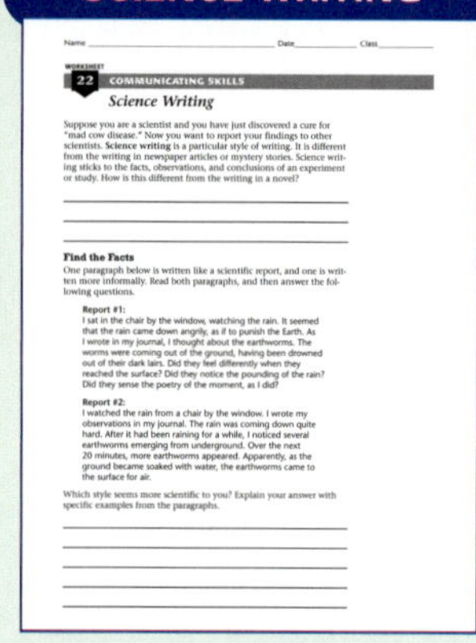

SCIENCE DRAWING

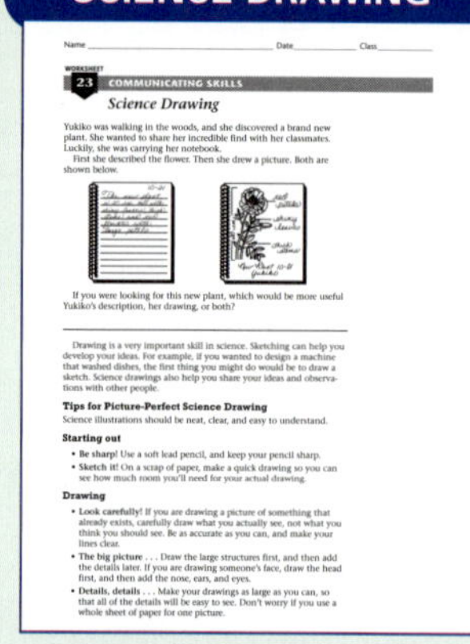

USING MODELS TO COMMUNICATE

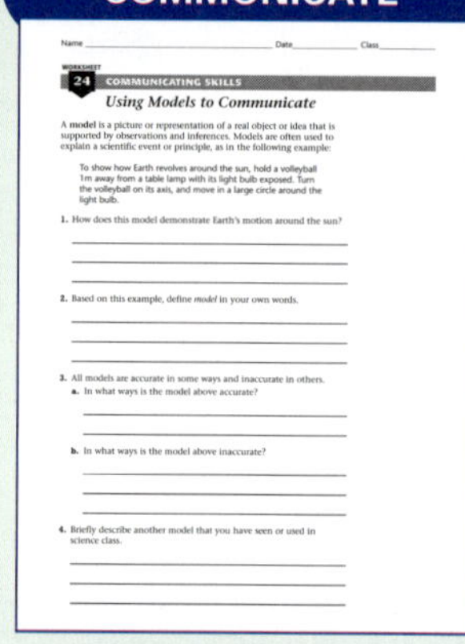

INTRODUCTION TO GRAPHS

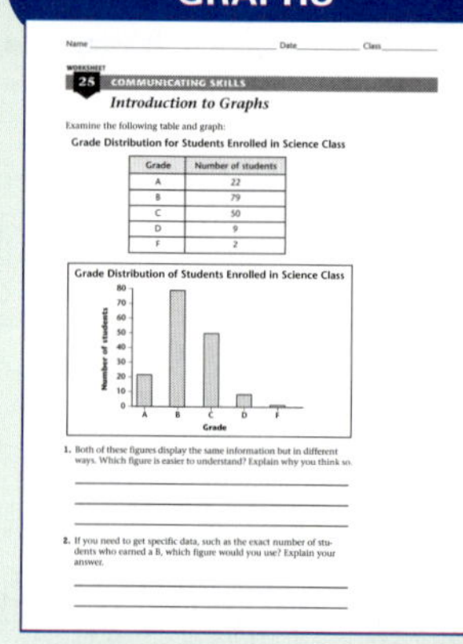

GRASPING GRAPHING

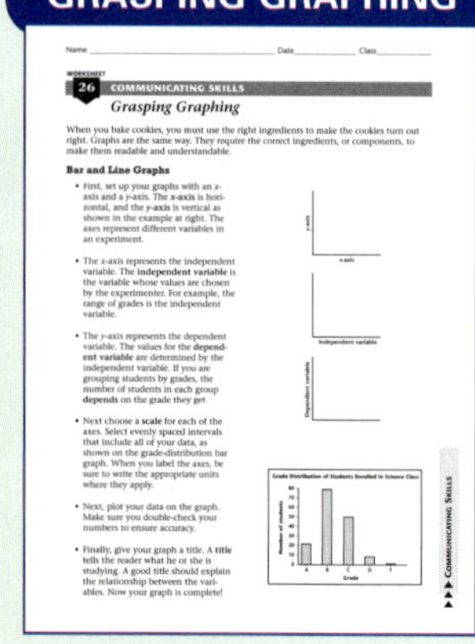

INTERPRETING YOUR DATA

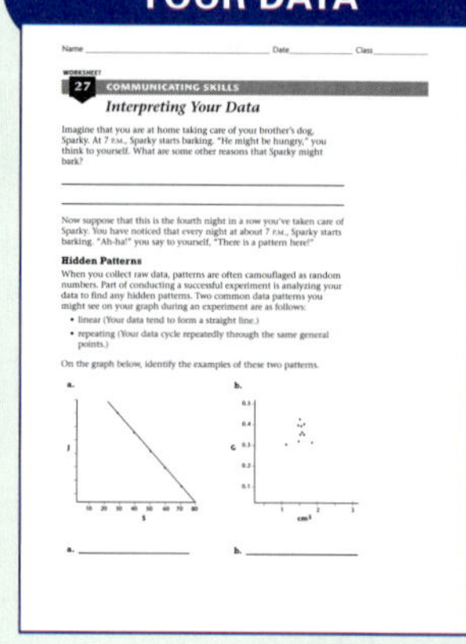

RECOGNIZING BIAS IN GRAPHS

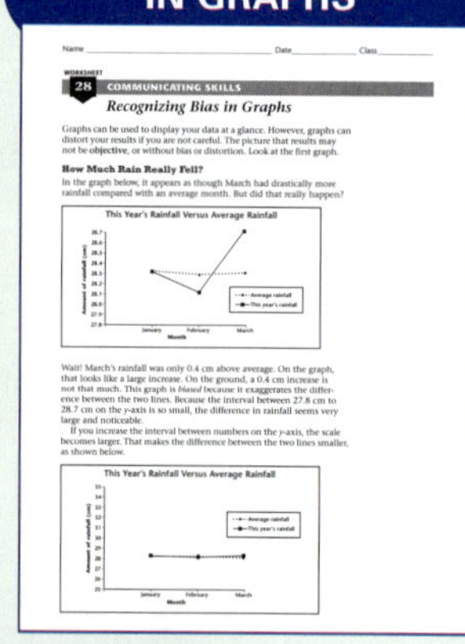

MAKING DATA MEANINGFUL

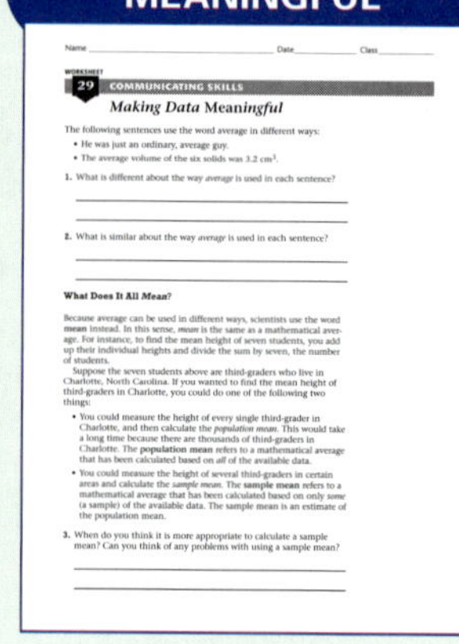

HINTS FOR ORAL PRESENTATIONS

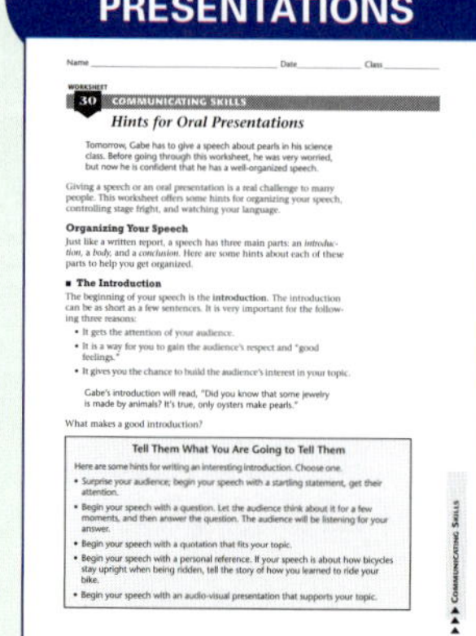

Math Skills for Science

ADDITION AND SUBTRACTION

Worksheet 1 — Math Skills: *Addition Review*

Worksheet 2 — Math Skills: *Subtraction Review*

MULTIPLICATION

Worksheet 3 — Math Skills: *Multiplying Whole Numbers*

Worksheet 4 — Math Skills: *A Shortcut for Multiplying Large Numbers*

DIVISION

Worksheet 5 — Math Skills: *Dividing Whole Numbers with Long Division*

Worksheet 6 — Math Skills: *Checking Division with Multiplication*

AVERAGES

Worksheet 7 — Math Skills: *What Is an Average?*

Worksheet 8 — Math Skills: *Average, Mode, and Median*

POSITIVE AND NEGATIVE NUMBERS

Worksheet 9 — Math Skills: *Comparing Integers on a Number Line*

Worksheet 10 — Math Skills: *Arithmetic with Positive and Negative Numbers*

FRACTIONS

Worksheet 11 — Math Skills: *What Is a Fraction?*

Worksheet 13 — Math Skills: *Improper Fractions and Mixed Numbers*

Worksheet 15 — Math Skills: *Multiplying and Dividing Fractions*

Worksheet 12 — Math Skills: *Reducing Fractions to Lowest Terms*

Worksheet 14 — Math Skills: *Adding and Subtracting Fractions*

Math Skills for Science (continued)

RATIOS AND PROPORTIONS

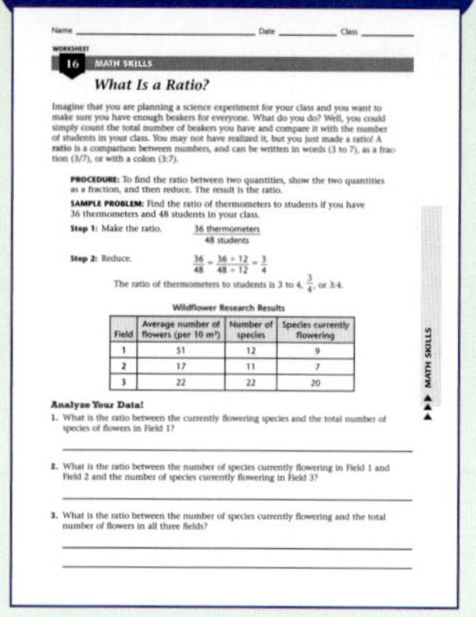
Worksheet 16 — Math Skills — What Is a Ratio?

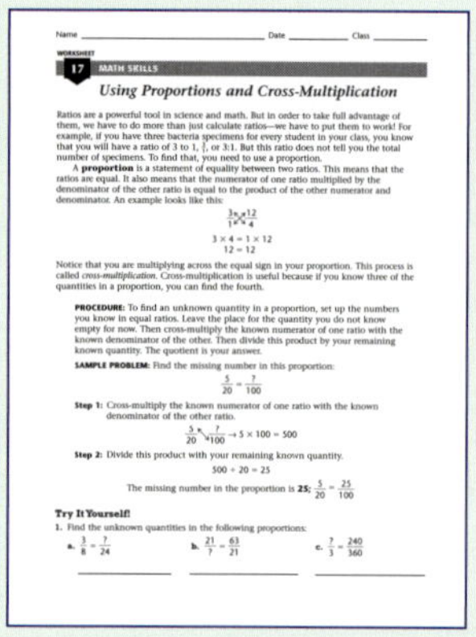
Worksheet 17 — Math Skills — Using Proportions and Cross-Multiplication

DECIMALS

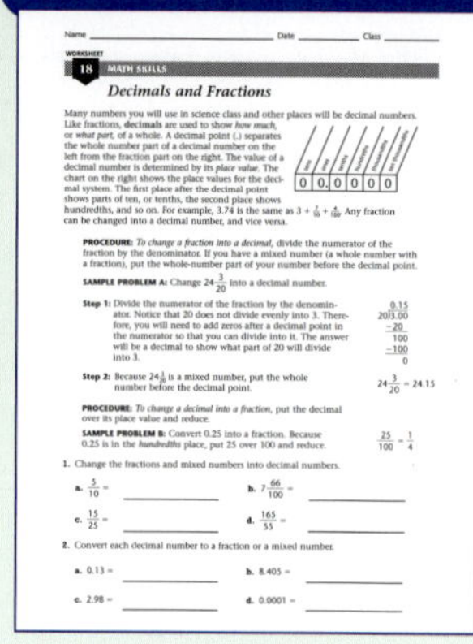
Worksheet 18 — Math Skills — Decimals and Fractions

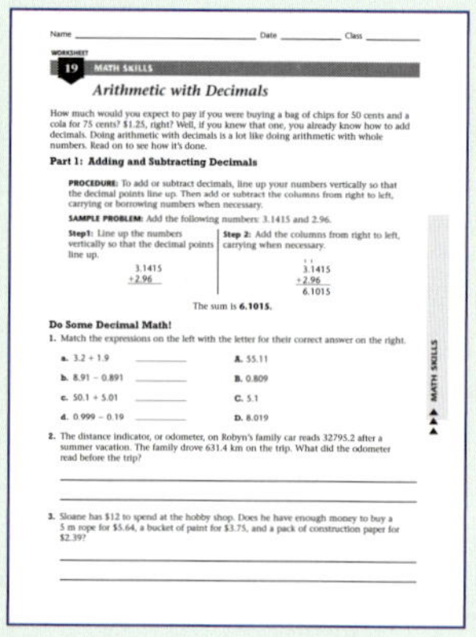
Worksheet 19 — Math Skills — Arithmetic with Decimals

PERCENTAGES

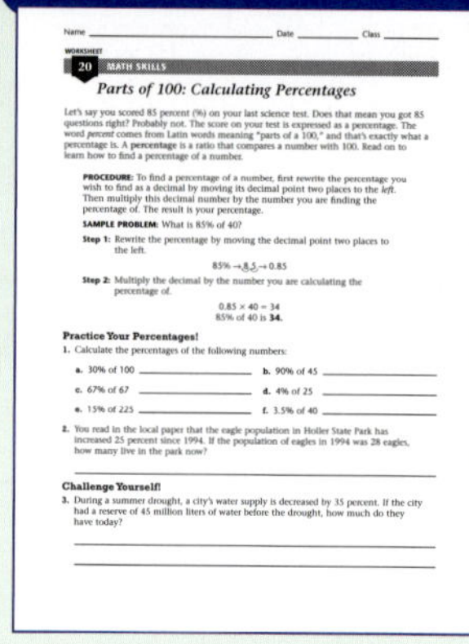
Worksheet 20 — Math Skills — Parts of 100: Calculating Percentages

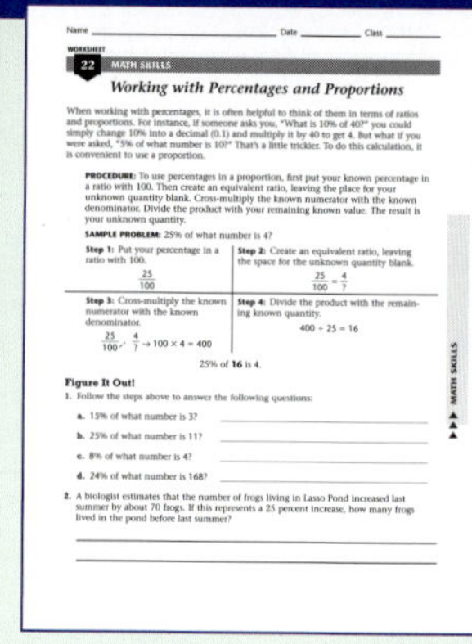
Worksheet 22 — Math Skills — Working with Percentages and Proportions

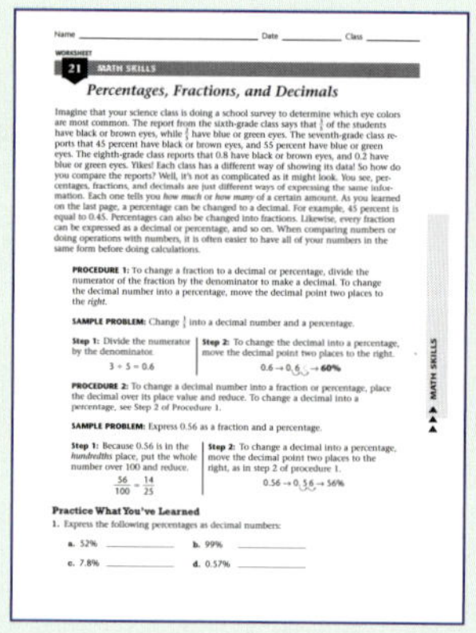
Worksheet 21 — Math Skills — Percentages, Fractions, and Decimals

POWERS OF 10

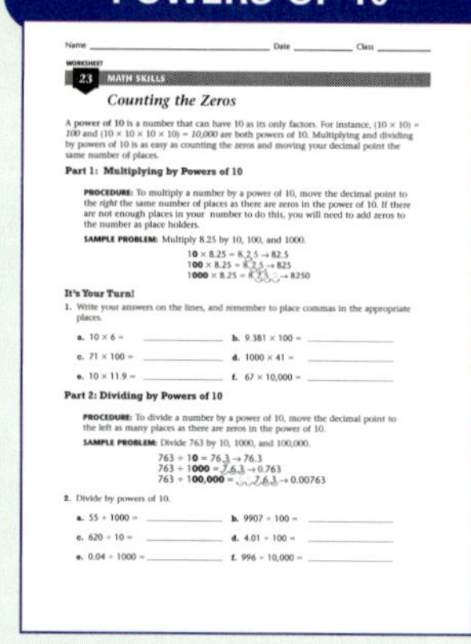
Worksheet 23 — Math Skills — Counting the Zeros

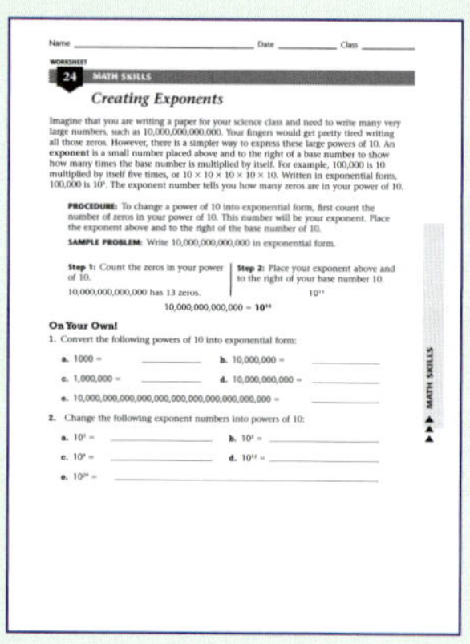
Worksheet 24 — Math Skills — Creating Exponents

SCIENTIFIC NOTATION

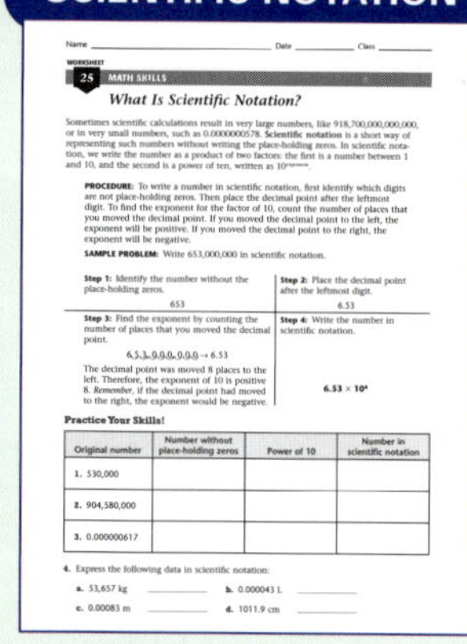
Worksheet 25 — Math Skills — What Is Scientific Notation?

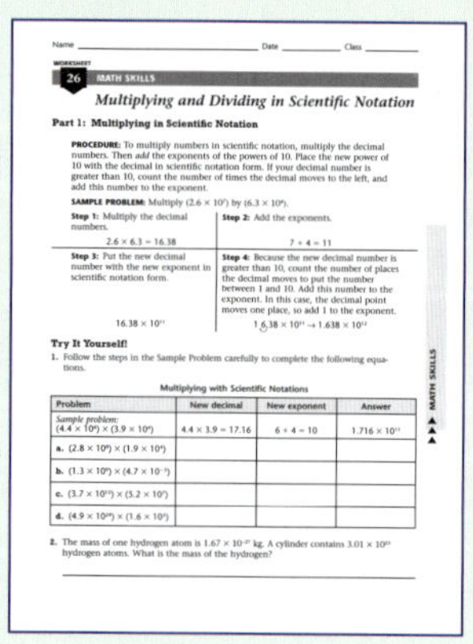
Worksheet 26 — Math Skills — Multiplying and Dividing in Scientific Notation

SI MEASUREMENT AND CONVERSION

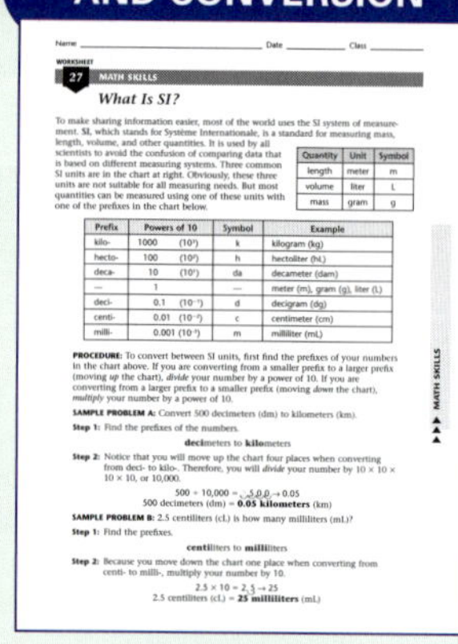
Worksheet 27 — Math Skills — What Is SI?

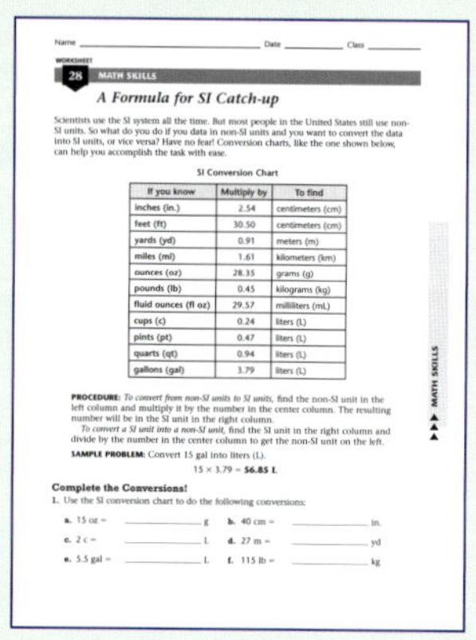
Worksheet 28 — Math Skills — A Formula for SI Catch-up

Math Skills for Science (continued)

GEOMETRY

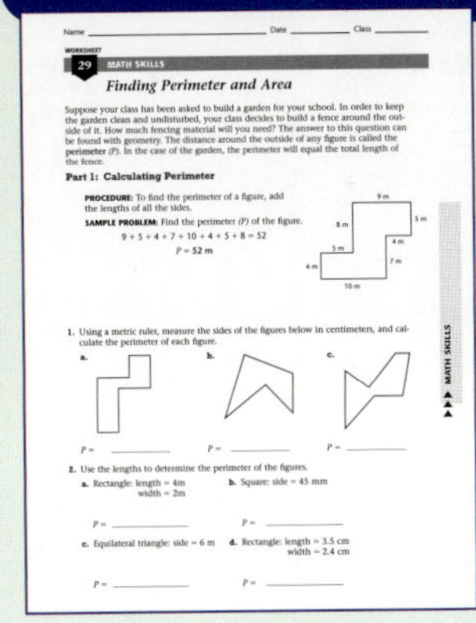

Worksheet 29 — Math Skills — Finding Perimeter and Area

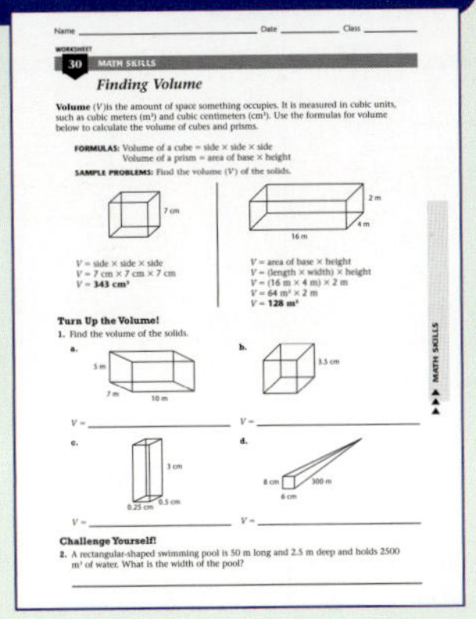

Worksheet 30 — Math Skills — Finding Volume

THE UNIT FACTOR AND DIMENSIONAL ANALYSIS

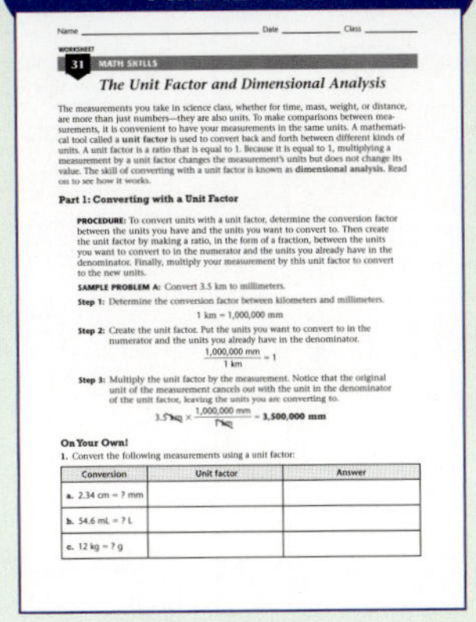

Worksheet 31 — Math Skills — The Unit Factor and Dimensional Analysis

MATH IN SCIENCE: INTEGRATED SCIENCE

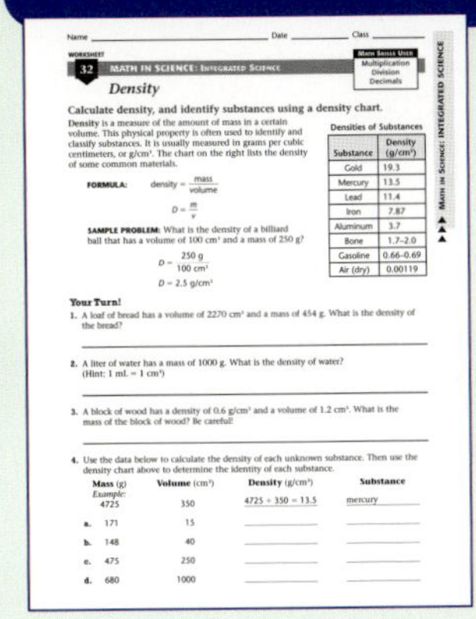

Worksheet 32 — Math in Science: Integrated Science — Density

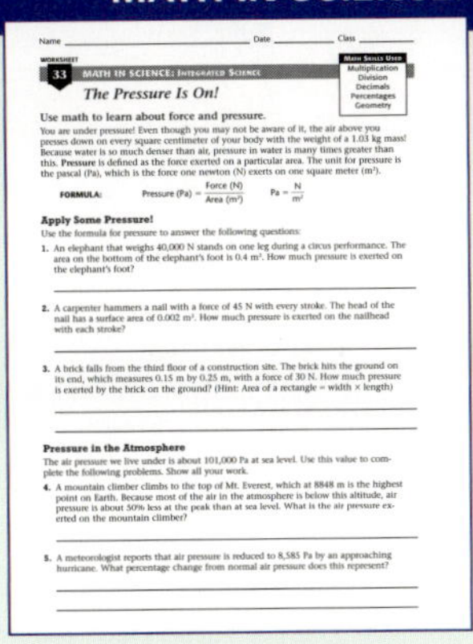

Worksheet 33 — Math in Science: Integrated Science — The Pressure Is On!

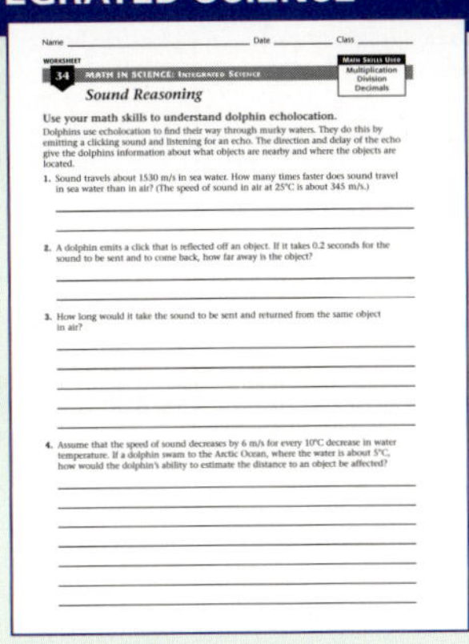

Worksheet 34 — Math in Science: Integrated Science — Sound Reasoning

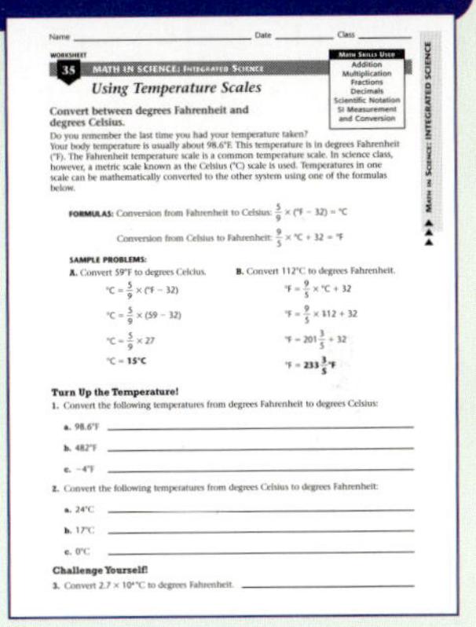

Worksheet 35 — Math in Science: Integrated Science — Using Temperature Scales

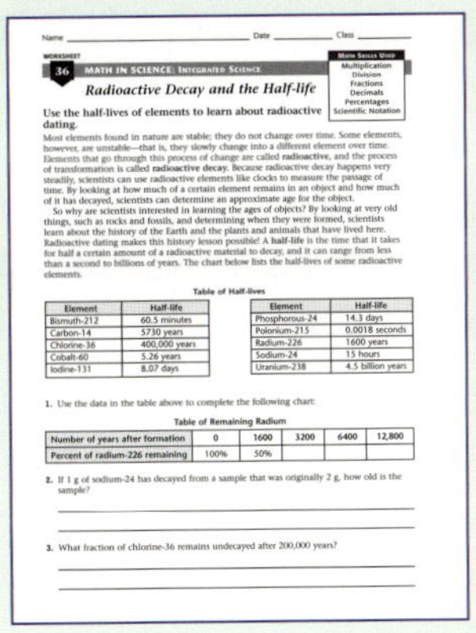

Worksheet 36 — Math in Science: Integrated Science — Radioactive Decay and the Half-life

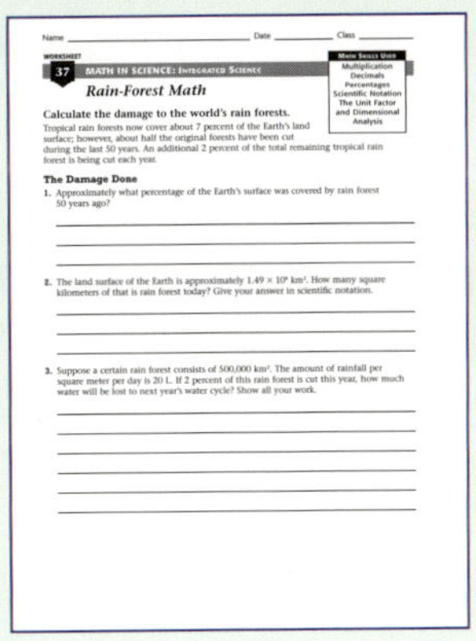

Worksheet 37 — Math in Science: Integrated Science — Rain-Forest Math

Math Skills for Science (continued)

MATH IN SCIENCE: LIFE SCIENCE

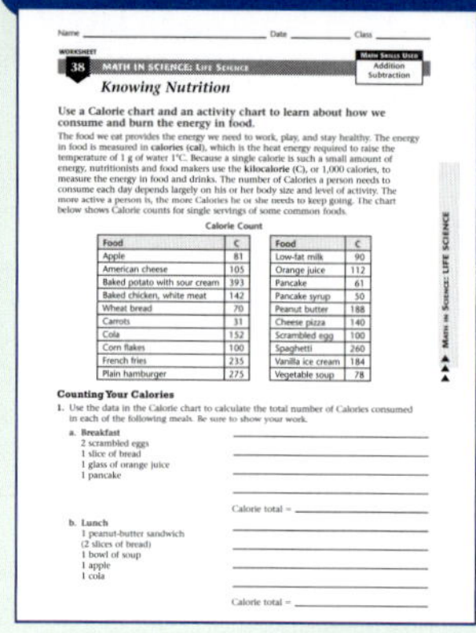

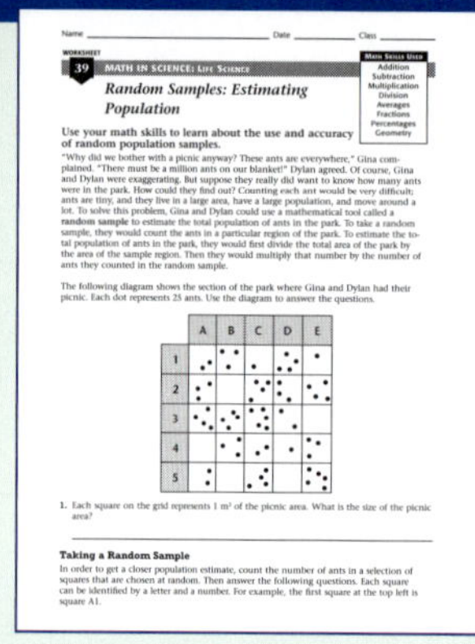

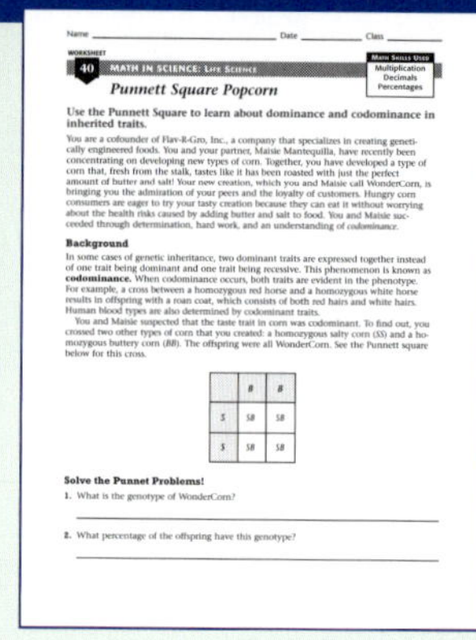

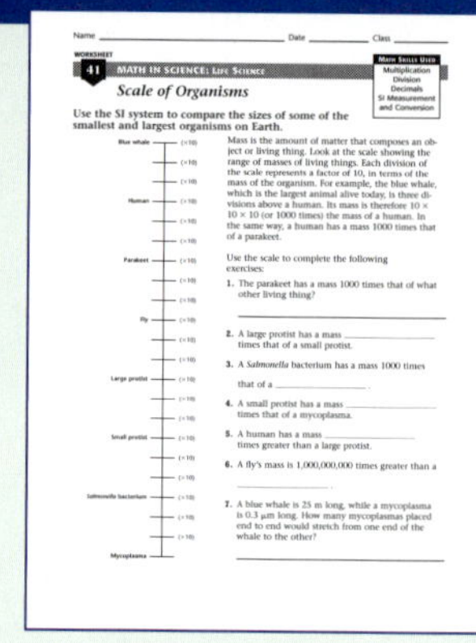

MATH IN SCIENCE: EARTH SCIENCE

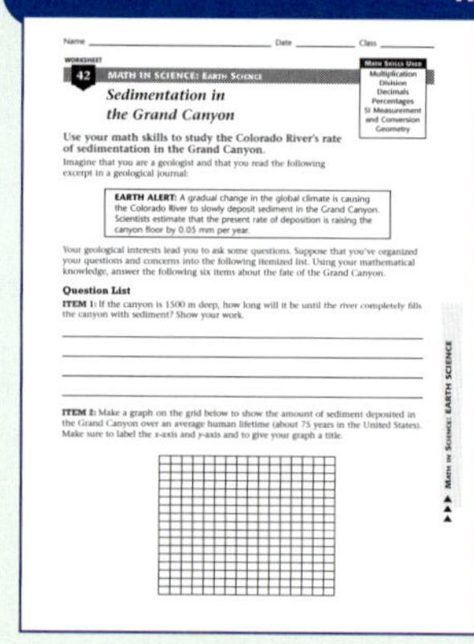

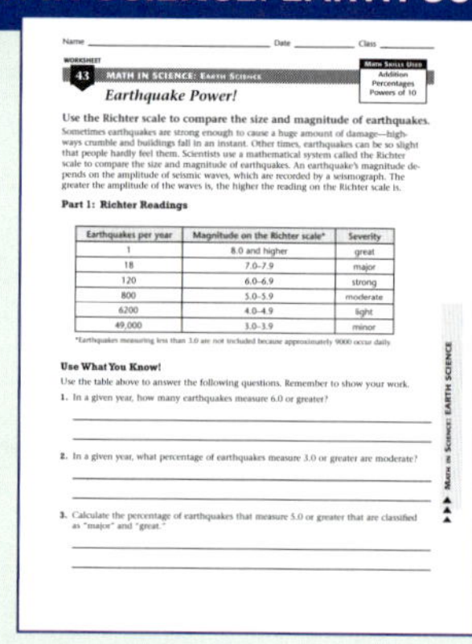

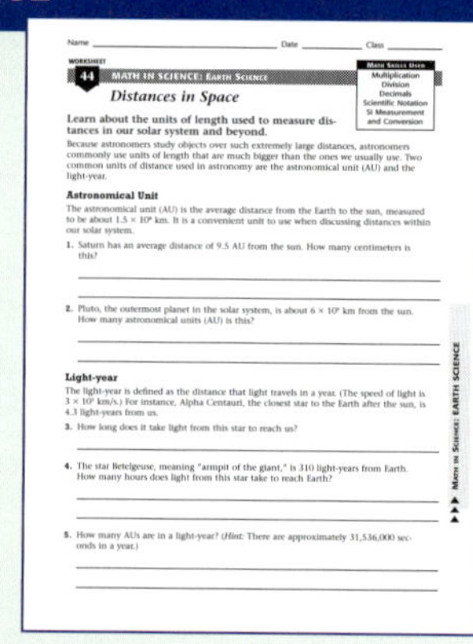

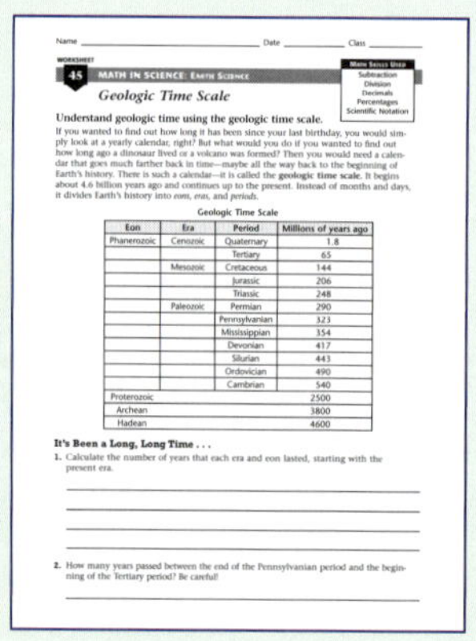

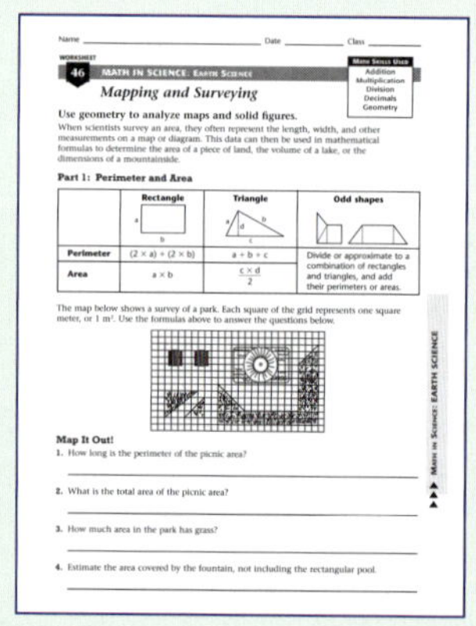

Math Skills for Science (continued)

MATH IN SCIENCE: PHYSICAL SCIENCE

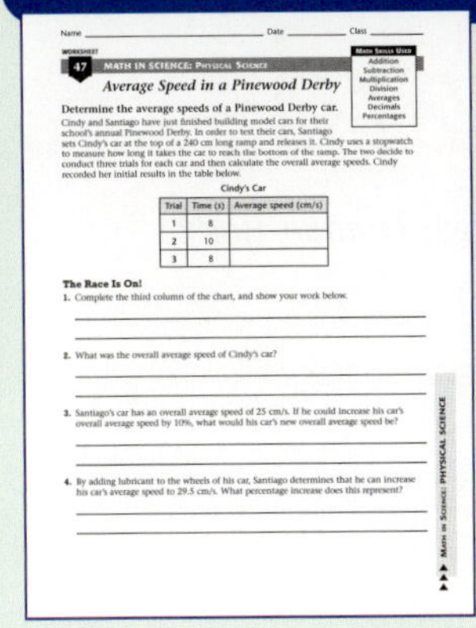
Worksheet 47 — Average Speed in a Pinewood Derby

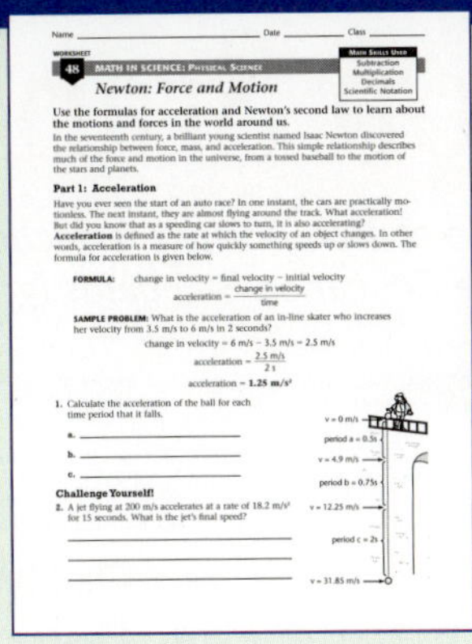
Worksheet 48 — Newton: Force and Motion

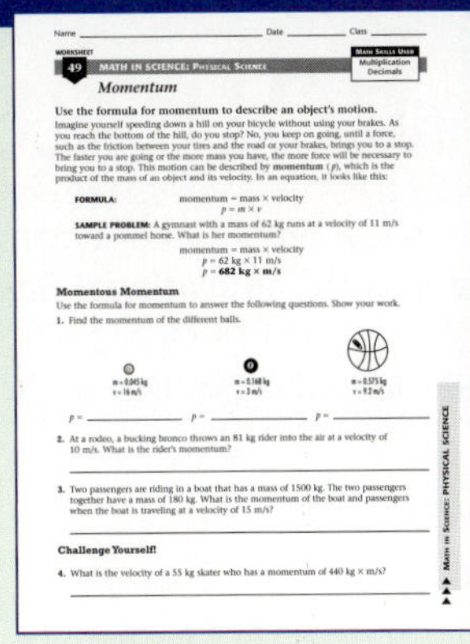
Worksheet 49 — Momentum

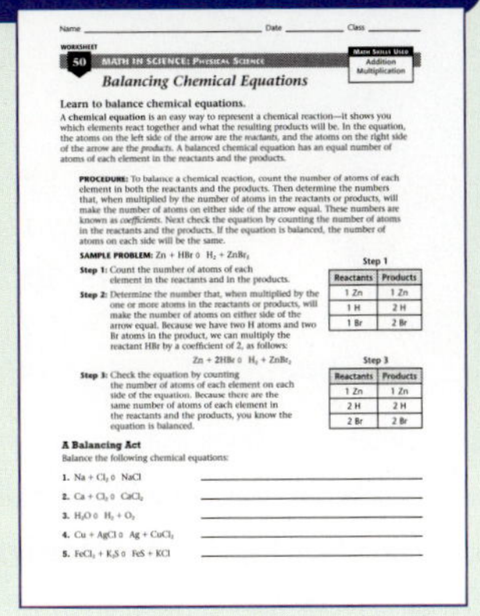
Worksheet 50 — Balancing Chemical Equations

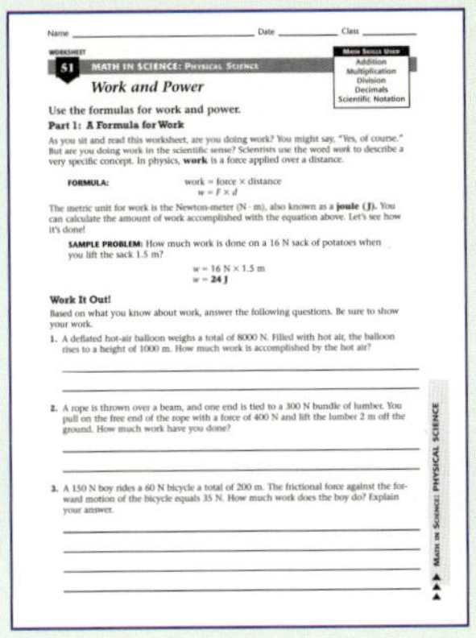
Worksheet 51 — Work and Power

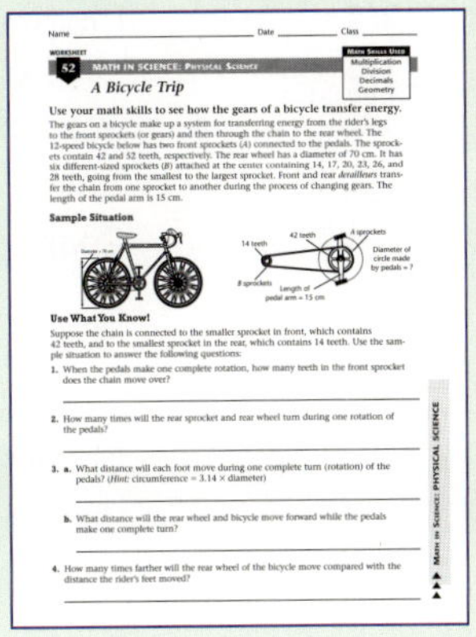
Worksheet 52 — A Bicycle Trip

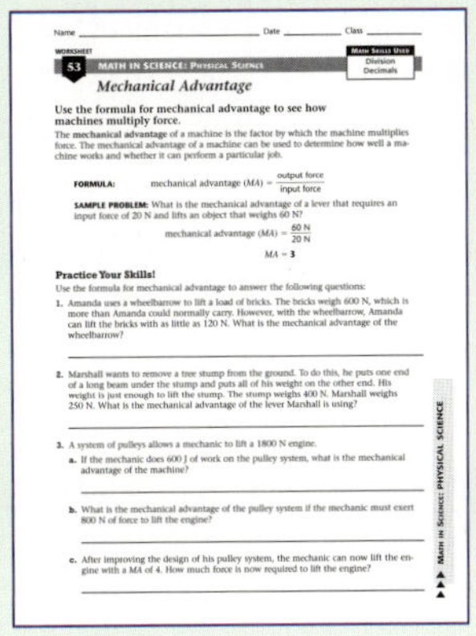
Worksheet 53 — Mechanical Advantage

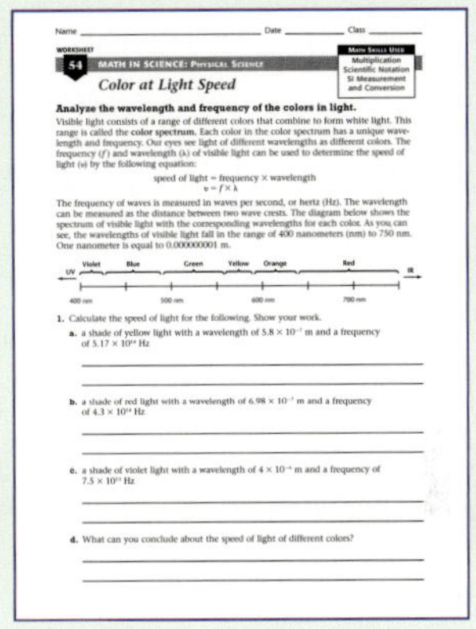
Worksheet 54 — Color at Light Speed

Assessment Checklist & Rubrics

The following is just a sample of over 50 checklists and rubrics contained in this booklet.

RUBRICS FOR WRITTEN WORK

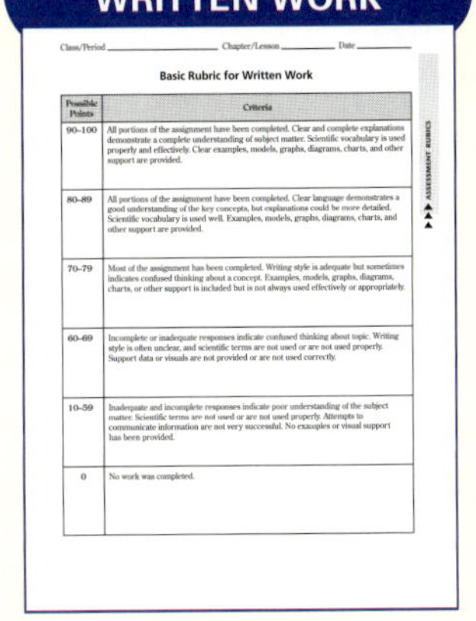
Basic Rubric for Written Work

RUBRIC FOR EXPERIMENTS

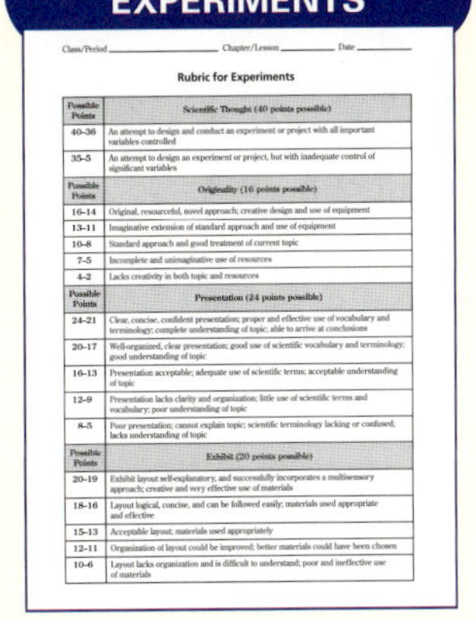
Rubric for Experiments

TEACHER EVALUATION OF COOPERATIVE LEARNING

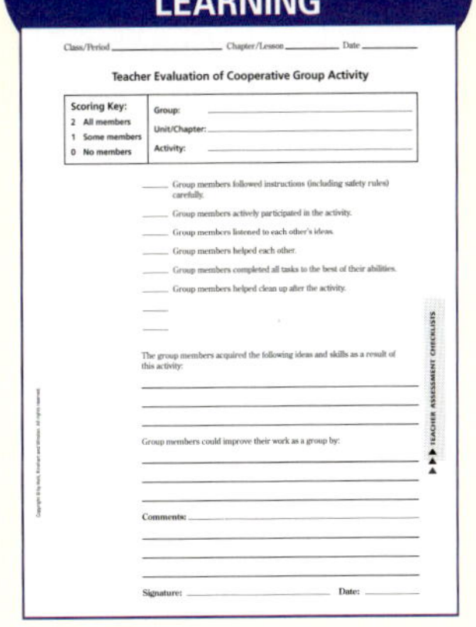
Teacher Evaluation of Cooperative Group Activity

TEACHER EVALUATION OF STUDENT PROGRESS

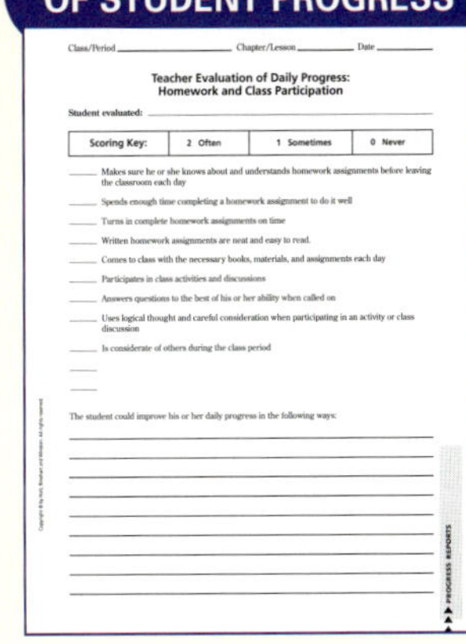
Teacher Evaluation of Daily Progress: Homework and Class Participation

National Science Education Standards

The following lists show the chapter correlation of ***Holt Science & Technology: Human Body Systems and Health*** with the ***National Science Education Standards*** (grades 5–8).

Standard	Chapter Correlation			
Unifying Concepts and Processes				
Systems, order, and organization Code: UCP 1	**Chapter 1** **Chapter 2** **Chapter 3** **Chapter 4**	1.1, 1.3 2.1, 2.2, 2.3, 2.4 3.1, 3.2 4.1, 4.3	**Chapter 5** **Chapter 6** **Chapter 7**	5.2, 5.3 6.2 7.1
Evidence, models, and explanation Code: UCP 2	**Chapter 1** **Chapter 2** **Chapter 3** **Chapter 4**	1.2, 1.3 2.1, 2.4 3.1, 3.2 4.1, 4.2	**Chapter 5** **Chapter 6**	5.3 6.2
Change, constancy, and measurement Code: UCP 3	**Chapter 1** **Chapter 2** **Chapter 3** **Chapter 4**	1.1, 1.2, 1.3, 1.4 2.1 3.2 4.1, 4.3	**Chapter 5** **Chapter 6** **Chapter 7**	5.3 6.2 7.3
Evolution and equilibrium Code: UCP 4	**Chapter 1** **Chapter 2** **Chapter 3**	1.1, 1.3 2.4 3.2	**Chapter 4** **Chapter 6** **Chapter 7**	4.1, 4.2, 4.3 6.2 7.1
Form and function Code: UCP 5	**Chapter 1** **Chapter 2** **Chapter 3**	1.2, 1.4 2.1, 2.3, 2.4 3.1, 3.2	**Chapter 4** **Chapter 5** **Chapter 6**	4.1, 4.2 5.2 6.2
Science as Inquiry				
Abilities necessary to do scientific inquiry Code: SAI 1	**Chapter 1** **Chapter 3** **Chapter 4**	1.2, 1.3 3.1, 3.2 4.1, 4.2	**Chapter 5** **Chapter 6** **Chapter 7**	5.2, 5.3 6.1, 6.2 7.1, 7.2, 7.3
Understandings about scientific inquiry Code: SAI 2	**Chapter 1** **Chapter 3**	1.3, 1.4 3.1, 3.2	**Chapter 4** **Chapter 6**	4.1 6.1
Science and Technology				
Understandings about science and technology Code: ST 2	**Chapter 5** **Chapter 6**	5.3 6.1		
Science in Personal Perspectives				
Personal Health Code: SPSP 1	**Chapter 1** **Chapter 2** **Chapter 3**	1.2, 1.3, 1.4 2.1, 2.2 3.1	**Chapter 4** **Chapter 5** **Chapter 7**	4.1 5.2 7.1, 7.2, 7.3

Science in Personal Perspectives (cont.)	
Standard	**Chapter Correlation**
Risks and benefits **Code: SPSP 4**	**Chapter 1** 1.2 **Chapter 2** 2.1 **Chapter 5** 5.2 **Chapter 7** 7.1, 7.2, 7.3
Science and technology in society **Code: SPSP 5**	**Chapter 2** 2.1, 2.2 **Chapter 6** 6.1 **Chapter 7** 7.2, 7.3
History and Nature of Science	
Science as a human endeavor **Code: HNS 1**	**Chapter 3** 3.1, 3.2 **Chapter 6** 6.1
History of science **Code: HNS 3**	**Chapter 6** 6.1, 6.2

Life Science Content Standards

Structure and Function in Living Systems	
Standard	**Chapter Correlation**
Living systems at all levels of organization demonstrate the complementary nature of structure and function. Important levels of organization for structure and function include cells, organs, tissues, organ systems, whole organisms, and ecosystems. Code: LS 1a	**Chapter 1** 1.1 **Chapter 2** 2.3 **Chapter 3** 3.1, 3.2 **Chapter 4** 4.1, 4.2, 4.3 **Chapter 5** 5.3
All organisms are composed of cells—the fundamental unit of life. Most organisms are single cells; other organisms, including humans, are multicellular. Code: LS 1b	**Chapter 6** 6.2
Cells carry on the many functions needed to sustain life. They grow and divide, thereby producing more cells. This requires that they take in nutrients, which they use to provide energy for the work that cells do and to make the materials that a cell or an organism needs. Code: LS 1c	**Chapter 1** 1.4 **Chapter 2** 2.3 **Chapter 3** 3.2 **Chapter 5** 5.3
Specialized cells perform specialized functions in multicellular organisms. Groups of specialized cells cooperate to form a tissue, such as a muscle. Different tissues are in turn grouped together and form larger functional units, called organs. Each type of cell, tissue, and organ has a distinct structure and set of functions that serve the organism as a whole. Code: LS 1d	**Chapter 1** 1.1, 1.2, 1.3, 1.4 **Chapter 2** 2.3 **Chapter 3** 3.1, 3.2 **Chapter 4** 4.1, 4.2, 4.3 **Chapter 5** 5.2, 5.3 **Chapter 6** 6.2
The human organism has systems for digestion, respiration, reproduction, circulation, excretion, movement, control and coordination, and protection from disease. These systems interact with one another. Code: LS 1e	**Chapter 1** 1.2, 1.3, 1.4 **Chapter 2** 2.3 **Chapter 3** 3.1, 3.2 **Chapter 4** 4.1, 4.2, 4.3 **Chapter 5** 5.2, 5.3 **Chapter 6** 6.2

Structure and Function in Living Systems (cont.)

Standard	Chapter Correlation
Disease is the breakdown in structures or functions of an organism. Some diseases are the result of intrinsic failures of the system. Others are the result of damage by infection by other organisms. Code: LS 1f	**Chapter 1** 1.4 **Chapter 2** 2.3 **Chapter 3** 3.1, 3.2 **Chapter 5** 5.2 **Chapter 6** 6.1, 6.2

Reproduction and Heredity

Standard	Chapter Correlation
In many species, including humans, females produce eggs and males produce sperm. Plants also reproduce sexually—the egg and sperm are produced in the flowers of flowering plants. An egg and sperm unite to begin development of a new individual. The individual receives genetic information from its mother (via the egg) and its father (via the sperm). Sexually produced offspring never are identical to either of their parents. Code: LS 2b	**Chapter 5** 5.3

Regulation and Behavior

Standard	Chapter Correlation
All organisms must be able to obtain and use resources, grow, reproduce, and maintain stable internal conditions while living in a constantly changing external environment. Code: LS 3a	**Chapter 1** 1.1, 1.4 **Chapter 2** 2.2 **Chapter 3** 3.2 **Chapter 4** 4.2 **Chapter 6** 6.2
Regulation of an organism's internal environment involves sensing the internal environment and changing physiological activities to keep conditions within the range required to survive. Code: LS 3b	**Chapter 1** 1.4 **Chapter 2** 2.1, 2.2 **Chapter 3** 3.1, 3.2 **Chapter 4** 4.1 **Chapter 6** 6.2
Behavior is one kind of response an organism can make to an internal or environmental stimulus. A behavioral response requires coordination and communication at many levels, including cells, organ systems, and whole organisms. Behavioral response is a set of actions determined in part by heredity and in part from experience. Code: LS 3c	**Chapter 4** 4.1

HOLT SCIENCE & TECHNOLOGY

Human Body Systems and Health

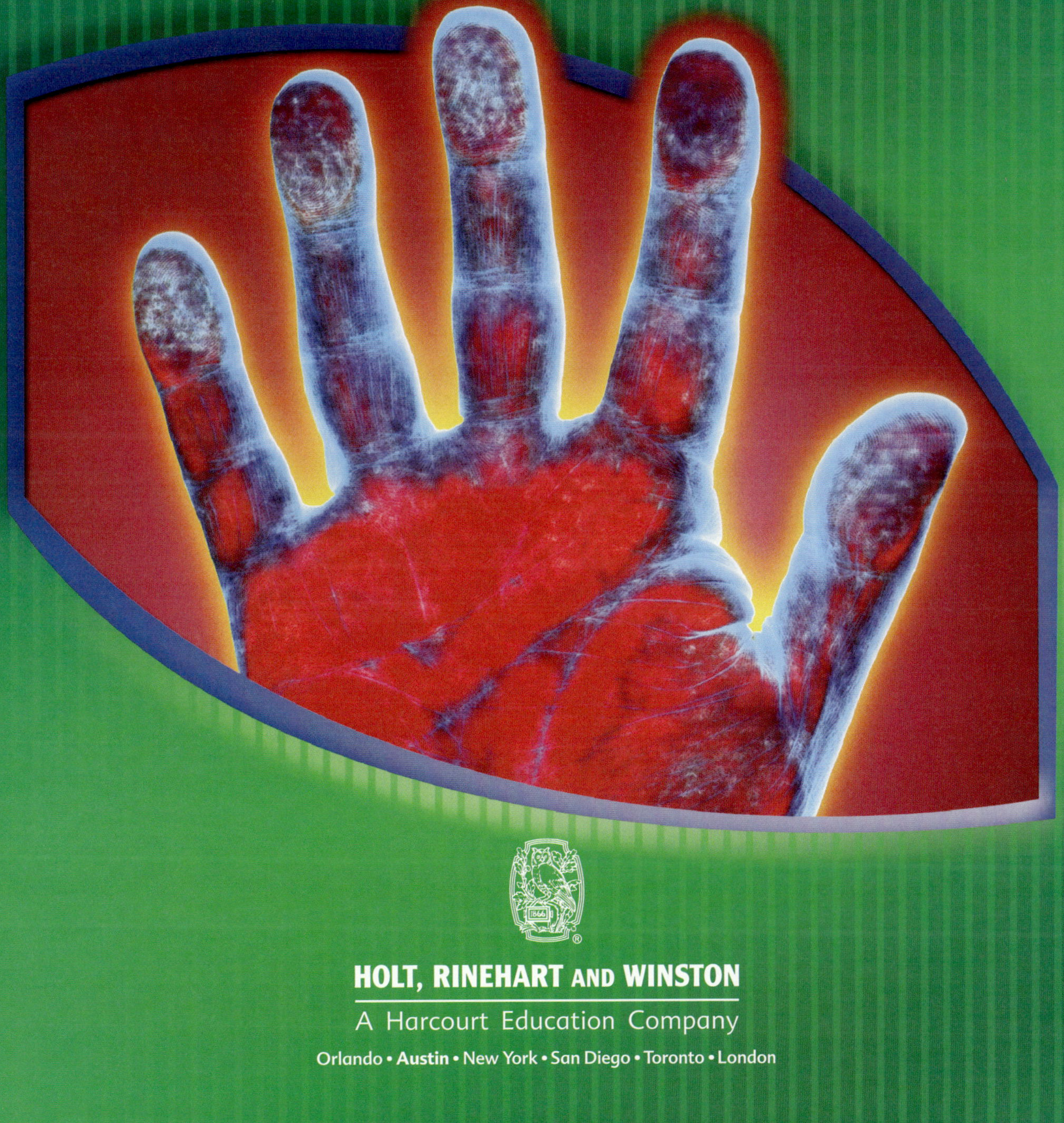

HOLT, RINEHART AND WINSTON
A Harcourt Education Company
Orlando • Austin • New York • San Diego • Toronto • London

Acknowledgments

Contributing Authors

Katy Z. Allen
Science Writer
Wayland, Massachusetts

Linda Ruth Berg, Ph.D.
Adjunct Professor
Natural Sciences
St. Petersburg College
St. Petersburg, Florida

Mark F. Taylor, Ph.D.
Associate Professor of Biology
Biology Department
Baylor University
Waco, Texas

Inclusion Specialist

Ellen McPeek Glisan
Special Needs Consultant
San Antonio, Texas

Safety Reviewer

Jack Gerlovich, Ph.D.
Associate Professor
School of Education
Drake University
Des Moines, Iowa

Academic Reviewers

William E. Dunscombe
Chairman
Biology Department
Union County College
Cranford, New Jersey

William Grisham, Ph.D.
Lecturer
Psychology Department
University of California, Los Angeles
Los Angeles, California

Ping H. Johnson, M.D., Ph.D., CHES
Assistant Professor of Health Education
Department of Health, Physical Education and Sport Science
Kennesaw State University
Kennesaw, Georgia

Linda Jones
Program Manager
Texas Department of Public Health
Austin, Texas

Eva Oberdoerster, Ph.D.
Lecturer
Department of Biology
Southern Methodist University
Dallas, Texas

Laurie Santos, Ph.D.
Assistant Professor
Department of Psychology
Yale University
New Haven, Connecticut

Teacher Reviewers

Diedre S. Adams
Physical Science Instructor
West Vigo Middle School
West Terre Haute, Indiana

Debra S. Kogelman, MAed.
Science Teacher
University of Chicago Laboratory Schools
Chicago, Illinois

Augie Maldonado
Science Teacher
Grisham Middle School
Round Rock, Texas

Jean Pletchette
Health Educator
Winterset Community Schools
Winterset, Iowa

Elizabeth Rustad
Science Teacher
Higley School District
Gilbert, Arizona

Helen P. Schiller
Instructional Coach
The School District of Greenville County
Greenville, South Carolina

Florence Vaughan
Science Teacher
University of Chicago Laboratory Schools
Chicago, Illinois

Carl Benson
General Science Teacher
Plains High School
Plains, Montana

Charlotte Blassingame
Technology Coordinator
White Station Middle School
Memphis, Tennessee

Marsha Carver
Science Teacher and Department Chair
McLean County High School
Calhoun, Kentucky

Kenneth E. Creese
Science Teacher
White Mountain Junior High School
Rock Springs, Wyoming

Linda Culp
Science Teacher and Department Chair
Thorndale High School
Thorndale, Texas

James Deaver
Science Teacher and Department Chair
West Point High School
West Point, Nebraska

Frank McKinney, Ph.D.
Professor of Geology
Appalachian State University
Boone, North Carolina

Lab Development

Diana Scheidle Bartos
Research Associate
School of Mines
Golden, Colorado

Printed in the United States of America

ISBN 0-03-025566-X

1 2 3 4 5 6 7 048 08 07 06 05 04

Alyson Mike
Science Teacher
East Valley Middle School
East Helena, Montana

C. Ford Morishita
Biology Teacher
Clackamas High School
Milwaukie, Oregon

Patricia D. Morrell, Ph.D.
Associate Professor
School of Education
University of Portland
Portland, Oregon

Hilary C. Olson, Ph.D.
Research Associate
Institute for Geophysics
The University of Texas at Austin
Austin, Texas

James B. Pulley
Science Editor and Former Science Teacher
North Kansas City, Missouri

Denice Lee Sandefur
Science Chairperson
Nucla High School
Nucla, Colorado

Patti Soderberg
Science Writer
The BioQUEST Curriculum Consortium
Biology Department
Beloit College
Beloit, Wisconsin

Phillip Vavala
Science Teacher and Department Chair
Salesianum School
Wilmington, Delaware

Albert C. Wartski, M.A.T.
Biology Teacher
Chapel Hill High School
Chapel Hill, North Carolina

Lynn Marie Wartski
Science Writer and Former Science Teacher
Hillsborough, North Carolina

Ivora D. Washington
Science Teacher and Department Chair
Hyattsville Middle School
Washington, D.C.

Lab Testing

Yvonne Brannum
Science Teacher and Department Chair
Hine Junior High School
Washington, D.C.

James Chin
Science Teacher
Frank A. Day Middle School
Newtonville, Massachusetts

Randy Christian
Science Teacher
Stovall Junior High School
Houston, Texas

M. R. Penny Kisiah
Science Teacher and Department Chair
Fairview Middle School
Tallahassee, Florida

Kathy LaRoe
Science Teacher
East Valley Middle School
East Helena, Montana

Edith C. McAlanis
Science Teacher and Department Chair
Socorro Middle School
El Paso, Texas

Elizabeth Rustad
Science Teacher
Crane Middle School
Yuma, Arizona

Ivora Washington
Science Teacher and Department Chair
Hyattsville Middle School
Washington, D.C.

Christopher Wood
Science Teacher
Western Rockingham Middle School
Madison, North Carolina

Feature Development

Hatim Belyamani
John A. Benner
David Bradford
Jennifer Childers
Mickey Coakley
Susan Feldkamp
Jane Gardner
Erik Hahn
Christopher Hess
Deena Kalai
Charlotte W. Luongo, MSc
Michael May
Persis Mehta, Ph.D.
Eileen Nehme, MPH
Catherine Podeszwa
Dennis Rathnaw
Daniel B. Sharp
April Smith West
John M. Stokes
Molly F. Wetterschneider

Answer Checking

Hatim Belyamani
Austin, Texas

D Human Body Systems and Health

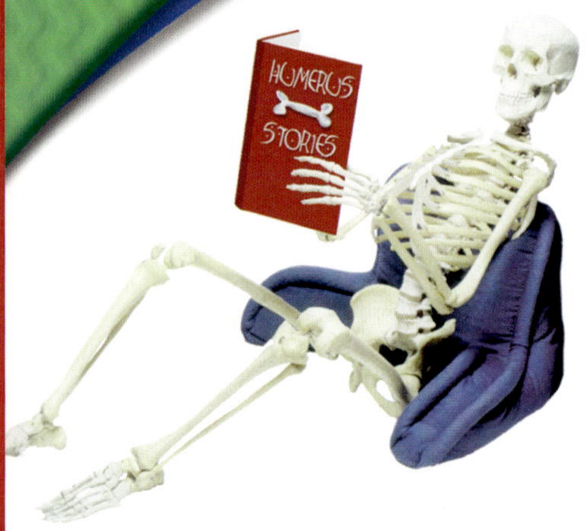

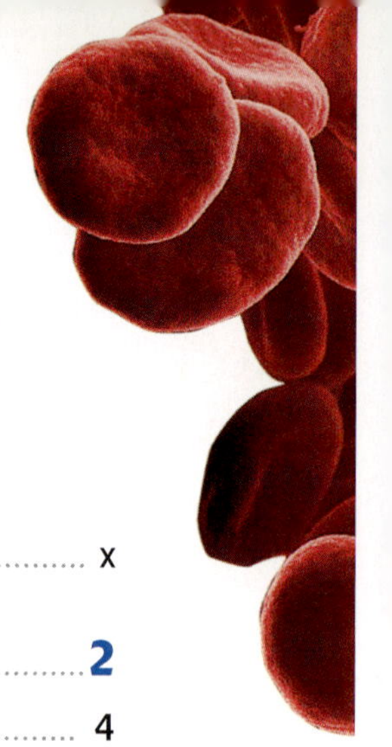

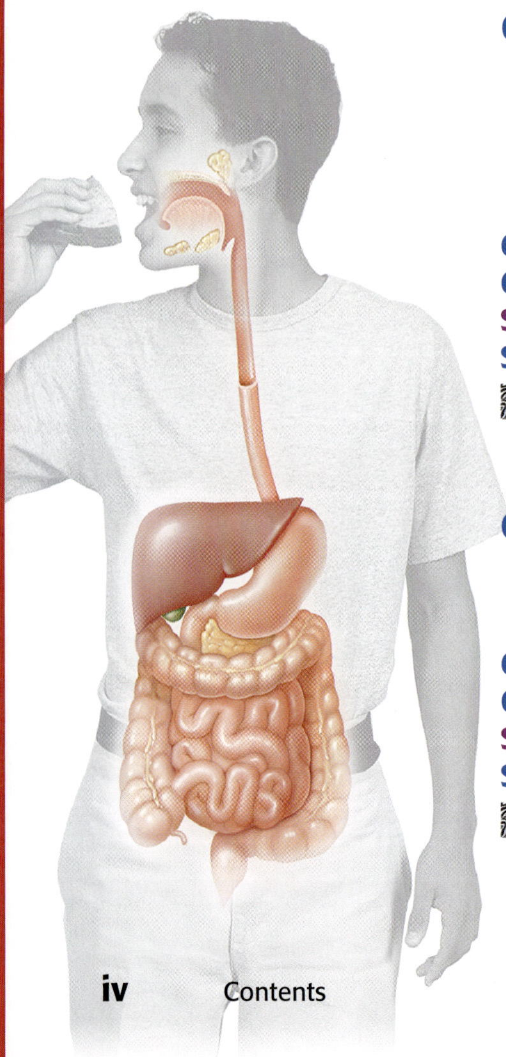

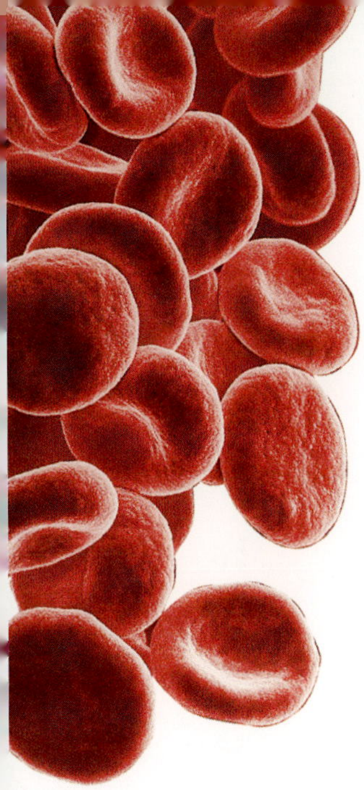

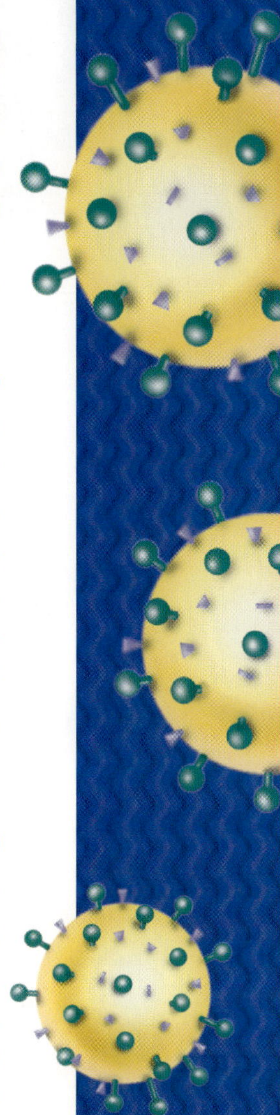

Skills Development

PRE-READING ACTIVITY

FOLDNOTES

Graphic Organizer

START-UP ACTIVITY

QUICK LAB

SCHOOL to HOME

READING STRATEGY

INTERNET ACTIVITY

Math Practice

Connection to . . .

Science in Action

How to Use Your Textbook

Your Roadmap for Success with Holt Science and Technology

Reading Warm-Up

A Reading Warm-Up at the beginning of every section provides you with the section's objectives and key terms. The objectives tell you what you'll need to know after you finish reading the section.

Key terms are listed for each section. Learn the definitions of these terms because you will most likely be tested on them. Each key term is highlighted in the text and is defined at point of use and in the margin. You can also use the glossary to locate definitions quickly.

STUDY TIP Reread the objectives and the definitions to the key terms when studying for a test to be sure you know the material.

SECTION 1

Body Organization

Imagine jumping into a lake. At first, your body feels very cold. You may even shiver. But eventually you get used to the cold water. How?

READING WARM-UP

Objectives

- Describe how tissues, organs, and organ systems are related.
- List 12 organ systems.
- Identify how organ systems work together to maintain homeostasis.

Terms to Learn

homeostasis organ
tissue

READING STRATEGY

Reading Organizer As you read this section, make a concept map by using the terms above.

Your body gets used to cold water because of homeostasis (HOH mee OH STAY sis). **Homeostasis** is the maintenance of a stable internal environment in the body. When you jump into a lake, homeostasis helps your body adapt to the cold water.

Cells, Tissues, and Organs

Maintaining homeostasis is not easy. Your internal environment is always changing. Your cells need nutrients and oxygen to survive. Your cells need wastes removed. If homeostasis is disrupted, cells may not get the materials they need. So, cells may be damaged or may die.

Your cells must do many jobs to maintain homeostasis. Fortunately, each of your cells does not have to do all of those jobs. Just as each person on a soccer team has a role during a game, each cell in your body has a job in maintaining homeostasis. Your cells are organized into groups. A group of similar cells working together forms a **tissue.** Your body has four main kinds of tissue. The four kinds of tissue are shown in **Figure 1.**

homeostasis the maintenance of a constant internal state in a changing environment

tissue a group of similar cells that perform a common function

Figure 1 Four Kinds of Tissue

Epithelial tissue covers and protects underlying tissue. When you look at the surface of your skin, you see epithelial tissue. The cells form a continuous sheet.

Nervous tissue sends electrical signals through the body. It is found in the brain, nerves, and sense organs.

580 Chapter 22 Body Organization and Structure

Get Organized

A Reading Strategy at the beginning of every section provides tips to help you organize and remember the information covered in the section. Keep a science notebook so that you are ready to take notes when your teacher reviews the material in class. Keep your assignments in this notebook so that you can review them when studying for the chapter test.

Be Resourceful — Use the Web

SciLinks®

Internet Connect boxes in your textbook take you to resources that you can use for science projects, reports, and research papers. Go to scilinks.org, and type in the SciLinks code to get information on a topic.

go.hrw.com

Visit go.hrw.com Find worksheets, **Current Science®** magazine articles online, and other materials that go with your textbook at **go.hrw.com.** Click on the textbook icon and the table of contents to see all of the resources for each chapter.

Figure 2 Organization of the Stomach

The stomach is an organ. The four kinds of tissue work together so that the stomach can carry out digestion.

Tissues Form Organs

One kind of tissue alone cannot do all of the things that several kinds of tissue working together can do. Two or more tissues working together form an **organ.** Your stomach, shown in **Figure 2,** uses all four kinds of tissue to carry out digestion.

organ a collection of tissues that carry out a specialized function of the body

Organs Form Systems

Your stomach does a lot to help you digest your food. But the stomach doesn't do it all. Your stomach works with other organs, such as the small and large intestines, to digest your food. Organs that work together make up an *organ system.*

Reading Check **How is the stomach part of an organ system?** (*See the Appendix for answers to Reading Checks.*)

Nervous System Your nervous system receives and sends electrical messages throughout your body.

Digestive System Your digestive system breaks down the food you eat into nutrients that your body can absorb.

Lymphatic System The lymphatic system returns leaked fluids to blood vessels and helps get rid of bacteria and viruses.

Endocrine System Your glands send out chemical messages. Ovaries and testes are part of this system.

SECTION Review

Summary

- A group of cells that work together is a tissue. Tissues form organs. Organs that work together form organ systems.
- There are four kinds of tissue in the human body.
- There are 12 major organ systems in the human body.
- Organ systems work together to help the body maintain homeostasis.

Using Key Terms

1. Use the following terms in the same sentence: *homeostasis, tissue,* and *organ.*

Understanding Key Ideas

2. Which of the following statements describes how tissues, organs, and organ systems are related?
   a. Organs form tissues, which form organ systems.
   b. Organ systems form organs, which form tissues.
   c. Tissues form organs, which form organ systems.
   d. None of the above
3. List the 12 organ systems.

Math Skills

4. The human skeleton has 206 bones. The human skull has 22 bones. What percentage of human bones are skull bones?

Critical Thinking

5. **Applying Concepts** Tanya went to a restaurant and ate a hamburger. Describe how Tanya used five organ systems to eat and digest her hamburger.
6. **Predicting Consequences** Predict what might happen if the human body did not have specialized cells, tissues, organs, and organ systems to maintain homeostasis.

SCILINKS. NSTA Developed and maintained by the National Science Teachers Association

For a variety of links related to this chapter, go to www.scilinks.org

Topic: Tissues and Organs; Body Systems
SciLinks code: HSM1530; HSM0184

583

Use the Illustrations and Photos

Art shows complex ideas and processes. Learn to analyze the art so that you better understand the material you read in the text.

Tables and graphs display important information in an organized way to help you see relationships.

A picture is worth a thousand words. Look at the photographs to see relevant examples of science concepts that you are reading about.

Answer the Section Reviews

Section Reviews test your knowledge of the main points of the section. Critical Thinking items challenge you to think about the material in greater depth and to find connections that you infer from the text.

STUDY TIP When you can't answer a question, reread the section. The answer is usually there.

Do Your Homework

Your teacher may assign worksheets to help you understand and remember the material in the chapter.

STUDY TIP Don't try to answer the questions without reading the text and reviewing your class notes. A little preparation up front will make your homework assignments a lot easier. Answering the items in the Chapter Review will help prepare you for the chapter test.

Visit Holt Online Learning

If your teacher gives you a special password to log onto the Holt Online Learning site, you'll find your complete textbook on the Web. In addition, you'll find some great learning tools and practice quizzes. You'll be able to see how well you know the material from your textbook.

Visit CNN Student News

You'll find up-to-date events in science at **cnnstudentnews.com.**

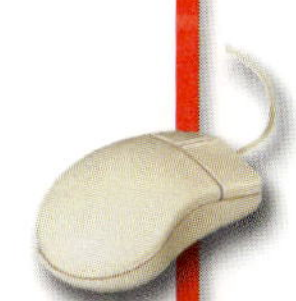

Exploring, inventing, and investigating are essential to the study of science. However, these activities can also be dangerous. To make sure that your experiments and explorations are safe, you must be aware of a variety of safety guidelines. You have probably heard of the saying, "It is better to be safe than sorry." This is particularly true in a science classroom where experiments and explorations are being performed. Being uninformed and careless can result in serious injuries. Don't take chances with your own safety or with anyone else's.

The following pages describe important guidelines for staying safe in the science classroom. Your teacher may also have safety guidelines and tips that are specific to your classroom and laboratory. Take the time to be safe.

Safety Rules!

Start Out Right

Always get your teacher's permission before attempting any laboratory exploration. Read the procedures carefully, and pay particular attention to safety information and caution statements. If you are unsure about what a safety symbol means, look it up or ask your teacher. You cannot be too careful when it comes to safety. If an accident does occur, inform your teacher immediately regardless of how minor you think the accident is.

If you are instructed to note the odor of a substance, wave the fumes toward your nose with your hand. Never put your nose close to the source.

Safety Symbols

All of the experiments and investigations in this book and their related worksheets include important safety symbols to alert you to particular safety concerns. Become familiar with these symbols so that when you see them, you will know what they mean and what to do. It is important that you read this entire safety section to learn about specific dangers in the laboratory.

Eye Safety

Wear safety goggles when working around chemicals, acids, bases, or any type of flame or heating device. Wear safety goggles any time there is even the slightest chance that harm could come to your eyes. If any substance gets into your eyes, notify your teacher immediately and flush your eyes with running water for at least 15 minutes. Treat any unknown chemical as if it were a dangerous chemical. Never look directly into the sun. Doing so could cause permanent blindness.

Avoid wearing contact lenses in a laboratory situation. Even if you are wearing safety goggles, chemicals can get between the contact lenses and your eyes. If your doctor requires that you wear contact lenses instead of glasses, wear eye-cup safety goggles in the lab.

Safety Equipment

Know the locations of the nearest fire alarms and any other safety equipment, such as fire blankets and eyewash fountains, as identified by your teacher, and know the procedures for using the equipment.

Neatness

Keep your work area free of all unnecessary books and papers. Tie back long hair, and secure loose sleeves or other loose articles of clothing, such as ties and bows. Remove dangling jewelry. Don't wear open-toed shoes or sandals in the laboratory. Never eat, drink, or apply cosmetics in a laboratory setting. Food, drink, and cosmetics can easily become contaminated with dangerous materials.

Certain hair products (such as aerosol hair spray) are flammable and should not be worn while working near an open flame. Avoid wearing hair spray or hair gel on lab days.

Sharp/Pointed Objects

Use knives and other sharp instruments with extreme care. Never cut objects while holding them in your hands. Place objects on a suitable work surface for cutting.

Be extra careful when using any glassware. When adding a heavy object to a graduated cylinder, tilt the cylinder so the object slides slowly to the bottom.

Heat

Wear safety goggles when using a heating device or a flame. Whenever possible, use an electric hot plate as a heat source instead of using an open flame. When heating materials in a test tube, always angle the test tube away from yourself and others. To avoid burns, wear heat-resistant gloves whenever instructed to do so.

Electricity

Be careful with electrical cords. When using a microscope with a lamp, do not place the cord where it could trip someone. Do not let cords hang over a table edge in a way that could cause equipment to fall if the cord is accidentally pulled. Do not use equipment with damaged cords. Be sure that your hands are dry and that the electrical equipment is in the "off" position before plugging it in. Turn off and unplug electrical equipment when you are finished.

Chemicals

Wear safety goggles when handling any potentially dangerous chemicals, acids, or bases. If a chemical is unknown, handle it as you would a dangerous chemical. Wear an apron and protective gloves when you work with acids or bases or whenever you are told to do so. If a spill gets on your skin or clothing, rinse it off immediately with water for at least 5 minutes while calling to your teacher.

Never mix chemicals unless your teacher tells you to do so. Never taste, touch, or smell chemicals unless you are specifically directed to do so. Before working with a flammable liquid or gas, check for the presence of any source of flame, spark, or heat.

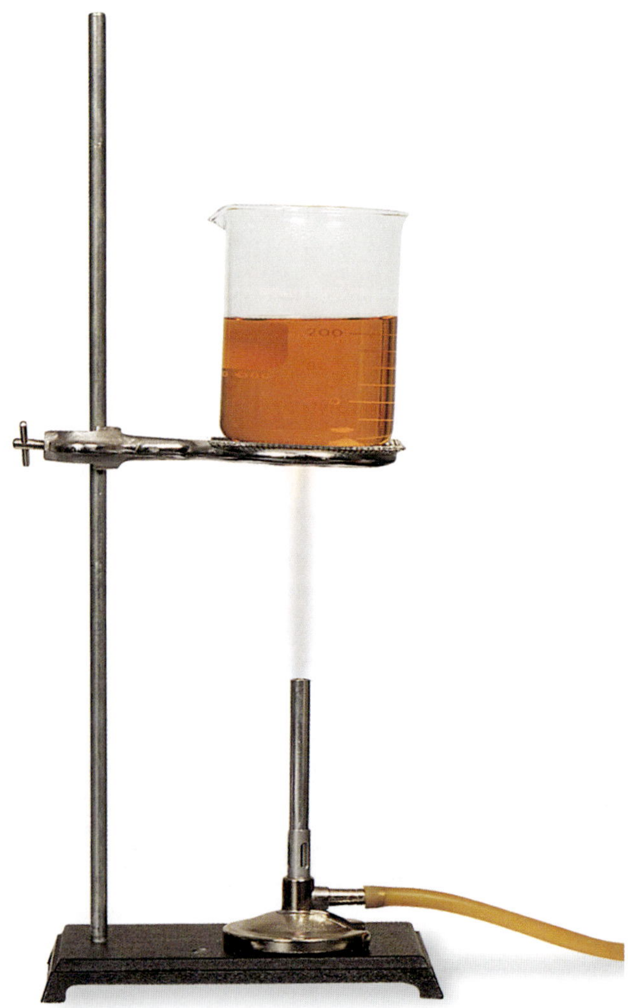

Animal Safety

Always obtain your teacher's permission before bringing any animal into the school building. Handle animals only as your teacher directs. Always treat animals carefully and respectfully. Wash your hands thoroughly after handling any animal.

Plant Safety

Do not eat any part of a plant or plant seed used in the laboratory. Wash your hands thoroughly after handling any part of a plant. When in nature, do not pick any wild plants unless your teacher instructs you to do so.

Glassware

Examine all glassware before use. Be sure that glassware is clean and free of chips and cracks. Report damaged glassware to your teacher. Glass containers used for heating should be made of heat-resistant glass.

Body Organization and Structure

Chapter Planning Guide

Compression guide: To shorten instruction because of time limitations, omit Section 1.

OBJECTIVES	LABS, DEMONSTRATIONS, AND ACTIVITIES	TECHNOLOGY RESOURCES
PACING • 90 min pp. 2–7 **Chapter Opener**	SE **Start-up Activity,** p. 3 GENERAL	OSP **Parent Letter** ■ GENERAL CD **Student Edition on CD-ROM** CD **Guided Reading Audio CD** ■ TR **Chapter Starter Transparency*** VID **Brain Food Video Quiz**
Section 1 Body Organization • Describe how tissues, organs, and organ systems are related. • List 12 organ systems. • Identify how organ systems work together to maintain homeostasis.	TE **Connection Activity** History, p. 5 ADVANCED	CRF **Lesson Plans*** TR **Bellringer Transparency*** TR Organ Systems* CRF **SciLinks Activity*** GENERAL
PACING • 45 min pp. 8–11 **Section 2 The Skeletal System** • Identify the major organs of the skeletal system. • Describe four functions of bones. • Describe three types of joints. • List three injuries and two diseases that affect bones and joints.	TE **Activity** Locating Bones, p. 8 GENERAL SE **Quick Lab** Pickled Bones, p. 9 ◆ GENERAL CRF **Datasheet for Quick Lab*** TE **Demonstration** Bone Dissection, p. 9 ◆ BASIC	CRF **Lesson Plans*** TR **Bellringer Transparency*** TR The Skeleton* TR *LINK TO PHYSICAL SCIENCE* Machines Change the Size and/or Direction of a Force*
PACING • 45 min pp. 12–15 **Section 3 The Muscular System** • List three kinds of muscle tissue. • Describe how skeletal muscles move bones. • Compare aerobic exercise with resistance exercise. • Describe two muscular system injuries.	TE **Group Activity** Poster Project, p. 12 GENERAL SE **School-to-Home Activity** Power in Pairs, p. 13 GENERAL TE **Demonstration** Muscle Contraction, p. 13 BASIC SE **Connection to Chemistry** Muscle Function, p. 14 GENERAL SE **Inquiry Lab** Muscles at Work, p. 180 ◆ GENERAL CRF **Datasheet for LabBook*** LB **Inquiry Labs** On a Wing and a Layer* ◆ GENERAL	CRF **Lesson Plans*** TR **Bellringer Transparency*** TR A Pair of Muscles in the Arm*
PACING • 90 min pp. 16–19 **Section 4 The Integumentary System** • List four functions of skin. • Describe the two layers of skin. • Describe the structure and function of hair and nails. • Describe two kinds of damage that can affect skin.	TE **Activity** Measuring Temperature, p. 17 ◆ GENERAL TE **Connection Activity** Real World, p. 18 GENERAL SE **Skills Practice Lab** Seeing Is Believing, p. 20 GENERAL CRF **Datasheet for Chapter Lab*** SE **Science in Action** Math, Social Studies, and Language Arts Activities, pp. 26–27 GENERAL LB **Long-Term Projects & Research Ideas** Mapping the Human Body* ADVANCED	CRF **Lesson Plans*** TR **Bellringer Transparency*** TR Structures of the Skin* VID **Lab Videos for Life Science** TE **Internet Activity,** p. 17 GENERAL

PACING • 90 min

CHAPTER REVIEW, ASSESSMENT, AND STANDARDIZED TEST PREPARATION

- CRF **Vocabulary Activity*** GENERAL
- SE **Chapter Review,** pp. 22–23 GENERAL
- CRF **Chapter Review*** ■ GENERAL
- CRF **Chapter Tests A*** ■ GENERAL, **B*** ADVANCED, **C*** SPECIAL NEEDS
- SE **Standardized Test Preparation,** pp. 24–25 GENERAL
- CRF **Standardized Test Preparation*** GENERAL
- CRF **Performance-Based Assessment*** GENERAL
- OSP **Test Generator** GENERAL
- CRF **Test Item Listing*** GENERAL

Online and Technology Resources

go.hrw.com — Visit **go.hrw.com** for a variety of free resources related to this textbook. Enter the keyword **HL5BD1.**

Holt Online Learning — Students can access interactive problem-solving help and active visual concept development with the *Holt Science and Technology* Online Edition available at **www.hrw.com.**

Guided Reading Audio CD — A direct reading of each chapter using instructional visuals as guideposts. For auditory learners, reluctant readers, and Spanish-speaking students. Available in English and Spanish.

KEY

SE Student Edition	**CRF** Chapter Resource File	**SS** Science Skills Worksheets	* Also on One-Stop Planner
TE Teacher Edition	**OSP** One-Stop Planner	**MS** Math Skills for Science Worksheets	◆ Requires advance prep
	LB Lab Bank	**CD** CD or CD-ROM	■ Also available in Spanish
	TR Transparencies	**VID** Classroom Video/DVD	

SKILLS DEVELOPMENT RESOURCES	SECTION REVIEW AND ASSESSMENT	STANDARDS CORRELATIONS
SE Pre-Reading Activity, p. 2 GENERAL **OSP Science Puzzlers, Twisters & Teasers** GENERAL		National Science Education Standards UCP 1, 3,4 5; SAI 1, 2; SPSP 1; LS 1d, 1e, 3a, 3b
CRF Directed Reading A* ■ BASIC, **B*** SPECIAL NEEDS **CRF Vocabulary and Section Summary*** ■ GENERAL **SE Reading Strategy** Reading Organizer, p. 4 GENERAL **TE Reading Strategy** Prediction Guide, p. 5 BASIC **TE Inclusion Strategies,** p. 6	**SE Reading Checks,** pp. 5, 6 GENERAL **TE Reteaching,** p. 6 BASIC **TE Quiz,** p. 6 GENERAL **TE Alternative Assessment,** p. 6 GENERAL **SE Section Review,*** p. 7 ■ GENERAL **CRF Section Quiz*** ■ GENERAL	UCP 1, 3, 4; LS 1a, 1d, 3a
CRF Directed Reading A* ■ BASIC, **B*** SPECIAL NEEDS **CRF Vocabulary and Section Summary*** ■ GENERAL **SE Reading Strategy** Reading Organizer, p. 8 GENERAL **SE Connection to Environmental Science** Bones from the Ocean, p. 10 GENERAL **MS Math Skills for Science** Mechanical Advantage* GENERAL **CRF Reinforcement Worksheet** The Hipbone's Connected to the… BASIC **CRF Critical Thinking** The Tissue Engineering Debate* ADVANCED	**SE Reading Checks,** pp. 9, 10 GENERAL **TE Reteaching,** p. 10 BASIC **TE Quiz,** p. 10 GENERAL **SE Section Review,*** p. 11 ■ GENERAL **TE Alternative Assessment,** p. 11 GENERAL **CRF Section Quiz*** ■ GENERAL	UCP 2, 3, 5; SAI 1; SPSP 1, 4; LS 1d, 1e
CRF Directed Reading A* ■ BASIC, **B*** SPECIAL NEEDS **CRF Vocabulary and Section Summary*** ■ GENERAL **SE Reading Strategy** Discussion, p. 12 GENERAL **SE Math Practice** Runner's Time, p. 15 GENERAL **MS Math Skills for Science** The Unit Factor and Dimensional Analysis* GENERAL **CRF Reinforcement Worksheet** Muscle Map* BASIC	**SE Reading Checks,** pp. 13, 15 GENERAL **TE Reteaching,** p. 14 BASIC **TE Quiz,** p. 14 GENERAL **TE Alternative Assessment,** p. 14 GENERAL **TE Homework,** p. 14 ADVANCED **SE Section Review,*** p. 15 ■ GENERAL **CRF Section Quiz*** ■ GENERAL	UCP 1, 2, 3, 4; SAI 1, 2; SPSP 1; LS 1d, 1e; *LabBook:* UCP 3; SAI 1, 2; LS 1a, 1d, 1e, 3b
CRF Directed Reading A* ■ BASIC, **B*** SPECIAL NEEDS **CRF Vocabulary and Section Summary*** ■ GENERAL **SE Reading Strategy** Paired Summarizing, p. 16 GENERAL **TE Inclusion Strategies,** p. 17 **SE Connection to Social Studies** Using Hair, p. 18 GENERAL	**SE Reading Checks,** pp. 17, 18 GENERAL **TE Reteaching,** p. 18 BASIC **TE Quiz,** p. 18 GENERAL **TE Alternative Assessment,** p. 18 ADVANCED **SE Section Review,*** p. 19 ■ GENERAL **CRF Section Quiz*** ■ GENERAL	UCP 3, 5; SAI 2; SPSP 1; LS 1c, 1d, 1e, 1f, 3a, 3b; *Chapter Lab:* UCP 1, 2, 3, 5; SAI 1, 2; SPSP 1; LS 1d

One-Stop Planner® CD-ROM

This convenient CD-ROM includes:

- **Lab Materials QuickList Software**
- **Holt Calendar Planner**
- **Customizable Lesson Plans**
- **Printable Worksheets**
- **ExamView® Test Generator**

CNN Student News

cnnstudentnews.com

Find the latest news, lesson plans, and activities related to important scientific events.

SciLinks NSTA

www.scilinks.org

Maintained by the **National Science Teachers Association.** See Chapter Enrichment pages for a complete list of topics.

Current Science®

Check out ***Current Science*** articles and activities by visiting the HRW Web site at **go.hrw.com.** Just type in the keyword **HL5CS22T.**

Classroom Videos

- **Lab Videos** demonstrate the chapter lab.
- **Brain Food Video Quizzes** help students review the chapter material.

Chapter Resources

Visual Resources

CHAPTER STARTER TRANSPARENCY

BELLRINGER TRANSPARENCIES

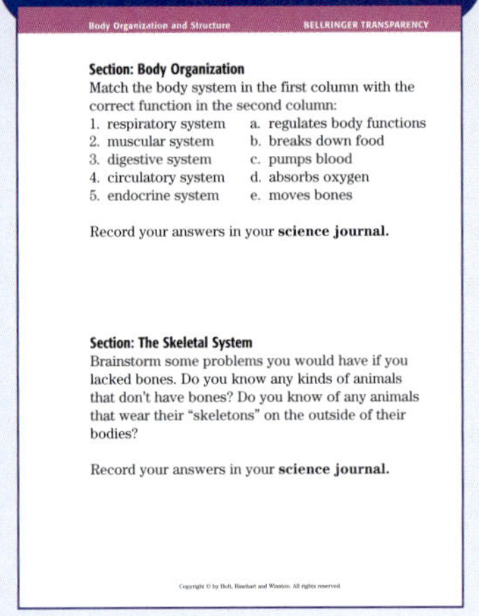

TEACHING TRANSPARENCIES

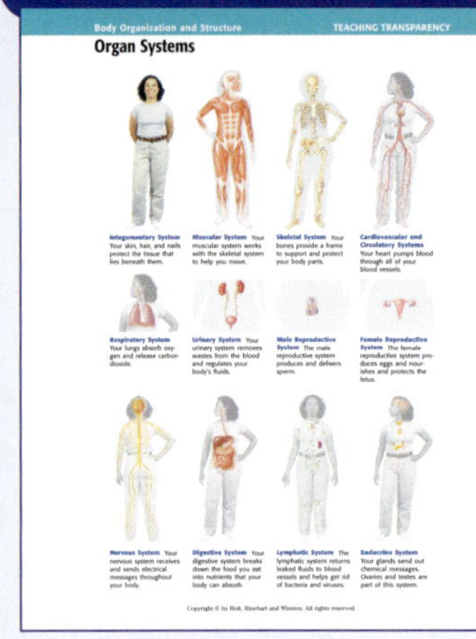

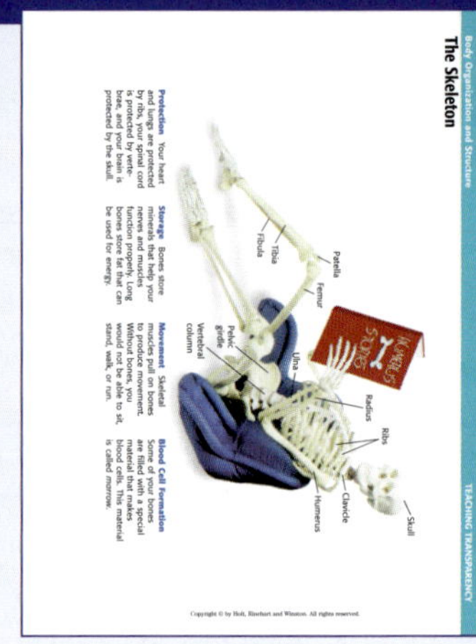

TEACHING TRANSPARENCIES

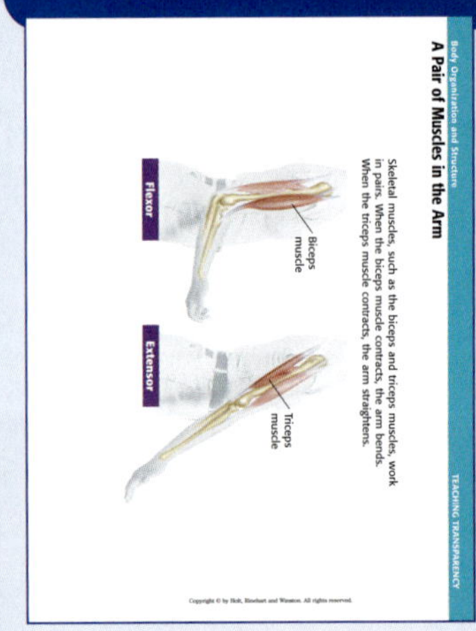

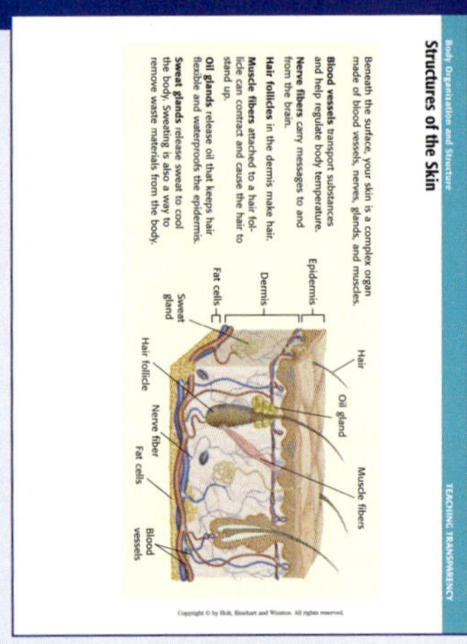

CONCEPT MAPPING TRANSPARENCY

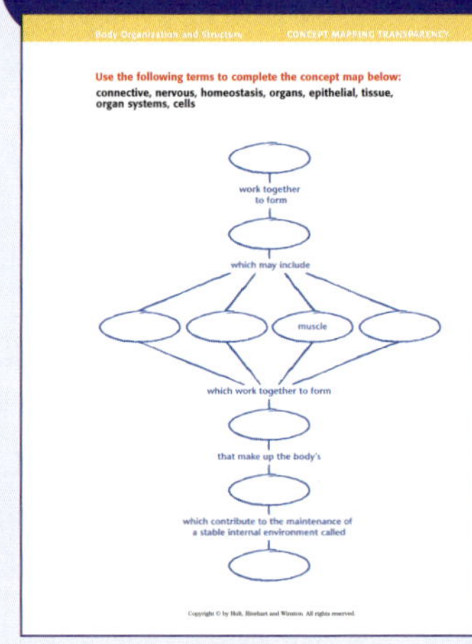

Planning Resources

LESSON PLANS

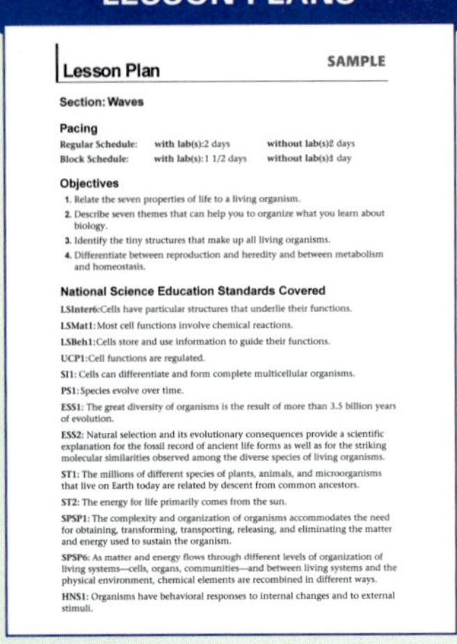

PARENT LETTER

TEST ITEM LISTING

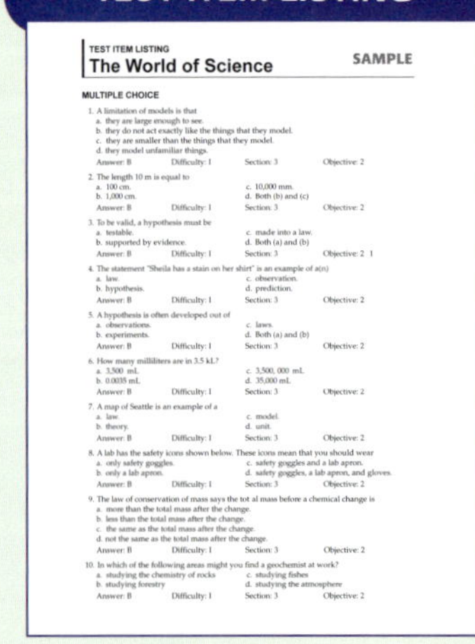

One-Stop Planner® CD-ROM

This CD-ROM includes all of the resources shown here and the following time-saving tools:

- *Lab Materials QuickList Software*
- *Customizable lesson plans*
- *Holt Calendar Planner*
- *The powerful ExamView® Test Generator*

For a preview of available worksheets covering math and science skills, see pages T12–T19. All of these resources are also on the One-Stop Planner®.

Meeting Individual Needs

DIRECTED READING A
Skills Worksheet — Directed Reading A — SAMPLE
BASIC · ALSO IN SPANISH

VOCABULARY ACTIVITY
Activity — Vocabulary Activity — SAMPLE
Getting the Dirt on the Soil
GENERAL

REINFORCEMENT
Skills Worksheet — Reinforcement — SAMPLE
The Plane Truth
BASIC

SCILINKS ACTIVITY
Activity — SciLinks Activity — SAMPLE
GENERAL

DIRECTED READING B
Skills Worksheet — Directed Reading B — SAMPLE
SPECIAL NEEDS

VOCABULARY AND SECTION SUMMARY
Skills Worksheet — Vocabulary & Notes — SAMPLE
GENERAL · ALSO IN SPANISH

CRITICAL THINKING
Skills Worksheet — Critical Thinking — SAMPLE
A Solar Solution
ADVANCED

SCIENCE PUZZLERS, TWISTERS & TEASERS
Chapter 22 — Science Puzzlers, Twisters & Teasers — Body Organization and Structure
GENERAL

Labs and Activities

LONG-TERM PROJECTS & RESEARCH IDEAS
Project 22 — Student Worksheet — Mapping the Human Body
ADVANCED

INQUIRY LABS
Lab 6 — Student Worksheet — On a Wing and a Layer
GENERAL

DATASHEETS FOR QUICK LABS
Quick Lab — Reaction to Stress — SAMPLE

DATASHEETS FOR CHAPTER LABS
Skills Practice Lab — Using Scientific Methods — SAMPLE

DATASHEETS FOR LABBOOK
Skills Practice Lab — Does It All Add Up? — SAMPLE

Review and Assessments

SECTION QUIZ
Assessment — Section Quiz — SAMPLE
GENERAL · ALSO IN SPANISH

CHAPTER REVIEW
Skills Worksheet — Chapter Review — SAMPLE
GENERAL · ALSO IN SPANISH

CHAPTER TEST B
Assessment — Chapter Test B — SAMPLE
ADVANCED

STANDARDIZED TEST PREPARATION
Assessment — Standardized Test Preparation — SAMPLE
GENERAL

SECTION REVIEW
Skills Worksheet — Section Review — SAMPLE
GENERAL · ALSO IN SPANISH

CHAPTER TEST A
Assessment — Chapter Test A — SAMPLE
GENERAL · ALSO IN SPANISH

CHAPTER TEST C
Assessment — Chapter Test C — SAMPLE
SPECIAL NEEDS

PERFORMANCE-BASED ASSESSMENT
Assessment — Performance-Based Assessment — SAMPLE
GENERAL

1 Chapter Enrichment

This Chapter Enrichment provides relevant and interesting information to expand and enhance your presentation of the chapter material.

Section 1

Body Organization

Tissues

- Tissues differ from each other in terms of the shape and size of cells, the amount and kind of material between the cells, and the special functions the tissues perform to maintain proper functioning of the body.

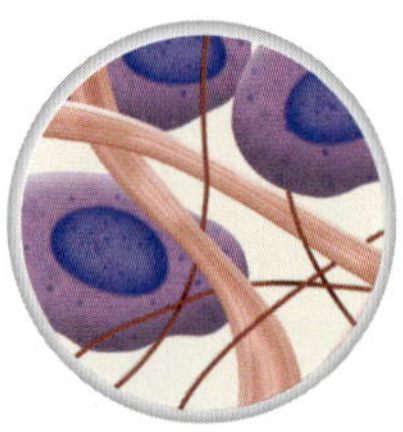

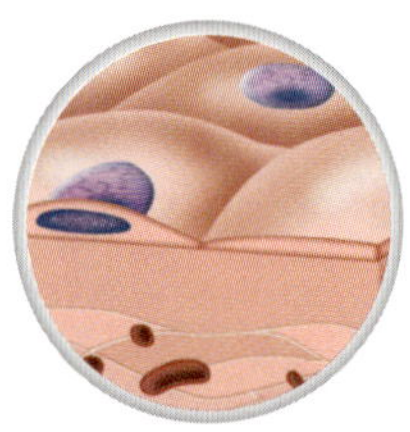

- Connective tissue is the most abundant tissue in the body. It displays the most variety in form and type. All connective tissue, however, can be classified into one of four types: dense connective tissue (cartilage and bone), loose connective tissue (found beneath the skin and around nerves, blood vessels, and organs), liquid connective tissue (blood and lymph), and adipose tissue (where the body stores energy as droplets of fat).
- Although bone is considerably harder than other body tissues, it accounts for only about 14% of a person's total body weight.

Section 2

The Skeletal System

The Human Skeleton

- The skeleton provides support for soft tissue. Also, it regulates body minerals and produces both red and white blood cells. There are typically 206 bones in the adult human body, but extra bones, particularly those in the hands and feet, can increase that number. The number of bones in children varies with age.
- The skeleton forms from more than 800 centers of ossification. All of the bony elements are generally not completely united to form an adult skeleton until a person reaches his or her mid-20s.
- The skeletons of male and female humans are slightly different. The most pronounced differences are in the pelvis. A female's pelvis is adapted for childbearing and thus has a larger pelvic inlet. Women who are malnourished during childhood typically do not develop the wider pelvis, which can make natural childbirth dangerous or even fatal for them.
- An individual's age can be determined by looking at the skeleton alone. A younger individual's dentition and bone fusion patterns indicate his or her age. In adults, age determination is more difficult because one must rely solely on signs of skeletal deterioration.

Bones

- Each bone is surrounded by a strong fibrous covering called a *periosteum*. Articular surfaces are covered in cartilage.
- Bones are made of three types of cells: osteoblasts, osteocytes, and osteoclasts. Osteoblasts are bone-producing cells. Osteocytes are bone-maintaining cells. Osteoclasts are bone-destroying cells.
- For its weight, bone is 5 times stronger than steel.

Joints

- Doctors typically classify joints by structure rather than movement. The three types of joint structures are called *fibrous*, *cartilaginous*, and *synovial*. Fibrous joints (such as those in the skull) are immovable joints in which a fibrous tissue or a hyaline cartilage connects the bones. Cartilaginous joints (such as those in the rib cage) are slightly moveable joints in which cartilage connects the bones. Synovial joints (such as the knee) are freely moving joints in which synovial membranes cover the cartilage and ligaments connecting the bone.

Section 3

The Muscular System

Skeletal Muscles

- There are more than 600 skeletal muscles in the human body. They are often organized into the following groups: muscles of the head and the neck, muscles of the trunk, muscles of the upper limbs, and muscles of the lower limbs.

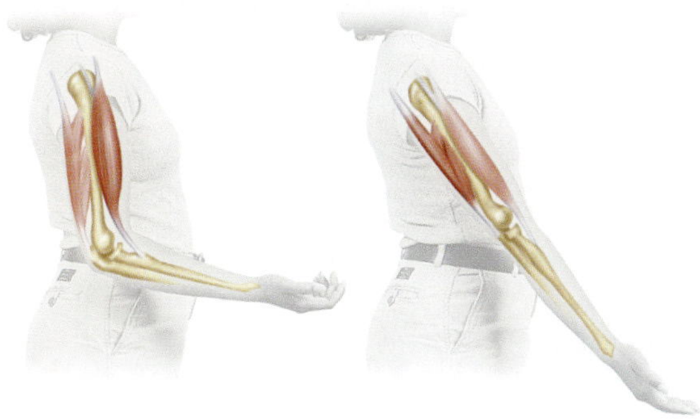

Types of Muscle Cells

- When observed through a microscope, the three types of muscles are clearly identifiable. Cells of smooth muscles have a long, tapered shape; no clearly defined striations; and a large central nucleus. Skeletal muscle cells are long and tapered and characterized by distinct light- and dark-colored bands. Each cell has multiple nuclei because several skeletal muscle cells merge, and the cell membranes become indistinct. The cells of cardiac muscle have one or more nuclei and have an irregular, branched shape.

Section 4

The Integumentary System

Skin

- One square inch of skin can hold as many as 650 sweat glands, 20 blood vessels, and more than 1,000 nerve endings.
- Each person has a unique series of ridges and indentations called *fingerprints* on the tips of his or her fingers. No two people have the same fingerprints. Fingerprints help the fingers to grip slippery surfaces. Each person also has unique patterns on the tips of his or her toes.

Is That a Fact!

- ◆ More than three-fourths of the dust in some homes is made up of dead skin cells!

Hair and Nails

- Only mammals have true hair. All mammals have hair somewhere on their bodies.
- The body's most visible signs of aging occur in the integumentary system. Skin becomes thin, dry, wrinkled, and less supple. Dark-colored age spots may develop. Hair turns gray or white and may begin to fall out. Hair follicles decrease in number. Sweat glands become less active, so older people are less tolerant to extremely hot weather.

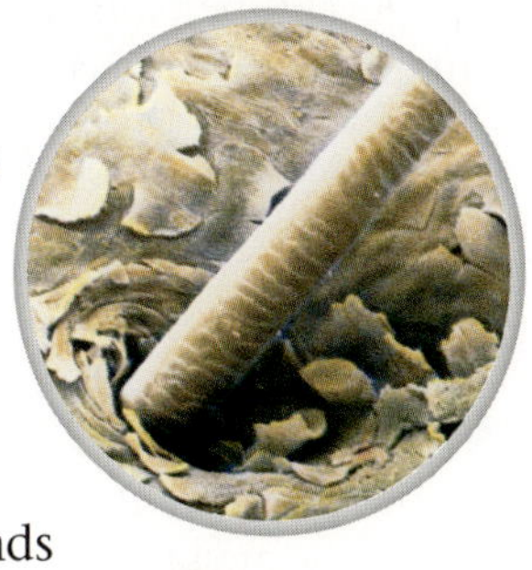

- Hair that is kept short grows an average of 2 cm per month. Growth slows to about 1 cm per month when hair reaches about 30 cm long. Fingernails grow about 2 cm each year. The fastest-growing nail is on the middle finger. Fingernails grow three to four times more quickly than toenails do.

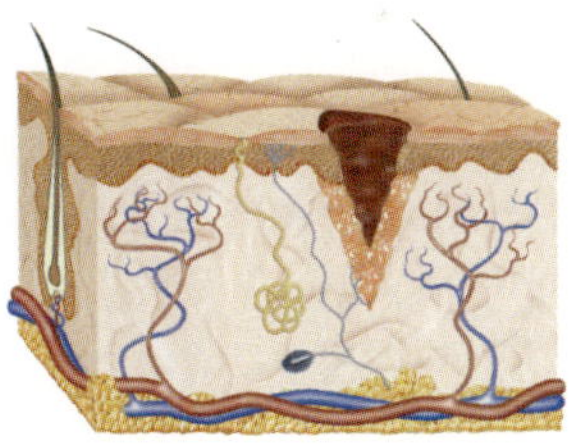

SciLinks — NSTA Developed and maintained by the National Science Teachers Association

SciLinks is maintained by the National Science Teachers Association to provide you and your students with interesting, up-to-date links that will enrich your classroom presentation of the chapter.

Visit www.scilinks.org and enter the SciLinks code for more information about the topic listed.

Topic: Tissues and Organs
SciLinks code: HSM1530

Topic: Body Systems
SciLinks code: HSM0184

Topic: Skeletal System
SciLinks code: HSM1399

Topic: Muscular System
SciLinks code: HSM1008

Topic: Integumentary System
SciLinks code: HSM0803

1

Overview

Tell students that this chapter will help them learn about human body systems. In particular, the chapter will introduce students to the skeletal, muscular, and integumentary systems.

Assessing Prior Knowledge

Students should be familiar with the following topics:

- cells
- body organization

Identifying Misconceptions

As students learn the material in this chapter, some of them may be confused about the concept of homeostasis. Students should understand that homeostasis is a state in which the internal environment of the human body is stable. Help students understand that because the external environment is always changing, the human body must adjust to these changes to maintain homeostasis within the body. Students should also understand that all cells in the body play a role in homeostasis, but because there are many kinds of cells, no single cell has to do all of the jobs necessary for homeostasis.

1

Body Organization and Structure

About the PHOTO

Lance Armstrong has won the Tour de France several times. These victories are especially remarkable because he was diagnosed with cancer in 1996. But with medicine and hard work, he grew strong enough to win one of the toughest events in all of sports.

PRE-READING ACTIVITY

FOLDNOTES **Four-Corner Fold** Before you read the chapter, create the FoldNote entitled "Four-Corner Fold" described in the **Study Skills** section of the Appendix. Label the flaps of the four-corner fold with "The skeletal system," "The muscular system," and "The integumentary system." Write what you know about each topic under the appropriate flap. As you read the chapter, add other information that you learn.

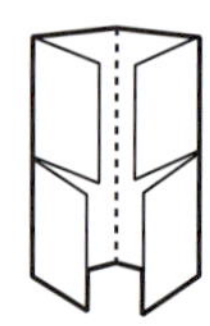

Standards Correlations

National Science Education Standards

The following codes indicate the National Science Education Standards that correlate to this chapter. The full text of the standards is at the front of the book.

Chapter Opener
UCP 1, 3, 4, 5; SAI 1, 2; SPSP 1; LS 1d, 1e, 3a, 3b

Section 1 Body Organization
UCP 1, 3, 4; LS 1a, 1d, 3a

Section 2 The Skeletal System
UCP 2, 3, 5; SAI 1; SPSP 1, 4; LS 1d, 1e

Section 3 The Muscular System
UCP 1, 2, 3, 4; SAI 1, 2; SPSP 1; LS 1d, 1e; *LabBook:* UCP 3; SAI 1, 2; LS 1a, 1d, 1e, 3b

Section 4 The Integumentary System
UCP 3, 5; SAI 2; SPSP 1; LS 1c, 1d, 1e, 1f, 3a, 3b

Chapter Lab
UCP 1, 2, 3, 5; SAI 1, 2; SPSP 1; LS 1d

Chapter Review
UCP 1, 2, 3, 4, 5; SAI 1, 2; SPSP 1, 4; LS 1a, 1c, 1d, 1e, 1f, 3a, 3b

Science in Action
UCP 1, 5; ST 1, 2; SPSP 1, 4, 5; LS 1d, 1e

START-UP ACTIVITY

Too Cold for Comfort

Your nervous system sends you messages about your body. For example, if someone steps on your toe, your nervous system sends you a message. The pain you feel is a message that tells you to move your toe to safety. Try this exercise to watch your nervous system in action.

Procedure

1. Hold **a few pieces of ice** in one hand. Allow the melting water to drip into a **dish.** Hold the ice until the cold is uncomfortable. Then, release the ice into the dish.
2. Compare the hand that held the ice with your other hand. Describe the changes you see.

Analysis

1. What message did you receive from your nervous system while you held the ice?
2. How quickly did the cold hand return to normal?
3. What organ systems do you think helped restore your hand to normal?
4. Think of a time when your nervous system sent you a message, such as an uncomfortable feeling of heat, cold, or pain. How did your body react?

START-UP ACTIVITY

MATERIALS

FOR EACH GROUP

- dish, waterproof
- ice

Safety Caution: Students should clean up the water that results from the melting ice, especially if it drips onto the floor.

Answers

1. Answers may vary, but students should feel discomfort that is intense enough to make them want to drop the ice.
2. Answers may vary. Students' hands will return to normal at varying rates, but most students should answer that it took a few minutes.
3. Students should recognize that the integumentary, circulatory, and cardiovascular systems played roles in returning their hands to normal. Point out that the redness reflects an increased blood supply to the cold area. By bringing warmth to the hand, the blood helps to restore the hand to normal.
4. Answers may vary, but students should recognize that the nervous system detected the discomfort and sent signals to other body systems to respond. Some students may note that they jerked away from the discomfort.

Body Organization and Structure — CHAPTER STARTER

This Really Happened!

On April 14, 1912, at 11:40 P.M., the British steamship *Titanic*, the largest and most luxurious steamship ever built, struck an iceberg. Jack Thayer, a 17-year-old from Pennsylvania, felt the impact in his cabin and went on deck to see what had happened. To his horror, he found that the ship was sinking and that there were not enough lifeboats for the 2,207 passengers and crew. Jack watched as the last of the lifeboats, filled with women and children, were launched. Just before the ship slid beneath the icy water, Jack jumped. The pain of the cold felt like knives stabbing every inch of his body. He swam to an overturned lifeboat and climbed on.

Jack Thayer was among the survivors who were able to keep themselves dry enough to survive until help arrived, at 4 A.M. An estimated 1,500 people died. Even those who were wearing life jackets couldn't survive the extreme temperature of the water. The freezing water caused their body systems to fail. Breathing became difficult, and muscles no longer functioned. Finally, they lost consciousness and their hearts stopped.

The water around the *Titanic* was just too cold for human survival.

Under normal circumstances, our bodies are able to maintain a constant internal temperature. Read on to find out about this and other interesting information on how our bodies work.

Chapter Starter Transparency
Use this transparency to help students begin thinking about human body systems.

CHAPTER RESOURCES

Technology

- **Transparencies** — READING SKILLS
  - Chapter Starter Transparency
- **Student Edition on CD-ROM**
- **Guided Reading Audio CD**
  - English or Spanish
- **Classroom Videos**
  - Brain Food Video Quiz

Workbooks

- **Science Puzzlers, Twisters & Teasers**
  - Body Organization and Structure GENERAL

SECTION 1

Focus

Overview

This section introduces the basic organization of the human body. Students will learn about the four major tissues of the body and that the body's organs are arranged by function into 12 organ systems.

Bellringer

Write the names of the following organ systems on the board or an overhead projector: *respiratory system, muscular system, digestive system, cardiovascular system,* and *endocrine system.* Then, write the following functions: *to pump blood, to enable movement, to send out chemical messages, to absorb oxygen,* and *to break down food.* Ask students to match each organ system with its correct function. (The respiratory system absorbs oxygen. The muscular system enables movement. The digestive system breaks down food. The cardiovascular system pumps blood. The endocrine system sends out chemical messages.)

SECTION 1

READING WARM-UP

Objectives

- Describe how tissues, organs, and organ systems are related.
- List 12 organ systems.
- Identify how organ systems work together to maintain homeostasis.

Terms to Learn

homeostasis organ
tissue

READING STRATEGY

Reading Organizer As you read this section, make a concept map by using the terms above.

Body Organization

Imagine jumping into a lake. At first, your body feels very cold. You may even shiver. But eventually you get used to the cold water. How?

Your body gets used to cold water because of homeostasis (HOH mee OH STAY sis). **Homeostasis** is the maintenance of a stable internal environment in the body. When you jump into a lake, homeostasis helps your body adapt to the cold water.

homeostasis the maintenance of a constant internal state in a changing environment

tissue a group of similar cells that perform a common function

Cells, Tissues, and Organs

Maintaining homeostasis is not easy. Your internal environment is always changing. Your cells need nutrients and oxygen to survive. Your cells need wastes removed. If homeostasis is disrupted, cells may not get the materials they need. So, cells may be damaged or may die.

Your cells must do many jobs to maintain homeostasis. Fortunately, each of your cells does not have to do all of those jobs. Just as each person on a soccer team has a role during a game, each cell in your body has a job in maintaining homeostasis. Your cells are organized into groups. A group of similar cells working together forms a **tissue.** Your body has four main kinds of tissue. The four kinds of tissue are shown in **Figure 1.**

Figure 1 Four Kinds of Tissue

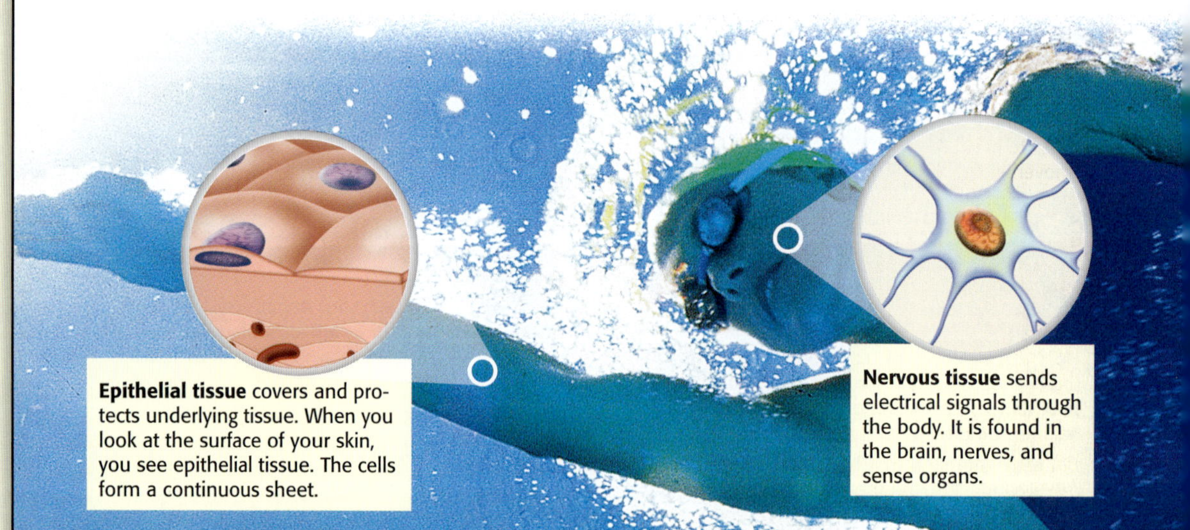

Epithelial tissue covers and protects underlying tissue. When you look at the surface of your skin, you see epithelial tissue. The cells form a continuous sheet.

Nervous tissue sends electrical signals through the body. It is found in the brain, nerves, and sense organs.

CHAPTER RESOURCES

Chapter Resource File

- Lesson Plan
- Directed Reading A BASIC
- Directed Reading B SPECIAL NEEDS

Technology

Transparencies
- Bellringer

Is That a Fact!

The Pompeii worm, *Alvinella pompejana,* can survive a temperature difference of 60°C between its head and its tail! Scientists theorize that a coating of furry bacteria living on the worm's back allows the worm to endure such extreme temperature differences.

Figure 2 **Organization of the Stomach**

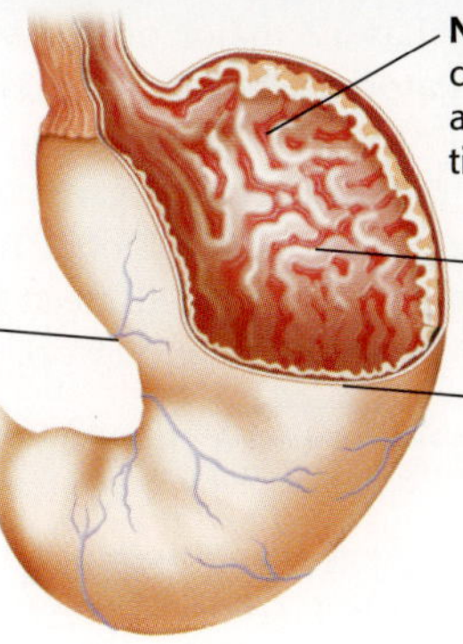

Tissues Form Organs

One kind of tissue alone cannot do all of the things that several kinds of tissue working together can do. Two or more tissues working together form an **organ.** Your stomach, shown in **Figure 2,** uses all four kinds of tissue to carry out digestion.

organ a collection of tissues that carry out a specialized function of the body

Organs Form Systems

Your stomach does a lot to help you digest your food. But the stomach doesn't do it all. Your stomach works with other organs, such as the small and large intestines, to digest your food. Organs that work together make up an *organ system.*

Reading Check **How is the stomach part of an organ system?** *(See the Appendix for answers to Reading Checks.)*

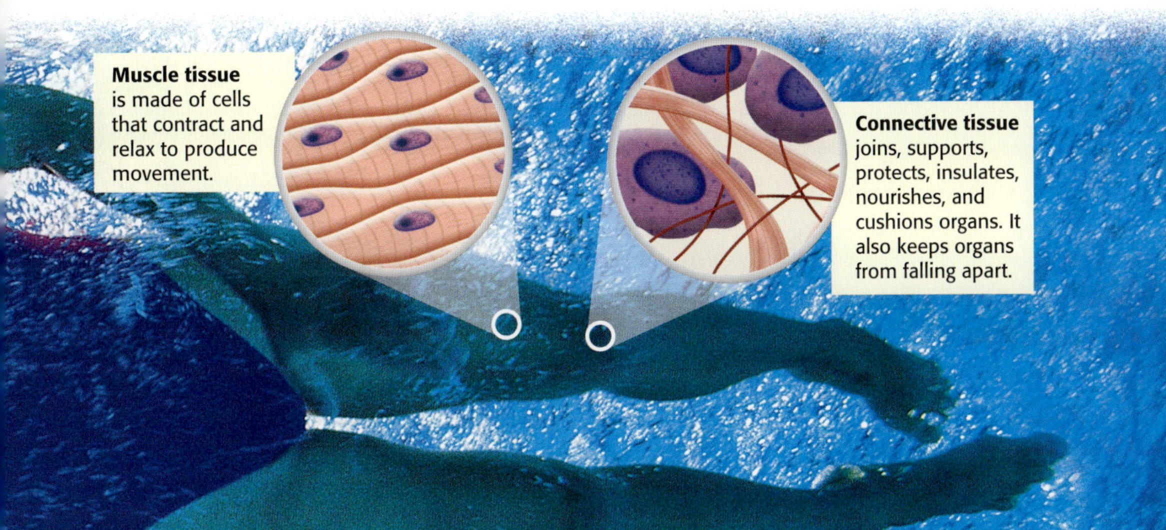

CONNECTION ACTIVITY
History — ADVANCED

Organ Transplants The first transplant of a human heart was performed in Cape Town, South Africa, on December 3, 1967, by Dr. Christiaan Barnard and a team of 30 physicians. Have interested students research when other organs, such as kidneys, livers, and lungs, were first transplanted. Have students make posters about their findings. LS **Visual**

Answer to Reading Check

The stomach works with other organs, such as the small and large intestines, to digest food.

Motivate

Discussion — GENERAL

Homeostasis Before students read this section, ask them to relate what they know about how different animals adjust to outside temperatures. (Most students will likely relate how ectotherms and endotherms adjust body temperature.) Ask students how the idea of homeostasis fits into what they know about animals and temperature regulation. (Students should recognize that temperature regulation is a function of homeostasis.) LS **Logical/Verbal**

Teach

READING STRATEGY — BASIC

Prediction Guide Before students read the pages that describe the body's tissues and organs, ask whether each of the following statements is true or false.

- Homeostasis is the body's ability to maintain a stable internal environment. (true)
- The human body has four main types of tissues. (true)
- An organ is a group of tissues that work together. (true)

LS **Verbal**

Debate — ADVANCED

Transplant Ethics Thousands of people in the United States are waiting for organ transplants. The average cost of an organ transplant is $120,000. Encourage students to research and debate the ethical issues surrounding transplants. The following are suggested topics:

- Should transplants happen at all?
- Who should get a transplant?
- Should a child receive a transplant before an older person does?

LS **Logcal/Interpersonal**

Close

Reteaching — Basic

Organizing Information To help students understand and identify the 12 major organ systems of the body, have them make a table with the following headings: "Name of organ system," "Function(s)," and "Main organs." Have students use the table to review the information presented on these pages. **LS Logical**

Quiz — General

Ask students whether each of the following statements is true or false. Have students correct false statements.

1. Homeostasis is the maintenance of a stable internal environment. (true)
2. Epithelial tissue sends electrical signals throughout the body. (false; Nervous tissue sends electrical signals throughout the body and epithelial tissue covers and protects underlying tissue.)
3. Blood is a type of connective tissue. (true)

Alternative Assessment — General

Writing **Organ Systems** In their **science journal,** have students describe three of the organ systems introduced in this section. Have them describe the functions and primary organs of each system and include drawings of each system. **LS Verbal**

Working Together

Organ systems work together to maintain homeostasis. Your body has 12 major organ systems, as shown in **Figure 3.** The circulatory and cardiovascular systems are shown together. The cardiovascular system includes your heart and blood vessels. Additionally, these organs are part of the circulatory system, which also includes blood. Together, these two systems deliver the materials your cells need to survive. This is just one example of how organ systems work together to keep you healthy.

Reading Check Give an example of how organ systems work together in the body.

Figure 3 Organ Systems

Integumentary System Your skin, hair, and nails protect the tissue that lies beneath them.

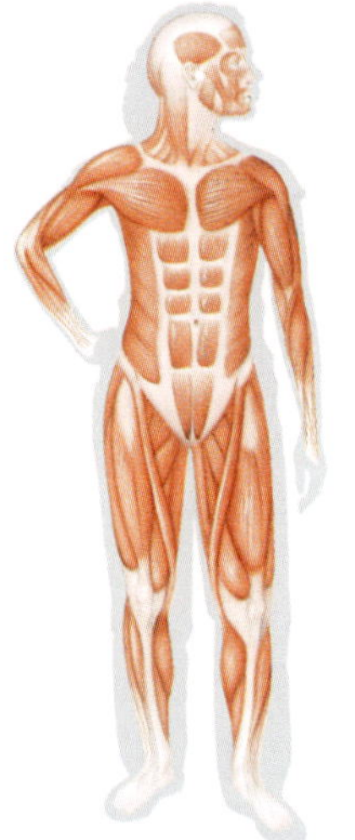

Muscular System Your muscular system works with the skeletal system to help you move.

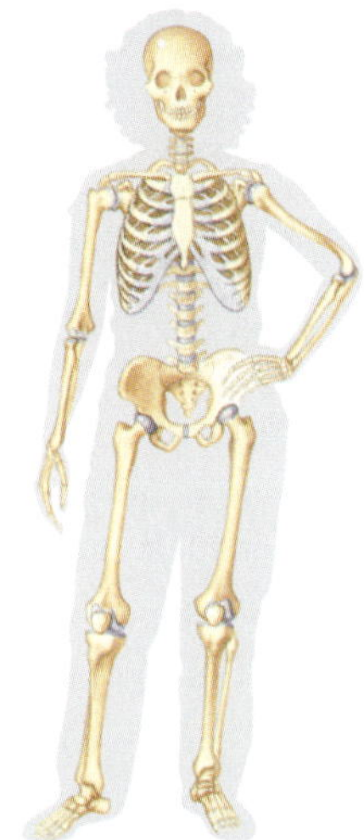

Skeletal System Your bones provide a frame to support and protect your body parts.

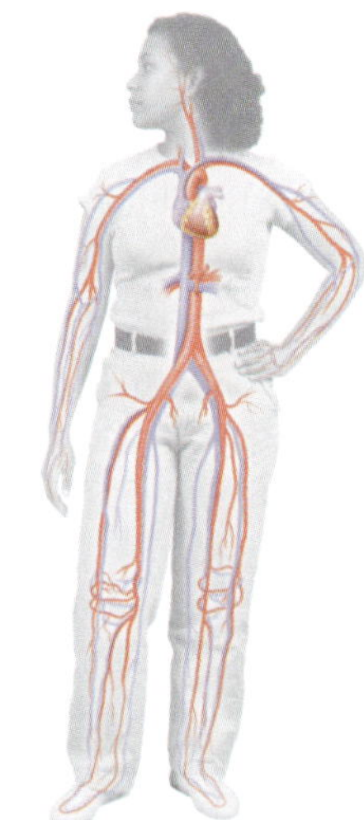

Cardiovascular and Circulatory Systems Your heart pumps blood through all of your blood vessels.

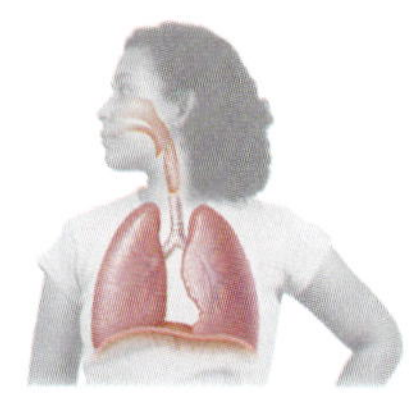

Respiratory System Your lungs absorb oxygen and release carbon dioxide.

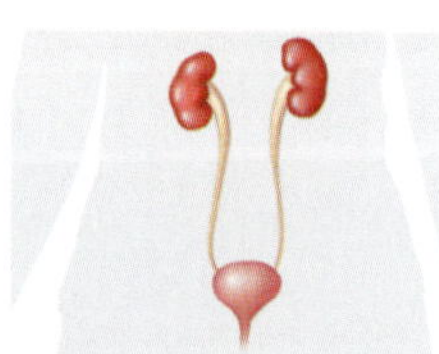

Urinary System Your urinary system removes wastes from the blood and regulates your body's fluids.

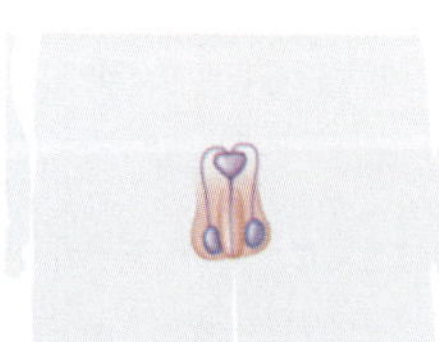

Male Reproductive System The male reproductive system produces and delivers sperm.

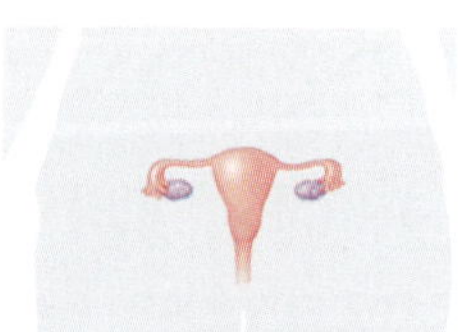

Female Reproductive System The female reproductive system produces eggs and nourishes and protects the fetus.

INCLUSION Strategies

- *Attention Deficit Disorder* • *Visually Impaired*
- *Developmentally Delayed*

Kinesthetic activities can help students learn more easily. Ask 11 students to hold signs identifying each student as an organ system (exclude the reproductive system). Ask the students to step forward if their system is involved in the following activities: snoring (respiratory and muscular), sweating (lymphatic and integumentary), jumping in fear (muscular, skeletal, nervous, and endocrine), drinking (muscular, digestive, and urinary), eating an apple (digestive and muscular), running (muscular, skeletal, circulatory, cardiovascular, and respiratory). You may want to tell students that some body systems, such as the nervous system and the endocrine system, play a role in nearly all of the body's functions. **LS Kinesthetic/Visual**

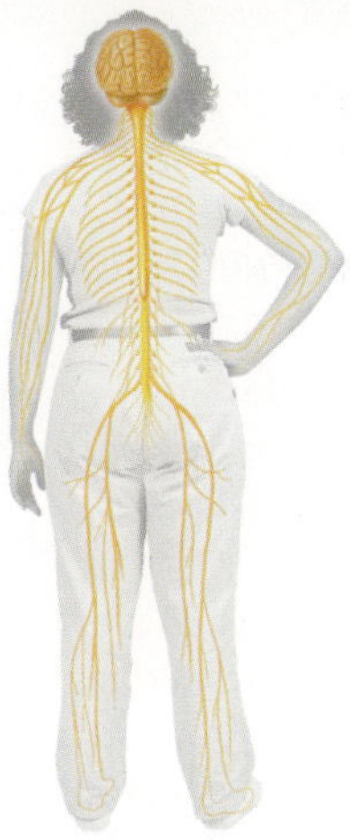

Nervous System Your nervous system receives and sends electrical messages throughout your body.

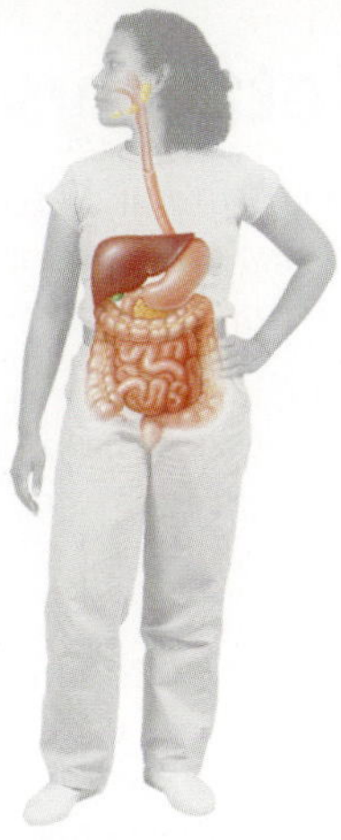

Digestive System Your digestive system breaks down the food you eat into nutrients that your body can absorb.

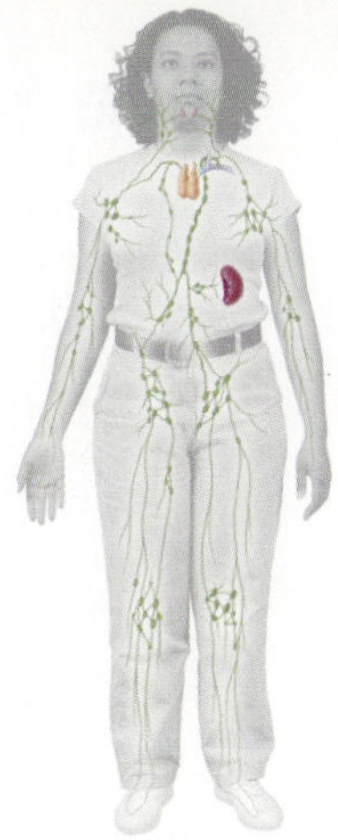

Lymphatic System The lymphatic system returns leaked fluids to blood vessels and helps get rid of bacteria and viruses.

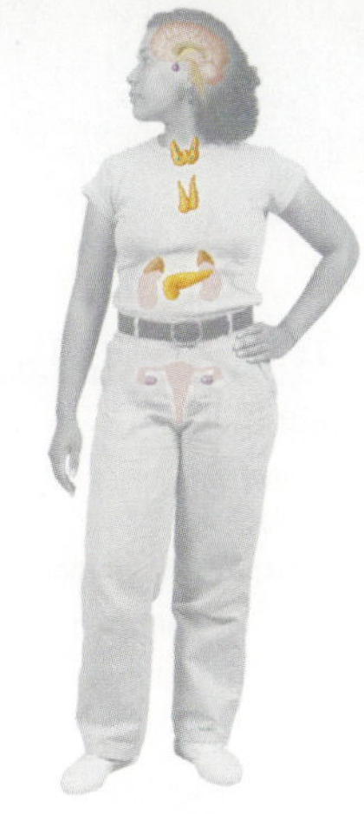

Endocrine System Your glands send out chemical messages. Ovaries and testes are part of this system.

SECTION Review

Summary

- A group of cells that work together is a tissue. Tissues form organs. Organs that work together form organ systems.
- There are four kinds of tissue in the human body.
- There are 12 major organ systems in the human body.
- Organ systems work together to help the body maintain homeostasis.

Using Key Terms

1. Use the following terms in the same sentence: *homeostasis*, *tissue*, and *organ*.

Understanding Key Ideas

2. Which of the following statements describes how tissues, organs, and organ systems are related?
   - **a.** Organs form tissues, which form organ systems.
   - **b.** Organ systems form organs, which form tissues.
   - **c.** Tissues form organs, which form organ systems.
   - **d.** None of the above
3. List the 12 organ systems.

Math Skills

4. The human skeleton has 206 bones. The human skull has 22 bones. What percentage of human bones are skull bones?

Critical Thinking

5. **Applying Concepts** Tanya went to a restaurant and ate a hamburger. Describe how Tanya used five organ systems to eat and digest her hamburger.
6. **Predicting Consequences** Predict what might happen if the human body did not have specialized cells, tissues, organs, and organ systems to maintain homeostasis.

SCILINKS. NSTA Developed and maintained by the National Science Teachers Association

For a variety of links related to this chapter, go to www.scilinks.org

Topic: Tissues and Organs; Body Systems
SciLinks code: HSM1530; HSM0184

Answers to Review

1. Sample answer: Homeostasis is maintained by cells, tissues, and organs.
2. c
3. integumentary system, muscular system, skeletal system, cardiovascular system, circulatory system, respiratory system, urinary system, reproductive system, nervous system, digestive system, lymphatic system, and endocrine system
4. about 11% (22 bones $\div$ 206 bones $\times$ 100 = 10.7%)
5. Sample answer: The nervous system sends messages to Tanya's muscular system to bite into the hamburger. Tanya's jaws, which are part of the skeletal system, help grind up the food. The food moves into Tanya's stomach, which is part of her digestive system. The digestive system breaks down the food, and the circulatory system picks up nutrients from the digestive system to take throughout the body.
6. Sample answer: The body would be unable to maintain homeostasis because the cells of the body would have too many jobs to do. The cells likely would not be able to do all of these jobs.

Answer to Reading Check

Sample answer: The cardiovascular system includes the heart and blood vessels. These organs are also part of the circulatory system, which includes blood. Together, these systems deliver the materials cells need to survive.

CHAPTER RESOURCES

Chapter Resource File

- Section Quiz GENERAL
- Section Review GENERAL
- Vocabulary and Section Summary GENERAL
- SciLinks Activity GENERAL

Technology

Transparencies
- Organ Systems

SECTION 2

Focus

Overview

This section introduces the skeletal system and describes the functions of bones. The section also illustrates the internal structure of bones and three types of joints.

Bellringer

Have students brainstorm problems that they would have if they lacked bones. (Students should understand that they would have no defined structure, mineral storage, organ protection, blood cells, or mobility.)

Motivate

ACTIVITY — GENERAL

Locating Bones Review with students that the skeletal system supports the body and protects delicate body parts. Encourage students to press the skin in various parts of their body to feel their bones. Ask students to describe any parts of their body where they cannot feel their bones. (Answers may vary but should include the abdomen, nose, and ears.) As you point to various parts of the body, ask students what organs the bones protect. (Sample answers: The skull protects the brain. The ribs protect the heart and lungs.) LS Kinesthetic/Visual

SECTION 2

The Skeletal System

When you hear the word skeleton, *you may think of the remains of something that has died. But your skeleton is not dead. It is very much alive.*

READING WARM-UP

Objectives

- Identify the major organs of the skeletal system.
- Describe four functions of bones.
- Describe three types of joints.
- List three injuries and two diseases that affect bones and joints.

Terms to Learn

skeletal system
joint

READING STRATEGY

Reading Organizer As you read this section, create an outline of the section. Use the headings from the section in your outline.

You may think your bones are dry and brittle. But they are alive and active. Bones, cartilage, and the connective tissue that holds bones together make up your **skeletal system.**

skeletal system the organ system whose primary function is to support and protect the body and to allow the body to move

Bones

The average adult human skeleton has 206 bones. Bones help support and protect parts of your body. They work with your muscles so you can move. Bones also help your body maintain homeostasis by storing minerals and making blood cells. **Figure 1** shows the functions of your skeleton.

Figure 1 The Skeleton

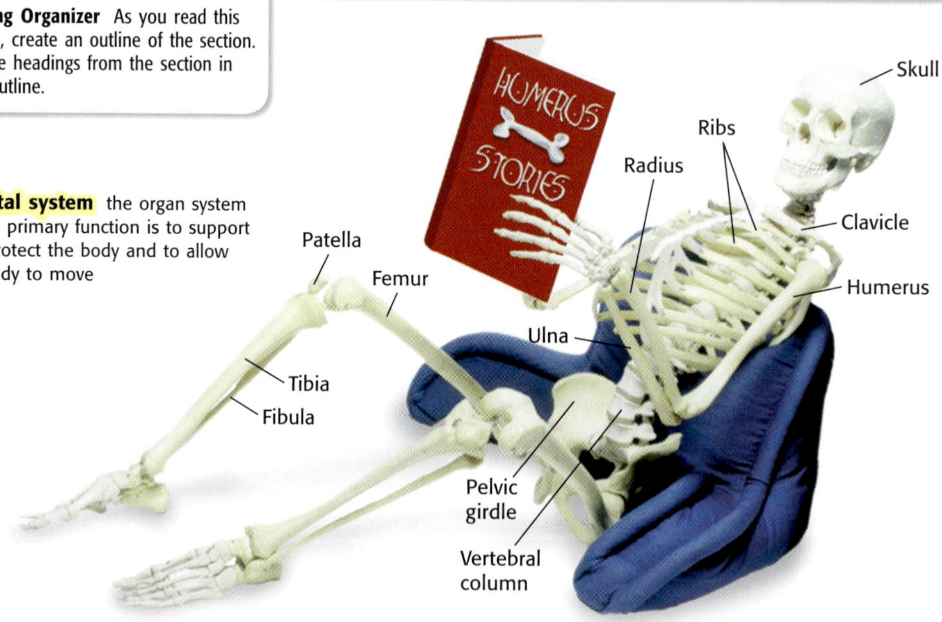

Protection Your heart and lungs are protected by ribs, your spinal cord is protected by vertebrae, and your brain is protected by the skull.

Storage Bones store minerals that help your nerves and muscles function properly. Long bones store fat that can be used for energy.

Movement Skeletal muscles pull on bones to produce movement. Without bones, you would not be able to sit, stand, walk, or run.

Blood Cell Formation Some of your bones are filled with a special material that makes blood cells. This material is called *marrow*.

CHAPTER RESOURCES

Chapter Resource File

- Lesson Plan
- Directed Reading A BASIC
- Directed Reading B SPECIAL NEEDS

Technology

Transparencies
- Bellringer
- The Skeleton

SCIENCE HUMOR

Q: Why didn't the skeleton cross the road?

A: It didn't have the guts.

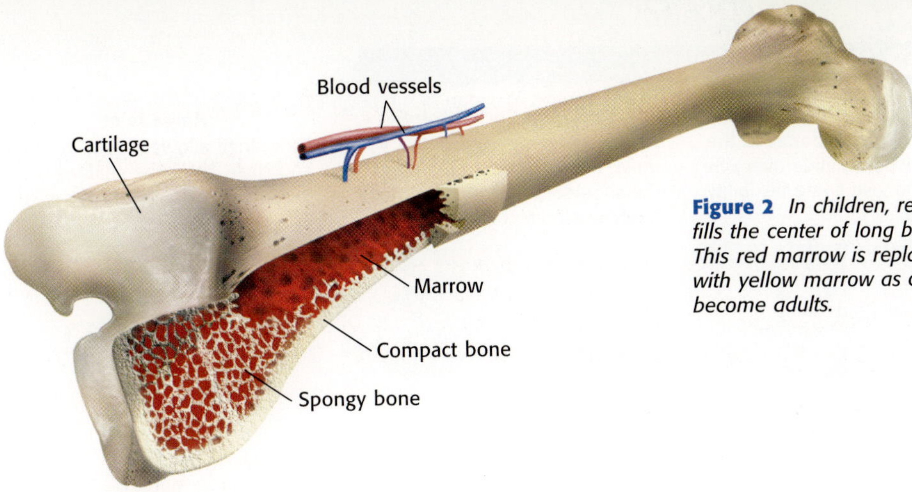

Figure 2 *In children, red marrow fills the center of long bones. This red marrow is replaced with yellow marrow as children become adults.*

Bone Structure

A bone may seem lifeless. But a bone is a living organ made of several different tissues. Bone is made of connective tissue and minerals. These minerals are deposited by living cells called *osteoblasts* (AHS tee oh BLASTS).

If you look inside a bone, you will notice two kinds of bone tissue. If the bone tissue does not have any visible open spaces, it is called *compact bone*. Compact bone is rigid and dense. Tiny canals within compact bone contain small blood vessels. Bone tissue that has many open spaces is called *spongy bone*. Spongy bone provides most of the strength and support for a bone.

Bones contain a soft tissue called *marrow*. There are two types of marrow. Red marrow produces both red and white blood cells. Yellow marrow, found in the central cavity of long bones, stores fat. **Figure 2** shows a cross section of a long bone, the femur.

Bone Growth

Did you know that most of your skeleton used to be soft and rubbery? Most bones start out as a flexible tissue called *cartilage*. When you were born, you didn't have much true bone. But as you grew, most of the cartilage was replaced by bone. During childhood, most bones still have growth plates of cartilage. These growth plates provide a place for bones to continue to grow.

Feel the end of your nose. Or bend the top of your ear. These areas are two places where cartilage is never replaced by bone. These areas stay flexible.

Reading Check How do bones grow? *(See the Appendix for answers to Reading Checks.)*

Quick Lab

Pickled Bones

1. Place a **clean chicken bone** in a **jar of vinegar.**
2. After 1 week, remove the bone and rinse it with **water.**
3. Describe the changes that you can see or feel.
4. How has the bone's strength changed?
5. What did the vinegar remove?

Teach

Demonstration — Basic

Bone Dissection Ask a local butcher to cut a long bone from a pig, a cow, or a sheep in half lengthwise to expose the bone's internal structure. If a bone is not available from a butcher, obtain a preserved long bone from a biological supply house. Point out to students the differences in structure and location of spongy bone and compact bone. Also point out any cartilage that might remain on the bone. Ask students what makes the bone relatively lightweight for its size. (Students should infer that the air spaces in the spongy bone make the bone relatively lightweight for its size.) English Language Learners
LS Logical/Visual

Quick Lab

MATERIALS

For Each Group
- chicken bone, clean
- jar of vinegar

Safety Caution: Remind students to review all safety cautions and icons before beginning this lab activity.

Answers

3. Sample answer: The bone becomes more flexible.
4. Sample answer: The bone has become weaker. Because it is no longer rigid, it will not be able to provide the support and protection it once did.
5. Sample answer: The hard minerals in the leg bone dissolved in the vinegar.

Cultural Awareness — General

El Día de los Muertos *Writing* On November 2 each year, people across Mexico celebrate El Día de los Muertos (the Day of the Dead). They set up elaborate shrines that are often decorated with figures of skeletons depicting a deceased person's job when the person was alive. Have interested students write a research paper on this holiday and share their findings with the class. LS Verbal

Answer to Reading Check

Sample answer: As people grow, most of the cartilage that they start out with is replaced with bone.

Teach, *continued*

Discussion — GENERAL

Joints After students read about the different kinds of joints, ask them the following questions.

- What kind of joint do you use when you bend your knee? (hinge joint)
- What kind of joint moves when you swing your arm back and forth? (ball-and-socket joint) Name another location in your body where this type of joint is located. (hip)

LS Verbal

Close

Reteaching — BASIC

Skeletal System Have students work in pairs to review the functions of the skeletal system. Students should stop each other if a concept is confusing or if they need clarification.

LS Interpersonal

Quiz — GENERAL

1. What is the difference between compact bone and spongy bone? (Sample answer: Compact bone has no visible, open spaces. Spongy bone has many visible spaces.)
2. Where in the body are ball-and-socket joints found? (hip and shoulder) Where are hinge joints? (knee) Where are gliding joints? (wrist and ankle)

Figure 3 Three Kinds of Joints

Gliding Joint
Gliding joints allow bones in the hand and wrist to glide over one another and give some flexibility to the area.

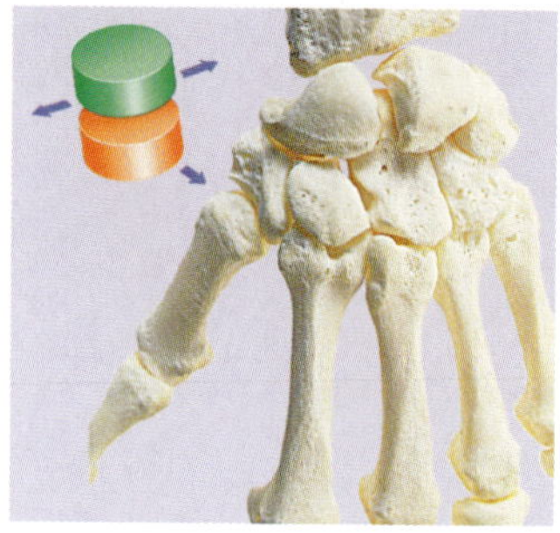

Ball-and-Socket Joint
As a video-game joystick lets you move your character all around, the shoulder lets your arm move freely in all directions.

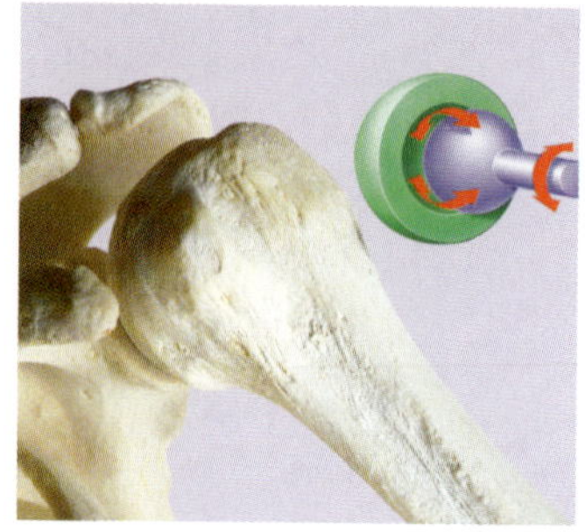

Hinge Joint
As a hinge allows a door to open and close, the knee enables you to flex and extend your lower leg.

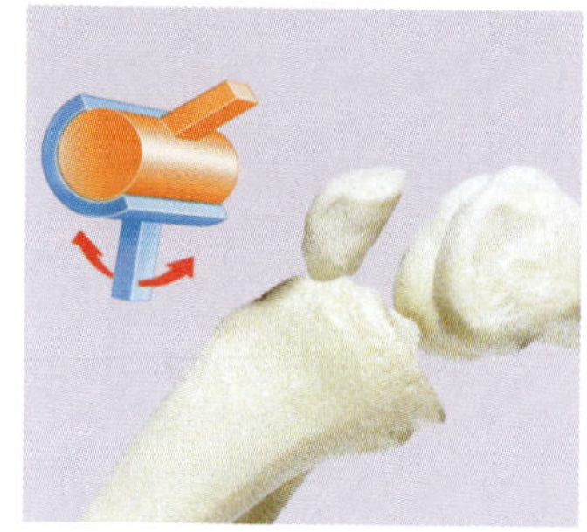

Joints

A place where two or more bones meet is called a **joint.** Your joints allow your body to move when your muscles contract. Some joints, such as fixed joints, allow little or no movement. Many of the joints in the skull are fixed joints. Other joints, such as your shoulder, allow a lot of movement. Joints can be classified based on how the bones in a joint move. For example, your shoulder is a ball-and-socket joint. Three kinds of joints are shown in **Figure 3.**

joint a place where two or more bones meet

Joints are held together by *ligaments* (LIG uh muhnts). Ligaments are strong elastic bands of connective tissue. They connect the bones in a joint. Also, cartilage covers the ends of many bones. Cartilage helps cushion the area in a joint where bones meet.

Reading Check Describe the basic structure of joints.

CONNECTION TO Environmental Science

WRITING SKILL **Bones from the Ocean** Sometimes, a bone or joint may become so damaged that it needs to be repaired or replaced with surgery. Often, replacement parts are made from a metal, such as titanium. However, some scientists have discovered that coral skeletons from coral reefs in the ocean can be used to replace human bone. Research bone surgery. Identify why doctors use metals such as titanium. Then, identify the advantages that coral may offer. Write a report discussing your findings.

Answer to Reading Check

Sample answer: Joints are held together by ligaments. Cartilage cushions the area in a joint where bones meet.

Answer to Connection to Environmental Science

Answers may vary. Students should understand that titanium is a relatively nonreactive metal, so it is used for bone replacement. Students should also demonstrate an understanding that coral easily binds to existing bone. You may want to ask students to examine the impact of coral bone replacement on the environment.

Skeletal System Injuries and Diseases

Sometimes, parts of the skeletal system are injured. As shown in **Figure 4,** bones may be fractured, or broken. Joints can also be injured. A dislocated joint is a joint in which one or more bones have been moved out of place. Another joint injury, called a *sprain,* happens if a ligament is stretched too far or torn.

There are also diseases of the skeletal system. *Osteoporosis* (AHS tee OH puh ROH sis) is a disease that causes bones to become less dense. Bones become weak and break more easily. Age and poor eating habits can make it more likely for people to develop osteoporosis. Other bone diseases affect the marrow or make bones soft. A disease that affects the joints is called *arthritis* (ahr THRIET is). Arthritis is painful. Joints may swell or stiffen. As they get older, some people are more likely to have some types of arthritis.

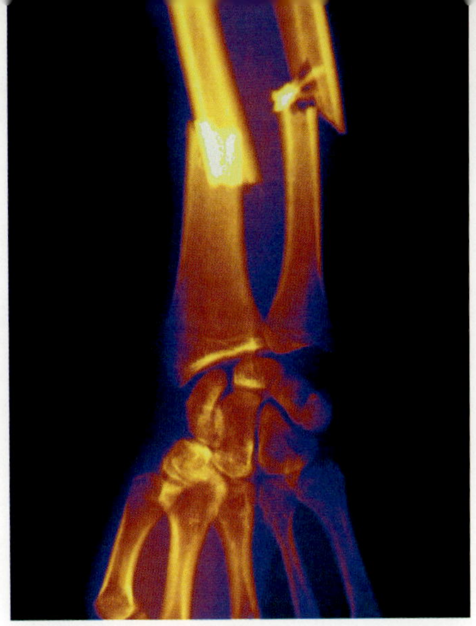

Figure 4 *This X ray shows that the two bones of the forearm have been fractured, or broken.*

SECTION Review

Summary

- The skeletal system includes bones, cartilage, and the connective tissue that connects bones.
- Bones protect the body, store minerals, allow movement, and make blood cells.
- Joints are places where two or more bones meet.
- Skeletal system injuries include fractures, dislocations, and sprains. Skeletal system diseases include osteoporosis and arthritis.

Using Key Terms

1. In your own words, write a definition for the term *skeletal system.*

Understanding Key Ideas

2. Which of the following is NOT an organ of the skeletal system?
   - **a.** bone
   - **b.** cartilage
   - **c.** muscle
   - **d.** None of the above
3. Describe four functions of bones.
4. What are three types of joints?
5. Describe two diseases that affect the skeletal system.

Math Skills

6. A broken bone usually heals in about six weeks. A mild sprain takes one-third as long to heal. In days, about how long does it take a mild sprain to heal?

Critical Thinking

7. **Identifying Relationships** Red bone marrow produces blood cells. Children have red bone marrow in their long bones, while adults have yellow bone marrow, which stores fat. Why might adults and children have different kinds of marrow?
8. **Predicting Consequences** What might happen if children's bones didn't have growth plates of cartilage?

SCILINKS. NSTA Developed and maintained by the National Science Teachers Association

For a variety of links related to this chapter, go to www.scilinks.org

Topic: Skeletal System
SciLinks code: HSM1399

CONNECTION to Physical Science — GENERAL

Levers Students may be surprised to learn that their arms and legs are machines. Human limbs are levers, and levers are the simplest kind of machine. Levers allow people to apply, increase, and change the direction of force. Use the teaching transparency titled "Machines Change the Size and/or Direction of a Force" to illustrate this point. LS Visual

CHAPTER RESOURCES

Chapter Resource File

- Section Quiz GENERAL
- Section Review GENERAL
- Vocabulary and Section Summary GENERAL
- Reinforcement Worksheet BASIC
- Critical Thinking ADVANCED
- Datasheet for Quick Lab

Technology

Transparencies

- LINK TO PHYSICAL SCIENCE Machines Change the Size and/or Direction of a Force

Alternative Assessment — GENERAL

Writing **Bone Essays** Have students write an essay about bones. Essays should address what bones do, how they are specialized, and how they are joined. LS Verbal

Answers to Section Review

1. Sample answer: The skeletal system includes bones, cartilage, and connective tissue, and it provides support to the body.
2. c
3. Sample answer: Bones protect organs. Bones store materials that help nerves and muscles function properly. Along with skeletal muscles, bones make it possible to move. Finally, marrow in bones produces blood cells.
4. gliding joints, hinge joints, and ball-and-socket joints
5. Sample answer: Osteoporosis is a disease that causes bones to become less dense. Bones become weak and break more easily. Arthritis is a painful joint disease that causes joints to swell or stiffen.
6. 14 days (6 weeks × 7 days/week = 42 days; 42 days × 1/3 = 14 days)
7. Sample answer: Children are still growing, so they need more blood cells as their bodies get bigger. After children become adults, they don't need as many new blood cells, so long bones store fat instead.
8. Sample answer: Growth plates are places where bones continue to grow. If children did not have growth plates, their bones would not continue to grow.

SECTION 3

Focus

Overview

This section introduces students to the major parts of the muscular system and describes three types of muscle. This section also describes movement, aerobic exercise, and resistance exercise. The section discusses muscular system injuries and ways to prevent them.

Bellringer

On the board or an overhead projector, write the following: "List at least five parts of your body that you use to drink a glass of water." (Sample answer: fingers, hands, arm, lips, and tongue) Remind students that all of the parts that they use, including the eyes that they use to see the glass, are controlled by muscles.

Motivate

Group Activity—GENERAL

Poster Project Have students create a poster illustrating the three types of muscle. Students should include information about where the muscle is found, what it looks like, and whether it is involuntary or voluntary. Have students present their posters to the class.

LS Verbal/Visual

SECTION 3

The Muscular System

Have you ever tried to sit still, without moving any muscles at all, for one minute? It's impossible! Somewhere in your body, muscles are always working.

Your heart is a muscle. Muscles make you breathe. And muscles hold you upright. If all of your muscles rested at the same time, you would collapse. The **muscular system** is made up of the muscles that let you move.

READING WARM-UP

Objectives

- List three kinds of muscle tissue.
- Describe how skeletal muscles move bones.
- Compare aerobic exercise with resistance exercise.
- Describe two muscular system injuries.

Terms to Learn

muscular system

READING STRATEGY

Discussion Read this section silently. Write down questions that you have about this section. Discuss your questions in a small group.

Kinds of Muscle

Figure 1 shows the three kinds of muscle in your body. *Smooth muscle* is found in the digestive tract and in the walls of blood vessels. *Cardiac muscle* is found only in your heart. *Skeletal muscle* is attached to your bones for movement. Skeletal muscle also helps protect your inner organs.

Muscle action can be voluntary or involuntary. Muscle action that is under your control is *voluntary.* Muscle action that is not under your control is *involuntary.* Smooth muscle and cardiac muscle are involuntary muscles. Skeletal muscles can be both voluntary and involuntary muscles. For example, you can blink your eyes anytime you want to. But your eyes will also blink automatically.

Figure 1 Three Kinds of Muscle

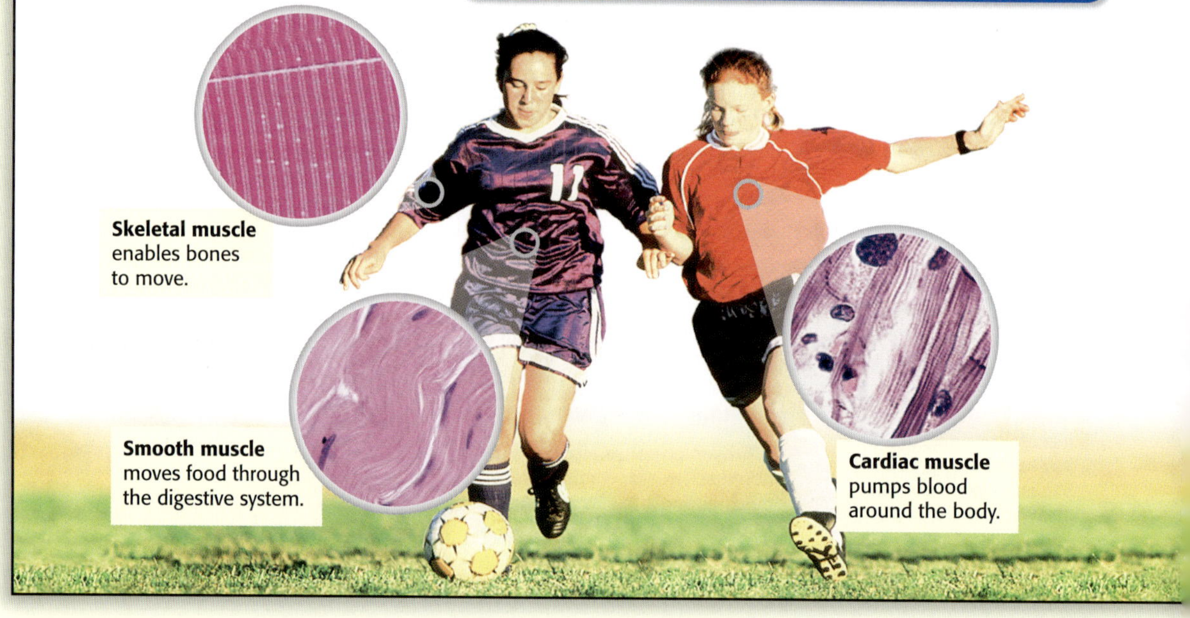

Skeletal muscle enables bones to move.

Smooth muscle moves food through the digestive system.

Cardiac muscle pumps blood around the body.

CHAPTER RESOURCES

Chapter Resource File

- Lesson Plan
- Directed Reading A BASIC
- Directed Reading B SPECIAL NEEDS

Technology

Transparencies

- Bellringer
- A Pair of Muscles in the Arm

Is That a Fact!

Horses can sleep standing up. Their legs can support their weight on their bones without the use of muscles. When horses fall asleep and their muscles relax, their leg bones lock in place underneath them and hold them upright for the duration of their nap.

Figure 2 **A Pair of Muscles in the Arm**

Skeletal muscles, such as the biceps and triceps muscles, work in pairs. When the biceps muscle contracts, the arm bends. When the triceps muscle contracts, the arm straightens.

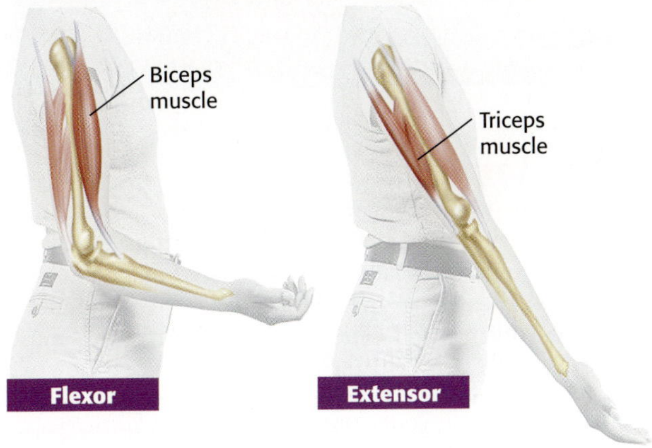

muscular system the organ system whose primary function is movement and flexibility

Movement

Skeletal muscles can make hundreds of movements. You can see many of these movements by watching a dancer, a swimmer, or even someone smiling or frowning. When you want to move, signals travel from your brain to your skeletal muscle cells. The muscle cells then contract, or get shorter.

Muscles Attach to Bones

Strands of tough connective tissue connect your skeletal muscles to your bones. These strands are called *tendons*. When a muscle that connects two bones gets shorter, the bones are pulled closer to each other. For example, tendons attach the biceps muscle to a bone in your shoulder and to a bone in your forearm. When the biceps muscle contracts, your forearm bends toward your shoulder.

Muscles Work in Pairs

Your skeletal muscles often work in pairs. Usually, one muscle in the pair bends part of the body. The other muscle straightens part of the body. A muscle that bends part of your body is called a *flexor* (FLEKS uhr). A muscle that straightens part of your body is an *extensor* (ek STEN suhr). As shown in **Figure 2,** the biceps muscle of the arm is a flexor. The triceps muscle of the arm is an extensor.

Reading Check Describe how muscles work in pairs. (*See the Appendix for answers to Reading Checks.*)

SCHOOL to HOME

Power in Pairs

Ask a parent to sit in a chair and place a hand palm up under the edge of a table. Tell your parent to apply gentle upward pressure. Feel the front and back of your parent's upper arm. Next, ask your parent to push down on top of the table. Feel your parent's arm again. What did you notice about the muscles in your parent's arm when he or she was pressing up? pushing down?

ACTIVITY

Answer to Reading Check

Sample answer: One muscle, the flexor, bends part of the body. Another muscle, the extensor, straightens part of the body.

Answer to School-to-Home Activity

Sample answer: While my parent pushes down, the muscles in the back of my parent's arm contract, and while my parent pushes up, the muscles in the front of the arm contract.

Teach

Demonstration — Basic

Muscle Contraction Ask a student to stand in a doorway with his or her arms and hands relaxed and palms turned inward. Ask the student to raise his or her hands against the door frame (backs of the hands on the frame) and press steadily against the frame for about 30 to 40 seconds. Then, ask the student to relax and step away from the door. Have the rest of the class observe what happens to the student's arms. (The student's arms should rise slowly without obvious effort by the student.) Explain that the student's arms rise because the muscles that were pushing against the door frame are still shortened, or contracted.

LS Kinesthetic/Visual — English Language Learners

CONNECTION to Real World — Advanced

PORTFOLIO

Polio In the early 1900s, tens of thousands of people, mostly children, were stricken with polio, a viral disease that paralyzes muscles. Often, polio would leave its victims unable to walk or move. An Australian nurse, Sister Elizabeth Kenny, treated patients by using flexible hot wraps and exercise instead of immobilizing casts. Using her treatments, patients avoided paralysis. Largely because Kenny had no formal medical education, the doctors and hospital administrators of the time fought against her practices. Eventually, her successes became well known, and her contributions began the field of physical therapy. Have students research Kenny's story and the opposition that she faced. Have them give a report to the class. LS Verbal

Close

Reteaching — Basic

Exercise Ask students to demonstrate different resistance and aerobic exercises. Then, ask students to explain the benefits of each exercise. Make sure students have warmed up before their demonstrations.
LS Kinesthetic

Quiz — General

1. What is the difference between voluntary muscle action and involuntary muscle action? (Sample answer: Voluntary muscle action is action that you can control. Involuntary muscle action is not under your control.)
2. What kind of muscle bends part of your body? (flexor) What kind of muscle straightens part of your body? (extensor)
3. What are the risks of using anabolic steroids? (Anabolic steroids can damage the heart, liver, and kidneys and can cause high blood pressure. They can cause bones to stop growing.)

Alternative Assessment — General

Crossword Puzzle Have students work in groups of four. Have groups make crossword puzzles using the vocabulary terms and other important terms in this section. Have groups exchange puzzles with each other.
LS Verbal English Language Learners

Figure 3 *This girl is strengthening her heart and improving her endurance by doing aerobic exercise. This boy is doing resistance exercise to build strong muscles.*

Use It or Lose It

What happens when someone wears a cast for a broken arm? Skeletal muscles around the broken bone become smaller and weaker. The muscles weaken because they are not exercised. Exercised muscles are stronger and larger. Strong muscles can help other organs, too. For example, contracting muscles squeeze blood vessels. This action increases blood flow without needing more work from the heart.

Certain exercises can give muscles more strength and endurance. More endurance lets muscles work longer without getting tired. Two kinds of exercise can increase muscle strength and endurance. They are resistance exercise and aerobic exercise. You can see an example of each kind in **Figure 3.**

Resistance Exercise

Resistance exercise is a great way to strengthen skeletal muscles. During resistance exercise, people work against the resistance, or weight, of an object. Some resistance exercises, such as curl-ups, use your own weight for resistance.

Aerobic Exercise

Steady, moderately intense activity is called *aerobic exercise.* Jogging, cycling, skating, swimming, and walking are aerobic exercises. This kind of exercise can increase muscle strength. However, aerobic exercise mostly strengthens the heart and increases endurance.

Connection to Chemistry

Muscle Function Body chemistry is very important for healthy muscle function. Spasms or cramps happen if too much sweating, poor diet, or illness causes a chemical imbalance in muscles. Identify three chemicals that the body needs for muscles to work properly. Make a poster explaining how people can make sure that they have enough of each chemical.
Activity

Homework — Advanced

Portfolio **Injuries** For one month, have students read the sports section in a local newspaper or look for articles in sports magazines about injuries sustained by athletes. Ask students to identify and count the types of injuries, such as sprains and strains. Have students compile their information on bar graphs in which they record the kinds of injuries on the x-axis and the number of injuries on the y-axis.
LS Logical

Answer to Connection to Chemistry

Students should discuss the importance of minerals, such as magnesium, calcium, potassium, or sodium, to muscle function. Students should identify important dietary sources of these minerals.

Muscle Injury

Any exercise program should be started slowly. Starting slowly means you are less likely to get hurt. You should also warm up for exercise. A *strain* is an injury in which a muscle or tendon is overstretched or torn. Strains often happen because a muscle has not been warmed up. Strains also happen when muscles are worked too hard.

People who exercise too much can hurt their tendons. The body can't repair an injured tendon before the next exercise session. So, the tendon becomes inflamed. This condition is called *tendinitis*. Often, a long rest is needed for the injured tendon to heal.

Some people try to make their muscles stronger by taking drugs. These drugs are called *anabolic steroids* (A nuh BAH lik STER OIDZ). They can cause long-term health problems. Anabolic steroids can damage the heart, liver, and kidneys. They can also cause high blood pressure. If taken before the skeleton is mature, anabolic steroids can cause bones to stop growing.

Reading Check What are the risks of using anabolic steroids?

MATH PRACTICE

Runner's Time

Jan has decided to enter a 5 km road race. She now runs 5 km in 30 min. She would like to decrease her time by 15% before the race. What will her time be when she reaches her goal?

SECTION Review

Summary

- The three kinds of muscle tissue are smooth muscle, cardiac muscle, and skeletal muscle.
- Skeletal muscles work in pairs. Skeletal muscles contract to move bones.
- Resistance exercise improves muscle strength. Aerobic exercise improves heart strength and muscle endurance.
- Strains are injuries that affect muscles and tendons. Tendinitis affects tendons.

Using Key Terms

1. In your own words, write a definition for the term *muscular system*.

Understanding Key Ideas

2. Muscles
   a. work in pairs.
   b. move bones by relaxing.
   c. get smaller when exercised.
   d. All of the above
3. Describe three kinds of muscle.
4. List two kinds of exercise. Give an example of each.
5. Describe two muscular system injuries.

Math Skills

6. If Trey can do one curl-up every 2.5 s, about how long will it take him to do 35 curl-ups?

Critical Thinking

7. **Applying Concepts** Describe some of the muscle action needed to pick up a book. Include flexors and extensors in your description.
8. **Predicting Consequences** If aerobic exercise improves heart strength, what likely happens to heart rate as the heart gets stronger? Explain your answer.

SCILINKS. NSTA Developed and maintained by the National Science Teachers Association

For a variety of links related to this chapter, go to www.scilinks.org

Topic: Muscular System
SciLinks code: HSM1008

Answer to Math Practice

25 min, 30 s (30 min × 0.85 = 25.5 min)

Answers to Section Review

1. Sample answer: The muscular system is the group of muscles that allow people to move.
2. a
3. Sample answer: Smooth muscle is found in the digestive tract and in the walls of blood vessels. Cardiac muscle is found only in the heart. Skeletal muscle is attached to bones for movement.
4. Sample answer: Curl-ups are a resistance exercise. Jogging is an aerobic exercise.
5. Sample answer: A strain is an injury in which a muscle or tendon is overstretched or torn. Tendinitis is a condition in which a tendon becomes inflamed when the body doesn't have enough time to repair the tendon between exercise sessions.
6. 87.5 s (35 curl-ups × 2.5 curl-ups/s = 87.5 s)
7. Sample answer: An extensor in the back of my arm straightens out my arm as I reach for the book. Flexors in my hand let me close my fingers on the book, and a flexor in my arm bends my arm as I pick up the book.
8. Sample answer: As the heart gets stronger, the heart likely will pump more blood with each beat. The heart will not need to work as hard to pump the same amount of blood. So, heart rate will likely decrease.

Answer to Reading Check

Sample answer: Anabolic steroids can damage the heart, liver, and kidneys. They can also cause high blood pressure. Anabolic steroids can cause bones to stop growing.

CHAPTER RESOURCES

Chapter Resource File

- Section Quiz GENERAL
- Section Review GENERAL
- Vocabulary and Section Summary GENERAL
- Reinforcement Worksheet BASIC

Workbooks

Math Skills for Science

- The Unit Factor and Dimensional Analysis GENERAL

SECTION 4

Focus

Overview

This section introduces students to the major functions of the integumentary system and describes the major parts of skin and their functions. Students will also learn about common skin injuries.

Bellringer

Write the following questions on the board or an overhead projector: "When do you see dogs panting?" (Sample answer: on hot days or after they have run) "Why do you think dogs pant?" (Sample answer: Dogs can't sweat. Dogs pant to regulate their body temperature.)

Motivate

Discussion — GENERAL

Homeostasis Relay the following story to students: "More than 200 years ago, Dr. Charles Blagden tested how mammals regulate body temperature. He spent 45 min in a room with an uncooked steak. The temperature in the room measured 126°C (260°F)." Ask students what they think happened to Dr. Blagden and the steak. (Dr. Blagden emerged from the room unharmed, but the steak was cooking! Living people and mammals can regulate their body temperature.)
LS Logical

SECTION 4

READING WARM-UP

Objectives

- List four functions of skin.
- Describe the two layers of skin.
- Describe the structure and function of hair and nails.
- Describe two kinds of damage that can affect skin.

Terms to Learn

integumentary system
epidermis
dermis

READING STRATEGY

Paired Summarizing Read this section silently. In pairs, take turns summarizing the material. Stop to discuss ideas that seem confusing.

The Integumentary System

What part of your body has to be partly dead to keep you alive? Here are some clues: It comes in many colors, it is the largest organ in the body, and it is showing right now!

Did you guess your skin? If you did, you guessed correctly. Your skin, hair, and nails make up your **integumentary system** (in TEG yoo MEN tuhr ee SIS tuhm). The integumentary system covers your body and helps you maintain homeostasis.

integumentary system the organ system that forms a protective covering on the outside of the body

Functions of Skin

Why do you need skin? Here are four good reasons:

- Skin protects you by keeping water in your body and foreign particles out of your body.
- Skin keeps you in touch with the outside world. Nerve endings in your skin let you feel things around you.
- Skin helps regulate your body temperature. Small organs in the skin called *sweat glands* make sweat. Sweat is a salty liquid that flows to the surface of the skin. As sweat evaporates, the skin cools.
- Skin helps get rid of wastes. Several kinds of waste chemicals can be removed in sweat.

As shown in **Figure 1,** skin comes in many colors. Skin color is determined by a chemical called *melanin.* If a lot of melanin is present, skin is very dark. If little melanin is present, skin is very light. Melanin absorbs ultraviolet light from the sun. So, melanin reduces damage that can lead to skin cancer. However, all skin, even dark skin, is vulnerable to cancer. Skin should be protected from sunlight whenever possible.

Figure 1 *Variety in skin color is caused by the pigment melanin. The amount of melanin varies from person to person.*

CHAPTER RESOURCES

Chapter Resource File

- Lesson Plan
- Directed Reading A BASIC
- Directed Reading B SPECIAL NEEDS

Technology

Transparencies

- Bellringer
- Structures of the Skin

Is That a Fact!

In an average adult, the skin has a surface area of about 2 m^2 and has a mass of about 4 kg. The skin on the human body varies in thickness from about 5 mm on the soles of the feet to about 0.5 mm on the eyelids.

Figure 2 **Structures of the Skin**

Beneath the surface, your skin is a complex organ made of blood vessels, nerves, glands, and muscles.

Blood vessels transport substances and help regulate body temperature.

Nerve fibers carry messages to and from the brain.

Hair follicles in the dermis make hair.

Muscle fibers attached to a hair follicle can contract and cause the hair to stand up.

Oil glands release oil that keeps hair flexible and waterproofs the epidermis.

Sweat glands release sweat to cool the body. Sweating is also a way to remove waste materials from the body.

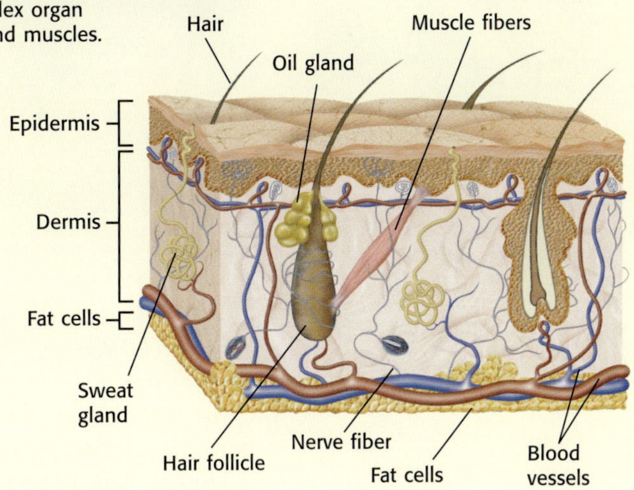

Layers of Skin

Skin is the largest organ of your body. In fact, the skin of an adult covers an area of about 2 m^2! However, there is more to skin than meets the eye. Skin has two main layers: the epidermis (EP uh DUHR mis) and the dermis. The **epidermis** is the outermost layer of skin. You see the epidermis when you look at your skin. The thicker layer of skin that lies beneath the epidermis is the **dermis.**

epidermis the surface layer of cells on a plant or animal

dermis the layer of skin below the epidermis

Epidermis

The epidermis is made of epithelial tissue. Even though the epidermis has many layers of cells, it is as thick as only two sheets of paper over most of the body. It is thicker on the palms of your hands and on the soles of your feet. Most cells in the epidermis are dead. These cells are filled with a protein called *keratin*. Keratin helps make the skin tough.

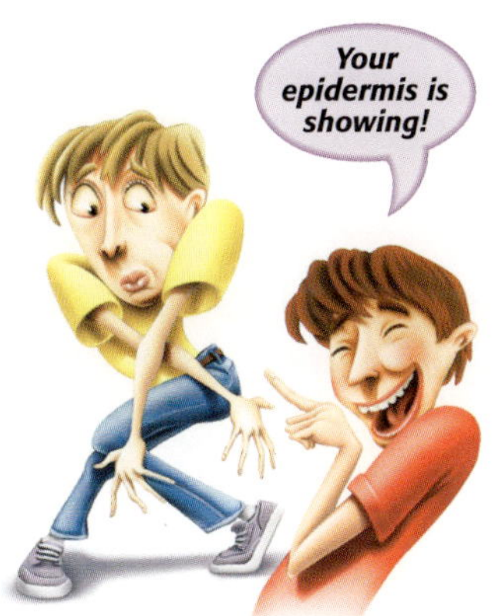

Dermis

The dermis lies beneath the epidermis. The dermis has many fibers made of a protein called *collagen*. These fibers provide strength. They also let skin bend without tearing. The dermis contains many small structures, as shown in **Figure 2.**

Reading Check **Describe the dermis. How does it differ from the epidermis?** (*See the Appendix for answers to Reading Checks.*)

Teach

ACTIVITY — GENERAL

MATERIALS

For Each Group
- clock or watch
- cotton balls
- fan
- thermometers, Celsius (2)
- water at room temperature

Measuring Temperature Have students work in groups of four students. Have students wrap the bulb of each thermometer with a cotton ball and then wet one of the cotton balls. Have students record the beginning temperature of each thermometer and then hold both thermometers in front of a fan. Students should record the temperature of each thermometer every minute for 5 min. Ask students: "How do the temperatures of the thermometers differ?" (Sample answer: The thermometer with the wet cotton has a lower temperature.) "Why?" (Sample answer: Evaporation lowers the temperature.) "How does this process relate to what happens when you sweat?" (Sample answer: As sweat evaporates from the body, the skin becomes cooler.) **LS Kinesthetic**

INTERNET ACTIVITY

Brochure — GENERAL

For an internet activity related to this chapter, have students go to **go.hrw.com** and type in the keyword **HL5BD1W.**

Answer to Reading Check

The dermis is the layer of skin that lies beneath the epidermis. It is composed of a protein called *collagen,* while the epidermis contains keratin.

INCLUSION *Strategies*

- ***Learning Disabled***
- ***Hearing Impaired***
- ***Developmentally Delayed***

Students can often better retain information if they create a visual image. Have students work in groups of four. Give each group a sheet of poster board. Have students divide the poster board into four sections and draw memory clues to help them remember the four functions of the skin. **LS Visual**

Close

Reteaching — BASIC

Hair and Nails Ask students to relate the functions of hair and nails. Then, ask students to describe how hair and nails form. LS Verbal

Quiz — GENERAL

1. What are four functions of skin? (to keep moisture in and foreign particles out, to provide information about the outside world, to help regulate body temperature, and to remove wastes)
2. Describe the two layers of skin. (The epidermis is the outermost layer of skin and contains epithelial tissue and mostly dead cells. The dermis lies beneath the epidermis and contains collagen.)

Alternative Assessment — ADVANCED

PORTFOLIO **Skin Art** Have students make a colorful drawing of a cross section of skin. Have students make their drawings from memory and label and describe the function of blood vessels, nerve fibers, muscle fibers, hair follicles, oil glands, and sweat glands.
LS Visual

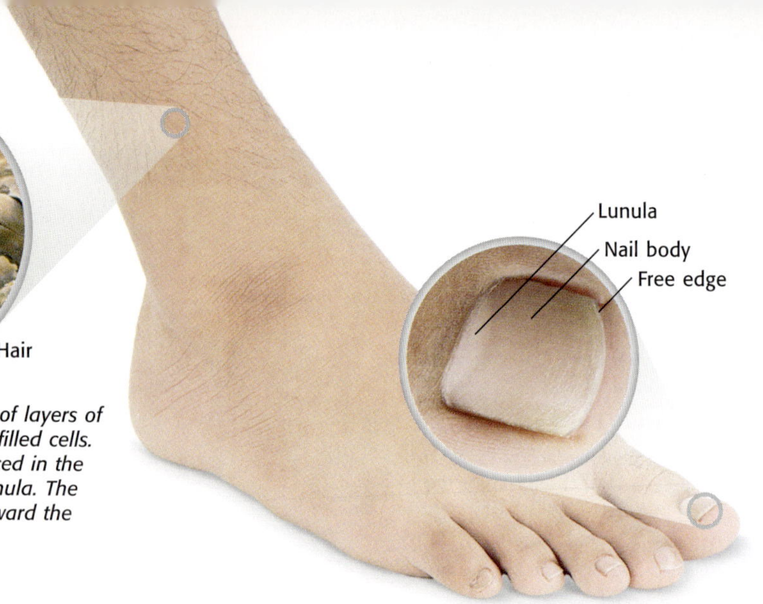

Figure 3 *A hair is made up of layers of dead, tightly packed, keratin-filled cells. In nails, new cells are produced in the nail root, just beneath the lunula. The new cells push older cells toward the outer edge of the nail.*

Hair and Nails

Hair and nails are important parts of the integumentary system. Like skin, hair and nails are made of living and dead cells. **Figure 3** shows hair and nails.

A hair forms at the bottom of a tiny sac called a *hair follicle.* The hair grows as new cells are added at the hair follicle. Older cells get pushed upward. The only living cells in a hair are in the hair follicle. Like skin, hair gets its color from melanin.

Hair helps protect skin from ultraviolet light. Hair also keeps particles, such as dust and insects, out of your eyes and nose. In most mammals, hair helps regulate body temperature. A tiny muscle attached to the hair follicle contracts. If the follicle contains a hair, the hair stands up. The lifted hairs work like a sweater. They trap warm air around the body.

A nail grows from living cells in the *nail root* at the base of the nail. As new cells form, the nail grows longer. Nails protect the tips of your fingers and toes. So, your fingers and toes can be soft and sensitive for a keen sense of touch.

Reading Check Describe how nails grow.

CONNECTION TO Social Studies

WRITING SKILL **Using Hair** Many traditional cultures use animal hair to make products, such as rugs and blankets. Identify a culture that uses animal hair. In your **science journal,** write a report describing how the culture uses animal hair.

Skin Injuries

Skin is often damaged. Fortunately, your skin can repair itself, as shown in **Figure 4.** Some damage to skin is very serious. Damage to the genetic material in skin cells can cause skin cancer. Skin may also be affected by hormones that cause oil glands in skin to make too much oil. This oil combines with dead skin cells and bacteria to clog hair follicles. The result is acne. Proper cleansing can help but often cannot prevent this problem.

CONNECTION ACTIVITY
Real World — GENERAL

PORTFOLIO **Protection from the Sun** Skin cancer is the most common kind of cancer. Hundreds of thousands of new cases of skin cancer are reported each year. Ask interested students to research ways people can prevent skin cancer. Have students create informative brochures about their findings. LS Verbal/Visual

Answer to Reading Check

Sample answer: A nail grows from living cells in the nail root at the base of the nail. As new cells form, the nail grows longer.

Figure 4 **How Skin Heals**

1. A blood clot forms over a cut to stop bleeding and to keep bacteria from entering the wound. Bacteria-fighting cells then come to the area to kill bacteria.

2. Damaged cells are replaced through cell division. Eventually, all that is left on the surface is a scar.

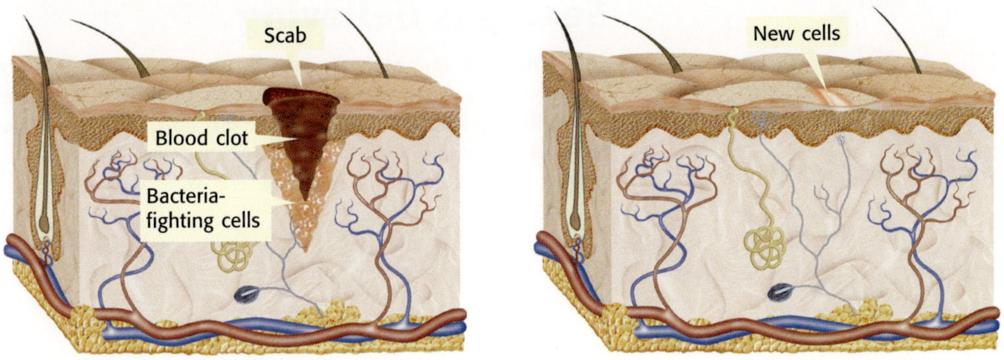

SECTION Review

Summary

- Skin keeps water in the body, keeps foreign particles out of the body, lets people feel things around them, regulates temperature, and removes wastes.
- The two layers of skin are the epidermis and the dermis.
- Hair grows from hair follicles. Nails grow from nail roots.
- Skin may develop skin cancer. Acne may develop if skin produces too much oil.

Using Key Terms

1. In your own words, write a definition for each of the following terms: *integumentary system*, *epidermis*, and *dermis*.

Understanding Key Ideas

2. Which of the following is NOT a function of skin?
   - **a.** to regulate body temperature
   - **b.** to keep water in the body
   - **c.** to move your body
   - **d.** to get rid of wastes
3. Describe the two layers of skin.
4. How do hair and nails develop?
5. Describe how a cut heals.

Math Skills

6. On average, hair grows 0.3 mm per day. How many millimeters does hair grow in 30 days? in a year?

Critical Thinking

7. **Making Inferences** Why do you feel pain when you pull on your hair or nails, but not when you cut them?
8. **Analyzing Ideas** The epidermis on the palms of your hands and on the soles of your feet is thicker than it is anywhere else on your body. Why might this skin need to be thicker?

Answers to Section Review

1. Sample answer: The integumentary system is the organ system that forms a protective covering over the body. The epidermis is the outer layer of skin. The dermis lies beneath the epidermis.
2. c
3. Sample answer: The epidermis is the outermost layer of skin. It is made of epithelial tissue. Most of the cells of the epidermis are dead cells that contain keratin. The dermis lies beneath the epidermis and contains sweat glands, hair follicles, nerve fibers, muscle fibers, and blood vessels. The dermis also contains collagen.
4. Sample answer: A hair grows as new cells are added in the hair follicle, which is where the only living cells are located. Nails grow from living cells in the nail root at the base of the nail.
5. Sample answer: A blood clot forms over a cut to stop bleeding and to keep bacteria out of the cut. Bacteria-fighting cells come to the area to kill bacteria. Damaged skin cells are replaced through cell division. Eventually, all that is left is a scar.
6. 9 mm (0.3 mm/day $\times$ 30 days $=$ 9 mm); 109.5 mm (0.3 mm/day $\times$ 365 days $=$ 109.5 mm)
7. Sample answer: Except in the hair follicle and in the nail root, hair and nails are made of dead cells, so cutting them does not hurt. Pulling on them, however, affects the living areas of hair and nails in the dermis, where nerve fibers are located.
8. Sample answer: The epidermis forms a protective covering of dead cells. The hands and the feet are areas that are often used, so the epidermis in these areas needs to be thicker for better protection.

CHAPTER RESOURCES

Chapter Resource File

- Section Quiz GENERAL
- Section Review GENERAL
- Vocabulary and Section Summary GENERAL

Seeing Is Believing

Teacher's Notes

Time Required

One 45-minute class period and 5 to 10 minutes every other day for 2 weeks

Lab Ratings

EASY ⟶ HARD

Teacher Prep 1
Student Set-Up 1
Concept Level 1
Clean Up 1

MATERIALS

The materials listed on the student page are enough for 1–2 students. This lab may be done with several different types of marking methods. The fingernail is very hard and is not very porous. Marking the nail permanently is a challenge. A permanent marker, such as a laundry-marking pen, may need to be refreshed only once a day. Fingernail polish may be an acceptable alternative. Acrylic paint may also be used.

Safety Caution

Remind students to review all safety cautions and icons before beginning this lab activity.

Skills Practice Lab

OBJECTIVES

Measure nail growth over time.

Draw a graph of nail growth.

MATERIALS

- graph paper (optional)
- metric ruler
- permanent marker

SAFETY

Seeing Is Believing

Like your hair and skin, fingernails are part of your body's integumentary system. Nails, shown in the figure below, are a modification of the outer layer of the skin. Nails grow from the nail bed and will grow continuously throughout your life. In this activity, you will measure the rate at which fingernails grow.

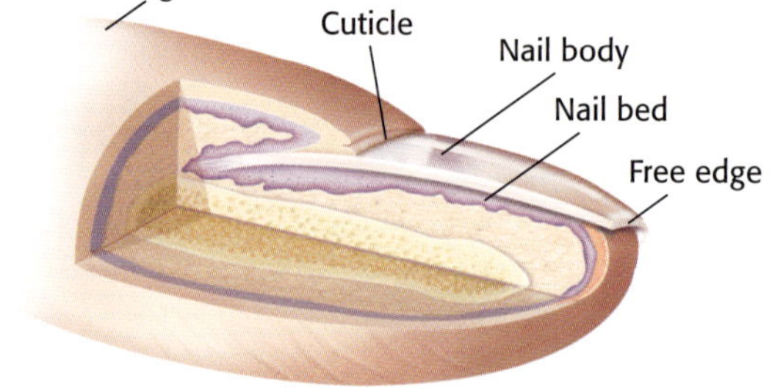

Procedure

1. Use a permanent marker to mark the center of the nail bed on your right index finger, as shown in the figure below. **Caution:** Do not get ink on your clothing.

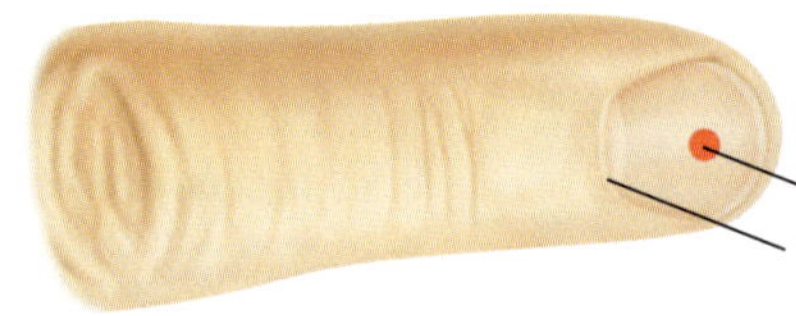

2. Measure from the mark to the base of your nail. Record the measurement, and label the measurement "Day 1."
3. Repeat steps 1 and 2 for your left index finger.
4. Let your fingernails grow for 2 days. Normal daily activity will not wash away the mark completely, but you may need to freshen the mark.
5. Measure the distance from the mark on your nail to the base of your nail. Record this distance, and label the measurement "Day 3."

Kathy LaRoe
East Valley Middle School
East Helena, Montana

CHAPTER RESOURCES

Chapter Resource File
- Datasheet for Chapter Lab
- Lab Notes and Answers

Technology

Classroom Videos
- Lab Video

LabBook
- Muscles at Work

6. Continue measuring and recording the growth of your nails every other day for 2 weeks. Refresh the mark as necessary. You may continue to file or trim your nails as usual throughout the course of the lab.

7. After you have completed your measurements, use them to create a graph similar to the graph below.

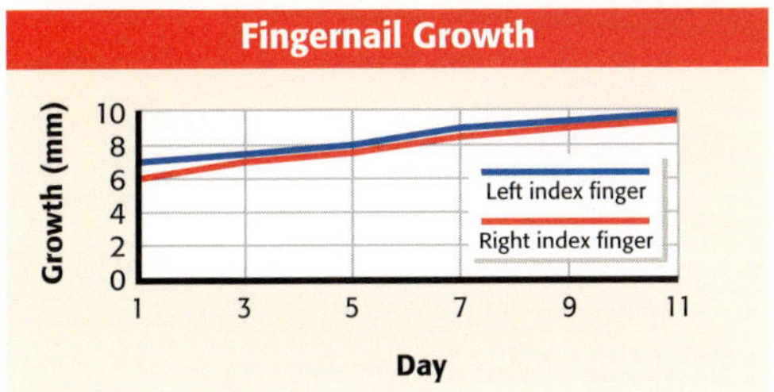

Analyze the Results

1. **Describing Events** Did the nail on one hand grow faster than the nail on the other hand?

2. **Examining Data** Did your nails grow at a constant rate, or did your nails grow more quickly at certain times?

Draw Conclusions

3. **Making Predictions** If one nail grew more quickly than the other nail, what might explain the difference in growth?

4. **Analyzing Graphs** Compare your graph with the graphs of your classmates. Do you notice any differences in the graphs based on gender or physical characteristics, such as height? If so, describe the difference.

Applying Your Data

Do additional research to find out how nails are important to you. Also, identify how nails can be used to indicate a person's health or nutrition. Based on what you learn, describe how your nail growth indicates your health or nutrition.

CHAPTER RESOURCES

Workbooks

Inquiry Labs
- On a Wing and a Layer GENERAL

Long-Term Projects & Research Ideas
- Mapping the Human Body ADVANCED

Applying Your Data

Students should discover that the appearance of a nail can reflect the health of an individual. Abnormalities in color, such as white, yellow, or red nail beds, can indicate serious problems such as liver, lung, and heart diseases. Other problems indicated by nail appearance are bacterial or fungal infections of the nails and poor nutrition.

Lab Notes

Few topics are as important to students as gaining knowledge and understanding of their own body. As they develop, students can't help but observe how they are changing physically. In this lab, students witness the growth of their own fingernails.

Tell students that the graphed data shown in this lab are only an example and will not be the same as their own data.

Analyze the Results

1. Answers may vary. Many students may note that rates of growth vary.
2. Answers may vary. Some students will answer that their nails did not grow at a constant rate.

Draw Conclusions

3. The nails of the dominant hand grow faster than those of the other hand. Damage to the nail root can affect how quickly a nail grows. Also, circulation to the area affects growth. If one hand receives poor circulation, the nail will grow at a slower rate.
4. Some students may note differences in nail growth. While these differences likely are not related to height, they are related to nutrition. Someone who does not have enough of certain minerals, such as calcium and magnesium, in his or her diet will have slower nail growth. Nail growth is also affected by age, time of year, and gender. Nail growth slows as people get older. Nails grow faster in the summer than they do in the winter. Often, men's nails grow faster than women's nails do.

Chapter Review

Assignment Guide

Section	Questions
1	2, 5, 7, 13, 17
2	1, 6, 12, 14, 18, 20–21
3	8, 10–11, 15
4	3–4, 9, 16, 19, 23–27
2 and 3	22

ANSWERS

Using Key Terms

1. joint
2. Homeostasis
3. epidermis
4. integumentary system
5. organ
6. skeletal system

Understanding Key Ideas

7. c
8. c
9. a
10. d
11. Sample answer: Muscles are connected to bones by tendons. When a muscle that connects two bones contracts, the bones are pulled closer together. Muscles often work in pairs.
12. Sample answer: The skeletal system includes the bones, cartilage, and connective tissue whose primary function is to support the body. The skeletal system protects organs, stores minerals, allows movement, and produces blood cells.
13. Sample answer: The circulatory and cardiovascular systems work closely together to maintain homeostasis. The cardiovascular system includes the heart and blood vessels. These structures are also part of the circulatory system, which includes blood. These two systems work together to deliver the materials that cells need to survive.
14. fractures, dislocations, sprains, osteoporosis, and arthritis

Chapter Review

USING KEY TERMS

Complete each of the following sentences by choosing the correct term from the word bank.

homeostasis	organ
joint	skeletal system
tissue	muscular system
epidermis	dermis
integumentary system	

1. A(n) ___ is a place where two or more bones meet.
2. ___ is the maintenance of a stable internal environment.
3. The outermost layer of skin is the ___.
4. The organ system that includes skin, hair, and nails is the ___.
5. A(n) ___ is made up of two or more tissues working together.
6. The ___ supports and protects the body, stores minerals, and allows movement.

UNDERSTANDING KEY IDEAS

Multiple Choice

7. Which of the following lists shows the way in which the body is organized?
   - **a.** cells, organs, organ systems, tissues
   - **b.** tissues, cells, organs, organ systems
   - **c.** cells, tissues, organs, organ systems
   - **d.** cells, tissues, organ systems, organs
8. Which muscle tissue can be both voluntary and involuntary?
   - **a.** smooth muscle
   - **b.** cardiac muscle
   - **c.** skeletal muscle
   - **d.** All of the above
9. The integumentary system
   - **a.** helps regulate body temperature.
   - **b.** helps the body move.
   - **c.** stores minerals.
   - **d.** None of the above
10. Muscles
    - **a.** work in pairs.
    - **b.** can be voluntary or involuntary.
    - **c.** become stronger if exercised.
    - **d.** All of the above

Short Answer

11. How do muscles move bones?
12. Describe the skeletal system, and list four functions of bones.
13. Give an example of how organ systems work together.
14. List three injuries and two diseases that affect the skeletal system.

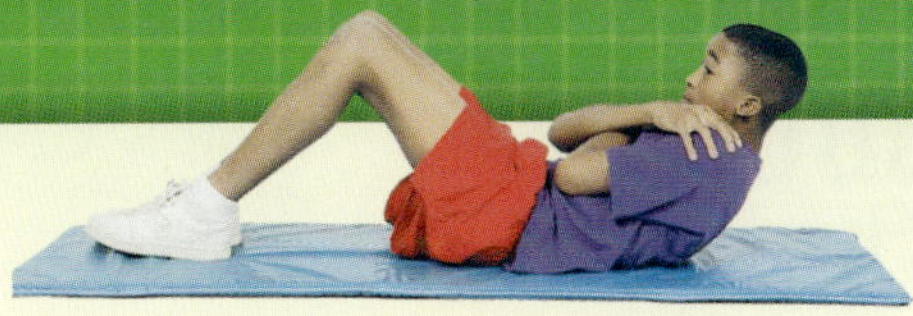

15 Compare aerobic exercise and resistance exercise.

16 What are two kinds of damage that may affect skin?

CRITICAL THINKING

17 **Concept Mapping** Use the following terms to create a concept map: *tissues, muscle tissue, connective tissue, cells, organ systems, organs, epithelial tissue,* and *nervous tissue.*

18 **Making Comparisons** Compare the shapes of the bones of the human skull with the shapes of the bones of the human leg. How do the shapes differ? Why are the shapes important?

19 **Making Inferences** Compare your elbows and fingertips in terms of the texture and sensitivity of the skin on these parts of your body. Why might the skin on these body parts differ?

20 **Making Inferences** Imagine that you are building a robot. Your robot will have a skeleton similar to a human skeleton. If the robot needs to be able to move a limb in all directions, what kind of joint would be needed? Explain your answer.

21 **Analyzing Ideas** Human bones are dense and are often filled with marrow. But many bones of birds are hollow. Why might birds have hollow bones?

22 **Identifying Relationships** Why might some muscles fail to work properly if a bone is broken?

INTERPRETING GRAPHICS

Use the cross section of skin below to answer the questions that follow.

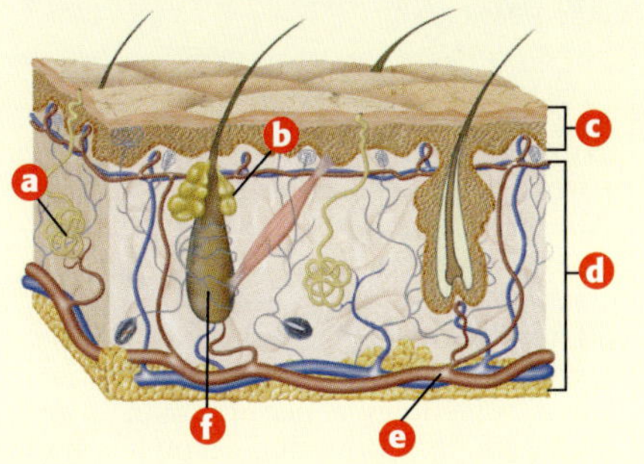

23 What is d called? What substance is most abundant in this layer?

24 What is the name and function of a?

25 What is the name and function of b?

26 Which letter corresponds to the part of the skin that is made up of epithelial tissue that contains dead cells?

27 Which letter corresponds to the part of the skin from which hair grows? What is this part called?

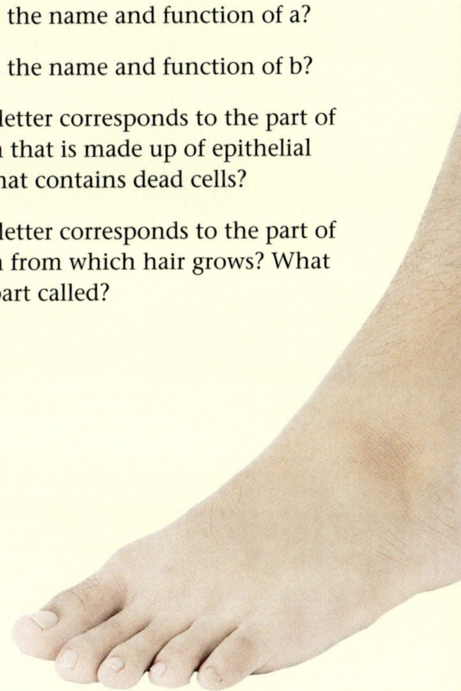

15. Sample answer: Aerobic exercise is steady, moderately intense activity that improves endurance. Resistance exercise strengthens skeletal muscles. During resistance exercise, muscles work against resistance, or weight.

16. Sample answer: Damage to genetic material in skin cells can cause skin cancer. If oil glands in skin produce too much oil, hair follicles may be clogged, which causes acne.

CHAPTER RESOURCES

Chapter Resource File

- Chapter Review GENERAL
- Chapter Test A GENERAL
- Chapter Test B ADVANCED
- Chapter Test C SPECIAL NEEDS
- Vocabulary Activity GENERAL

Workbooks

Study Guide
- Assessment Resources are also available in Spanish.

Critical Thinking

17. 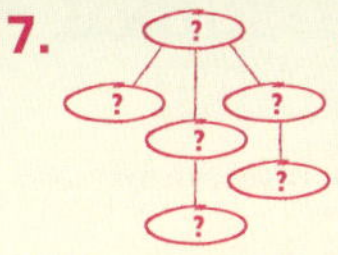 An answer to this exercise can be found at the end of the book.

18. Sample answer: Leg bones are long and skinny, while skull bones are thin and wide. Many skull bones are curved. Leg bones enable walking and standing, while skull bones are designed to protect the brain.

19. Sample answer: The skin on the elbows is thin, loose, and sometimes rough. The skin on the fingertips is thicker, but more sensitive to touch. The skin on elbows allows elbows to bend because it is loose and can stretch. The skin on the fingers is important for the sense of touch.

20. Sample answer: The robot would need a ball-and-socket joint. A ball-and-socket joint allows movement in all directions.

21. Sample answer: Dense bones that are filled with marrow weigh more than hollow bones. Because most birds fly, they need lighter bones. So, birds' bones are hollow.

22. Sample answer: If a bone is broken, flexors or extensors likely will not work properly. These muscles will not be able to pull on a bone in the way that they do when a bone is not broken. Also, if the area where the muscle attaches is damaged, the ability of the muscle to pull on the bone may be affected.

Interpreting Graphics

23. dermis; collagen

24. Sample answer: Sweat glands release sweat to cool the body.

25. Sample answer: Oil glands release oil to keep hair flexible and to waterproof the epidermis.

26. c

27. f; hair follicle

Standardized Test Preparation

Teacher's Note

To provide practice under more realistic testing conditions, give students 20 minutes to answer all of the questions in this Standardized Test Preparation.

MISCONCEPTION ALERT

Answers to the standardized test preparation can help you identify student misconceptions and misunderstandings.

READING

Passage 1

1. C
2. F
3. B

TEST DOCTOR

Question 2: Some students may answer that a skin graft is skin made of plastic because plastic bandages are discussed in the passage. Some students may think that a skin graft is damaged skin that has been removed, but it is undamaged skin that has been removed to replace damaged skin. Finally, some students may think that a skin graft is burned skin because burns are mentioned in the passage.

Standardized Test Preparation

READING

Read the passages below. Then, answer the questions that follow each passage.

Passage 1 Sometimes, doctors perform a skin graft to transfer some of a person's healthy skin to an area where skin has been damaged. Doctors perform skin grafts because skin is often the best "bandage" for a wound. Like cloth or plastic bandages, skin protects the wound. Skin allows the wound to breathe. Unlike cloth or plastic bandages, skin can regenerate itself as it covers a wound. But sometimes a person's skin is so severely damaged (by burns, for example) that the person doesn't have enough skin to spare.

1. Based on the passage, what can skin do that manufactured bandages can't do?

A Skin can protect a wound.
B Skin can stop more skin from being damaged.
C Skin can regenerate itself.
D Skin can prevent burns.

2. In the passage, what does the term *skin graft* most likely mean?

F a piece of skin transplanted from one part of the body to another
G a piece of skin made of plastic
H a piece of damaged skin that has been removed from the body
I burned skin

3. Based on the passage, why might a severe burn victim not receive a skin graft?

A Manufactured bandages are better.
B He or she doesn't have enough healthy skin.
C There isn't enough damaged skin to repair.
D Skin is the best bandage for a wound.

Passage 2 Making sure that your body maintains homeostasis is not an easy task. The task is difficult because your internal environment is always changing. Your body must do many different jobs to maintain homeostasis. Each cell in your body has a specific job in maintaining homeostasis. Your cells are organized into groups. A group of similar cells working together forms a tissue. Your body has four main kinds of tissue—epithelial tissue, connective tissue, muscle tissue, and nervous tissue. These tissues work together to form organs, which help maintain homeostasis.

1. Based on the passage, which of the following statements about tissues is true?

A Tissues do not help maintain homeostasis.
B Tissues form organ systems.
C Tissues are changing because the body's internal environment is always changing.
D There are four kinds of tissue.

2. According to the passage, which of the following statements about homeostasis is true?

F It is easy for the body to maintain homeostasis.
G The body must do different jobs to maintain homeostasis.
H Your internal environment rarely changes.
I Organs and organ systems do not help maintain homeostasis.

3. Which of the following statements about cells is false?

A Cells are organized into different groups.
B Cells form tissues.
C Cells work together.
D Cells don't maintain homeostasis.

Passage 2

1. D
2. G
3. D

TEST DOCTOR

Question 1: Some students may answer that tissues do not help maintain homeostasis. However, because all cells in the body have a role in homeostasis, so do tissues. Some students may answer that tissues form organ systems, but tissues form organs. Some students may answer that tissues are always changing, but this is not mentioned in the passage.

INTERPRETING GRAPHICS

The line graph below shows hair growth over time. Use the graph to answer the questions that follow.

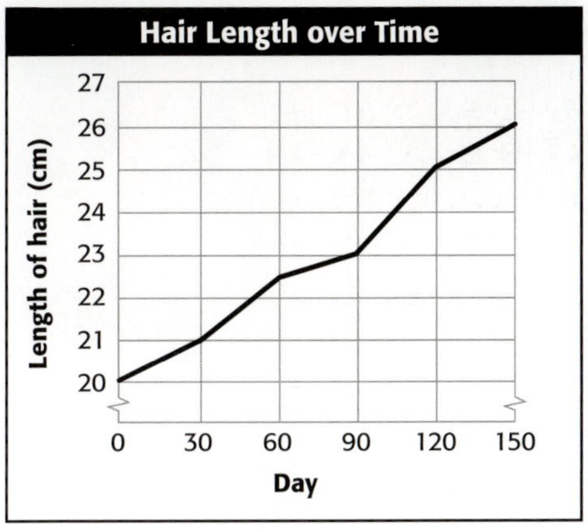

1. How long was the hair on day 60?

A 20.0 cm
B 21.0 cm
C 22.5 cm
D 23.0 cm

2. On which day was hair length 23 cm?

F day 60
G day 90
H day 120
I day 150

3. From day 0 to day 150, what is the average amount that hair grows every 30 days?

A 0.5 cm
B 1.2 cm
C 1.5 cm
D 2.0 cm

4. Based on the average amount of hair growth per 30-day period, how long would it take the hair to grow another 3.6 cm?

F 30 days
G 60 days
H 90 days
I 120 days

MATH

Read each question below, and choose the best answer.

1. About 40% of a person's mass is muscle tissue. If Max has a mass of 40 kg, about how much muscle tissue does he have?

A 16 kg
B 20 kg
C 24 kg
D 30 kg

2. When running, an adult inhales about 72 L of air per minute. That amount is 12 times the amount that an adult needs while resting. How much air does an adult inhale while resting?

F 6 L/min
G 12 L/min
H 60 L/min
I 64 L/min

3. Maggie likes to do bench presses, a resistance exercise. She bench presses 10 kg. If Maggie added 2 kg every 2 weeks, how long would it take her to reach 20 kg?

A 4 weeks
B 5 weeks
C 10 weeks
D 12 weeks

4. A box of 25 bandages costs $4.00. A roll of tape costs $1.50. Troy needs 125 bandages and 3 rolls of tape for a first-aid kit. Which of the following equations shows the cost of first-aid supplies, *x*?

F $x = (125 \times 4.00) + (3 \times 1.50)$
G $x = (25 \times 4.00) + (3 \times 1.50)$
H $x = [(25 \times 4.00) \div 125] + (3 \times 1.50)$
I $x = [(125 \div 25) \times 4.00] + (3 \times 1.50)$

5. Stephen wants to run a 10 K race. Right now, he can run 5 K. What is the percentage increase from 5 K to 10 K?

A 50%
B 100%
C 200%
D 500%

CHAPTER RESOURCES

Chapter Resource File

• Standardized Test Preparation GENERAL

State Resources

For specific resources for your state, visit **go.hrw.com** and type in the keyword **HSMSTR.**

INTERPRETING GRAPHICS

1. C
2. G
3. B
4. H

TEST DOCTOR

Question 3: Students can calculate average hair growth by examining the total hair growth (6 cm) and the number of 30-day periods (5). In calculating the average growth, students should find that hair grows about 1.2 cm every 30 days (6 cm ÷ 5 = 1.2 cm). If students round their answers, they may incorrectly answer 1.0 cm or 1.5 cm.

MATH

1. A
2. F
3. C
4. I
5. B

TEST DOCTOR

Question 2: Students may answer 12 L because the number 12 is introduced in the question. Students may answer 60 L if they subtract 12 from 72 L. Students may answer 64 L if they assume that people need 12% less air rather than 12 times less air.

Question 5: To find percentage increase, students must find the difference between 10 K and 5 K, divide it by 5 K, and multiply by 100. If students divide 5 K by 10 K, they will get the incorrect answer of 50%. If they divide 10 K by 5 K, they will get the incorrect answer of 200%. Finding the difference between 10 K and 5 K and multiplying the answer by 100 without dividing first will yield the incorrect answer of 500%.

Science in Action

Weird Science

Background

The way that a wound heals when engineered skin is used is far better than the way that a wound heals when engineered skin is not used—more scar tissue forms. Scar tissue is weaker and more brittle than the skin that the scar tissue replaces. Scar tissue does not stretch and grow, which is a particularly difficult problem for children suffering from burns. Engineered skin also helps reduce the disfigurement associated with scarring. One significant limitation of the engineered skin is that when it is new, it lacks sweat glands. Patients who have large skin grafts need to be cautious about overexercising and exposure to the sun.

Science, Technology, and Society

Activity — General

Have students model having a prosthetic hand by using a clothespin to pick up papers, hold a pencil to write their name, and tie their shoes. Have them discuss how it feels, and have them try to imagine what having an artificial hand would be like. Then, ask students to brainstorm ways that they might improve their "prosthetic hand" (the clothespin).

Science in Action

Weird Science

Engineered Skin

Your skin is your first line of defense against the outside world. Your skin keeps you safe from dehydration and infection, helps regulate body temperature, and helps remove some wastes. But what happens if a large portion of skin is damaged? Skin may not be able to function properly. For someone who has a serious burn, a doctor often uses skin from an undamaged part of the person's body to repair the damaged skin. But some burn victims don't have enough undamaged skin to spare. Doctors have discovered ways to engineer skin that can be used in place of human skin.

Math Activity

A doctor repaired 0.35 m^2 of an adult patient's skin with engineered skin. If an adult has about 2 m^2 of skin, what percentage of the patient's skin was repaired?

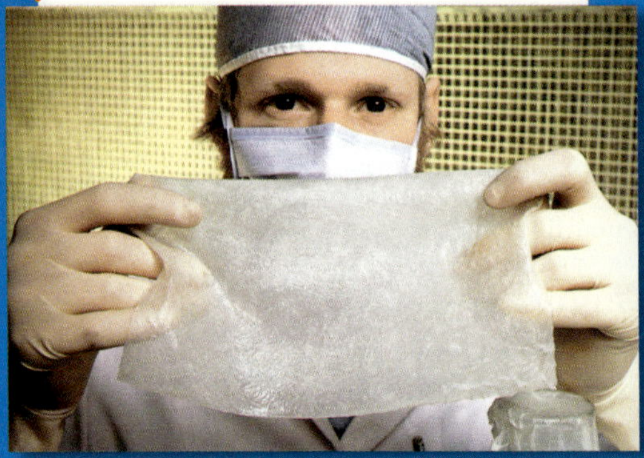

Science, Technology, and Society

Beating the Odds

Sometimes, people are born without limbs or lose limbs in accidents. Many of these people have prostheses (prahs THEE SEEZ), or human-made replacements for the body parts. Until recently, many of these prostheses made it more difficult for many people to participate in physical activities, such as sports. But new designs have led to lighter, more comfortable prostheses that move the way that a human limb does. These new designs have allowed athletes with physical disabilities to compete at higher levels.

Social Studies Activity

Research the use of prostheses throughout history. Create a timeline showing major advances in prosthesis use and design.

Answer to Math Activity

17.5% ($0.35\ m^2 \div 2\ m^2 \times 100 = 17.5\%$)

Answer to Social Studies Activity

Students' timelines should display an understanding of how modern technology has improved prosthesis design. Timelines should include the use of wooden prostheses, expand into the use of metal and the development of plastics and other synthetics in prosthesis construction, and discuss modern ergonomic designs.

Careers

Zahra Beheshti

Physical Therapist A physical therapist is a licensed professional who helps people recover from injuries by using hands-on treatment instead of medicines. Dr. Zahra Beheshti is a physical therapist at the Princeton Physical Therapy Center in New Jersey. She often helps athletes who suffer from sports injuries.

After an injury, a person may go through a process called *rehabilitation* to regain the use of the injured body part. The most common mistake made by athletes is that they play sports before completely recovering from injuries. Dr. Beheshti explains, "Going back to their usual pre-injury routine could result in another injury."

Dr. Beheshti also teaches patients about preventing future sports injuries. "Most injuries happen when an individual engages in strenuous activities without a proper warm-up or cool-down period." Being a physical therapist is rewarding work. Dr. Beheshti says, "I get a lot of satisfaction when treating patients and see them regain their function and independence and return to their normal life."

Language Arts ACTIVITY

WRITING SKILL Interview a physical therapist who works in or near your community. Write a newspaper article about your interview.

go.hrw.com

To learn more about these Science in Action topics, visit go.hrw.com and type in the keyword HL5BD1F.

Current Science

Check out Current Science® articles related to this chapter by visiting go.hrw.com. Just type in the keyword HL5CS22.

Careers

Teaching Strategy—BASIC

Invite a local physical therapist to the classroom. Ask the physical therapist to bring some of the tools that he or she uses to treat injuries. Ask the physical therapist to demonstrate each tool by having student volunteers participate in a physical therapy exercise. Ask the physical therapist to describe how each exercise helps someone recover from an injury and explain about how long recovery will take.

Answer to Language Arts Activity

Students should be able to relate the importance of physical therapy as told to them by the physical therapist who was interviewed. Students should also be able to relate some reasons that people require physical therapy.

2 Circulation and Respiration

Chapter Planning Guide

Compression guide:
To shorten instruction because of time limitations, omit the Chapter Lab.

OBJECTIVES	LABS, DEMONSTRATIONS, AND ACTIVITIES	TECHNOLOGY RESOURCES
PACING • 90 min pp. 28–35 **Chapter Opener**	SE **Start-up Activity,** p. 29 GENERAL	OSP **Parent Letter** ■ GENERAL CD **Student Edition on CD-ROM** CD **Guided Reading Audio CD** ■ TR **Chapter Starter Transparency*** VID **Brain Food Video Quiz**
Section 1 The Cardiovascular System • List four main parts of the cardiovascular system, and describe their functions. • Describe the two types of circulation of blood in the body. • List four cardiovascular problems.	TE **Connection Activity** Math, p. 31 GENERAL TE **Activity** Viewing Blood Vessels, p. 32 ◆ GENERAL TE **Group Activity** Circulate!, p. 33 ◆ GENERAL LB **Whiz-Bang Demonstrations** Get the Beat* ◆ BASIC	CRF **Lesson Plans*** TR **Bellringer Transparency*** TR The Flow of Blood Through the Heart* TR The Flow of Blood Through the Body* TE **Internet Activity,** p. 35 GENERAL
PACING • 45 min pp. 36–39 **Section 2 Blood** • Identify the four main components of blood. • Describe three functions of blood. • Explain how blood pressure is measured. • Explain what the ABO blood types are and why they are important.	TE **Activity** Making a Model of Blood, p. 37 ◆ BASIC SE **Science in Action** Math, Social Studies, and Language Arts Activities, pp. 54–55 GENERAL LB **Calculator-Based Labs** A Hot Hand* ◆ ADVANCED	CRF **Lesson Plans*** TR **Bellringer Transparency***
PACING • 45 min pp. 40–43 **Section 3 The Lymphatic System** • Describe the relationship between the lymphatic system and the circulatory system. • Identify six parts of the lymphatic system, and describe their functions.	SE **Connection to Social Studies** Vent Your Spleen, p. 42 GENERAL	CRF **Lesson Plans*** TR **Bellringer Transparency*** CRF **SciLinks Activity*** GENERAL
PACING • 90 min pp. 44–47 **Section 4 The Respiratory System** • Describe the parts of the respiratory system and their functions. • Explain how breathing happens. • Discuss the relationship between the respiratory system and the cardiovascular system. • Identify two respiratory disorders.	TE **Activity** Deep Breathing, p. 44 GENERAL TE **Activity** Investigating Speech, p. 45 GENERAL SE **Connection to Chemistry** Oxygen and Blood, p. 46 GENERAL SE **Quick Lab** Why Do People Snore?, p. 47 GENERAL SE **Skills Practice Labs** Carbon Dioxide Breath, p. 48 ◆ GENERAL SE **Model-Making Labs** Build a Lung, p. 181 ◆ GENERAL LB **Whiz-Bang Demonstrations** Take a Deep Breath* ◆ GENERAL LB **EcoLabs & Field Activities** There's Something in the Air* ◆ GENERAL LB **Long-Term Projects & Research Ideas** Getting to the Heart* ◆ ADVANCED	CRF **Lesson Plans*** TR **Bellringer Transparency*** TR The Role of Blood in Respiration* TR *LINK TO PHYSICAL SCIENCE* Exhaling, Pressure, and Fluid Flow* VID **Lab Videos for Life Science**

PACING • 90 min

CHAPTER REVIEW, ASSESSMENT, AND STANDARDIZED TEST PREPARATION

CRF **Vocabulary Activity*** GENERAL
SE **Chapter Review,** pp. 50–51 GENERAL
CRF **Chapter Review*** ■ GENERAL
CRF **Chapter Tests A*** ■ GENERAL, **B*** ADVANCED, **C*** SPECIAL NEEDS
SE **Standardized Test Preparation,** pp. 52–53 GENERAL
CRF **Standardized Test Preparation*** GENERAL
CRF **Performance-Based Assessment*** GENERAL
OSP **Test Generator** GENERAL
CRF **Test Item Listing*** GENERAL

Online and Technology Resources

go.hrw.com

Visit **go.hrw.com** for a variety of free resources related to this textbook. Enter the keyword **HL5BD2.**

Holt Online Learning

Students can access interactive problem-solving help and active visual concept development with the *Holt Science and Technology* Online Edition available at **www.hrw.com.**

Guided Reading Audio CD

A direct reading of each chapter using instructional visuals as guideposts. For auditory learners, reluctant readers, and Spanish-speaking students. Available in English and Spanish.

KEY

SE Student Edition	**CRF** Chapter Resource File	**SS** Science Skills Worksheets	* Also on One-Stop Planner
TE Teacher Edition	**OSP** One-Stop Planner	**MS** Math Skills for Science Worksheets	◆ Requires advance prep
	LB Lab Bank	**CD** CD or CD-ROM	■ Also available in Spanish
	TR Transparencies	**VID** Classroom Video/DVD	

SKILLS DEVELOPMENT RESOURCES	SECTION REVIEW AND ASSESSMENT	STANDARDS CORRELATIONS
SE **Pre-Reading Activity,** p. 28 GENERAL OSP **Science Puzzlers, Twisters & Teasers** GENERAL		National Science Education Standards UCP 1, 2; SAI 1; ST 2; SPSP 5; LS 1e, 3b
CRF **Directed Reading A*** ■ BASIC, **B*** SPECIAL NEEDS CRF **Vocabulary and Section Summary*** ■ GENERAL SE **Reading Strategy** Paired Summarizing, p. 30 GENERAL TE **Reading Strategy** Mnemonics, p. 32 GENERAL SE **Math Practice** The Beat Goes On, p. 34 GENERAL MS **Math Skills for Science** Multiplying Whole Numbers* BASIC CRF **Reinforcement Worksheet** Matchmaker, Matchmaker* GENERAL CRF **Reinforcement Worksheet** Colors of the Heart* BASIC CRF **Critical Thinking** Doctor for a Day* ADVANCED	SE **Reading Checks,** pp. 30, 32, 34 GENERAL TE **Homework,** p. 31 GENERAL TE **Homework,** p. 31 GENERAL TE **Reteaching,** p. 34 BASIC TE **Quiz,** p. 34 GENERAL TE **Alternative Assessment,** p. 34 GENERAL SE **Section Review,*** p. 35 ■ GENERAL CRF **Section Quiz*** ■ GENERAL	UCP 1, 2, 3, 5; SPSP 1, 4, 5; LS 1a, 1b, 1c, 1d, 1e, 1f, 3b
CRF **Directed Reading A*** ■ BASIC, **B*** SPECIAL NEEDS CRF **Vocabulary and Section Summary*** ■ GENERAL SE **Reading Strategy** Reading Organizer, p. 36 GENERAL TE **Discussion** Blood, p. 36 GENERAL TE **Reading Strategy** Asking Questions, p. 37 GENERAL TE **Inclusion Strategies,** p. 38	SE **Reading Checks,** pp. 36, 37, 38, 39 GENERAL TE **Reteaching,** p. 38 BASIC TE **Quiz,** p. 38 GENERAL TE **Alternative Assessment,** p. 38 GENERAL SE **Section Review,*** p. 39 ■ GENERAL CRF **Section Quiz*** ■ GENERAL	UCP 1; SPSP 1, 5; LS 1a, 1b, 1c, 1d, 1e, 1f, 3a, 3b
CRF **Directed Reading A*** ■ BASIC, **B*** SPECIAL NEEDS CRF **Vocabulary and Section Summary*** ■ GENERAL SE **Reading Strategy** Prediction Guide, p. 40 GENERAL TE **Research** Lymphatic System and Disease, p. 41 ADVANCED TE **Reading Strategy** Paired Summarizing, p. 41 GENERAL	SE **Reading Checks,** pp. 40, 42 GENERAL TE **Reteaching,** p. 42 BASIC TE **Quiz,** p. 42 GENERAL TE **Alternative Assessment,** p. 42 ADVANCED SE **Section Review,*** p. 43 ■ GENERAL CRF **Section Quiz*** ■ GENERAL	UCP 1, 5; LS 1a, 1c, 1d, 1e, 1f
CRF **Directed Reading A*** ■ BASIC, **B*** SPECIAL NEEDS CRF **Vocabulary and Section Summary*** ■ GENERAL SE **Reading Strategy** Reading Organizer, p. 44 GENERAL TE **Inclusion Strategies,** p. 45	SE **Reading Checks,** pp. 45, 46 GENERAL TE **Reteaching,** p. 46 BASIC TE **Quiz,** p. 46 GENERAL TE **Alternative Assessment,** p. 46 ADVANCED SE **Section Review,*** p. 47 ■ GENERAL CRF **Section Quiz*** ■ GENERAL	UCP 1, 2, 4, 5; LS 1a, 1b, 1c, 1d, 1e, 1f; UCP 2, 5; SAI 1, 2; LS 1a; *Chapter Lab:* UCP 2, 5; SAI 1, 2; LS 1a; *LabBook:* UCP 2, 5; SAI 1, 2; LS 1a

One-Stop Planner® CD-ROM

This convenient CD-ROM includes:
- **Lab Materials QuickList Software**
- **Holt Calendar Planner**
- **Customizable Lesson Plans**
- **Printable Worksheets**
- **ExamView® Test Generator**

CNN Student News

cnnstudentnews.com

Find the latest news, lesson plans, and activities related to important scientific events.

SciLinks NSTA

www.scilinks.org

Maintained by the **National Science Teachers Association.** See Chapter Enrichment pages for a complete list of topics.

Current Science®

Check out ***Current Science*** articles and activities by visiting the HRW Web site at **go.hrw.com.** Just type in the keyword **HL5CS23T.**

Classroom Videos

- **Lab Videos** demonstrate the chapter lab.
- **Brain Food Video Quizzes** help students review the chapter material.

2 Chapter Resources

Visual Resources

CHAPTER STARTER TRANSPARENCY

BELLRINGER TRANSPARENCIES

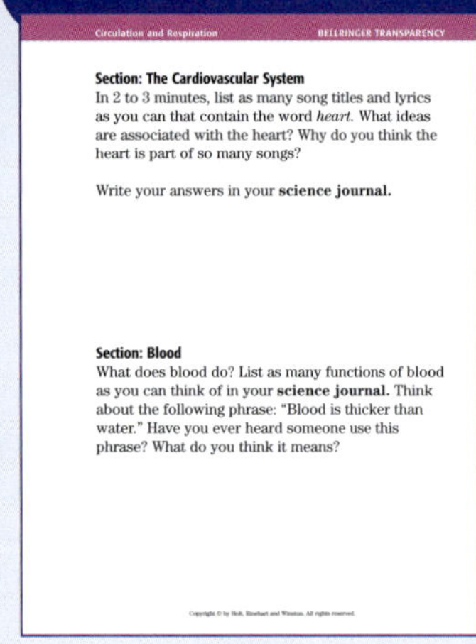

TEACHING TRANSPARENCIES

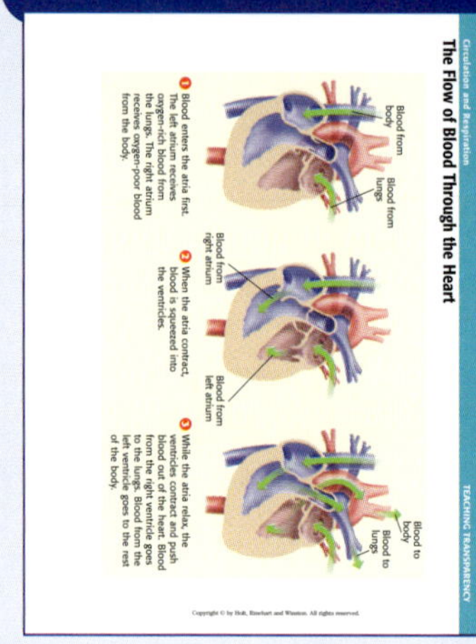

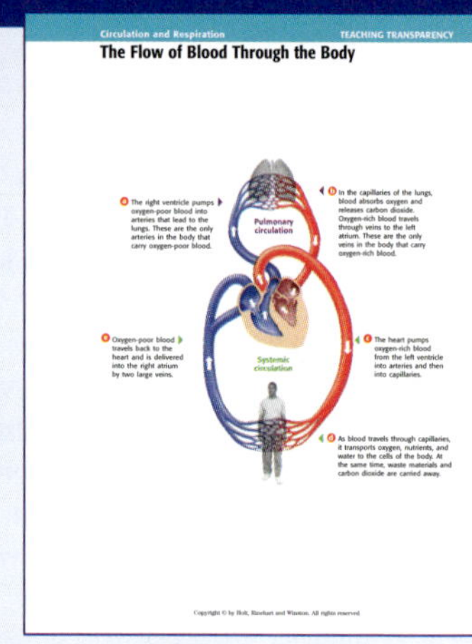

TEACHING TRANSPARENCIES

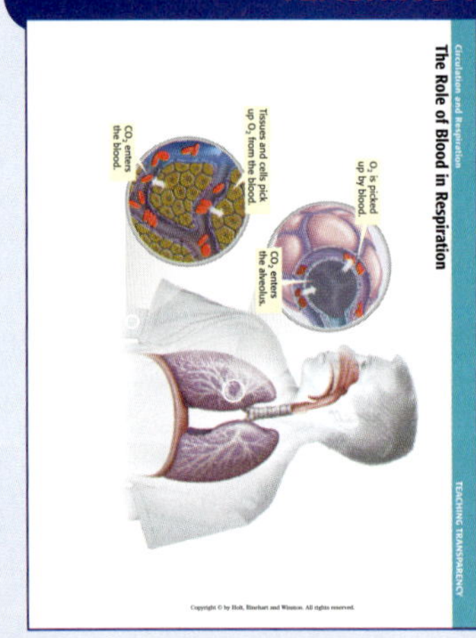

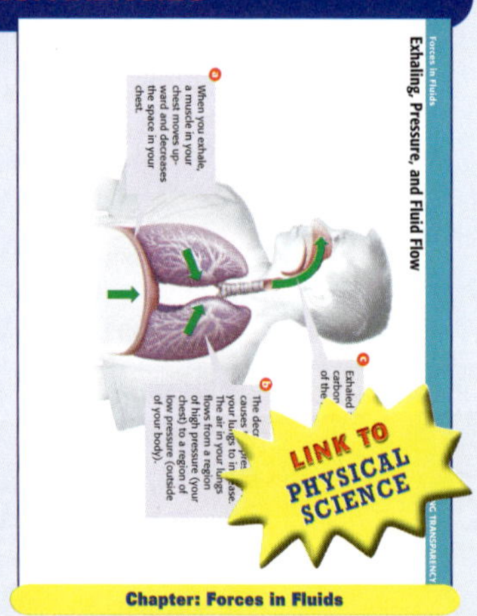

CONCEPT MAPPING TRANSPARENCY

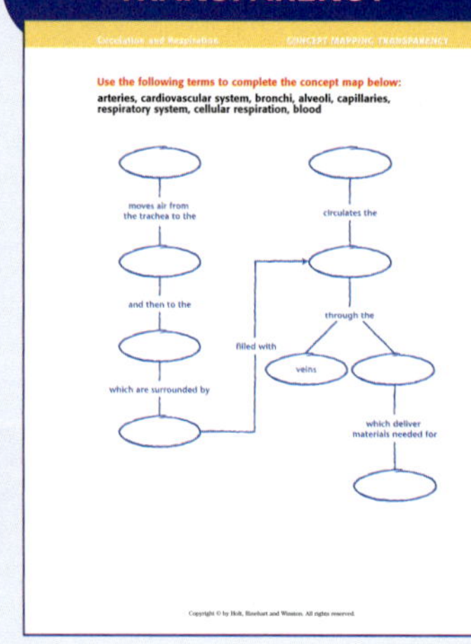

Planning Resources

LESSON PLANS

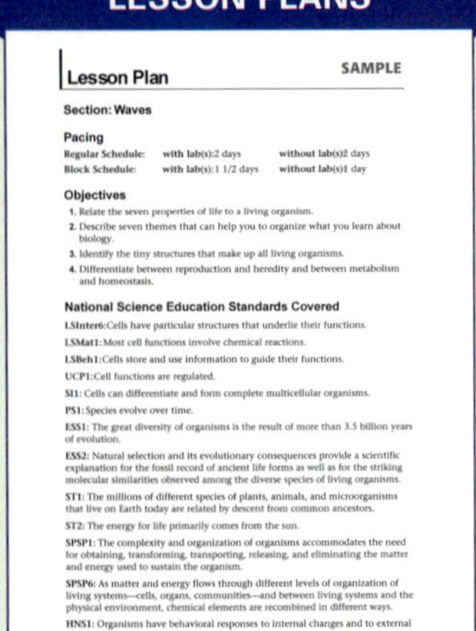

PARENT LETTER

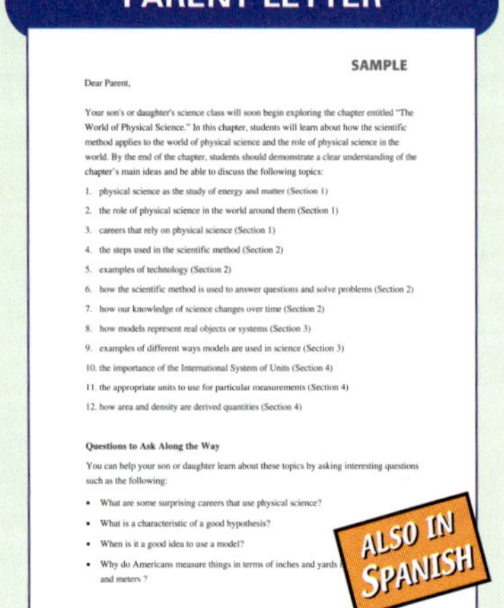

TEST ITEM LISTING

One-Stop Planner® CD-ROM

This CD-ROM includes all of the resources shown here and the following time-saving tools:

- *Lab Materials QuickList Software*
- *Customizable lesson plans*
- *Holt Calendar Planner*
- *The powerful ExamView® Test Generator*

For a preview of available worksheets covering math and science skills, see pages T12–T19. All of these resources are also on the One-Stop Planner®.

Meeting Individual Needs

DIRECTED READING A — Directed Reading A — SAMPLE — BASIC — ALSO IN SPANISH

DIRECTED READING B — Directed Reading B — SAMPLE — SPECIAL NEEDS

VOCABULARY ACTIVITY — Vocabulary Activity — SAMPLE — GENERAL

VOCABULARY AND SECTION SUMMARY — Vocabulary & Notes — SAMPLE — GENERAL — ALSO IN SPANISH

REINFORCEMENT — Reinforcement — SAMPLE — BASIC

CRITICAL THINKING — Critical Thinking — SAMPLE — ADVANCED

SCILINKS ACTIVITY — SciLinks Activity — SAMPLE — GENERAL

SCIENCE PUZZLERS, TWISTERS & TEASERS — Circulation and Respiration — GENERAL

Labs and Activities

ECOLABS & FIELD ACTIVITIES — There's Something in the Air — GENERAL

WHIZ-BANG DEMONSTRATIONS — Get the Beat! — BASIC

WHIZ-BANG DEMONSTRATIONS — Take a Deep Breath — GENERAL

LONG-TERM PROJECTS & RESEARCH IDEAS — Getting to the Heart — ADVANCED

CALCULATOR-BASED LABS — A Hot Hand — ADVANCED

DATASHEETS FOR QUICK LABS — Reaction to Stress — SAMPLE

DATASHEETS FOR CHAPTER LABS — Using Scientific Methods — SAMPLE

DATASHEETS FOR LABBOOK — Does It All Add Up? — SAMPLE

Review and Assessments

SECTION QUIZ — Section Quiz — SAMPLE — GENERAL — ALSO IN SPANISH

SECTION REVIEW — Section Review — SAMPLE — GENERAL — ALSO IN SPANISH

CHAPTER REVIEW — Chapter Review — SAMPLE — GENERAL — ALSO IN SPANISH

CHAPTER TEST A — Chapter Test A — SAMPLE — GENERAL — ALSO IN SPANISH

CHAPTER TEST B — Chapter Test B — SAMPLE — ADVANCED

CHAPTER TEST C — Chapter Test C — SAMPLE — SPECIAL NEEDS

STANDARDIZED TEST PREPARATION — Standardized Test Preparation — SAMPLE — GENERAL

PERFORMANCE-BASED ASSESSMENT — Performance-Based Assessment — SAMPLE — GENERAL

Chapter Enrichment

This Chapter Enrichment provides relevant and interesting information to expand and enhance your presentation of the chapter material.

Section 1

The Cardiovascular System

The Flow of Blood

- William Harvey (1578–1657) is credited with being the first European to discover the circulation of the blood through the body.

- Based on his dissections of animals, Harvey rightly concluded that the heart was a muscle that served to pump blood through the body. Harvey correctly maintained that arteries carry blood away from the heart and veins carry blood toward the heart. His views, for which many other physicians ridiculed him, contradicted conventional wisdom and theories of the famous physician Galen (129–c. 201 CE).

Atherosclerosis

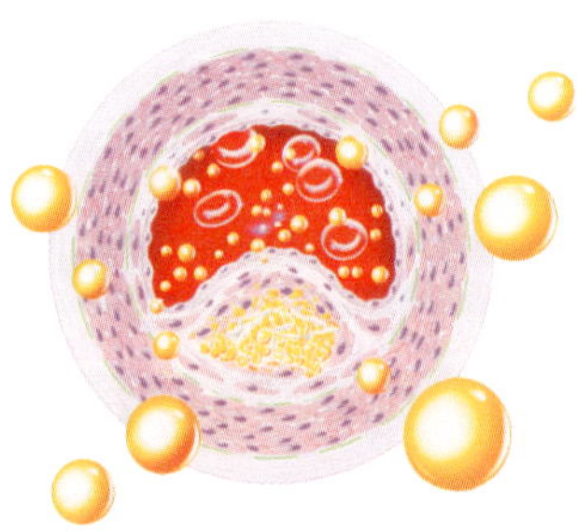

- Atherosclerosis is a disease of the arteries in which the inside layer of the arterial walls thickens with plaque. The plaque forms in areas of turbulent blood flow. As the walls of an artery thicken, the diameter of the vessel narrows. This narrowing of the vessel impedes blood flow.

- Cigarette smoking, high blood pressure, obesity, inactivity, high cholesterol level, and a family history of heart disease are risk factors for atherosclerosis.

Hypertension

- Like people who have atherosclerosis, people afflicted with hypertension may not exhibit symptoms of the disease for years. In fact, hypertension is often called "the silent killer." Although blood pressure varies within a wide range across the population, a person whose resting blood pressure is consistently at the high end of that range is said to have hypertension.

- Smoking, obesity, stress, excessive consumption of alcohol, and diabetes mellitus exacerbate high blood pressure.

Stroke

- A stroke occurs when the brain is damaged because of an interruption in the blood flow or the leakage of blood from the blood vessels. Atherosclerosis and hypertension are some of the causes of strokes.

Heart Attack

- A heart attack occurs when part of the heart muscle dies because of blood and oxygen deprivation. About 1 million people in the United States have a heart attack each year.

Section 2

Blood

Blood Types and Rh Factor

- Human blood has four ABO types (A, B, AB, and O), which are determined by surface antigens on red blood cells (RBCs). Blood is also either Rh+ or Rh–. If RBCs have an Rh antigen (a protein on the surface of the cells), the blood is Rh+. If an Rh antigen is not present, the blood is Rh–. People have one of the following blood types: A+, A–, B+, B–, AB+, AB–, O+, or O–.

- People make antibodies against the antigens that their RBCs do not have. People who have Rh– blood (no Rh antigens) make antibodies against the Rh antigen only after being exposed to Rh+ blood. For example, people who have type B– blood make A antibodies that attack any blood cell that has an A antigen. They may also make Rh antibodies that will attack any blood cell that has an Rh antigen on it. So, people who have type B– blood who have been exposed to Rh+ blood cannot be given A+, A–, B+, AB+, or AB– blood.

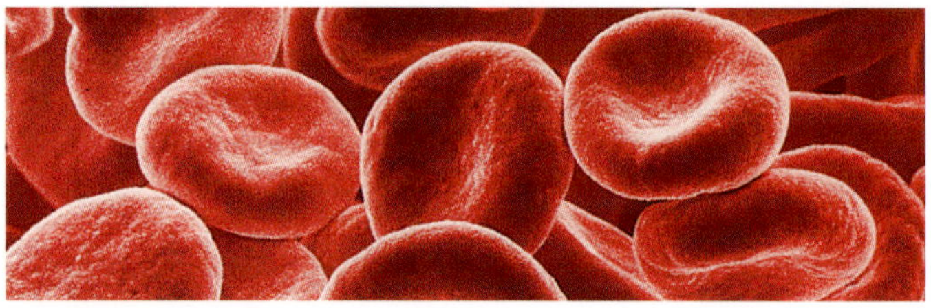

- Type O– blood can be given to anyone because its RBCs do not have any A, B, or Rh antigens that a recipient's antibodies could attack. For this reason, a type O– person is said to be a universal donor. Type AB+ people are called *universal recipients;* they can be given any type of blood because they do not make any antibodies against A, B, or Rh antigens.

For background information about teaching strategies and issues, refer to the ***Professional Reference for Teachers.***

Section 3

The Lymphatic System

Lymphocytes

- Lymphocytes are white blood cells that specialize in fighting pathogens. The two main kinds of lymphocytes are B cells and T cells.
- About 10 % of lymphocytes are B cells. When confronted with foreign antigens, B cells produce antibodies that destroy the antigens. This process is called *humoral immunity.*
- About 90 % of lymphocytes are T cells, which form in the bones and mature in the thymus. Cells called *killer T cells* locate and attack cells on whose surface are foreign antigens. This type of immunity is called *cell-mediated immunity.*

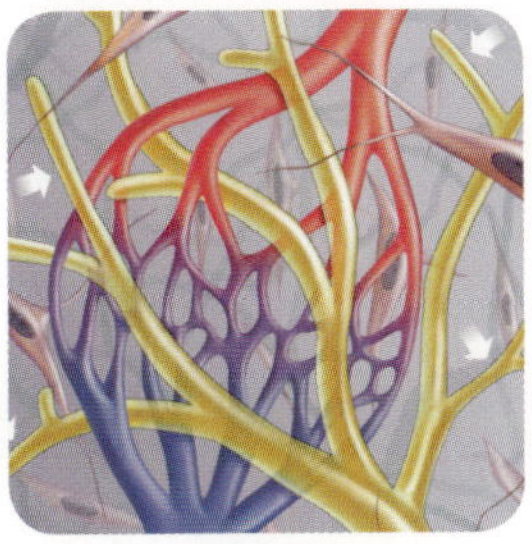

- HIV infects and destroys lymphocytes called *helper T cells.* When a person's helper T cell count falls below 200 cells per cubic millimeter of blood, the person is diagnosed with AIDS.

Is That a Fact!

- The tonsils reach their largest size when a person is about seven years old. Then, the tonsils begin to shrink.

Section 4

The Respiratory System

Control of Breathing

- Unless a person consciously holds his or her breath or changes the rate of his or her breathing, breathing is controlled automatically by breathing control centers in the base of the brain (in the medulla oblongata and in the pons).
- Hiccups are a result of a sudden, jerky contraction of the diaphragm. When a person eats too much food, the full stomach may irritate the diaphragm muscle and cause the muscle to contract jerkily. However, other causes of hiccups are unknown.

Is That a Fact!

- Each day, about 2,000 adolescents in the United States become regular daily smokers Each day. The habit will eventually kill one-third of these children.
- Children who have asthma are particularly at risk from their parents' second-hand smoke. Each year, second-hand smoke increases the number of asthma attacks and the severity of the symptoms in 200,000 to 1 million children who have asthma.

Smoking

- Tobacco smoking has been implicated in more than 90% of lung cancers among men. Among people who do not smoke, 3,000 cases of lung cancer are linked to secondhand cigarette smoke each year.

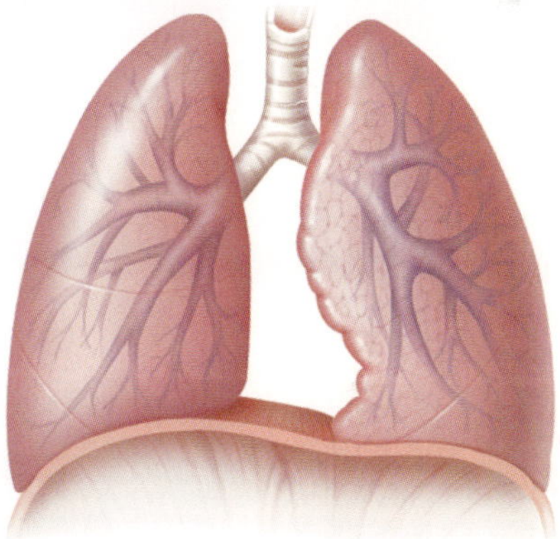

SCILINKS. NSTA Developed and maintained by the National Science Teachers Association

SciLinks is maintained by the National Science Teachers Association to provide you and your students with interesting, up-to-date links that will enrich your classroom presentation of the chapter.

Visit www.scilinks.org and enter the SciLinks code for more information about the topic listed.

Topic: The Cardiovascular System
SciLinks code: HSM0221

Topic: Cardiovascular Problems
SciLinks code: HSM0220

Topic: Blood
SciLinks code: HSM0175

Topic: Blood Donations
SciLinks code: HSM0178

Topic: The Lymphatic System
SciLinks code: HSM0891

Topic: The Immune System
SciLinks code: HSM0786

Topic: The Respiratory System
SciLinks code: HSM1307

Topic: Respiratory Disorders
SciLinks code: HSM1306

Overview

Tell students that this chapter will help them learn about their cardiovascular, lymphatic, and respiratory systems. The chapter has a separate section about blood and a clear description of the way that the cardiovascular system and blood make up the circulatory system.

Assessing Prior Knowledge

Students should be familiar with the following topics:

- the basic life processes
- cells

Identifying Misconceptions

As students learn the material in this chapter, some of them may be confused about the difference between cells and molecules. When asked to draw a molecule, many students will draw something that resembles a cell. Instruction should describe the relationship between molecules and cells. Also, many students may think that proteins and molecules are bigger than cells. The concepts of the cell and the molecule are important when studying the way that blood transports materials to cells and the way that gases are exchanged in the lungs. Use a Venn diagram to help clarify the relationship between atoms, molecules and cells.

2 Circulation and Respiration

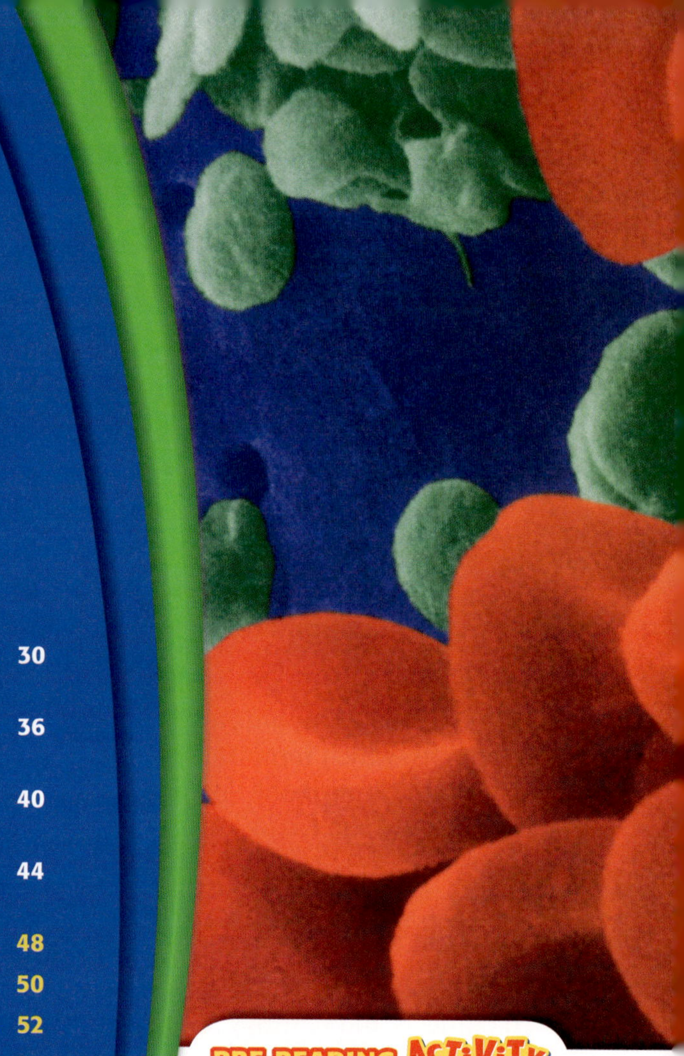

About the PHOTO

Your circulatory system is made up of the heart, blood vessels, and blood. This picture is a colored scanning electron micrograph of red and white blood cells and cell fragments called *platelets.* Red blood cells are disk shaped, white blood cells are rounded, and platelets are the small green fragments. There are millions of blood cells in a drop of blood. Blood cells are so important that your body makes about 200 billion red blood cells every day.

PRE-READING ACTIVITY

FOLDNOTES **Four-Corner Fold**

Before you read the chapter, create the FoldNote entitled "Four-Corner Fold" described in the **Study Skills** section of the Appendix. Label the flaps of the four-corner fold with the section titles "Cardiovascular system," "Blood," Lymphatic system," and "Respiratory system." Write what you know about each topic under the appropriate flap. As you read the chapter, add other information that you learn.

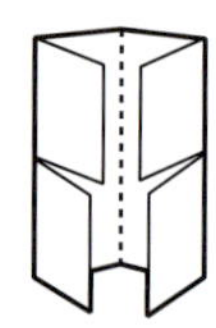

Standards Correlations

National Science Education Standards

The following codes indicate the National Science Education Standards that correlate to this chapter. The full text of the standards is at the front of the book.

Chapter Opener
UCP 1, 2; SAI 1; ST 2; SPSP 5; LS 1e, 3b

Section 1 The Cardiovascular System
UCP 1, 2, 3, 5; SPSP 1, 4, 5; LS 1a, 1b, 1c, 1d, 1e, 1f, 3b

Section 2 Blood
UCP 1; SPSP 1, 5; LS 1a, 1b, 1c, 1d, 1e, 1f, 3a, 3b

Section 3 The Lymphatic System
UCP 1, 5; LS 1a, 1c, 1d, 1e, 1f

Section 4 The Respiratory System
UCP 1, 2, 4, 5; LS 1a, 1b, 1c, 1d, 1e, 1f; *LabBook:* UCP 2, 5; SAI 1, 2; LS 1a

Chapter Lab
UCP 2, 5; SAI 1, 2; LS 1a

Chapter Review
UCP 1; SAI 1; LS 1a, 1c, 1d, 1e, 3a

Science in Action
SAI 1, 2; ST 2; SPSP 4, 5; HNS 1; LS 1e

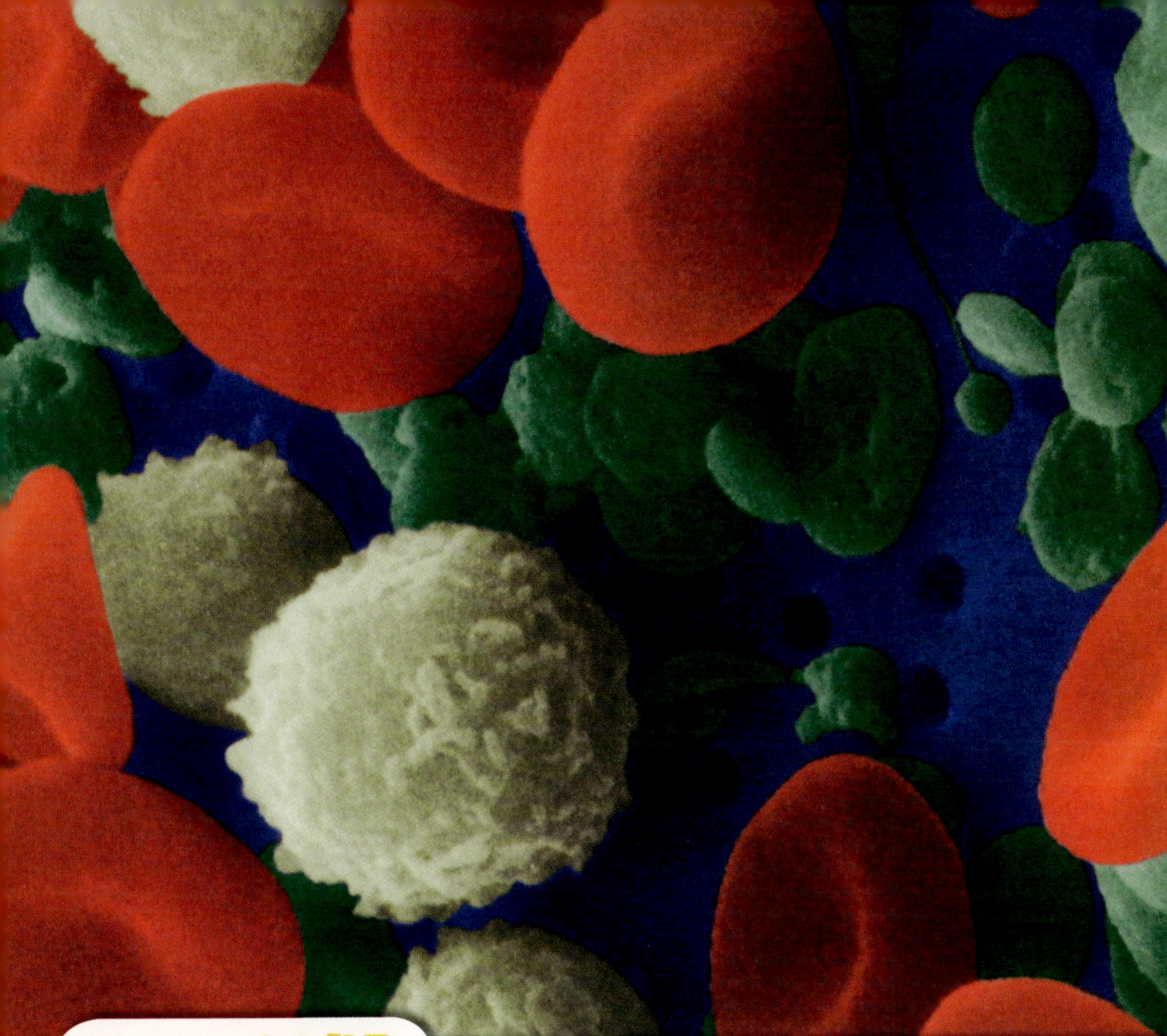

START-UP ACTIVITY

Exercise Your Heart

How does your heart respond to exercise? You can see this reaction by measuring your pulse.

Procedure

1. Take your pulse while remaining still. (Take your pulse by placing your fingers on the inside of your wrist just below your thumb.)
2. Using a **watch with a second hand,** count the number of heart beats in 15 s. Then, multiply this number by 4 to calculate the number of beats in 1 minute.
3. Do some moderate physical activity, such as jumping jacks or jogging in place, for 30 s.
4. Stop and calculate your heart rate again.
   Caution: Do not perform this exercise if you have difficulty breathing, if you have high blood pressure or asthma, or if you get dizzy easily.
5. Rest for 5 min.
6. Take your pulse again.

Analysis

1. How did exercise affect your heart rate? Why do you think this happened?
2. How does your heart rate affect the rate at which red blood cells travel throughout your body?
3. Did your heart rate return to normal (or almost normal) after you rested? Why or why not?

START-UP ACTIVITY

MATERIALS

For Each Group

- watch with a second hand

Safety Caution: Have students bring in a signed permission slip for this activity. Any students who have a health problem that may be worsened by exercise should be excused from the exercise portion of this activity. Instruct students who feel pain or become dizzy or tired to stop exercising immediately. Some students may feel embarrassed to exercise in front of their peers. You may want to invite a few volunteers to perform the exercise portion instead of having all of the students exercise.

Answers

1. Students should notice that their heart rate goes up when they are exercising. Explanations may vary, but students should understand that during exercise an increase in circulation rate is needed to ensure delivery of sufficient oxygen and fuel to the body.
2. Sample answer: An increased heart rate increases the rate of blood flow and delivers red blood cells to the body more quickly.
3. Sample answer: When I stop exercising, my body no longer needs extra oxygen and fuel, so my heart rate returns to normal after a period of rest.

Circulation and Respiration — CHAPTER STARTER

This Really Happened!

Chapter Starter Transparency
Use this transparency to help students begin thinking about their circulatory system and its functions.

CHAPTER RESOURCES

Technology

- **Transparencies** — READING SKILLS
  - Chapter Starter Transparency
- **Student Edition on CD-ROM**
- **Guided Reading Audio CD**
  - English or Spanish
- **Classroom Videos**
  - Brain Food Video Quiz

Workbooks

- **Science Puzzlers, Twisters & Teasers**
  - Circulation and Respiration GENERAL

SECTION 1

Focus

Overview

This section introduces the structures and functions of the cardiovascular and circulatory systems. Students study the three types of blood vessels, trace the path of blood in the body, and learn about blood types.

Bellringer

Ask students to list as many song titles, phrases, and slogans that contain the word *heart* as they can in two to three minutes. Ask for examples, and list the examples on the board. Ask for reasons that the word *heart* is the focus of so many songs and slogans.

Motivate

Discussion — GENERAL

Cardiovascular System Ask students to describe or diagram the flow of blood through their body. Remind students that their body is made up of trillions of cells. Ask students how they think that oxygen and nutrients get to each cell. Give a general description of the roles of the heart and the blood vessels, and discuss how sturdy these structures must be to work for a lifetime. LS Verbal/Logical

SECTION 1

The Cardiovascular System

READING WARM-UP

Objectives

- List four main parts of the cardiovascular system, and describe their functions.
- Describe the two types of circulation of blood in the body.
- List four cardiovascular problems.

Terms to Learn

cardiovascular system
artery
capillary
vein
pulmonary circulation
systemic circulation

READING STRATEGY

Paired Summarizing Read this section silently. In pairs, take turns summarizing the material. Stop to discuss ideas that seem confusing.

When you hear the word heart, *what do you think of first? Many people think of romance. Some people think of courage. But the heart is much more than a symbol of love or bravery. Your heart is an amazing pump.*

The heart is an organ that is part of your circulatory system. The *circulatory system* includes your heart; your blood; your veins, capillaries, and arteries; and your lymphatic system.

Your Cardiovascular System

Your heart creates pressure every time it beats. This pressure moves blood to every cell in your body through your cardiovascular system (KAR dee OH VAS kyoo luhr SIS tuhm). The **cardiovascular system** consists of the heart and the three types of blood vessels that carry blood throughout your body. The word *cardio* means "heart," and *vascular* means "blood vessel." The blood vessels—arteries, capillaries, and veins—carry blood pumped by the heart. **Figure 1** shows the major arteries and veins.

cardiovascular system a collection of organs that transport blood throughout the body

Reading Check What are the four main parts of the cardiovascular system? (*See the Appendix for answers to Reading Checks.*)

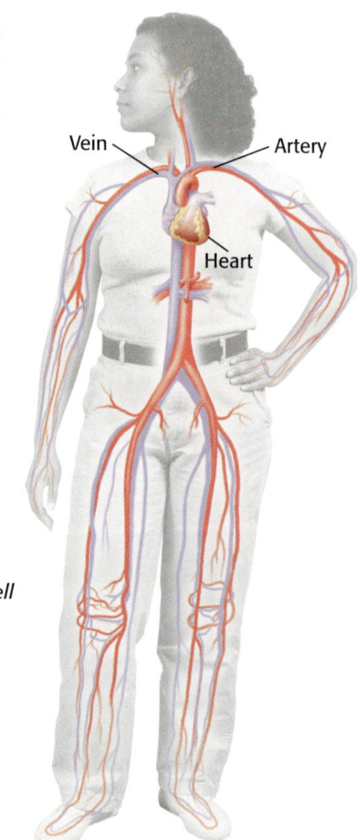

Figure 1 *The cardiovascular system carries blood to every cell in your body.*

CHAPTER RESOURCES

Chapter Resource File

- Lesson Plan
- Directed Reading A BASIC
- Directed Reading B SPECIAL NEEDS

Technology

Transparencies
- Bellringer
- The Flow of Blood Through the Heart

Workbooks

Math Skills for Science
- The Unit Factor and Dimensional Analysis

Answer to Reading Check

The four main parts of the cardiovascular system are the heart and the arteries, capillaries, and veins.

The Heart

Your *heart* is an organ made mostly of cardiac muscle tissue. It is about the size of your fist and is almost in the center of your chest cavity. Like hearts of all mammals, your heart has a left side and a right side that are separated by a thick wall. The right side of the heart pumps oxygen-poor blood to the lungs. The left side pumps oxygen-rich blood to the body. As you can see in **Figure 2,** each side has an upper chamber and a lower chamber. Each upper chamber is called an *atrium* (plural, *atria*). Each lower chamber is called a *ventricle*.

Flaplike structures called *valves* are located between the atria and ventricles and in places where large arteries are attached to the heart. As blood moves through the heart, these valves close to prevent blood from going backward. The "lub-dub, lub-dub" sound of a beating heart is caused by the valves closing. **Figure 3** shows the flow of blood through the heart.

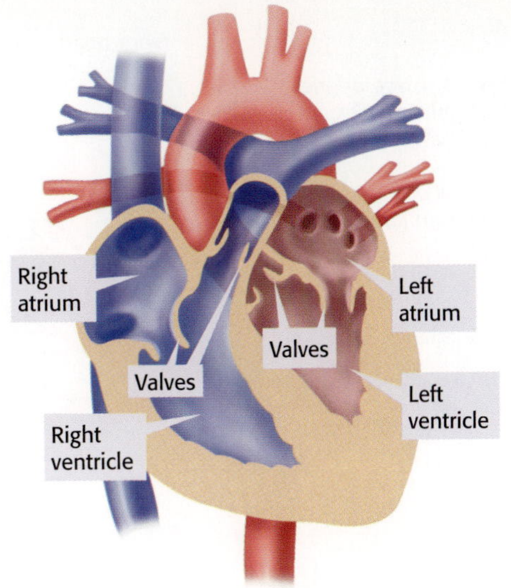

Figure 2 *The heart pumps blood through blood vessels. The vessels carrying oxygen-rich blood are shown in red. The vessels carrying oxygen-poor blood are shown in blue.*

Figure 3 The Flow of Blood Through the Heart

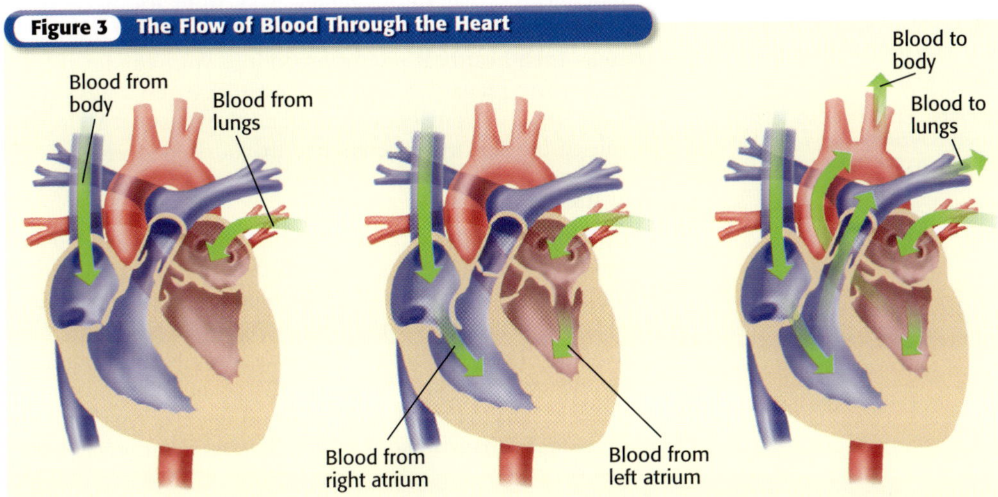

1. Blood enters the atria first. The left atrium receives oxygen-rich blood from the lungs. The right atrium receives oxygen-poor blood from the body.
2. When the atria contract, blood is squeezed into the ventricles.
3. While the atria relax, the ventricles contract and push blood out of the heart. Blood from the right ventricle goes to the lungs. Blood from the left ventricle goes to the rest of the body.

MISCONCEPTION ALERT

Left Side, Right Side The words *left* and *right* as used to diagram the anatomy of the heart might confuse students. When students are looking at a picture of the heart, the left atrium appears on their right and the right atrium appears on their left. Help students understand anatomical left and right by facing them and asking them to identify your left and right hands in relation to their own.

Homework — GENERAL

Poster Project Have students make a poster showing a diagram of the heart. Remind students to label each chamber of the heart properly and to use arrows to indicate the flow of blood. Students can also include diagrams of the lungs and of the rest of the body to show how blood flows to and from the heart. LS Visual

Teach

CONNECTION ACTIVITY
Math — GENERAL

The Pump in Your Chest Your heart beats about 100,800 times per day. With every beat, about 70 mL of blood is pumped out of your heart. In 1 h, how many liters of blood does your heart pump out? (294 L)

About how many liters of blood does your heart pump out in a day? (7,056 L) Help students visualize this amount by showing them a liter of water. LS Logical/Visual

Homework — GENERAL

Making Models Have students make a model of the human heart that shows the path of blood through the heart. Models can be drawings or three-dimensional constructions. Have students present their completed models to the class. Provide yarn or a pen light so that students can demonstrate to the class the flow of blood through their model heart. LS Visual English Language Learners

CONNECTION to Chemistry — ADVANCED

Pacemaker Contraction of the heart is started by a structure called the *sinoatrial node,* a small cluster of cells embedded in the upper wall of the right atrium. Have students research the sinoatrial node (and the atrioventricular node, which stimulates the ventricles to contract) and make a model or a poster showing how these special cells control the heartbeat.

(Students should find that the heartbeat is not triggered by the nervous system, that the atria contract almost simultaneously, and that the contraction spreads to the ventricles but that there is a short delay during which the atria empty blood into the ventricles.)

Teach, *continued*

READING STRATEGY — GENERAL

Mnemonics Challenge students to create mnemonic devices to help them recall the flow of blood in the body. For example, arteries carry blood away from the heart, and both *artery* and *away* begin with the letter *a*.
LS Verbal

CONNECTION to Real World — BASIC

Blood Vessels Have students rank blood vessels in order from thickest vessel wall to thinnest vessel wall. Have students explain why wall thicknesses vary. (arteries—blood is under pressure; veins—blood is not under as much pressure as in arteries, but walls do not need to be as thin as capillaries; capillaries—their walls are thinnest because the walls must allow for gas, nutrient, and waste exchange with cells) English Language Learners
LS Logical

ACTIVITY — GENERAL

Viewing Blood Vessels Set up microscope stations that have prepared slides of an artery and a vein. Have students view the slides and match them with the descriptions of arteries and veins in the textbook. Have students sketch what they see under the microscope and write descriptive words beside each sketch. English Language Learners
LS Visual

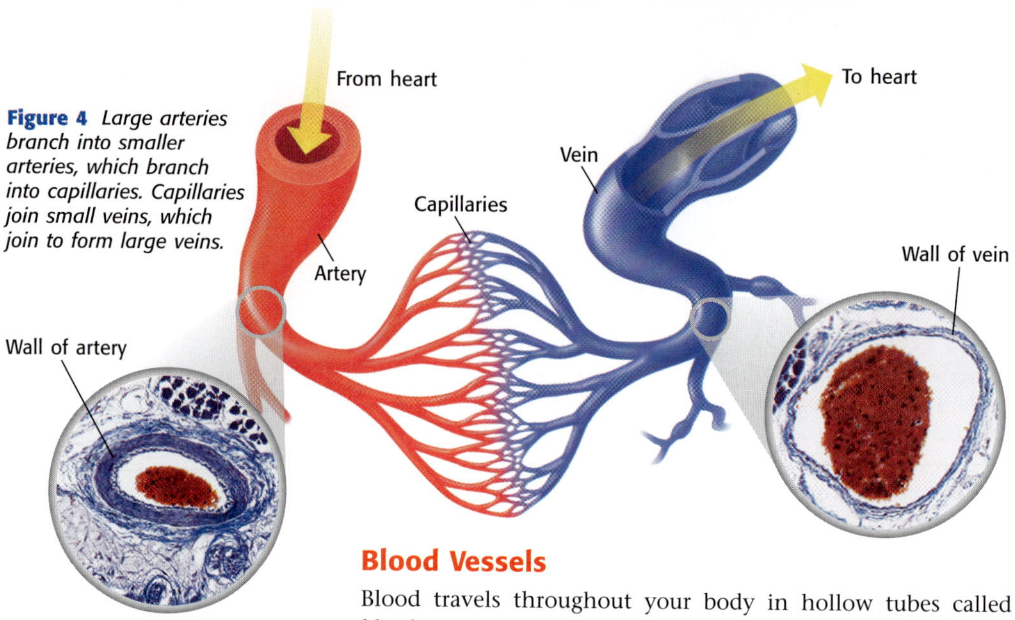

Figure 4 *Large arteries branch into smaller arteries, which branch into capillaries. Capillaries join small veins, which join to form large veins.*

Blood Vessels

Blood travels throughout your body in hollow tubes called *blood vessels*. The three types of blood vessels—arteries, capillaries, and veins—are shown in **Figure 4.**

Arteries

A blood vessel that carries blood away from the heart is an **artery.** Arteries have thick walls, which contain a layer of smooth muscle. Each heartbeat pumps blood into your arteries at high pressure. This pressure is your *blood pressure*. Artery walls stretch and are usually strong enough to stand the pressure. Your *pulse* is caused by the rhythmic change in your blood pressure.

artery a blood vessel that carries blood away from the heart to the body's organs

capillary a tiny blood vessel that allows an exchange between blood and cells in other tissue

vein in biology, a vessel that carries blood to the heart

Capillaries

Nutrients, oxygen, and other substances must leave blood and get to your body's cells. Carbon dioxide and other wastes leave body cells and are carried away by blood. A **capillary** is a tiny blood vessel that allows these exchanges between body cells and blood. These exchanges can take place because capillary walls are only one cell thick. Capillaries are so narrow that blood cells must pass through them in single file. No cell in the body is more than three or four cells away from a capillary.

Veins

After leaving capillaries, blood enters veins. A **vein** is a blood vessel that carries blood back to the heart. As blood travels through veins, valves in the veins keep the blood from flowing backward. When skeletal muscles contract, they squeeze nearby veins and help push blood toward the heart.

✓ **Reading Check** **Describe the three types of blood vessels.**

WEIRD SCIENCE

Babies that have certain congenital heart defects are blue at birth. The condition, called *cyanosis*, can be caused by low levels of oxygen in the blood. Oxygen-poor blood may be a result of defects in the heart. These defects cause oxygen-poor blood to be pumped to the body before it is pumped to the lungs. These defects can be repaired surgically.

Answer to Reading Check

Arteries have thick, stretchy walls and carry blood away from the heart. Capillaries are tiny blood vessels that allow the exchange of oxygen, carbon dioxide, and nutrients between cells and blood. Veins are blood vessels that carry blood back to the heart.

Two Types of Circulation

Where does blood get the oxygen to deliver to your body? From your lungs! Your heart pumps blood to the lungs. In the lungs, carbon dioxide leaves the blood and oxygen enters the blood. The oxygen-rich blood then flows back to the heart. This circulation of blood between your heart and lungs is called **pulmonary circulation** (PUL muh NER ee SUHR kyoo LAY shuhn).

The oxygen-rich blood returning to the heart from the lungs is then pumped to the rest of the body. The circulation of blood between the heart and the rest of the body is called **systemic circulation** (sis TEM ik SUHR kyoo LAY shuhn). Both types of circulation are shown in **Figure 5.**

pulmonary circulation the flow of blood from the heart to the lungs and back to the heart through the pulmonary arteries, capillaries, and veins

systemic circulation the flow of blood from the heart to all parts of the body and back to the heart

Figure 5 **The Flow of Blood Through the Body**

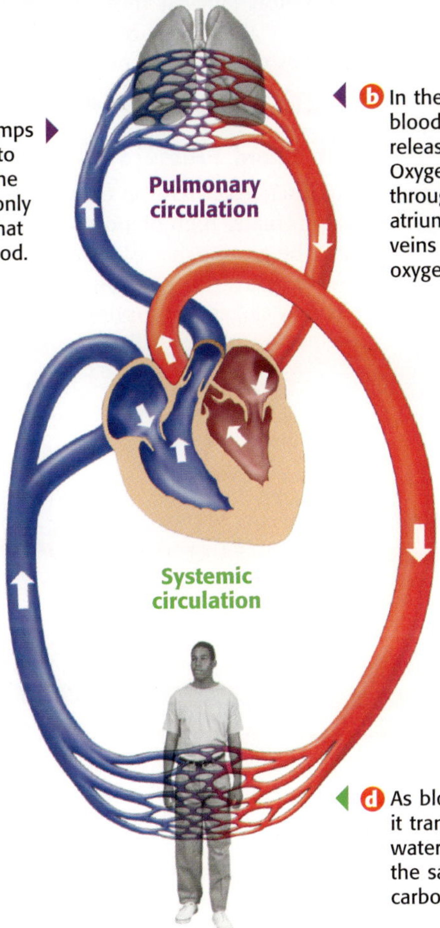

a The right ventricle pumps oxygen-poor blood into arteries that lead to the lungs. These are the only arteries in the body that carry oxygen-poor blood.

b In the capillaries of the lungs, blood absorbs oxygen and releases carbon dioxide. Oxygen-rich blood travels through veins to the left atrium. These are the only veins in the body that carry oxygen-rich blood.

c The heart pumps oxygen-rich blood from the left ventricle into arteries and then into capillaries.

d As blood travels through capillaries, it transports oxygen, nutrients, and water to the cells of the body. At the same time, waste materials and carbon dioxide are carried away.

e Oxygen-poor blood travels back to the heart and is delivered into the right atrium by two large veins.

MISCONCEPTION ALERT

Pulmonary Circulation Students may think that blood from the heart enters one lung and leaves from the other. However, each lung is serviced by its own vessels that carry blood to and from the heart.

CHAPTER RESOURCES

Technology

Transparencies
- The Flow of Blood Through the Body

Group ACTIVITY—GENERAL

MATERIALS

For Each Class
- balloon, blue, inflated (or blue paper disk) (5)
- balloon, red, inflated (or red paper disk) (5)
- diagram, "The Flow of Blood Through the Body"
- station, marked by (cardboard boxes,flags, or other markers) (8)

Circulate! Organize the class into five teams for a relay race. The race should proceed as follows:

1. Students begin in the left ventricle carrying a red balloon, which represents an oxygenated blood cell.
2. They travel through the aorta.
3. After passing through the aorta, students carry oxygenated blood to the muscles and exchange the red balloon for a blue one.
4. From the muscles, students carry blood loaded with CO_2 to the right atrium.
5. From the right atrium, students travel into the right ventricle.
6. Students travel through the pulmonary artery.
7. From the pulmonary artery, students travel into the lungs, where they exchange their CO_2 for oxygen (they exchange blue balloons for red ones).
8. Carrying oxygenated blood (red balloons), students enter the left atrium and are ready to begin again or hand off their balloons.

Walk one student at a time through the pathway; then, have the teams send students through in a relay race. LS Kinesthetic/Interpersonal

Co-op Learning

English Language Learners

Close

Answer to Math Practice

1. 100,800 beats per day
2. 2,759,400,000 beats per 75 years
3. 864,000 fewer beats

Reteaching — Basic

Writing **Owner's Guide** Have students make an owner's guide for their cardiovascular system. The guide should include a description of the various components of the system as well as information about the care and maintenance of the system. Encourage students to include a diagram or flowchart to illustrate their guide. **LS Verbal**

Quiz — General

Ask students whether each of the statements below is true or false. Have students correct false statements.

1. High blood pressure makes arteries stronger. (false)
2. Capillaries and veins have the same function. (false)
3. A major cause of heart disease is atherosclerosis. (true)
4. Arteries carry blood to the body and to the lungs. (true)

Alternative Assessment — General

Spider Map Have students make a spider map using the key terms in this chapter. Have them label the circle "Cardiovascular system," and have them make a leg for each of the other key terms. **LS Logical**

MATH PRACTICE

The Beat Goes On

A person's heart averages about 70 beats per minute.

1. Calculate how many times a heart beats in a day.
2. If a person lives for 75 years, how many times will his or her heart beat?
3. If an athlete's heart beats 50 times a minute, how many fewer times than an average heart will his or her heart beat in 30 days?

Cardiovascular Problems

More than just your heart and blood vessels are at risk if you have cardiovascular problems. Your whole body may be harmed. Cardiovascular problems can be caused by smoking, high levels of cholesterol in the blood, stress, physical inactivity, or heredity. Eating a healthy diet and getting plenty of exercise can reduce the risk of having cardiovascular problems.

Atherosclerosis

Heart diseases are the leading cause of death in the United States. A major cause of heart diseases is a cardiovascular disease called *atherosclerosis* (ATH uhr OH skluh ROH sis). Atherosclerosis happens when cholesterol (kuh LES tuhr AWL) builds up inside of blood vessels. This cholesterol buildup causes the blood vessels to become narrower and less elastic. **Figure 6** shows how clogged the pathway through a blood vessel can become. When an artery that supplies blood to the heart becomes blocked, the person may have a heart attack.

Reading Check Why is atherosclerosis dangerous?

High Blood Pressure

Atherosclerosis may be caused by hypertension. *Hypertension* is abnormally high blood pressure. The higher the blood pressure, the greater the risk of a heart attack, heart failure, kidney disease, and stroke. A *stroke* is when a blood vessel in the brain becomes clogged or ruptures. As a result, that part of the brain receives no oxygen. Without oxygen, brain cells die.

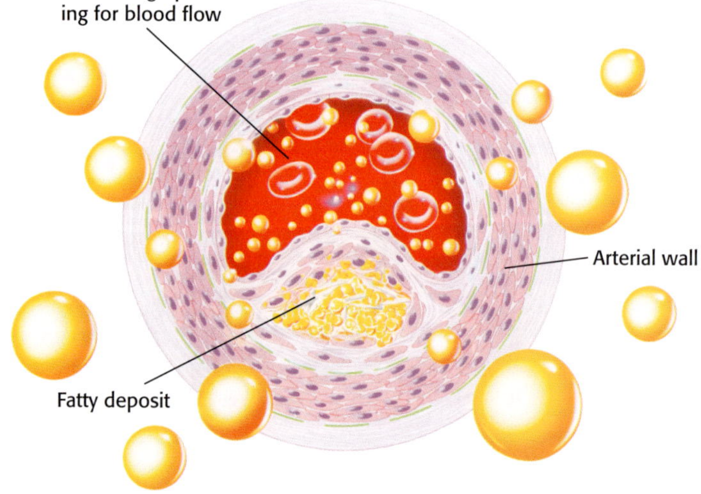

Figure 6 *This illustration shows the narrowing of an artery as the result of high levels of cholesterol in the blood. Lipid deposits (yellow) build up inside the blood vessel walls and block the flow of blood. Red blood cells and lipid particles (yellow balls) are shown escaping.*

Answer to Reading Check

Atherosclerosis is dangerous because it can cause the buildup of material inside an artery. When the artery becomes blocked, blood can't flow and can't reach the cells. In some cases, a person can have a heart attack from a blocked artery.

CHAPTER RESOURCES

Workbooks

Math Skills for Science
- Multiplying Whole Numbers Basic
- Counting the Zeros General
- Creating Exponents General
- Multiplying and Dividing in Scientific Notation Advanced

Heart Attacks and Heart Failure

Two cardiovascular problems are heart attacks and heart failure. A *heart attack* happens when heart muscle cells die and part of the heart muscle is damaged. As shown in **Figure 7,** arteries that deliver oxygen to the heart may be blocked. Without oxygen, heart muscle cells die quickly. When enough heart muscle cells die, the heart may stop.

Heart failure is different. *Heart failure* happens when the heart cannot pump enough blood to meet the body's needs. Organs, such as the brain, lungs, and kidneys, may be damaged by lack of oxygen or nutrients, or by the buildup of fluids or wastes.

Figure 7 Heart Attack

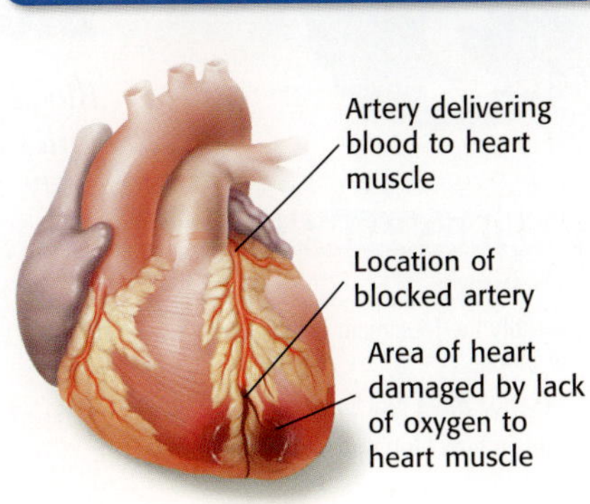

SECTION Review

Summary

- The cardiovascular system is made up of the heart and three types of blood vessels.
- The three types of blood vessels are arteries, veins, and capillaries.
- Oxygen-poor blood flows from the heart through the lungs, where it picks up oxygen.
- Oxygen-rich blood flows from the heart to the rest of the body.
- Cardiovascular problems include atherosclerosis, hypertension, heart attacks, and strokes.

Using Key Terms

For each pair of terms, explain how the meanings of the terms differ.

1. *artery* and *vein*
2. *systemic circulation* and *pulmonary circulation*

Understanding Key Ideas

3. Which of the following is true of blood in the pulmonary veins?
   a. The blood is going to the body.
   b. The blood is oxygen poor.
   c. The blood is going to the lungs.
   d. The blood is oxygen rich.
4. What are the four parts of the cardiovascular system? Describe the functions of each part.
5. What is the difference between a heart attack and heart failure?

Math Skills

6. An adult male's heart pumps about 2.8 million liters of blood a year. If his heart beats 70 times a minute, how much blood does his heart pump with each beat?

Critical Thinking

7. **Identifying Relationships** How is the structure of capillaries related to their function?
8. **Making Inferences** One of aspirin's effects is that it prevents platelets from being too "sticky." Why might doctors prescribe aspirin for patients who have had a heart attack?
9. **Analyzing Ideas** Veins and arteries are everywhere in your body. When a pulse is taken, it is usually taken at an artery in the neck or wrist. Explain why.
10. **Making Comparisons** Why is the structure of arteries different from the structure of capillaries?

SCILINKS. NSTA Developed and maintained by the National Science Teachers Association

For a variety of links related to this chapter, go to www.scilinks.org

Topic: The Cardiovascular System; Cardiovascular Problems

SciLinks code: HSM0221; HSM0220

INTERNET ACTIVITY

Short Story — GENERAL

For an Internet activity related to this chapter, have students go to **go.hrw.com** and type in the keyword **HL5BD2W.**

CHAPTER RESOURCES

Chapter Resource File

- Section Quiz GENERAL
- Section Review GENERAL
- Vocabulary and Section Summary GENERAL
- Reinforcement Worksheet BASIC
- Critical Thinking ADVANCED

Answers to Section Review

1. Arteries have stretchy, thick walls and carry blood away from the heart. Veins have valves and carry blood toward the heart.
2. Blood travels to and from the lungs in pulmonary circulation. Blood travels to and from the body in systemic circulation.
3. d
4. Sample answer: heart: to pump blood to lungs and to all parts of the body; arteries: to carry blood from the heart to the lungs and to all parts of the body; capillaries: to allow the exchange of gases, nutrients, and wastes between cells blood; veins: to carry blood back to the heart
5. Sample answer: A heart attack happens when heart muscle cells get insufficient oxygen and die, which may cause the heart to stop beating. Heart failure is when the heart cannot pump enough blood to meet the needs of the body.
6. about 76 mL per beat
7. Sample answer: Capillaries are very small and have very thin walls. This structure lets capillaries be near every cell and allows oxygen, nutrients, and carbon dioxide and other wastes to pass into and out of the capillaries and to and from cells.
8. Sample answer: A doctor might prescribe aspirin to keep platelets from sticking together and blocking an artery or vein, especially one in the heart. Such a blockage might cause another heart attack.
9. The wrist and the neck are two places where a large artery is close to the surface of the skin and the pulse can be felt.
10. Arteries and capillaries have different functions: arteries carry blood under pressure away from the heart, and capillaries allow the exchange of gases, nutrients and wastes in cells.

SECTION 2

Focus

Overview

In this section, students learn about blood, the parts of blood, and blood types.

Bellringer

Ask students to list three things that they know about blood, such as the parts of blood, the places where blood cells form, the function of blood in the body, and the way in which blood is donated.

Motivate

Discussion — GENERAL

Blood Have students describe a time when their skin was cut and describe what their blood looked like. List the words that students use to describe their blood. Based on students' experiences with blood and bleeding, lead a discussion about the structure and functions of blood. Have students read these pages and compare their descriptions of blood with the one in the textbook. Ask students the following questions: "Can you see individual blood cells when you bleed?" (no) "Why or why not?" (They are too small.) "How are red blood cells different from other cells in the body?" (They have hemoglobin, and they carry oxygen to other cells in the body.) LS Verbal

SECTION 2

Blood

READING WARM-UP

Objectives

- Identify the four main components of blood.
- Describe three functions of blood.
- Explain how blood pressure is measured.
- Explain what the ABO blood types are and why they are important.

Terms to Learn

blood
blood pressure

READING STRATEGY

Reading Organizer As you read this section, create an outline of the section. Use the headings from the section in your outline.

Blood is part of the circulatory system. It travels through miles and miles of blood vessels to reach every cell in your body. So, you must have a lot of blood, right?

Well, actually, an adult human body has about 5 L of blood. Your body probably has a little less than that. All the blood in your body would not fill two 3 L soda bottles.

What Is Blood?

Your *circulatory system* is made up of your heart, your blood vessels, and blood. **Blood** is a connective tissue made up of plasma, red blood cells, platelets, and white blood cells. Blood carries oxygen and nutrients to all parts of your body.

blood the fluid that carries gases, nutrients, and wastes through the body and that is made up of plasma, red blood cells, platelets, and white blood cells

Reading Check What are the four main components of blood? (*See the Appendix for answers to Reading Checks.*)

Plasma

The fluid part of the blood is called plasma (PLAZ muh). *Plasma* is a mixture of water, minerals, nutrients, sugars, proteins, and other substances. Red blood cells, white blood cells, and platelets are found in plasma.

Red Blood Cells

Most blood cells are *red blood cells,* or RBCs. RBCs, such as the ones shown in **Figure 1,** take oxygen to every cell in your body. Cells need oxygen to carry out their functions. Each RBC has hemoglobin (HEE moh GLOH bin). *Hemoglobin* is an oxygen-carrying protein. Hemoglobin clings to the oxygen you inhale. RBCs can then transport oxygen throughout the body. Hemoglobin also gives RBCs their red color.

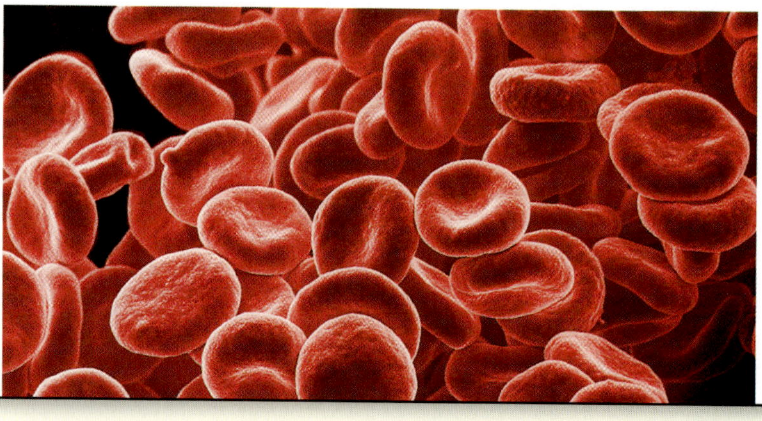

Figure 1 *Red blood cells are made in the bone marrow of certain bones. As red blood cells mature, they lose their nucleus and their DNA.*

CHAPTER RESOURCES

Chapter Resource File

- Lesson Plan
- Directed Reading A BASIC
- Directed Reading B SPECIAL NEEDS

Technology

Transparencies

- Bellringer

Answer to Reading Check

plasma, red blood cells, white blood cells, and platelets

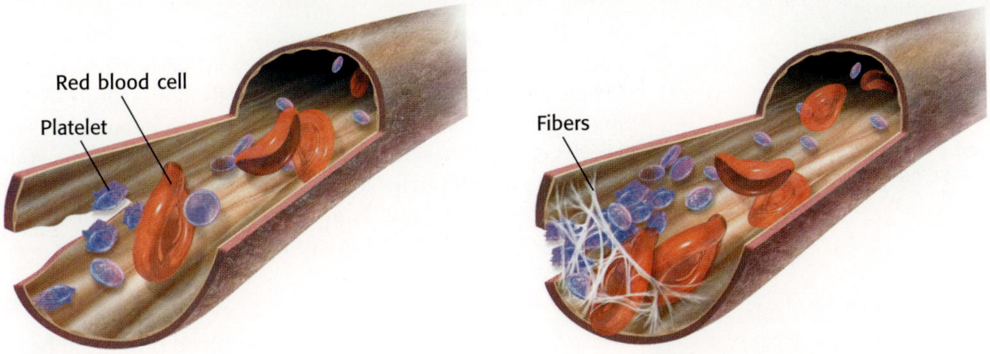

Figure 2 *Platelets release chemicals in damaged vessels and cause fibers to form. The fibers make a "net" that traps blood cells and stops bleeding.*

Platelets

Drifting among the blood cells are tiny particles called platelets. *Platelets* are pieces of larger cells found in bone marrow. These larger cells remain in the bone marrow, but fragments are pinched off and enter the bloodstream as platelets. Platelets last for only 5 to 10 days, but they are an important part of blood. When you cut or scrape your skin, you bleed because blood vessels have been opened. As soon as bleeding starts, platelets begin to clump together in the damaged area. They form a plug that helps reduce blood loss, as shown in **Figure 2.** Platelets also release chemicals that react with proteins in plasma. The reaction causes tiny fibers to form. The fibers help create a blood clot.

White Blood Cells

Sometimes *pathogens* (PATH uh juhnz)—bacteria, viruses, and other microscopic particles that can make you sick—enter your body. When they do, they often meet *white blood cells,* or WBCs. WBCs, shown in **Figure 3,** help keep you healthy by destroying pathogens. WBCs also help clean wounds.

WBCs fight pathogens in several ways. Some WBCs squeeze out of blood vessels and move around in tissues, searching for pathogens. When they find a pathogen, they destroy it. Other WBCs release antibodies. *Antibodies* are chemicals that identify or destroy pathogens. WBCs also keep you healthy by destroying body cells that have died or been damaged. Most WBCs are made in bone marrow. Some WBCs mature in the lymphatic system.

Figure 3 *White blood cells defend the body against pathogens. These white blood cells have been colored yellow to make their shape easier to see.*

Reading Check **Why are WBCs important to your health?**

MISCONCEPTION ALERT

Blue Blood? Students may think that deoxygenated blood is blue because it is often depicted that way in illustrations and because blood appears blue in veins through the skin. The color of blood does change depending on the amount of oxygen in the blood; however, human blood is not blue. It varies in color from scarlet to deep red.

Answer to Reading Check

White blood cells identify and attack pathogens that may make you sick.

Teach

READING STRATEGY — GENERAL

Writing **Asking Questions** Before students read this page, have them read the headings aloud. Then, ask students to formulate and write one question that they expect the text under each heading to answer.
LS Logical

Using the Figure — GENERAL

Red Cells, White Cells Have students compare **Figure 1,** which shows red blood cells, with **Figure 3,** which shows white blood cells. Ask students to relate any differences in shape to the different functions of these two types of blood cells. Help students understand that the flat shape and small size of a red blood cell enables the red blood cell to carry oxygen through tiny capillaries to cells throughout the body. Help students link the larger size and sticky-looking surface of a white blood cell to the white blood cell's ability to fight pathogens. English Language Learners
LS Visual

ACTIVITY — BASIC

Making a Model of Blood To help students visualize the components of blood, have them make a model of blood. Provide students with a variety of materials, such as a resealable plastic sandwich bag, thick corn syrup, plastic-foam balls, and other craft materials. Then, have students use the materials to make a model of the parts of blood. Encourage students to be creative in their representations of the parts of blood. English Language Learners
LS Kinesthetic

Close

Reteaching — Basic

Blood Types The subject of blood types may be confusing. Review the material in **Figure 4** to reinforce the material presented in the text. Help students compare the shapes of the antigens on each red blood cell with the shapes of the antibodies that bind to the antigens. LS Verbal

Quiz — General

1. Which antibodies will a person who has type AB blood produce? (none)
2. Can a person who has type O blood receive blood from a person who has type AB blood? Why or why not? (No, the type O blood will have antibodies that attack the A and B antigens in the AB blood.)
3. In your own words, define the term *transfusion*. (A transfusion occurs when a person is given an injection of blood to replace blood that has been lost.)

Alternative Assessment — General

Blood Groups Have students write a report about Karl Landsteiner (1868–1943), a Nobel Prize-winning scientist. (Landsteiner discovered that some mixtures of blood are compatible and others are not and that he could divide the population into different groups based on how blood from one group of people reacted with blood from another group.) LS Verbal

blood pressure the force that blood exerts on the walls of the arteries

Body Temperature Regulation

Your blood does more than supply your cells with oxygen and nutrients. It also helps regulate your body temperature. When your brain senses that your body temperature is rising, it signals blood vessels in your skin to enlarge. As the vessels enlarge, heat from your blood is transferred to your skin. This transfer helps lower your temperature. When your brain senses that your temperature is normal, it instructs your blood vessels to return to their normal size.

Blood Pressure

Every time your heart beats, it pushes blood out of the heart and into your arteries. The force exerted by blood on the inside walls of arteries is called **blood pressure.**

Blood pressure is expressed in millimeters of mercury (mm Hg). For example, a blood pressure of 110 mm Hg means the pressure on the artery walls can push a narrow column of mercury to a height of 110 mm.

Blood pressure is usually given as two numbers, such as 110/70 mm Hg. Systolic (sis TAHL ik) pressure is the first number. *Systolic pressure* is the pressure inside large arteries when the ventricles contract. The surge of blood causes the arteries to bulge and produce a pulse. The second number, *diastolic* (DIE uh STAHL ik) *pressure,* is the pressure inside arteries when the ventricles relax. For adults, blood pressure around 120/80 mm Hg is normal. High blood pressure (above 140/90 mm Hg) can cause heart or kidney damage.

Reading Check **What is the difference between systolic pressure and diastolic pressure?**

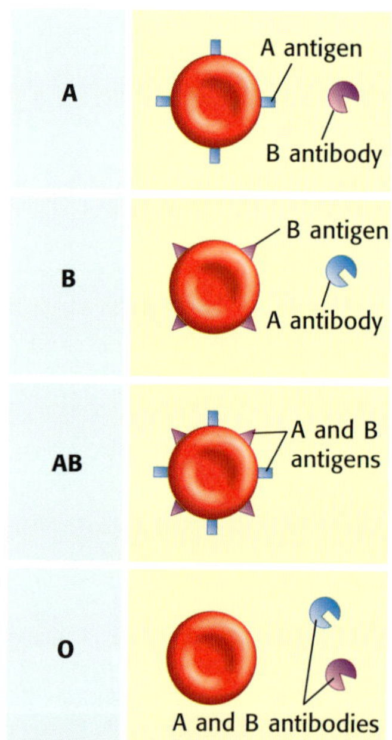

Figure 4 *This figure shows which antigens and antibodies may be present in each blood type.*

Blood Types

Every person has one of four blood types: A, B, AB, or O. Your blood type refers to the type of chemicals you have on the surface of your RBCs. These surface chemicals are called *antigens* (AN tuh juhnz). Type A blood has A antigens; type B has B antigens; and type AB has both A and B antigens. Type O blood has neither the A nor the B antigen.

The different blood types have different antigens on their RBCs. They may also have different antibodies in the plasma. These antibodies react to antigens of other blood types as if the antigens were pathogens. As shown in **Figure 4,** type A blood has antibodies that react to type B blood. If a person with type A blood receives type B blood, the type B antibodies attach themselves to the type B RBCs. These RBCs begin to clump together, and the clumps may block blood vessels. A reaction to the wrong blood type may be fatal.

Inclusion Strategies

• *Learning Disabled* • *Developmentally Delayed*

Some students may have difficulty remembering a variety of details, such as the functions of different types of blood cells. Help students develop hints or clues, such as mnemonic or visual clues, that will help them remember the main jobs of red and white blood cells. (Examples of clues include REd = REspiration and WhIte = WIpe out problems.) LS Verbal

Answer to Reading Check

Systolic pressure is the pressure inside arteries when the ventricles contract. Diastolic pressure is the pressure inside the arteries when the ventricles are relaxed.

Blood Types and Transfusions

Sometimes, a person must be given a blood transfusion. A *transfusion* is the injection of blood or blood components into a person to replace blood that has been lost because of surgery or an injury. **Figure 5** shows bags of blood that may be given in a transfusion. The blood type is clearly marked. Because the ABO blood types have different antigen-antibody reactions, a person receiving blood cannot receive blood from just anyone. **Table 1** shows blood transfusion possibilities.

Table 1 Blood Transfusion Possibilities

Type	Can receive	Can donate to
A	A, O	A, AB
B	B, O	B, AB
AB	all	AB only
O	O	all

Reading Check People with type O blood are sometimes called universal donors. Why might this be true?

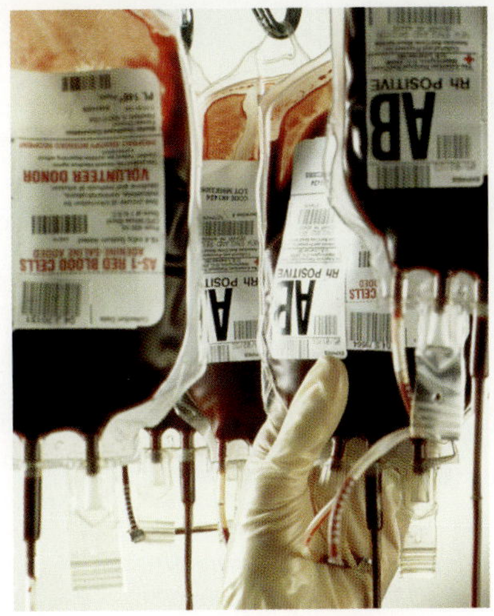

Figure 5 *The blood type must be clearly labeled on blood stored for transfusions.*

SECTION Review

Summary

- Blood's four main components are plasma, red blood cells, platelets, and white blood cells.
- Blood carries oxygen and nutrients to cells, helps protect against disease, and helps regulate body temperature.
- Blood pressure is the force blood exerts on the inside walls of arteries.
- Every person has one of four ABO blood types.
- Mixing blood types may be fatal.

Using Key Terms

1. Use each of the following terms in a separate sentence: *blood* and *blood pressure*.

Understanding Key Ideas

2. A person with type B blood can donate blood to people with which type(s) of blood?
   a. B, AB
   b. A, AB
   c. AB only
   d. All types
3. List the four main components of blood and tell what each component does.
4. Why is it important for a doctor to know a patient's blood type?

Math Skills

5. A person has a systolic pressure of 174 mm Hg. What percentage of normal (120 mm Hg) is this?

Critical Thinking

6. **Identifying Relationships** How does the body use blood and blood vessels to help maintain proper body temperature?
7. **Predicting Consequences** Some blood conditions and diseases affect the ability of red blood cells to deliver oxygen to cells of the body. Predict what might happen to a person with a disease of that type.

SCILINKS. NSTA Developed and maintained by the National Science Teachers Association

For a variety of links related to this chapter, go to www.scilinks.org

Topic: Blood; Blood Donations
SciLinks code: HSM0175; HSM0178

Answers to Section Review

1. Sample answer: Blood carries oxygen to all the cells. Sometimes, high blood pressure can cause damage to other body organs.
2. a
3. plasma, red blood cells, platelets, and white blood cells; plasma: it contains water, minerals, vitamins, and other substances; red blood cells: they carry oxygen to the body's cells; platelets: they clump together to form blood clots; white blood cells: they help fight disease by identifying and destroying pathogens
4. Sample answer: A doctor must know a person's blood type so that the doctor can order the right type of blood if the person needs a transfusion.
5. $174 \div 120 = 1.45 \times 100 = 145\%$
6. Sample answer: As blood circulates in the body, it picks up heat from the cells and carries it through blood vessels close to the surface of the skin. As blood circulates close to the skin's surface, the heat leaves the blood and goes into one's surroundings.
7. Sample answer: A person whose red blood cells cannot deliver oxygen to the body's cells might feel tired or weak all the time. In addition, the lack of oxygen may damage some of the person's cells, tissues, or organs.

Answer to Reading Check

The red blood cells of a person who has type O blood have no A or B antigens. The A or B antibodies in another person's blood will not react to the type O cells. It is safe for anyone to receive type O blood.

CHAPTER RESOURCES

Chapter Resource File

- Section Quiz GENERAL
- Section Review GENERAL
- Vocabulary and Section Summary GENERAL
- Reinforcement Worksheet GENERAL

Workbooks

Math Skills for Science

- Parts of 100: Calculating Percentages BASIC
- Percentages, Fractions, and Decimals GENERAL

SECTION 3

Focus

Overview

This section introduces students to the lymphatic system. Students will also learn about the relationship between blood and lymph.

Bellringer

Ask students, "Do you know people who have had their tonsils out? What are tonsils? What is their function? Where are they located in the body?" Have students write the answers to these questions in their **science journal.** Tell students that if they are not sure of the answers to make an educated guess at them: (Tonsils are small masses of lymphatic tissue at the back of the tongue and in the throat. Their function is to trap pathogens that might otherwise get into the body and cause disease.)

Motivate

Discussion — GENERAL

Swollen Glands Ask students if a doctor has ever felt their neck or under their jaw when they were sick. Ask them if they know what the doctor was checking. Encourage students who have had this experience to share it with the class. Then, invite students to explore the purpose of this type of examination. English Language Learners
LS Verbal

SECTION 3

The Lymphatic System

Every time your heart pumps, a little fluid is forced out of the thin walls of the capillaries. Some of this fluid collects in the spaces around your cells. What happens to this fluid?

Most of the fluid is reabsorbed through the capillaries into your blood. But some is not. Your body has a second circulatory system called the lymphatic (lim FAT ik) system.

The **lymphatic system** is the group of organs and tissues that collect the excess fluid and return it to your blood. The lymphatic system also helps your body fight pathogens.

READING WARM-UP

Objectives

- Describe the relationship between the lymphatic system and the circulatory system.
- Identify six parts of the lymphatic system, and describe their functions.

Terms to Learn

lymphatic system, lymph, lymph node, thymus, spleen, tonsils

READING STRATEGY

Prediction Guide Before reading this section, write the title of each heading in this section. Next, under each heading, write what you think you will learn.

Vessels of the Lymphatic System

The fluid collected by the lymphatic system is carried through vessels. The smallest vessels of the lymphatic system are *lymph capillaries.* Lymph capillaries absorb some of the fluid and particles from between the cells. These particles are too large to enter blood capillaries. Some of these particles are dead cells or pathogens. The fluid and particles absorbed into lymph capillaries are called **lymph.**

As shown in **Figure 1,** lymph capillaries carry lymph into larger vessels called *lymphatic vessels.* Skeletal muscles squeeze these vessels to force lymph through the lymphatic system. Valves inside lymphatic vessels stop backflow. Lymph drains into the large neck veins of the cardiovascular system.

Reading Check How is the lymphatic system related to the cardiovascular system? (*See the Appendix for answers to Reading Checks.*)

lymphatic system a collection of organs whose primary function is to collect extracellular fluid and return it to the blood

lymph the fluid that is collected by the lymphatic vessels and nodes

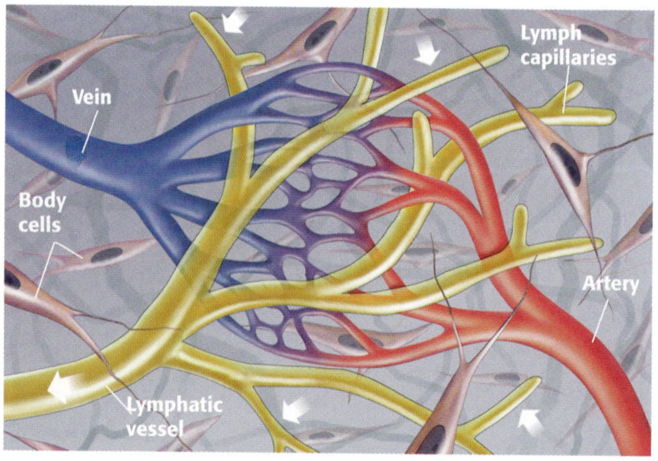

Figure 1 *The white arrows show the movement of lymph into lymph capillaries and through lymphatic vessels.*

CHAPTER RESOURCES

Chapter Resource File

- Lesson Plan
- Directed Reading A BASIC
- Directed Reading B SPECIAL NEEDS

Technology

Transparencies
- Bellringer

Answer to Reading Check

The lymphatic system is a secondary circulatory system in the body. The lymphatic system collects fluid and particles from between the cells and returns them to the cardiovascular system.

Other Parts of the Lymphatic System

In addition to vessels and capillaries, several organs and tissues are part of the lymphatic system. These organs and tissues are shown in **Figure 2.** Bone marrow plays an important role in your lymphatic system. The other parts of the lymphatic system are the lymph nodes, the thymus gland, the spleen, and the tonsils.

Bone Marrow

Bones—part of your skeletal system—are very important to your lymphatic system. *Bone marrow* is the soft tissue inside of bones. Bone marrow is where most red and white blood cells, including lymphocytes (LIM foh SIETS), are produced. *Lymphocytes* are a type of white blood cell that helps your body fight pathogens.

Lymph Nodes

As lymph travels through lymphatic vessels, it passes through lymph nodes. **Lymph nodes** are small, bean-shaped masses of tissue that remove pathogens and dead cells from the lymph. Lymph nodes are concentrated in the armpits, neck, and groin.

Lymph nodes contain lymphocytes. Some lymphocytes—called *killer T cells*—surround and destroy pathogens. Other lymphocytes—called *B cells*—produce antibodies that attach to pathogens. These marked pathogens clump together and are then destroyed by other cells.

When bacteria or other pathogens cause an infection, WBCs may multiply greatly. The lymph nodes fill with WBCs that are fighting the infection. As a result, some lymph nodes may become swollen and painful. Your doctor may feel these swollen lymph nodes to see if you have an infection. In fact, if your lymph nodes are swollen and sore, you or your parent can feel them, too. Swollen lymph nodes are sometimes an early clue that you have an infection.

Thymus

T cells develop from immature lymphocytes produced in the bone marrow. Before these cells are ready to fight infections, they develop further in the thymus. The **thymus** is the gland that produces T cells that are ready to fight infection. The thymus is located behind the breastbone, just above the heart. Mature lymphocytes from the thymus travel through the lymphatic system to other areas of your body.

lymph node oval mass of lymphatic tissue found in the lymphatic vessels that filter lymph

thymus the main gland of the lymphatic system; it produces mature T lymphocytes

Figure 2 The Lymphatic System

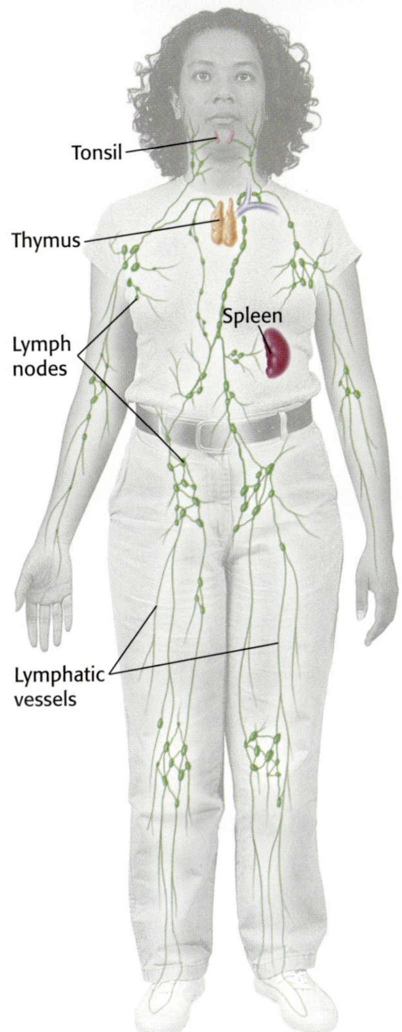

SCIENTISTS AT ODDS

First to Publish? The Danish physician Thomas Bartholin (1616–1680) is often credited with being the first person to describe the lymphatic system. His Swedish contemporary Olof Rudbeck (1630–1702), known as Rudbeck the Elder, also studied the lymphatic system and published his own description of the lymphatic system. For several years, Bartholin and Rudbeck engaged in a bitter dispute over who deserved the credit for discovering lymph glands and the lymphatic system.

Teach

READING STRATEGY — GENERAL

Paired Summarizing Have students read this section silently. Then, have students work in pairs to summarize the information in the section. Have the pairs of students review their summaries to make sure that they have included the important information. LS Intrapersonal/Interpersonal

Research — ADVANCED

Writing **Lymphatic System and Disease** Have students use library and on line resources to find out about the role of the lymphatic system in protecting the body against disease. Have students make a poster or a model of some part of the lymphatic system, showing where that part is located in the body and how it functions to help fight disease. To accompany their poster or model, students should write and present short reports to the class based on their findings. LS Verbal/Visual

Using the Figure — BASIC

Parts of the Lymphatic System
Use **Figure 2** to help students understand the material on this page and on the next page. It is important for students to understand that the lymphatic system is separate from the circulatory system. But the lymphatic system connects to and shares organs with the circulatory sytem. For example, bone marrow and the spleen are important to both systems.
LS Visual/Logical

Close

Reteaching — BASIC

Planning a Brochure Ask students to outline and plan a brochure or pamphlet explaining the lymphatic system. Students do not have to make the brochure; they must only plan it. The brochure or pamphlet should be similar to one that you might find in a doctor's office. **LS Verbal**

Quiz — GENERAL

1. What is one function of the tonsils? (The tonsils contain white blood cells, which help fight pathogens.)
2. How is the lymphatic system a circulatory system? (The lymphatic system collects fluid that is not reabsorbed by the capillaries and returns it to the bloodstream.)

Alternative Assessment — ADVANCED

Model of Lymphatic System Have students make a life-sized model of the lymphatic system. Provide students with butcher paper, dried beans, glue, and markers. Then, have students use their model to demonstrate the way that the lymphatic system and the cardiovascular system interact. **LS Visual/Kinesthetic** English Language Learners

spleen the largest lymphatic organ in the body

CONNECTION TO Social Studies

WRITING SKILL **Vent Your Spleen** Why do we say that someone is "venting his spleen"? What does it mean? Conduct library or Internet research about this phrase. Write a report on what you have learned.

Spleen

Your spleen is the largest lymphatic organ. The **spleen** stores and produces lymphocytes. It is a purplish organ about the size of your fist. Your spleen is soft and spongy. It is located in the upper left side of your abdomen. As blood flows through the spleen, lymphocytes attack or mark pathogens in the blood. If pathogens cause an infection, the spleen may also release lymphocytes into the bloodstream.

In addition to being part of the lymphatic system, the spleen produces, monitors, stores, and destroys blood cells. When red blood cells (RBCs) are squeezed through the spleen's capillaries, the older and more fragile cells burst. These damaged RBCs are then taken apart by some of the cells in the spleen. Some parts of these RBCs may be reused. For this reason, you can think of the spleen as the red-blood-cell recycling center.

The spleen has two important functions. The *white pulp*, shown in **Figure 3**, is part of the lymphatic system. It helps to fight infections. The *red pulp*, also shown in **Figure 3**, removes unwanted material, such as defective red blood cells, from the blood. However, it is possible to lead a healthy life without your spleen. If the spleen is damaged or removed, other organs in the body take over many of its functions.

Reading Check What are two important functions of the spleen?

Figure 3 White and Red Pulp in the Spleen

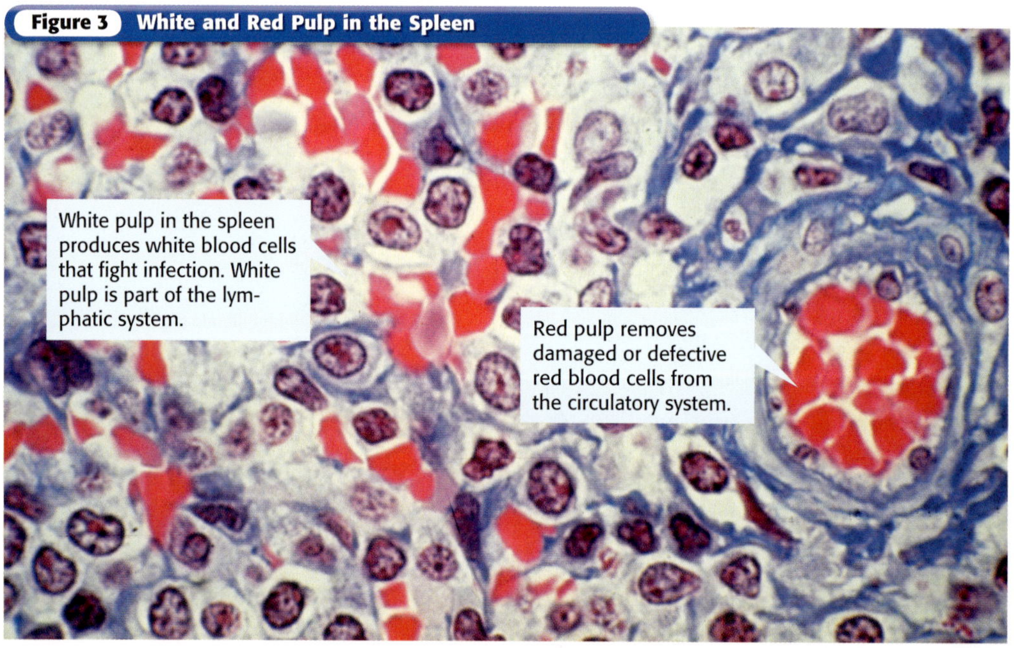

White pulp in the spleen produces white blood cells that fight infection. White pulp is part of the lymphatic system.

Red pulp removes damaged or defective red blood cells from the circulatory system.

Answer to Reading Check

The white pulp of the spleen is part of the lymphatic system. It helps fight infections by storing and producing lymphocytes. The red pulp of the spleen removes unwanted material, such as defective red blood cells, from the circulatory system.

Tonsils

The lymphatic system includes your tonsils. **Tonsils** are lymphatic tissue in the nasal cavity and at the back of the mouth on either side of the tongue. Each tonsil is about the size of a large olive.

Tonsils help defend the body against infection. Lymphocytes in the tonsils trap pathogens that enter the throat. Sometimes, tonsils become infected and are red, swollen, and very sore. Severely infected tonsils may be covered with patches of white, infected tissue. Sore, swollen tonsils, such as those in **Figure 4,** make swallowing difficult.

Sometimes, a doctor will suggest surgery to remove the tonsils. In the past, this surgery was frequently done in childhood. It is less common today. Surgery is now done only if a child has frequent, severe tonsil infections or if a child's tonsils are so enlarged that breathing is difficult.

tonsils small, rounded masses of lymphatic tissue located in the pharynx and in the passage from the mouth to the pharynx

Figure 4 *Tonsils help protect your throat and lungs from infection by trapping pathogens.*

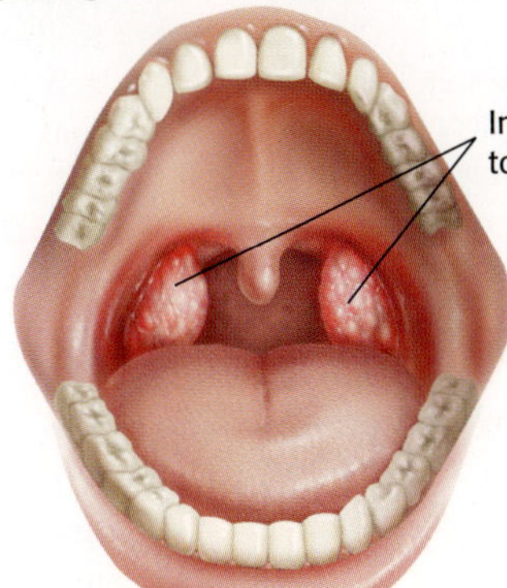

SECTION Review

Summary

- The lymphatic system collects fluid from between the cells and returns it to the blood.
- The lymphatic system contains cells that help the body fight disease.
- The lymphatic system consists of lymphatic vessels, lymph, and tissues and organs throughout the body.
- The thymus, spleen, and tonsils contain lymphocytes that help fight pathogens.

Using Key Terms

1. Use each of the following terms in a separate sentence: *lymph nodes, spleen,* and *tonsils.*

Understanding Key Ideas

2. Lymph
   a. is the same as blood.
   b. is fluid in the cells.
   c. drains into your muscles.
   d. is fluid collected by lymphatic vessels.
3. Name six parts of the lymphatic system. Tell what each part does.
4. How are your cardiovascular and lymphatic systems related?

Math Skills

5. One cubic millimeter of blood contains 5 million RBCs and 10,000 WBCs. How many times more RBCs are there than WBCs?

Critical Thinking

6. **Expressing Opinions** Some people have frequent, severe tonsil infections. These infections can be treated with medicine, and the infections usually go away after a few days. Do you think removing tonsils in such a case is a good idea? Explain.
7. **Analyzing Ideas** Why is it important that lymphatic tissue is spread throughout the body?

SCILINKS. NSTA Developed and maintained by the National Science Teachers Association

For a variety of links related to this chapter, go to www.scilinks.org

Topic: The Lymphatic System
SciLinks code: HSM0891

CHAPTER RESOURCES

Chapter Resource File

- Section Quiz GENERAL
- Section Review GENERAL
- Vocabulary and Section Summary GENERAL
- SciLinks Activity GENERAL

Answers to Section Review

1. Sample answer: Sometimes, when you are sick, your lymph nodes get swollen. Your spleen is an organ that is part of the circulatory system and the lymphatic system. Tonsils are lymphatic tissue at the back of your throat.
2. d
3. lymph vessels and capillaries, bone marrow, lymph nodes, thymus, spleen, and tonsils; lymph vessels and capillaries: they carry lymph from the body to the cardiovascular system; bone marrow: it is the tissue in which white blood cells are produced; lymph nodes: they remove pathogens and dead cells from lymph; thymus: it is the gland in which killer T cells mature; spleen: it produces and stores lymphocytes; tonsils: they trap pathogens
4. The lymphatic system collects particles and excess fluid from around your cells and returns them to the bloodstream, which is part of your cardiovascular system. The two systems work together to help the body fight pathogens: both lymph and blood carry white blood cells around the body to fight pathogens.
5. There are 500 times as many RBCs as WBCs.
6. Answers may vary as students offer their opinions. The most important part of the answer is the reasoning and the way in which the ideas that lead to the opinion are supported.
7. Sample answer: It is important for lymphatic tissue to be spread throughout the body so that fluid and particles from around every cell can be removed and pathogens in any part of the body can be identified and attacked.

SECTION 4

Focus

Overview

This section introduces students to the respiratory system. Students will learn about the flow of air in the respiratory system and about the way that the respiratory and the circulatory systems are related.

Bellringer

Ask students, "Are breathing and respiration the same thing?" (No, breathing is only one part of respiration. Respiration also includes cellular respiration.)

Motivate

Activity — General

Deep Breathing Have students place their hands on either side of their rib cage and breathe deeply several times. Then, ask students to describe what they felt while they breathed in and out. (Students should feel their rib cage moving up and expanding during inhalation and moving down and returning to its initial size during exhalation.) English Language Learners

LS Kinesthetic

SECTION 4

The Respiratory System

Breathing—you do it all the time. You're doing it right now. You hardly ever think about it, though, unless you suddenly can't breathe.

Then, it becomes very clear that you have to breathe in order to live. But why is breathing important? Your body needs oxygen in order to get energy from the foods you eat. Breathing makes this process possible.

READING WARM-UP

Objectives

- Describe the parts of the respiratory system and their functions.
- Explain how breathing happens.
- Discuss the relationship between the respiratory system and the cardiovascular system.
- Identify two respiratory disorders.

Terms to Learn

respiration
respiratory system
pharynx
larynx
trachea
bronchus
alveoli

READING STRATEGY

Reading Organizer As you read this section, make a flowchart of the steps of the process of respiration.

Respiration and the Respiratory System

The words *breathing* and *respiration* are often used to mean the same thing. However, breathing is only one part of respiration. **Respiration** is the process by which a body gets and uses oxygen and releases carbon dioxide and water. Respiration is divided into two parts. The first part is breathing, which involves inhaling and exhaling. The second part is cellular respiration, which involves chemical reactions that release energy from food.

Breathing is made possible by your respiratory system. The **respiratory system** is the group of organs that take in oxygen and get rid of carbon dioxide. The nose, throat, lungs, and passageways that lead to the lungs make up the respiratory system. **Figure 1** shows the parts of the respiratory system.

respiration the exchange of oxygen and carbon dioxide between living cells and their environment; includes breathing and cellular respiration

respiratory system a collection of organs whose primary function is to take in oxygen and expel carbon dioxide

Figure 1 *Air moves into and out of the body through the respiratory system.*

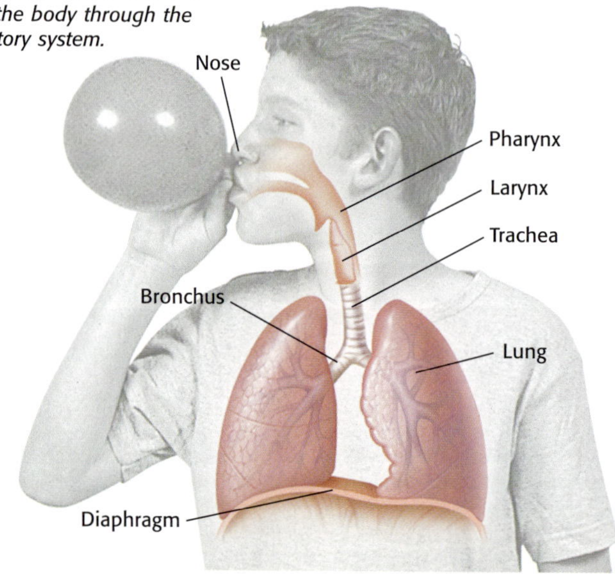

CHAPTER RESOURCES

Chapter Resource File

- Lesson Plan
- Directed Reading A BASIC
- Directed Reading B SPECIAL NEEDS

Technology

Transparencies

- Bellringer
- The Role of Blood in Respiration

Cultural Awareness — General

Adapting to High Altitudes Newcomers to Peruvian villages in the Andes Mountains may get headaches, feel nauseated, and be short of breath. Why? The newcomer is suffering from a lack of oxygen. Villagers don't have these problems; they have adapted to the elevation. Over time, villagers develop lungs that are larger than those of a newcomer. Villagers also have more red blood cells than newcomers have.

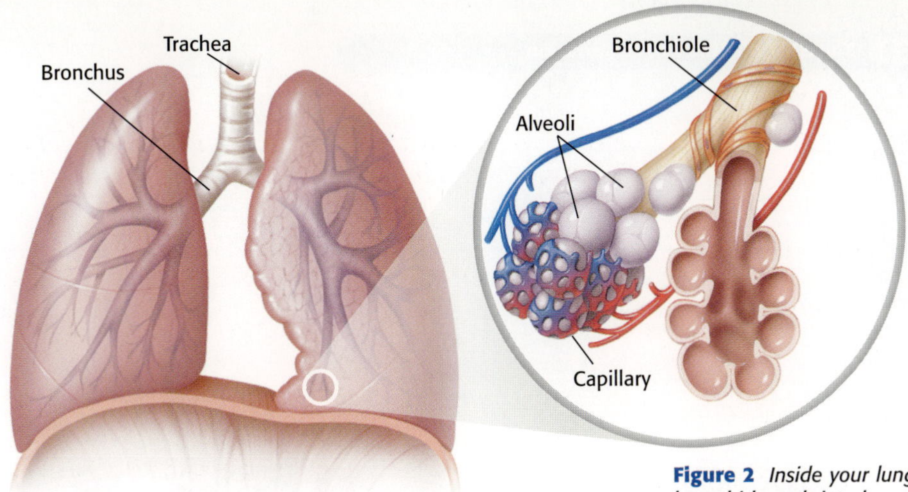

Figure 2 *Inside your lungs, the bronchi branch into bronchioles. The bronchioles lead to tiny sacs called alveoli.*

Nose, Pharynx, and Larynx

Your *nose* is the main passageway into and out of the respiratory system. Air can be breathed in through and out of the nose. Air can also enter and leave through the mouth.

From the nose, air flows into the **pharynx** (FAR ingks), or throat. Food and drink also travel through the pharynx on the way to the stomach. The pharynx branches into two tubes. One tube, the *esophagus,* leads to the stomach. The other tube is the larynx (LAR ingks). The larynx leads to the lungs.

The **larynx** is the part of the throat that contains the vocal cords. The *vocal cords* are a pair of elastic bands that stretch across the larynx. Muscles connected to the larynx control how much the vocal cords are stretched. When air flows between the vocal cords, the cords vibrate. These vibrations make sound.

Trachea

The larynx guards the entrance to a large tube called the **trachea** (TRAY kee uh), or windpipe. Your body has two large, spongelike lungs. The trachea, shown in **Figure 2,** is the passageway for air traveling from the larynx to the lungs.

Bronchi and Alveoli

The trachea splits into two branches called **bronchi** (BRAHNG KIE) (singular, *bronchus*). One bronchus connects to each lung. Each bronchus branches into thousands of smaller tubes that are called *bronchioles* (BRAHNG kee OHLZ). In the lungs, each bronchiole branches to form thousands of tiny sacs that are called **alveoli** (al VEE uh LIE) (singular, *alveolus*).

Reading Check Describe the flow of air from your nose to your alveoli. (*See the Appendix for answers to Reading Checks.*)

pharynx the passage from the mouth to the larynx and esophagus

larynx the area of the throat that contains the vocal cords and produces vocal sounds

trachea the tube that connects the larynx to the lungs

bronchus one of the two tubes that connect the lungs with the trachea

alveoli any of the tiny air sacs of the lungs where oxygen and carbon dioxide are exchanged

Teach

ACTIVITY — GENERAL

Investigating Speech Ask students to place their hands lightly on their neck near the larynx and to say, "ah." Have students keep their hands in place and alternate between blowing as they would blow out candles on a birthday cake and saying "ah." Then, ask students the following questions:

- What happened when you said "ah"? (The neck vibrated.)
- What happened when you blew without saying anything? (The neck stopped vibrating, but air still rushed out of the mouth.)
- What caused the sound and the vibrations when you said "ah"? (Air rushing past the vocal cord muscles in the larynx caused the muscles to vibrate. This vibration caused the sound.)

Point out that speech is made up of sounds that are voiced and sounds that are not voiced. If students are unconvinced, have them say "ssssssss" (the snake sound) while touching their neck near the larynx. Then, have them say "zzzzz" (buzz like a bee) while touching their neck. Students should find that their neck vibrates when they pronounce "z," which is voiced, but does not vibrate when they pronounce "s," which is not voiced. English Language Learners
LS Verbal

Answer to Reading Check

nose, pharynx, larynx, trachea, bronchi, bronchioles, alveoli

INCLUSION *Strategies*

- *Learning Disabled* • *Behavior Control Issues*

Do an oral matching activity to clarify the functions of the tubelike parts of the respiratory system. Write these words on the board: *nose, pharynx, larynx, trachea,* and *bronchi*. Call out the following numbered descriptions, and ask students to write the number of the description next to the term that matches the description.

1. two Y-shaped tubes that go into the lungs (bronchi)
2. a tube that holds the vocal cords (larynx)
3. a tube that connects the vocal cord area to the Y-shaped tubes that go into the lungs (trachea)
4. one of the main ways air enters and leaves the body (nose)
5. a tube that carries both air and food into the body (pharynx)

LS Verbal/Auditory

Close

Reteaching — BASIC

Section Outline Have students make an outline of this section by using the chapter title, subheads, and key terms as topics. Help them fill in their outline to make a useful study tool.
LS Logical

Quiz — GENERAL

Ask students whether each of the statements below is true or false. Have students correct false statements.

1. SARS is caused by air pollution. (false)
2. There are two main bronchi —one for each lung— in the human body. (true)
3. The lungs are made of muscle tissue. (false)

Alternative Assessment — ADVANCED

Lung Models Have students make models of healthy lungs and lungs damaged by smoking. Photographs of healthy lungs and damaged lungs can be found in literature from the American Lung Association and the American Cancer Society and in various science and health textbooks. English Language Learners
LS Visual/Kinesthetic

Answer to Connection to Chemistry

Because there is less oxygen in the air at high altitudes, it is difficult for the body to get the oxygen needed.

Figure 3 The Role of Blood in Respiration

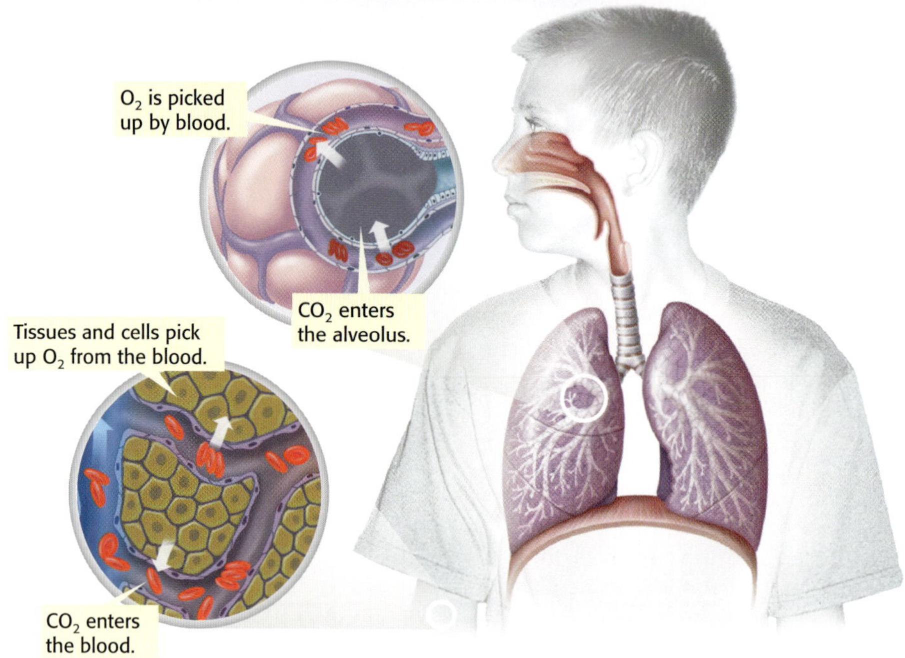

Breathing

When you breathe, air is sucked into or forced out of your lungs. However, your lungs have no muscles of their own. Instead, breathing is done by the diaphragm (DIE uh FRAM) and rib muscles. The *diaphragm* is a dome-shaped muscle beneath the lungs. When you inhale, the diaphragm contracts and moves down. The chest cavity's volume increases. At the same time, some of your rib muscles contract and lift your rib cage. As a result, your chest cavity gets bigger and a vacuum is created. Air is sucked in. Exhaling is this process in reverse.

CONNECTION TO Chemistry

Oxygen and Blood When people who live at low elevations travel up into the mountains, they may find themselves breathing heavily even when they are not exerting themselves. Why might this happen?

Breathing and Cellular Respiration

In *cellular respiration,* oxygen is used by cells to release energy stored in molecules of glucose. Where does the oxygen come from? When you breathe, you take in oxygen. This oxygen diffuses into red blood cells and is carried to tissue cells. The oxygen then diffuses out of the red blood cells and into each cell. Cells use the oxygen to release chemical energy. During the process, carbon dioxide (CO_2) and water are produced. Carbon dioxide is exhaled from the lungs. **Figure 3** shows how breathing and blood circulation are related.

✓ *Reading Check* **What is cellular respiration?**

Answer to Reading Check

Cellular respiration is the process inside a cell in which oxygen is used to release energy stored in molecules of glucose. During the process, carbon dioxide (CO_2) and water are released.

CONNECTION to *Physical Science* — GENERAL

Air Pressure and Breathing Lead a discussion about the ways in which the body creates changes in air pressure to make breathing possible. The teaching transparency "Exhaling, Pressure, and Fluid Flow" is a helpful illustration of the process of breathing. **LS Visual/Logical**

Respiratory Disorders

Millions of people suffer from respiratory disorders. Respiratory disorders include asthma, emphysema, and severe acute respiratory syndrome (SARS). Asthma causes the bronchioles to narrow. A person who has asthma has difficulty breathing. An asthma attack may be triggered by irritants such as dust or pollen. SARS is caused by a virus. A person who has SARS may have a fever and difficulty breathing. Emphysema happens when the alveoli have been damaged. People who have emphysema have trouble getting the oxygen they need. **Figure 4** shows a lung damaged by emphysema.

Figure 4 *The photo on the left shows a healthy lung. The photo on the right shows the lung of a person who had emphysema.*

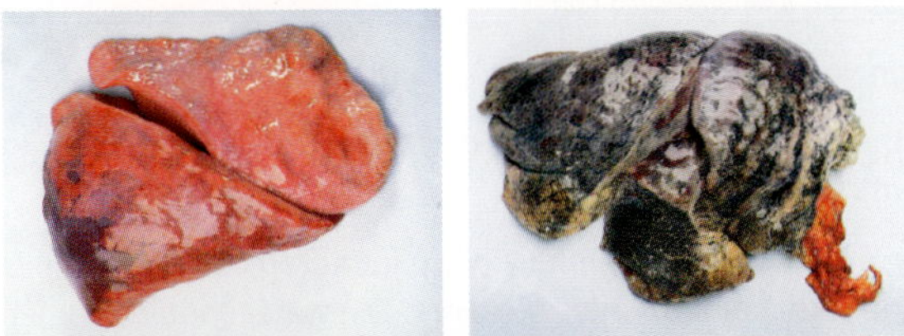

Quick Lab

Why Do People Snore?

1. Get a **15 cm² sheet of wax paper.**
2. Hum your favorite song.
3. Then, take the wax paper and press it against your lips. Hum the song again.
4. How was your humming different when wax paper was pressed to your mouth?
5. Use your observations to guess what might cause snoring.

SECTION Review

Summary

- Air travels to the lungs through the nose or mouth, pharynx, larynx, trachea, and bronchi.
- In the lungs, the bronchi branch into thousands of bronchioles and alveoli.
- Breathing involves lungs, muscles in the rib cage, and the diaphragm.
- Oxygen enters the blood through the alveoli in the lungs. Carbon dioxide leaves the blood and is exhaled.
- Respiratory disorders include asthma, SARS, and emphysema.

Using Key Terms

For each pair of terms, explain how the meanings of the terms differ.

1. *pharynx* and *larynx*

Understanding Key Ideas

2. Which of the following are respiratory disorders?
   - a. SARS, alveoli, and asthma
   - b. alveoli, emphysema, and SARS
   - c. larynx, asthma, and SARS
   - d. SARS, emphysema, and asthma
3. Explain how breathing happens.
4. Describe how your cardiovascular and respiratory systems work together.

Math Skills

5. Total lung capacity (TLC) is about 6 L. A person can exhale about 3.6 L. What percentage of TLC cannot be exhaled?

Critical Thinking

6. **Interpreting Statistics** About 6.3 million children in the United States have asthma. About 4 million of them had an asthma attack last year. What do these statistics tell you about the relationship between asthma and asthma attacks?
7. **Identifying Relationships** If a respiratory disorder causes lungs to fill with fluid, how might this affect a person's health?

SCILINKS. NSTA Developed and maintained by the National Science Teachers Association

For a variety of links related to this chapter, go to www.scilinks.org

Topic: The Respiratory System; Respiratory Disorders

SciLinks code: HSM1307; HSM1306

Is That a Fact!

The lungs contain about 300 million alveoli. The alveoli provide a tremendous surface area for gas exchange. In fact, because of this large capacity for gas exchange, a person can breathe easily with only one lung.

CHAPTER RESOURCES

Chapter Resource File

- Section Quiz GENERAL
- Section Review GENERAL
- Vocabulary and Section Summary GENERAL
- Datasheet for Quick Lab

Technology

Transparencies

- LINK TO PHYSICAL SCIENCE Exhaling, Pressure, and Fluid Flow

Quick Lab

MATERIALS

FOR EACH GROUP

- wax paper, 15 cm² sheet

Answers

4. A strong vibrating sound was made when the paper was pressed against the lips.
5. Sample answer: Most snoring happens when soft tissues in the mouth block a person's airway and vibrate.

Answers to Section Review

1. The pharynx is the passage from the mouth to the larynx and esophagus, while the larynx is the part of the throat that contains the vocal cords.
2. d
3. When you breathe, air is sucked into or forced out of your lungs. A muscle called the *diaphragm* contracts and creates a vacuum in your chest cavity. The vaccum pulls air in through the nose or mouth. Then, the air travels through the pharynx, larynx, trachea, and bronchi to reach the lungs.
4. The respiratory system brings in oxygen and expels carbon dioxide, and the cardiovascular system transports those gases to and from the lungs.
5. 6 L − 3.6 L = 2.4 L unexhaled
   2.4 L ÷ 6 L = 0.40, or 40% of TLC cannot be exhaled
6. Sample answer: These statistics tell us that last year not every child who had asthma had an asthma attack.
7. Sample answer: The alveoli in the lungs are made for the exhange of gases, not liquids. Fluids in the lungs would prevent a person from getting all of the oxygen needed to maintain regular activity. The person might feel weak or tired. Also, the person could not get rid of the carbon dioxide from the cells, which might cause a problem, too.

Skills Practice Lab

Carbon Dioxide Breath

Teacher's Notes

Time Required
One 45-minute class period

Lab Ratings
EASY → HARD

Teacher Prep 2

Student Set-Up 2

Concept Level 2

Clean Up 2

MATERIALS

You may wish to substitute bromthymol blue indicator solution for the phenol red indicator. The bromthymol blue will turn green in the presence of CO_2. Clear plastic cups (6 oz or 8 oz) may be used instead of 150 mL flasks if glassware is in short supply or if you have concerns about breakage.

Safety Caution
Remind students to review all safety cautions and icons before beginning this lab activity.

Lab Notes
Tell students that carbon dioxide is in the air in the classroom. They may need to cover their indicator solution to delay the reaction with the air. Tell them not to leave the indicator solution sitting exposed for several minutes before it is used.

Skills Practice Lab

Carbon Dioxide Breath

OBJECTIVES

Detect the presence of carbon dioxide in your breath.

Compare the data for carbon dioxide in your breath with the data from your classmates.

MATERIALS

- calculator (optional)
- clock with a second hand, or a stopwatch
- Erlenmeyer flask, 150 mL
- eyedropper
- gloves, protective
- graduated cylinder, 150 mL
- paper towels
- phenol red indicator solution
- plastic drinking straw
- water, 100 mL

SAFETY

Carbon dioxide is important to both plants and animals. Plants take in carbon dioxide during photosynthesis and give off oxygen as a byproduct of the process. Animals—including you—take in oxygen during respiration and give off carbon dioxide as a byproduct of the process.

Procedure

1. Put on your gloves, safety goggles, and apron.
2. Use the graduated cylinder to pour 100 mL of water into a 150 mL flask.
3. Using an eyedropper, carefully place four drops of phenol red indicator solution into the water. The water should turn orange.
4. Place a plastic drinking straw into the solution of phenol red and water. Drape a paper towel over the flask to prevent splashing.
5. Carefully blow through the straw into the solution. **Caution:** Do not inhale through the straw. Do not drink the solution, and do not share a straw with anyone.

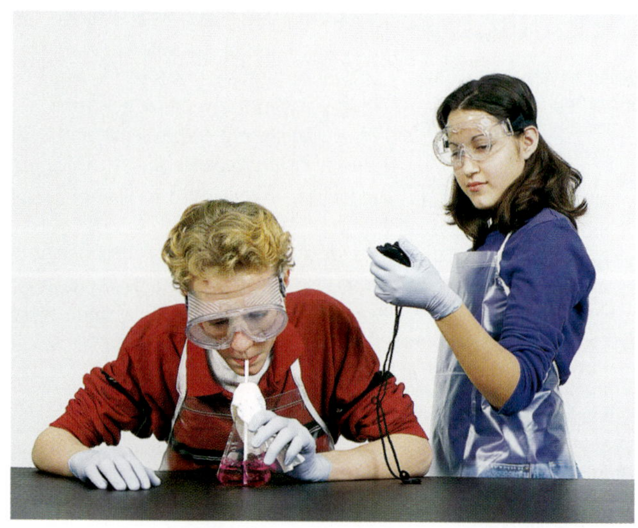

CLASSROOM TESTED & APPROVED

Yvonne Brannum
Hine Junior High School
Washington, D.C.

CHAPTER RESOURCES

Chapter Resource File

- Datasheet for Chapter Lab
- Lab Notes and Answers

Technology

Classroom Videos
- Lab Video

LabBook
- Build a Lung

6. Your lab partner should begin keeping time as soon as you start to blow through the straw. Have your lab partner time how long the solution takes to change color. Record the time.

Analyze the Results

1. **Describing Events** Describe what happens to the indicator solution.
2. **Examining Data** Compare your data with those of your classmates. What was the longest length of time it took to see a color change? What was the shortest? How do you account for the difference?
3. **Constructing Graphs** Make a bar graph that compares your data with the data of your classmates.

Draw Conclusions

4. **Interpreting Information** Do you think that there is a relationship between the length of time the solution takes to change color and the person's physical characteristics, such as which gender the tester is or whether the tester has an athletic build? Explain your answer.
5. **Making Predictions** Predict how exercise might affect the results of your experiment. For example, would you predict that the level of carbon dioxide in the breath of someone who was exercising would be higher or lower than the carbon dioxide level in the breath of someone who was sitting quietly? Would you predict that the level of carbon dioxide in the breath would affect the timing of any color change in the phenol solution?

Applying Your Data

Do jumping jacks or sit-ups for 3 minutes, and then repeat the experiment. Did the phenol solution still change color? Did your exercising change the timing? Describe and explain any change.

Analyze the Results

1. The phenol red solution turns yellowish.
2. Answers will depend on student observations. It is typical for the solution to change color faster when the student is breathing fast after exercise.
3. Bar graphs will vary according to the students' observations and the data collected.

Drawing Conclusions

4. In general, there should be little difference observed between genders. An athlete at rest will usually take the longest time to generate a color change in the indicator solution. There are exceptions, and all answers will depend on students' observations.
5. Students should predict that the carbon dioxide level in the breath of a person who is exercising is higher than that of the person at rest. This prediction should lead to a prediction that exercise would shorten the time it takes to make a color change in the solution.

Applying Your Data

Students should see that the breath of students who do the exercise changes the color of the solution in less time than the breath of students who did not do the exercise does. The breath of the students who exercised made the color change happen faster because exercise increases the level of carbon dioxide in the blood and in the exhaled breath.

CHAPTER RESOURCES

Workbooks

- **Whiz-Bang Demonstrations**
  - Get the Beat BASIC
  - Take a Deep Breath GENERAL
- **EcoLabs & Field Activities**
  - There's Something in the Air GENERAL
- **Long-Term Projects & Research Ideas**
  - Getting to the Heart ADVANCED
- **Calculator-Based Labs**
  - A Hot Hand ADVANCED

Chapter Review

Assignment Guide	
Section	**Questions**
1	2, 6, 8, 10, 24–26
2	1, 7, 13, 21–22
3	3, 8, 16, 23
4	4–5, 9, 11–12, 14, 18, 20
1, 2, and 3	15, 19
1, 2, and 4	17

ANSWERS

Using Key Terms

1. Red blood cells
2. Arteries
3. lymphatic system
4. larynx
5. alveoli

Understanding Key Ideas

6. b
7. d
8. a
9. c
10. b
11. a
12. Pulmonary circulation carries blood to the lungs and back to the heart. Systemic circulation carries blood from the heart to the rest of the body and then returns the blood to the heart.
13. The first number, systolic pressure, is pressure in arteries when ventricles contract. The second number, diastolic pressure, is pressure in arteries when ventricles relax.

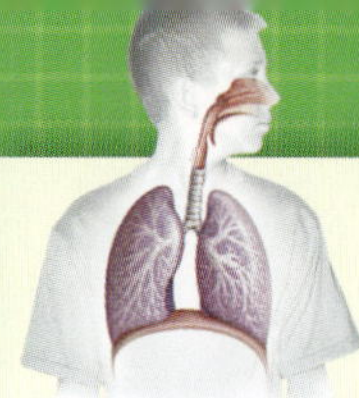

Chapter Review

USING KEY TERMS

Complete each of the following sentences by choosing the correct term from the word bank.

red blood cells	veins
white blood cells	arteries
lymphatic system	larynx
alveoli	bronchi
respiratory system	trachea

1. ___ deliver oxygen to the cells of the body.
2. ___ carry blood away from the heart.
3. The ___ helps the body fight pathogens.
4. The ___ contains the vocal cords.
5. The pathway of air through the respiratory system ends at the tiny sacs called ___.

UNDERSTANDING KEY IDEAS

Multiple Choice

6. Blood from the lungs enters the heart at the
   a. left ventricle.
   b. left atrium.
   c. right atrium.
   d. right ventricle.
7. Blood cells are made
   a. in the heart.
   b. from plasma.
   c. from lymph.
   d. in the bones.
8. Which of the following activities is a function of the lymphatic system?
   a. returning excess fluid to the circulatory system
   b. delivering nutrients to the cells
   c. bringing oxygen to the blood
   d. pumping blood to all parts of the body
9. Alveoli are surrounded by
   a. veins.
   b. muscles.
   c. capillaries.
   d. lymph nodes.
10. What prevents blood from flowing backward in veins?
    a. platelets
    b. valves
    c. muscles
    d. cartilage
11. Air moves into the lungs when the diaphragm muscle
    a. contracts and moves down.
    b. contracts and moves up.
    c. relaxes and moves down.
    d. relaxes and moves up.

Short Answer

12. What is the difference between pulmonary circulation and systemic circulation in the cardiovascular system?
13. Walton's blood pressure is 110/65. What do the two numbers mean?

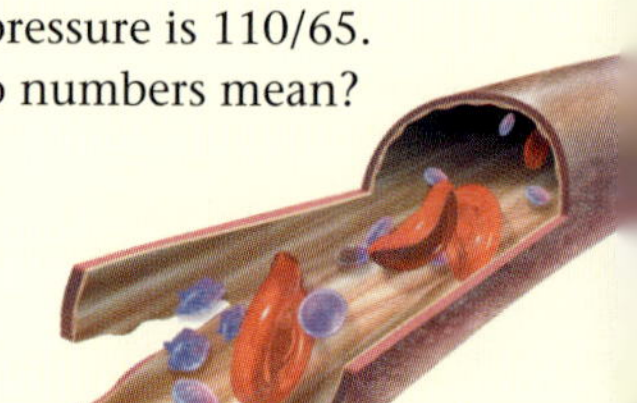

14. What body process produces the carbon dioxide you exhale?

15. Describe how the circulatory system and the lymphatic system work together to keep your body healthy.

16. How is the spleen important to both the lymphatic system and the circulatory system?

17. Briefly describe the path that oxygen follows in your respiratory system and your circulatory system.

CRITICAL THINKING

18. **Concept Mapping** Use the following terms to create a concept map: *blood, oxygen, alveoli, capillaries,* and *carbon dioxide.*

19. **Making Comparisons** Compare and contrast the functions of the circulatory system and the lymphatic system.

20. **Identifying Relationships** Why do you think there are hairs in your nose?

21. **Applying Concepts** After a person donates blood, the blood is stored in one-pint bags until it is needed for a transfusion. A healthy person has about 5 million RBCs in each cubic millimeter (1 mm^3) of blood.
    a. How many RBCs are in 1 mL of blood? (One milliliter is equal to 1 cm^3 and to 1,000 mm^3.)
    b. How many RBCs are there in 1 pt? (One pint is equal to 473 mL.)

22. **Predicting Consequences** What would happen if all of the red blood cells in your blood disappeared?

23. **Identifying Relationships** When a person is not feeling well, a doctor may examine samples of the person's blood to see how many white blood cells are present. Why would this information be useful?

INTERPRETING GRAPHICS

The diagram below shows how the human heart would look in cross section. Use the diagram to answer the questions that follow.

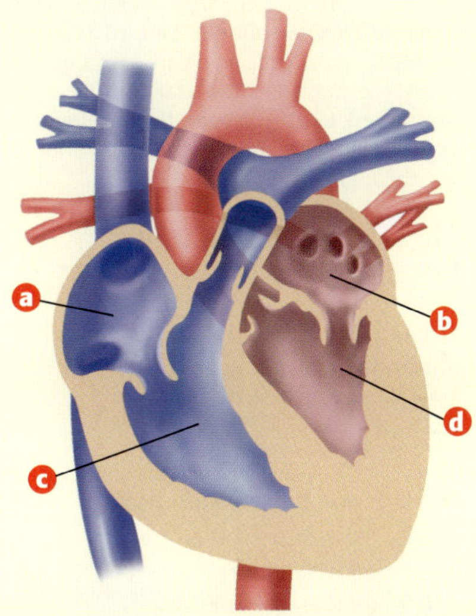

24. Which letter identifies the chamber that receives blood from systemic circulation? What is this chamber's name?

25. Which letter identifies the chamber that receives blood from the lungs? What is this chamber's name?

26. Which letter identifies the chamber that pumps blood to the lungs? What is this chamber's name?

14. Carbon dioxide is a product of cellular respiration, which occurs in the body's cells.
15. Sample answer: The circulatory system delivers oxygen, nutrients, and white blood cells to all parts of the body. The lymphatic system returns excess fluid from between the cells to the circulatory system.
16. The spleen is important to the circulatory system because it removes old or damaged red blood cells. It is important to the lymphatic system because it stores white blood cells, which help fight disease.
17. Oxygen comes into the body through the nose or mouth, travels to the lungs, and enters the alveoli. In the alveoli, oxygen moves through the capillary walls and enters the blood. The circulatory system carries the blood from the lungs to the heart, which pumps the blood to all parts of the body. Oxygen is carried by red blood cells to every body cell.

Critical Thinking

18. 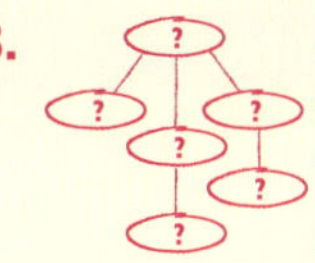 An answer to this exercise can be found at the end of the book.
19. Sample answer: The circulatory system and the lymphatic system use fluids to carry substances to and from all parts of the body. The circulatory system delivers oxygen and nutrients to the cells and removes wastes from the cells. The lymphatic system helps fight disease and infection.
20. The hairs catch dust and other foreign particles, which helps keep your lungs clean.
21. **a.** 5 billion (5,000,000,000) cells
    b. 2.365 trillion (2,365,000,000,000) cells
22. Sample answer: In a short time, I would probably die because my body could not get the oxygen it needs.
23. The immune system produces white blood cells to fight pathogens. A high white blood cell count tells the doctor that the person may have an infection.

Interpreting Graphics

24. a, the right atrium
25. b, the left atrium
26. c, the right ventricle

CHAPTER RESOURCES

Chapter Resource File

- Chapter Review GENERAL
- Chapter Test A GENERAL
- Chapter Test B ADVANCED
- Chapter Test C SPECIAL NEEDS
- Vocabulary Activity GENERAL

Workbooks

Study Guide
- Assessment resources are also available in Spanish.

Standardized Test Preparation

Teacher's Note

To provide practice under more realistic testing conditions, give students 20 minutes to answer all of the questions in this Standardized Test Preparation.

MISCONCEPTION ALERT

Answers to the standardized test preparation can help you identify student misconceptions and misunderstandings.

READING

Passage 1

1. C
2. I
3. C

TEST DOCTOR

Question 2: Students may select the incorrect answers F or H because the passage is about sneezing and mentions heredity. Students may select incorrect answer G because they may recall that some plants show phototropism, or a response to light.

Standardized Test Preparation

READING

Read each of the passages below. Then, answer the questions that follow each passage.

Passage 1 For some reason, about one in five people sneeze when they step from a dimly lit area into a brightly lit area. In fact, some may sneeze a dozen times or more! Fortunately, the sneezing usually stops relatively quickly. This sneeze reaction is called a photic sneeze reflex (FOHT ik SNEEZ REE FLEKS). No one knows for certain why it happens. A few years ago, some geneticists studied the photic sneeze reflex. They named it the *ACHOO syndrome*. Scientists know that the ACHOO syndrome runs in families. So, the photic sneeze may be hereditary and can be passed from parent to child. Sometimes, even the number of times in a row that each person sneezes is the same throughout a family.

1. According to the passage, the ACHOO syndrome is most likely to be which of the following?
- **A** contagious
- **B** photosynthetic
- **C** hereditary
- **D** allergic

2. In the passage, what does *photic* mean?
- **F** having to do with sneezing
- **G** having to do with plants
- **H** having to do with genetics
- **I** having to do with light

3. Which of the following statements is one clue that the photic sneeze reflex can be passed from parent to child?
- **A** The reflex is triggered by bright light.
- **B** Sneezing usually stops after a few sneezes.
- **C** Family members even sneeze the same number of times.
- **D** Scientists do not know what causes the ACHOO syndrome.

Passage 2 The two main functions of blood are transporting nutrients and oxygen from the lungs to cells and carrying carbon dioxide and other waste materials away from cells to the lungs or other organs. Blood also transfers body heat to the body surface and plays a role in defending the body against disease. The respiratory system transports gases to and from blood. The respiratory system and blood work together to carry out external respiration and internal respiration. External respiration is the exchange of gases between the atmosphere and blood. Internal respiration is the exchange of gases between blood and the cells of the body.

1. In the passage, what does *external respiration* mean?
- **A** the exchange of gases outdoors
- **B** the inhalation of gases as you breathe in
- **C** the exchange of gases between blood and the atmosphere
- **D** the exhalation of gases as you breathe out

2. Which of the following statements is a fact in the passage?
- **F** The respiratory system transports oxygen to all the cells of the body.
- **G** The respiratory system is part of the circulatory system.
- **H** Blood is a kind of cardiac tissue.
- **I** Blood transports oxygen to cells.

3. According to the passage, what are two of the roles blood plays in the human body?
- **A** transferring body heat and defending against disease
- **B** defending against disease and transporting gases to the circulatory system
- **C** transporting carbon dioxide to body cells and transferring body heat
- **D** external respiration and atmosphere

Passage 2

1. C
2. I
3. A

TEST DOCTOR

Question 3: Students may select incorrect answers B or C if they do not read the passage carefully. Blood does not transport gases to the circulatory system (answer B) or transport carbon dioxide to the body cells (answer C). Students may select incorrect answer D if they do not see that the question asks for two roles.

INTERPRETING GRAPHICS

Use the graph below to answer the questions that follow.

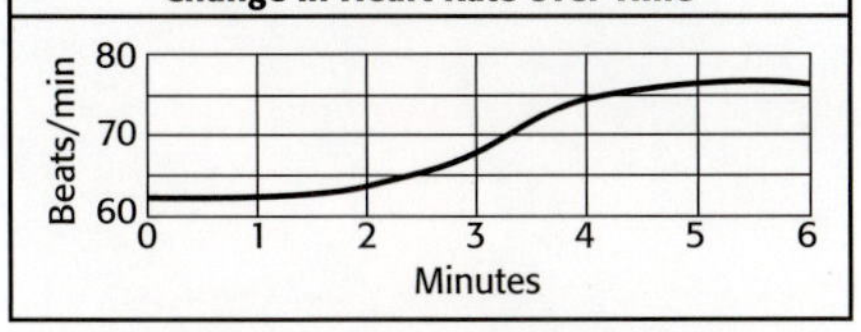

Change in Heart Rate over Time

1. What is the most likely explanation for the change seen after the two-minute mark?

A The person started exercising.
B The person fell asleep.
C The person inhaled.
D The person sat down.

2. How much faster is the heart beating during minute 5 than during minute 2?

F 10 beats per minute more
G 12 beats per minute more
H 15 beats per minute more
I 17 beats per minute more

3. About how many minutes did it take for this person's heart rate to go from 65 beats per minute to 75 beats per minute?

A 0.7 minute
B 1.0 minute
C 1.7 minutes
D 4.0 minutes

4. After how many minutes does this person's heart rate return to its resting rate?

F 1.0 minute
G 2.0 minutes
H 5.0 minutes
I There is not enough information to determine the answer.

MATH

Read each question below, and choose the best answer.

1. If Jim's heart beats 73 times every minute, Jen's heart beats 68 times every minute, and Leigh's heart beats 81 times every minute, what is the average heart rate for these 3 people?

A 73 beats per minute
B 74 beats per minute
C 141 beats per minute
D 222 beats per minute

2. The Griffith family has 4 dogs. Each of the dogs eats between 0.9 kg and 1.3 kg of food every day. Which is a reasonable estimate of the total amount of food all 4 dogs eat every day?

F 1 kg of food
G 3 kg of food
H 4 kg of food
I 8 kg of food

3. Assume that the average person's resting heart rate is 70 beats per minute. The resting heart rate of a particular person is 10 beats per minute more than the average person's. If a person with the higher heart rate lives 75 years, about how many more times will his or her heart beat than the average person's heart in that time?

A 3,942
B 394,200
C 3,942,000
D 394,200,000

4. At rest, the cells of the human body use about 250 mL of oxygen per minute. At that rate, how much oxygen would the cells of the human body use every 24 hours?

F about 36 L
G about 360 L
H about 36,000 L
I about 360,000 L

INTERPRETING GRAPHICS

1. A
2. G
3. C
4. I

TEST DOCTOR

Question 2: Students may select any of the incorrect answers if they do not look closely at the graph. The heart rate increases from about 64 beats per minute to 76 beats per minute.

Question 4: Students may select answer F, G, or H by guessing because they think that they should be able to find on the graph the point at which the heart rate returns to normal. However, the heart rate does not return to normal during the time period shown on the graph.

MATH

1. B
2. H
3. D
4. G

TEST DOCTOR

Question 3: Students may select any of the incorrect answers because they have not kept track of the proper number of decimal places in this two-step problem.

Question 4: Students may select incorrect answer I if they forget to convert milliliters to liters in their final answer.

CHAPTER RESOURCES

Chapter Resource File

- **Standardized Test Preparation** GENERAL

State Resources

For specific resources for your state, visit **go.hrw.com** and type in the keyword **HSMSTR.**

Science in Action

Science, Technology, and Society

Background

Because of the dangers associated with giving someone blood of the wrong type, people have searched for safe blood substitutes for hundreds of years. At one time, even wine was proposed as a substitute! Four substances are in advanced clinical trials in the United States. PolyHeme and Hemolink are derived from modified human hemoglobin. Hemopure is made from modified bovine hemoglobin. Oxygent is entirely synthetic and is made from an emulsion of perfluorocarbon molecules.

Weird Science

Background

Drawings of men playing an instrument that appears to be a didjeridu appear in caves in Australia. Evidence indicates that the didjeridu has been in Australia for about 1,000 years. Aboriginals, though, trace the history of the instrument back to the "Dreamtime," the very beginning of their people. The first reference by Westerners comes from a man named T. B. Wilson in 1835. Wilson described an aboriginal man playing an instrument that made of bamboo and was about 1 m long.

Science in Action

Science, Technology, and Society

Artificial Blood

What happens when someone loses blood rapidly? Loss of blood can be fatal in a very short time, so lost blood must be replaced as quickly as possible. But what if enough blood, or blood of the right type, is not immediately available? Scientists are developing different types of artificial blood—including one based on cow hemoglobin—that may soon be used to save lives that would otherwise be lost.

Language Arts ACTIVITY

WRITING SKILL Imagine that you are a doctor and one of your patients needs surgery. Create a pamphlet or brochure that explains what artificial blood is and how it may be used in surgical procedures.

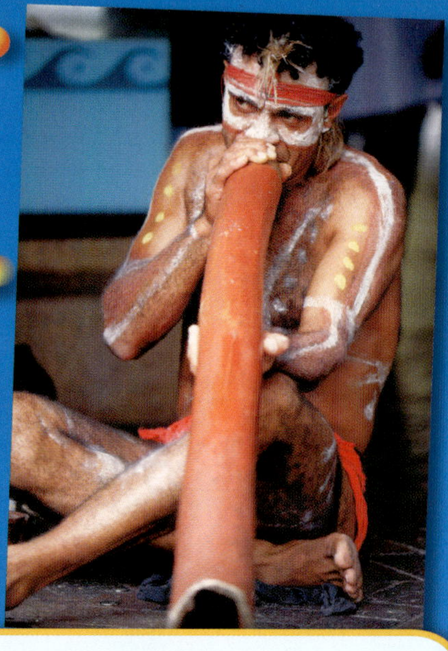

Weird Science

Circular Breathing and the Didgeridoo

Do you play a musical instrument such as a clarinet, flute, or tuba? How long can you blow into it before you have to take a breath? Can you blow into it for one minute? two minutes? And what happens when you stop to breathe? The Aboriginal people of Australia have a musical instrument called the *didgeridoo* (DIJ uh ree DOO). Didgeridoo players can play for hours without stopping to take a breath. They use a technique called *circular breathing* that lets them inhale and exhale at the same time. Circular breathing lets a musician play music without having to take breaths as often. With a little practice, maybe you can do it, too.

Social Studies ACTIVITY

WRITING SKILL Select a country from Africa or Asia. Research that country's traditional musical instruments or singing style. Write a description of how the instruments or singing style of that country differs from those of the United States. Illustrate your report.

Answer to Language Arts Activity

Students' brochures should be attractive, well illustrated, and informative. Brochures should explain which type of artificial blood the patient may receive and whether the patient or the doctor (or student) recommends the type of artificial blood to be used.

Answer to Social Studies Activity

Students' reports will vary depending on the country or region, the instrument, and the musical or singing style that the students have selected. Reports should be clear, interesting, and informative. Students should provide sources and references for their report, and their report should include illustrations that enhance the report.

People in Science

Anthony Roberts, Jr.

Leader in Training Anthony Roberts, Jr., has asthma. When he was in the 5th grade, his school counselor told him about a summer camp—The Boggy Creek Gang Camp—that was just being built. His counselor said that the camp was designed to serve kids who have asthma or other disabilities and diseases, such as AIDS, cancer, diabetes, epilepsy, hemophilia, heart disease, kidney disease, rheumatic diseases, and sickle cell anemia. Kids, in other words, who might otherwise never go to summer camp. Anthony jumped at the chance to go. Now, Anthony is too old to be a camper, and he is too young to be a regular counselor. But he can be a *Leader in Training* (LIT). Some camps have LIT programs that help young people make the transition from camper to counselor.

For Anthony, the chance to be an LIT fit perfectly with his love of camping and with his desire to work with kids with disabilities. Anthony remembers the fun he had and wants to help other kids have the same summer fun he did.

Math ACTIVITY

Research how many children under 17 years of age in the United States have asthma. Make a bar graph that shows how the number of children who have asthma has changed since 1981. What does this graph tell you about rates of asthma among children in the United States?

go.hrw.com

To learn more about these Science in Action topics, visit go.hrw.com and type in the keyword HL5BD2F.

Current Science

Check out Current Science® articles related to this chapter by visiting go.hrw.com. Just type in the keyword HL5CS23.

Answers to Math Activity

Students should find that asthma rates for children under 17 have increased from 3.2% in 1981 to 5.7% in 2001. Students should be able to update the figures each year as more-current data is published. Bar graphs should reflect the growth in the rate from the 1980s to the present. As an extension, students can research the costs associated with asthma in children or can research the ways in which asthma rates differ between races.

People in Science

Teaching Strategy-GENERAL

If appropriate for your class, begin by talking in a general way about summer vacation: what students like to do, how they spend their time, whether they go to summer camp (day camp or residence camp), and whether they know of anyone (no names) who has asthma or another disability that might make summer camp activities difficult. Then, discuss the Boggy Creek Gang Camp and Anthony Roberts, Jr., and his role as a leader in training. Your class may be interested to know that the Boggy Creek Gang Camp is one of the Hole in the Wall camps started by actor Paul Newman. The Hole in the Wall Gang was the outlaw gang of which Butch Cassidy was a member. Newman portrayed Cassidy in the 1969 film *Butch Cassidy and the Sundance Kid.*

The Digestive and Urinary Systems

Chapter Planning Guide

Compression guide: To shorten instruction because of time limitations, omit Section 2.

OBJECTIVES	LABS, DEMONSTRATIONS, AND ACTIVITIES	TECHNOLOGY RESOURCES
PACING • 90 min pp. 56–65 **Chapter Opener**	SE **Start-up Activity,** p. 57 ◆ GENERAL	OSP **Parent Letter** ■ GENERAL CD **Student Edition on CD-ROM** CD **Guided Reading Audio CD** ■ TR **Chapter Starter Transparency*** VID **Brain Food Video Quiz**
Section 1 The Digestive System • Compare mechanical digestion with chemical digestion. • Describe the parts and functions of the digestive system.	SE **Quick Lab** Break It Up!, p. 59 ◆ GENERAL CRF **Datasheet for Quick Lab*** TE **Demonstration** Measuring Saliva, p. 59 ◆ BASIC TE **Demonstration** Peristalsis, p. 61 ◆ GENERAL TE **Connection Activity** Chemistry, p. 62 ◆ GENERAL TE **Activity** Examining Tissue, p. 62 ◆ ADVANCED TE **Activity** Digestive Tract, p. 63 BASIC SE **School-to-Home Activity** Bile Model, p. 63 GENERAL SE **Connection to Environmental Science** Waste Away, p. 64 GENERAL SE **Skills Practice Lab** As the Stomach Churns, p. 70 ◆ GENERAL CRF **Datasheet for Chapter Lab*** SE **Skills Practice Lab** Enzymes in Action, p. 186 ◆ GENERAL CRF **Datasheet for LabBook*** LB **Whiz-Bang Demonstrations** Liver Let Live* ◆ GENERAL	CRF **Lesson Plans*** TR **Bellringer Transparency*** TR The Digestive System* TR The Role of Enzymes in Protein Digestion* TR The Stomach; The Small Intestine and Villi* TR *LINK TO EARTH SCIENCE* Mohs' Hardness Scale* CRF **SciLinks Activity*** GENERAL TE **Internet Activity,** p. 63 GENERAL VID **Lab Videos for Life Science**
PACING • 45 min pp. 66–69 **Section 2 The Urinary System** • Explain how the kidneys filter blood. • Describe three disorders of the urinary system.	SE **Science in Action** Math, Social Studies, and Language Arts Activities, pp. 76–77 GENERAL TE **Activity** Defining Terms, p. 67 BASIC TE **Demonstration** Kidney Structure, p. 67 ◆ GENERAL LB **Long-Term Projects & Research Ideas** Copying the Kidney* ADVANCED	CRF **Lesson Plans*** TR **Bellringer Transparency*** TR How the Kidneys Filter Blood*

PACING • 90 min

CHAPTER REVIEW, ASSESSMENT, AND STANDARDIZED TEST PREPARATION

CRF **Vocabulary Activity*** GENERAL
SE **Chapter Review,** pp. 72–73 GENERAL
CRF **Chapter Review*** ■ GENERAL
CRF **Chapter Tests A*** ■ GENERAL, **B*** ADVANCED, **C*** SPECIAL NEEDS
SE **Standardized Test Preparation,** pp. 74–75 GENERAL
CRF **Standardized Test Preparation*** GENERAL
CRF **Performance-Based Assessment*** GENERAL
OSP **Test Generator** GENERAL
CRF **Test Item Listing*** GENERAL

Online and Technology Resources

go.hrw.com

Visit **go.hrw.com** for a variety of free resources related to this textbook. Enter the keyword **HL5BD3.**

Holt Online Learning

Students can access interactive problem-solving help and active visual concept development with the *Holt Science and Technology* Online Edition available at **www.hrw.com.**

Guided Reading Audio CD

A direct reading of each chapter using instructional visuals as guideposts. For auditory learners, reluctant readers, and Spanish-speaking students. Available in English and Spanish.

KEY

SE Student Edition	**CRF** Chapter Resource File	**SS** Science Skills Worksheets	* Also on One-Stop Planner
TE Teacher Edition	**OSP** One-Stop Planner	**MS** Math Skills for Science Worksheets	◆ Requires advance prep
	LB Lab Bank	**CD** CD or CD-ROM	■ Also available in Spanish
	TR Transparencies	**VID** Classroom Video/DVD	

SKILLS DEVELOPMENT RESOURCES	SECTION REVIEW AND ASSESSMENT	STANDARDS CORRELATIONS
SE Pre-Reading Activity, p. 56 GENERAL **OSP Science Puzzlers, Twisters & Teasers** GENERAL		National Science Education Standards UCP2, SAI 1; SPSP 5; HNS 2; LS 1e
CRF Directed Reading A* ■ BASIC, **B*** SPECIAL NEEDS **CRF Vocabulary and Section Summary*** ■ GENERAL **SE Reading Strategy** Prediction Guide, p. 58 GENERAL **TE Reading Strategy** Activity, p. 60 BASIC **SE Math Practice** Tooth Truth, p. 61 GENERAL **TE Inclusion Strategies,** p. 61 **TE Connection to Math** Liver Regeneration, p. 63 GENERAL **CRF Critical Thinking** Frankenstein's Food* ADVANCED	**SE Reading Checks,** pp. 59, 61, 63, 64 GENERAL **TE Homework,** p. 59 GENERAL **TE Homework,** p. 63 GENERAL **TE Reteaching,** p. 64 BASIC **TE Quiz,** p. 64 GENERAL **TE Alternative Assessment,** p. 64 ADVANCED **SE Section Review,*** p. 65 ■ GENERAL **CRF Section Quiz*** ■ GENERAL	UCP 1, 2, 5; SAI 1; SPSP1, 5; HNS 1; LS 1a, 1d, 1e, 1f, 3b; *Chapter Lab:* UCP 2, 5; SAI 1; LS 1a, 1e; *LabBook:* UCP 2, 5; SAI 1; LS 1a, 1e
CRF Directed Reading A* ■ BASIC, **B*** SPECIAL NEEDS **CRF Vocabulary and Section Summary*** ■ GENERAL **SE Reading Strategy** Reading Organizer, p. 66 GENERAL **TE Inclusion Strategies,** p. 68 **CRF Reinforcement Worksheet** Annie Apple's Amazing Adventure* BASIC	**SE Reading Checks,** pp. 67, 68 GENERAL **TE Reteaching,** p. 68 BASIC **TE Quiz,** p. 68 GENERAL **TE Alternative Assessment,** p. 68 GENERAL **SE Section Review,*** p. 69 ■ GENERAL **CRF Section Quiz*** ■ GENERAL	UCP 1, 2, 3, 4, 5; HNS 1; LS la, 1c, 1d, 1e, 1f, 3a, 3b

One-Stop Planner® CD-ROM

This convenient CD-ROM includes:
- **Lab Materials QuickList Software**
- **Holt Calendar Planner**
- **Customizable Lesson Plans**
- **Printable Worksheets**
- **ExamView® Test Generator**

CNN Student News

cnnstudentnews.com

Find the latest news, lesson plans, and activities related to important scientific events.

SciLinks NSTA

www.scilinks.org

Maintained by the **National Science Teachers Association.** See Chapter Enrichment pages for a complete list of topics.

Current Science®

Check out ***Current Science*** articles and activities by visiting the HRW Web site at **go.hrw.com.** Just type in the keyword **HL4CS24T.**

Classroom Videos

- **Lab Videos** demonstrate the chapter lab.
- **Brain Food Video Quizzes** help students review the chapter material.

Chapter Resources

Visual Resources

CHAPTER STARTER TRANSPARENCY

BELLRINGER TRANSPARENCIES

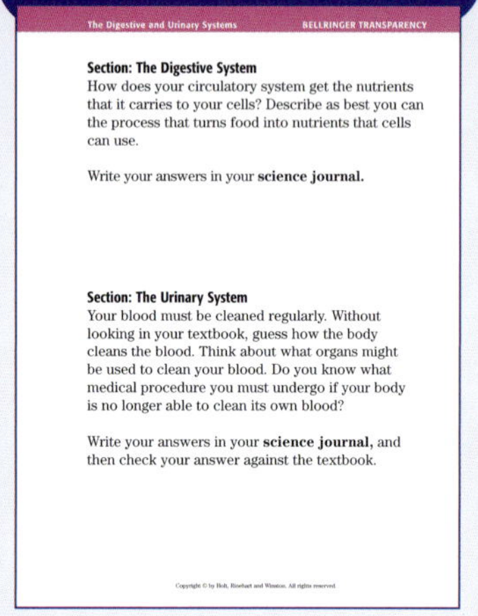

TEACHING TRANSPARENCIES

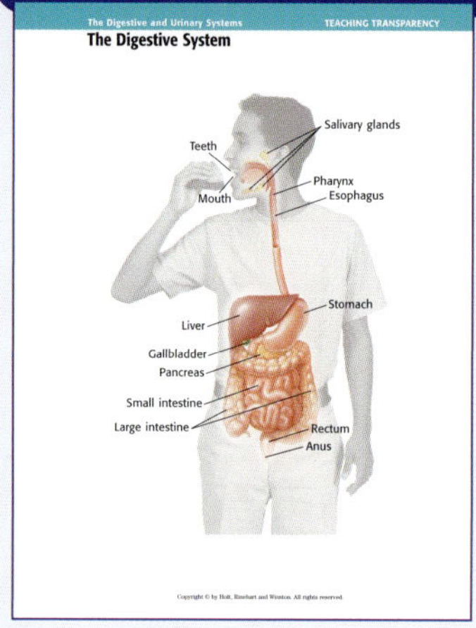

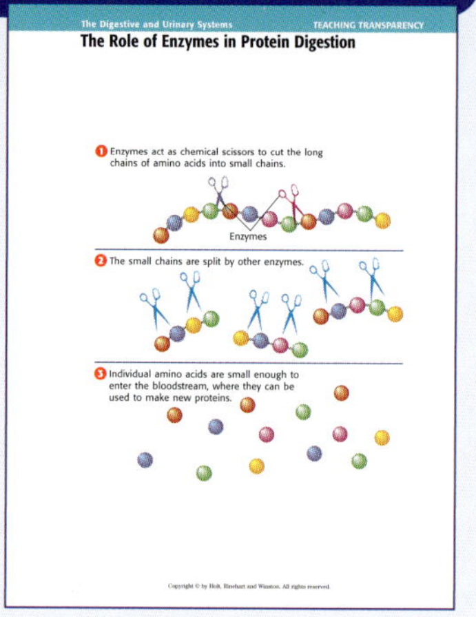

TEACHING TRANSPARENCIES

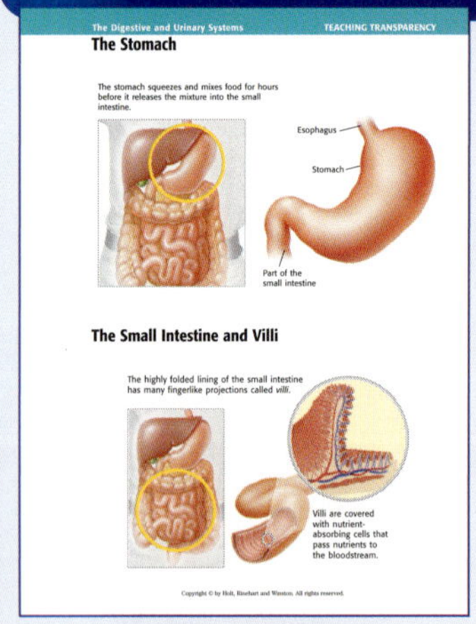

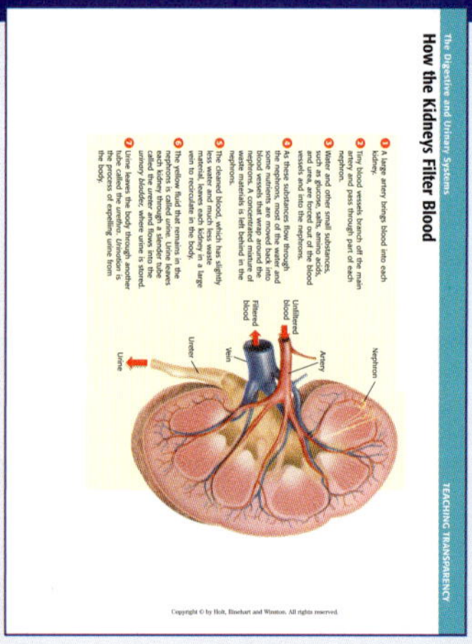

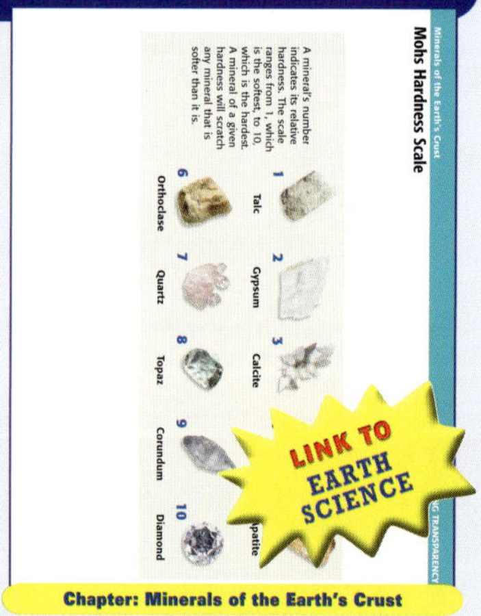

Chapter: Minerals of the Earth's Crust

CONCEPT MAPPING TRANSPARENCY

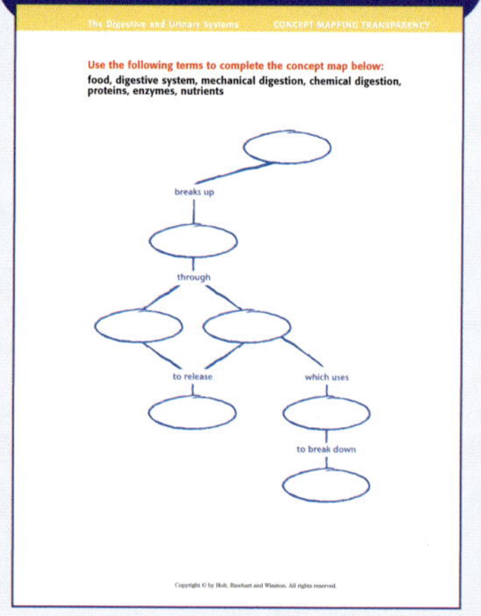

Planning Resources

LESSON PLANS

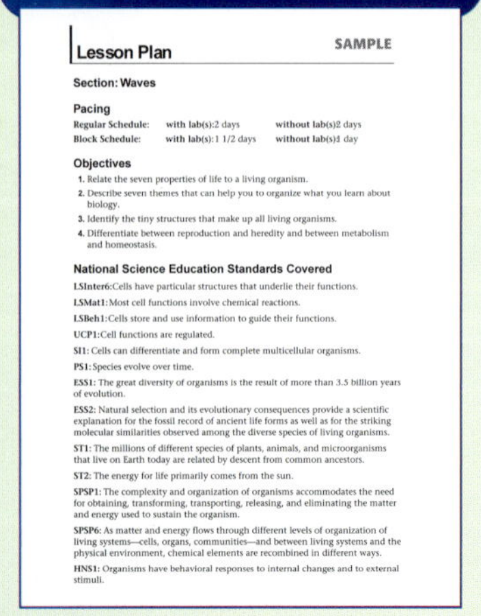

PARENT LETTER

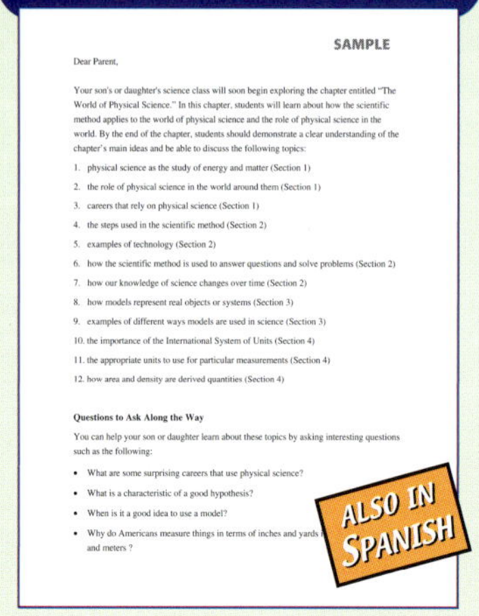

TEST ITEM LISTING

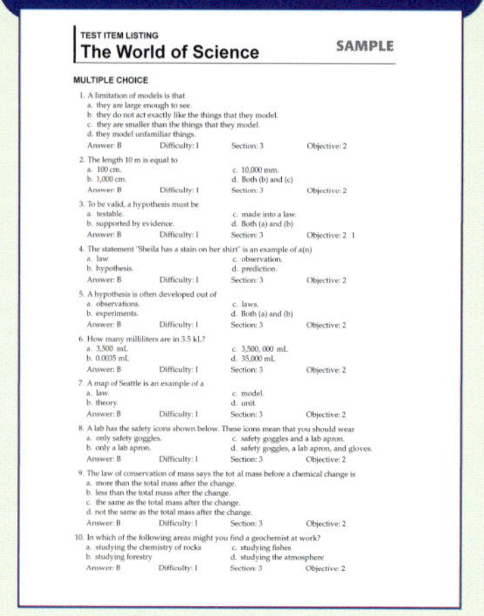

One-Stop Planner® CD-ROM

This CD-ROM includes all of the resources shown here and the following time-saving tools:

- *Lab Materials QuickList Software*
- *Customizable lesson plans*
- *Holt Calendar Planner*
- *The powerful ExamView® Test Generator*

For a preview of available worksheets covering math and science skills, see pages T12–T19. All of these resources are also on the One-Stop Planner®.

Meeting Individual Needs

DIRECTED READING A

Directed Reading A — SAMPLE — ALSO IN SPANISH — BASIC

DIRECTED READING B

Directed Reading B — SAMPLE — SPECIAL NEEDS

VOCABULARY ACTIVITY

Vocabulary Activity — SAMPLE — GENERAL

VOCABULARY AND SECTION SUMMARY

Vocabulary & Notes — SAMPLE — ALSO IN SPANISH — GENERAL

REINFORCEMENT

Reinforcement — SAMPLE — BASIC

CRITICAL THINKING

Critical Thinking — SAMPLE — ADVANCED

SCILINKS ACTIVITY

SciLinks Activity — SAMPLE — GENERAL

SCIENCE PUZZLERS, TWISTERS & TEASERS

The Digestive and Urinary Systems — GENERAL

Labs and Activities

LONG-TERM PROJECTS & RESEARCH IDEAS

Copying the Kidney — ADVANCED

WHIZ-BANG DEMONSTRATIONS

Liver Let Live — GENERAL

DATASHEETS FOR QUICK LABS

Reaction to Stress — SAMPLE

DATASHEETS FOR CHAPTER LABS

Using Scientific Methods — SAMPLE

DATASHEETS FOR LABBOOK

Does It All Add Up? — SAMPLE

Review and Assessments

SECTION QUIZ

Section Quiz — SAMPLE — ALSO IN SPANISH — GENERAL

SECTION REVIEW

Section Review — SAMPLE — ALSO IN SPANISH — GENERAL

CHAPTER REVIEW

Chapter Review — SAMPLE — ALSO IN SPANISH — GENERAL

CHAPTER TEST A

Chapter Test A — SAMPLE — ALSO IN SPANISH — GENERAL

CHAPTER TEST B

Chapter Test B — SAMPLE — ADVANCED

CHAPTER TEST C

Chapter Test C — SAMPLE — SPECIAL NEEDS

STANDARDIZED TEST PREPARATION

Standardized Test Preparation — SAMPLE — GENERAL

PERFORMANCE-BASED ASSESSMENT

Performanced-Based Assessment — SAMPLE — GENERAL

Chapter Enrichment

This Chapter Enrichment provides relevant and interesting information to expand and enhance your presentation of the chapter material.

Section 1

The Digestive System

Parts of the System

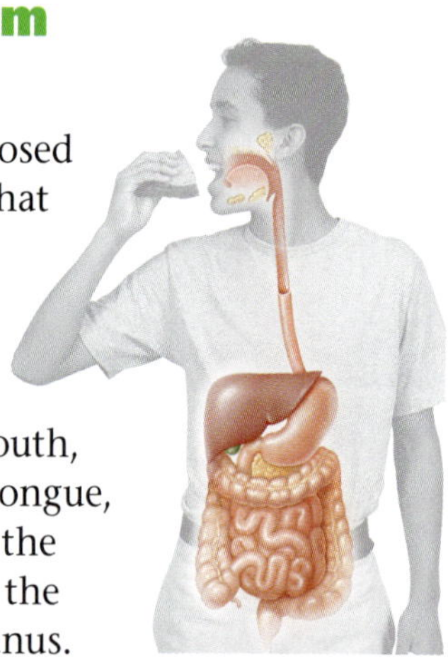

- The digestive system is composed of two sets of organs: those that make up the digestive tract and those that are called *accessory organs*. The digestive tract is a continuous tube consisting of the mouth, including the teeth and the tongue, the pharynx, the esophagus, the stomach, the small intestine, the large intestine, rectum, and anus.
- The accessory digestive organs include the salivary glands, the gallbladder, the liver, and the pancreas.

Is That a Fact!

- The digestive tract in a cadaver is about 9 m long. However, because of muscle tone, the digestive tract of a living person is considerably shorter.

The Stomach

- In short people, the stomach is found high in the abdomen and runs horizontally. In tall people, the stomach tends to run vertically, forming a **J** shape.

Is That a Fact!

- It takes from about four to eight seconds for food to pass from the top of the throat to the stomach.
- It takes less than two seconds for liquids to pass from the top of the throat to the stomach.

The Intestines

- The small intestine is the longest portion of the digestive tract. It is about 6 m long and 2.5 cm in diameter.
- Unlike the small intestine, the large intestine does not have villi and does not secrete digestive enzymes.

The Appendix

- The appendix is an offshoot of the large intestine and is 5 to 10 cm long. Like the tonsils, the appendix contains a large amount of lymphoid tissue.
- Occasionally, doctors begin an appendectomy on patients misdiagnosed with appendicitis. Instead of having appendicitis, however, the patients have had holes in the lining of their intestines made by parasitic worms. The incidence of intestinal parasitic worms is increasing as eating raw fish, such as sushi and sashimi, becomes more popular.

Food Poisoning

- Food poisoning is characterized by nausea, vomiting, abdominal cramps, and diarrhea. The two most common culprits are the bacteria *Salmonella* and *Shigella*.
- *Salmonella* is typically spread by poorly cooked meat and feces-contaminated hands. *Salmonella* infects the microvilli in the intestinal lining, causing blisters. If *Salmonella* poisoning is not treated promptly, the bacteria may spread throughout the body.
- *Shigella* is commonly found among people who have visited developing countries. It is spread via food, feces, fingers, flies, and contaminated public bodies of water, such as swimming pools.

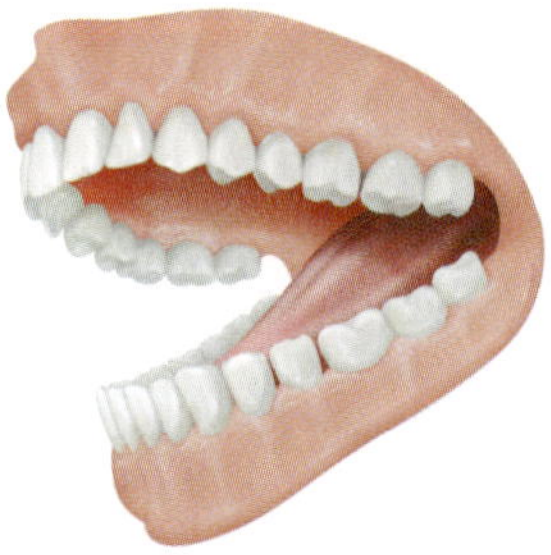

Eating Disorders

- Anorexia nervosa and bulimia nervosa are two common eating disorders that are characterized by an obsessive fear of gaining weight. Often, underlying psychological factors initiate these disorders.
- Anorexia nervosa is an eating disorder characterized by self-induced starvation. People with this condition refuse to eat and often exercise obsessively. Although the disorder can affect anyone, adolescent girls and young women are commonly afflicted. Boys, especially those involved in weight-conscious sports such as wrestling, can also suffer from anorexia nervosa.

For background information about teaching strategies and issues, refer to the ***Professional Reference for Teachers.***

- Bulimia nervosa is a binge-purge syndrome in which those afflicted binge on enormous amounts of food and then force themselves to vomit or use laxatives to eliminate the food. The binge-purge cycle may be repeated several times per day or only a couple of times per week.
- Although bulimics generally maintain a normal body weight and appear to be healthy, they are not healthy. They tend to have swollen salivary glands, pancreatitis, and liver and kidney problems. In addition, they are at risk of heart failure and stomach rupture, both of which can result in death. Excessive vomiting damages the esophagus and the stomach and wears away tooth enamel.
- Treatment for eating disorders may include diet control, often by means of hospitalization, behavior modification, nutrition education, and use of antidepressants.

Section 2

The Urinary System

Filtering Wastes

- The urinary system functions primarily to maintain the correct balance of salts and water and to remove metabolic, nitrogen-containing wastes, such as urea, from the body in the form of urine.
- The digestive system, the circulatory system, and the respiratory system also excrete wastes. The digestive system excretes food wastes in the form of feces. The circulatory and respiratory systems work together to rid the body of carbon dioxide.
- The yellow color of urine comes from the yellow pigment urochrome.

The Kidneys

- The kidneys are surrounded and kept in their proper place in the body by fat. People who become too thin risk damage to the kidneys and other related urinary problems. As much as 1,200 mL of blood passes through the nephrons each minute.

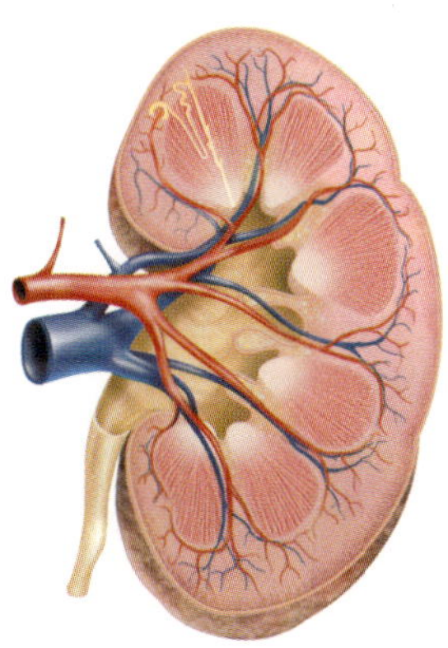

Is That a Fact!

- Urinary bladders of average size can contain as much as 1 L of urine!
- The kidneys require up to one-fourth of the body's oxygen supply to carry out their functions.
- The kidneys filter about 180 L of fluid from the blood each day. About 99% of this fluid is returned to the bloodstream. The other 1% leaves the body in the form of urine.
- A person who donates a kidney can maintain normal kidney functions. The remaining kidney enlarges and carries out the functions previously performed by two kidneys.

The Urinary Bladder

- Like the stomach, the size of the bladder varies with the amount of its contents. The bladder can hold 300 mL of urine without increasing its internal pressure significantly. At 500 mL of urine, the bladder is fairly full and may be 12.5 cm in length.

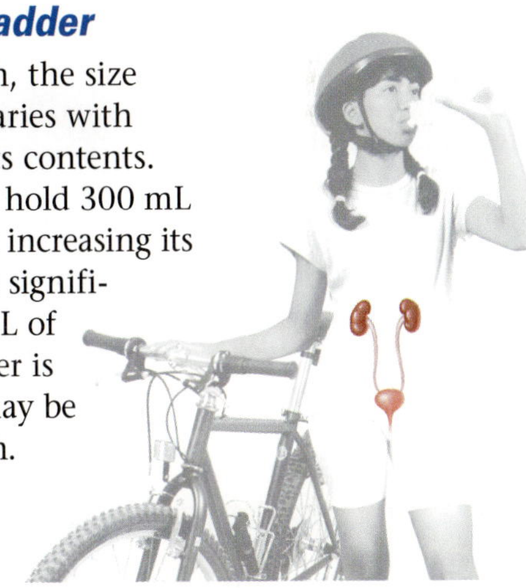

SCILINKS. NSTA Developed and maintained by the National Science Teachers Association

SciLinks is maintained by the National Science Teachers Association to provide you and your students with interesting, up-to-date links that will enrich your classroom presentation of the chapter.

Visit www.scilinks.org and enter the SciLinks code for more information about the topic listed.

Topic: The Digestive System
SciLinks code: HSM0409

Topic: Problems in the Digestive System
SciLinks code: HSM1218

Topic: The Urinary System
SciLinks code: HSM1583

Topic: Urinary System Ailments
SciLinks code: HSM1584

Topic: Tapeworms
SciLinks code: HSM1492

Overview

In this chapter, students will learn about the digestive and urinary systems of the human body. Students will learn how the digestive system breaks down food into building blocks the body can use and how the digestive and urinary systems eliminate wastes from the body.

Assessing Prior Knowledge

Students should be familiar with the following topics:

- cellular structure and function
- body organization

Identifying Misconceptions

Students may have misconceptions regarding the digestive and urinary systems. Many people commonly refer only to the stomach when discussing eating and digestion. So, some students may not know other organs are also part of the digestive system. Students may also mistakenly think that all digestion is mechanical. Though mechanical digestion is important, chemical digestion reduces food into subunits the body can use.

3

The Digestive and Urinary Systems

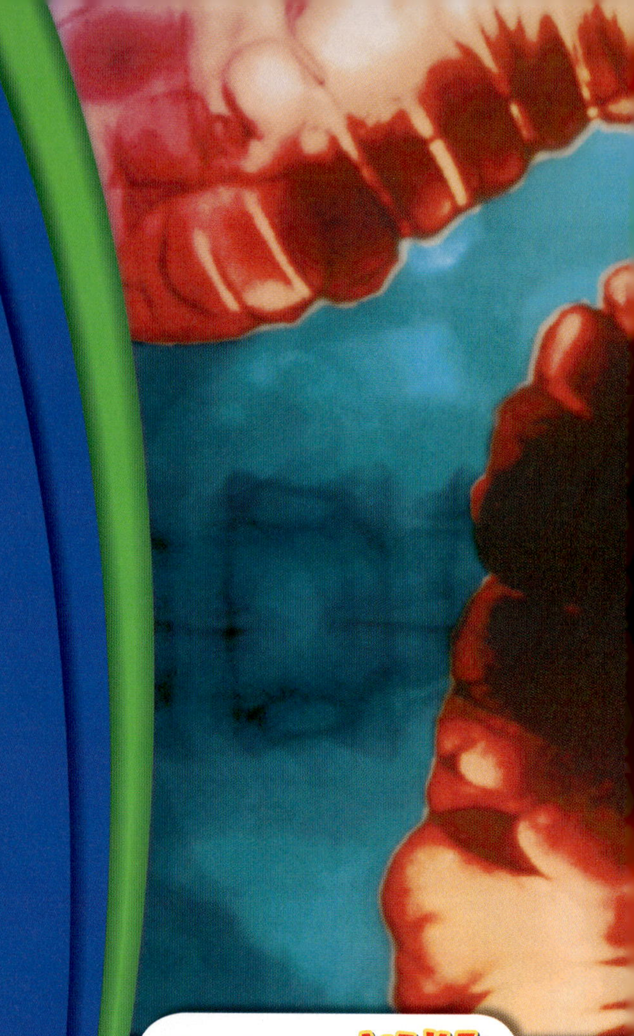

About the PHOTO

Is this a giant worm? No, it's an X ray of a healthy large intestine! Your large intestine helps your body preserve water. As mostly digested food passes through your large intestine, water is drawn out of the food. This water is returned to the bloodstream. The gray shadow behind the intestine is the spinal column. The areas that look empty are actually filled with organs. A special liquid helps this large intestine show up on the X ray.

PRE-READING ACTIVITY

Graphic Organizer

Chain-of-Events Chart Before you read the chapter, create the graphic organizer entitled "Chain-of-Events Chart" described in the **Study Skills** section of the Appendix. As you read the chapter, fill in the chart with details about each step of the processes that your body uses to digest food.

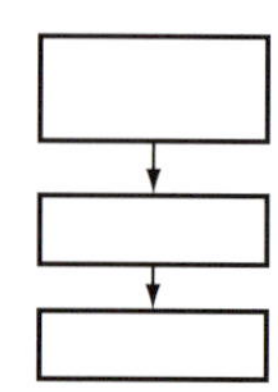

Standards Correlations

National Science Education Standards

The following codes indicate the National Science Education Standards that correlate to this chapter. The full text of the standards is at the front of the book.

Chapter Opener
UCP2, SAI 1; SPSP 5; HNS 2, 3; LS 1e

Section 1 The Digestive System
UCP 1, 2, 5; SAI 1; SPSP1, 5; HNS 1, 3; LS 1a, 1d, 1e, 1f, 3b; *LabBook:* UCP 2, 5; SAI 1; LS 1a, 1e

Section 2 The Urinary System
UCP 1, 2, 3, 4, 5; HNS 1; LS 1a, 1c, 1d, 1e, 1f, 3a, 3b

Chapter Lab
UCP 2, 5; SAI 1; LS 1a, 1e

Chapter Review
LS1 a, 1c, 1d, 1e, 1f, 3a, 3b

Science in Action
HNS 1; SPSP 5; LS 1f

START-UP ACTIVITY

MATERIALS

FOR EACH STUDENT

- bag, plastic, sealable
- flour, 200 mL
- vegetable oil, 100 mL
- water, 100 mL

Safety Caution: Tell students not to ingest any of the materials used in this investigation. Using strong bags or doubling the bags works best. Have paper towels on hand, and clean up any spills immediately. Spilled liquids are a slipping hazard.

Answers

1. Sample answer: Before I squeezed the bag, the contents were well dispersed. Once I began squeezing the bag, the oil was mixed into the flour mixture.
2. Sample answer: The stomach is a muscular, baglike organ that is involved in the physical and chemical digestion of food. By squeezing the bag, I am modeling the muscular contractions of the stomach. (Note: This activity does not model the chemical digestion that occurs in the stomach.)
3. Sample answer: This is a good model of how the stomach works. Squeezing the bag helps show how the stomach mechanically mashes food. But the model does not have acids and enzymes that also help a real stomach break down food.

START-UP ACTIVITY

Changing Foods

The stomach breaks down food by, in part, squeezing the food. You can model the action of the stomach in the following activity.

Procedure

1. Add **200 mL of flour** and **100 mL of water** to a **resealable plastic bag.**
2. Mix **100 mL of vegetable oil** with the flour and water.
3. Seal the plastic bag.
4. Shake the bag until the flour, water, and oil are well mixed.
5. Remove as much air from the bag as you can, and reseal the bag carefully.
6. Knead the bag carefully with your hands for 5 min. Be careful to keep the bag sealed.

Analysis

1. Describe the mixture before and after you kneaded the bag.
2. How might the changes you saw in the mixture relate to how your stomach digests food?
3. Do you think this activity is a good model of how your stomach works? Explain your answer.

The Digestive and Urinary Systems CHAPTER STARTER

This Really Happened!

In 1822, a Canadian fur trader named Alexis St. Martin accidentally shot himself in his left side. An army surgeon named Dr. William Beaumont was able to save St. Martin's life, but the wound never completely healed. A hole approximately 6 cm in diameter remained open in his side, leading through skin and muscle right into his stomach.

Dr. Beaumont seized the opportunity

stomach squeeze and relax as it digested the bits of food. When he removed the food, he could see that it had changed. He also found that food changed when he placed it in a test tube with juices from St. Martin's stomach. Beaumont was able to determine that the stomach's juices, not just its squeezing action, were responsible for turning the food to mush.

What began as an unfortunate acci-

Chapter Starter Transparency
Use this transparency to help students begin thinking about how the human body digests food.

CHAPTER RESOURCES

Technology

Transparencies READING SKILLS
- Chapter Starter Transparency

Student Edition on CD-ROM

Guided Reading Audio CD
- English or Spanish

Classroom Videos
- Brain Food Video Quiz

Workbooks

Science Puzzlers, Twisters & Teasers
- The Digestive and Urinary Systems GENERAL

SECTION 1

Focus

Overview

This section introduces the structures and functions of the digestive system. Students will compare mechanical digestion with chemical digestion and will learn to trace the path of food through the digestive system. Students will also learn about some ailments of the digestive system and their causes.

Bellringer

Ask students to answer the following questions:

- How does your circulatory system obtain the nutrients that it brings to your cells? (from the digestive system)
- Describe as best you can the process that turns food into nutrients that cells can use. (Sample answer: Food is broken down into smaller pieces by chewing and by the muscles of the digestive system. Chemicals in the digestive system then break these smaller pieces into nutrients that cells can use.)

SECTION 1

The Digestive System

It's your last class before lunch, and you're starving! Finally, the bell rings, and you get to eat!

READING WARM-UP

Objectives

- Compare mechanical digestion with chemical digestion.
- Describe the parts and functions of the digestive system.

Terms to Learn

digestive system
esophagus
stomach
pancreas
small intestine
liver
gallbladder
large intestine

READING STRATEGY

Prediction Guide Before reading this section, write the title of each heading in this section. Next, under each heading, write what you think you will learn.

You feel hungry because your brain receives signals that your cells need energy. But eating is only the beginning of the story. Your body must change a meal into substances that you can use. Your **digestive system,** shown in **Figure 1,** is a group of organs that work together to digest food so that it can be used by the body.

digestive system the organs that break down food so that it can be used by the body

Digestive System at a Glance

The most obvious part of your digestive system is a series of tubelike organs called the *digestive tract.* Food passes through the digestive tract. The digestive tract includes your mouth, pharynx, esophagus, stomach, small intestine, large intestine, rectum, and anus. The human digestive tract can be more than 9 m long! The liver, gallbladder, pancreas, and salivary glands are also part of the digestive system. But food does not pass through these organs.

Figure 1 The Digestive System

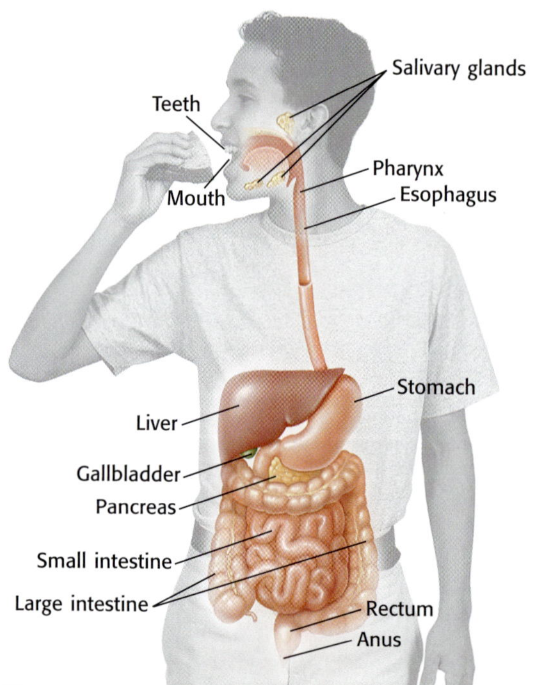

CHAPTER RESOURCES

Chapter Resource File

- Lesson Plan
- Directed Reading A BASIC
- Directed Reading B SPECIAL NEEDS

Technology

Transparencies

- Bellringer
- The Role of Enzymes in Protein Digestion

Cultural Awareness GENERAL

World Diets Have students research the diets of people in countries such as Japan, India, Israel, Egypt, Mexico, and Russia. Encourage students to discover strategies that people in these countries use to obtain a nutritious diet, and have students compare the diets of people in each country with the diets of people in the United States. You may also want to have students compare the diets based on fat or cholesterol intake. LS Verbal/Logical

Breaking Down Food

Digestion is the process of breaking down food, such as a peanut butter and jelly sandwich, into a form that can pass from the digestive tract into the bloodstream. There are two types of digestion—mechanical and chemical. The breaking, crushing, and mashing of food is called *mechanical digestion*. In *chemical digestion,* large molecules are broken down into nutrients. Nutrients are substances in food that the body needs for normal growth, maintenance, and repair.

Three major types of nutrients—carbohydrates, proteins, and fats—make up most of the food you eat. In fact, a peanut butter and jelly sandwich contains all three of these nutrients. Substances called *enzymes* break some nutrients into smaller particles that the body can use. For example, proteins are chains of smaller molecules called *amino acids*. Proteins are too large to be absorbed into the bloodstream. So, enzymes cut up the chain of amino acids. The amino acids are small enough to pass into the bloodstream. This process is shown in **Figure 2.**

Reading Check How do enzymes help digestion? (*See the Appendix for answers to Reading Checks.*)

Figure 2 The Role of Enzymes in Protein Digestion

1 Enzymes act as chemical scissors to cut the long chains of amino acids into small chains.

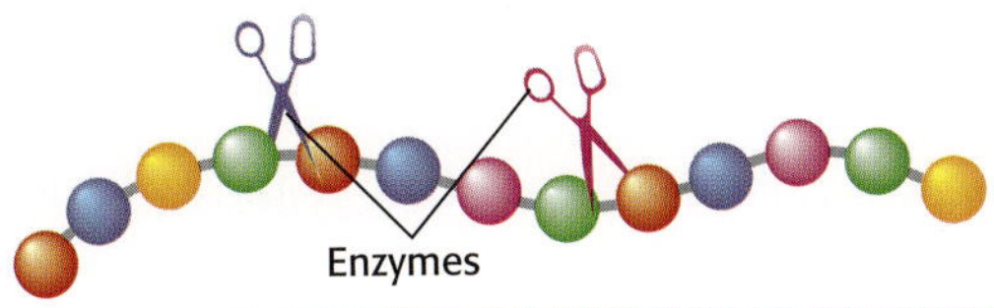

2 The small chains are split by other enzymes.

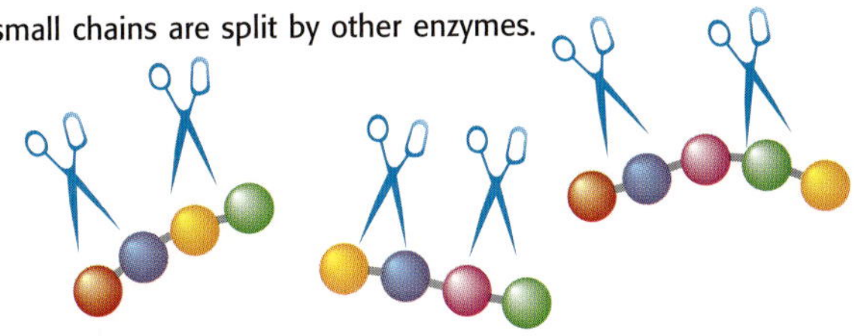

3 Individual amino acids are small enough to enter the bloodstream, where they can be used to make new proteins.

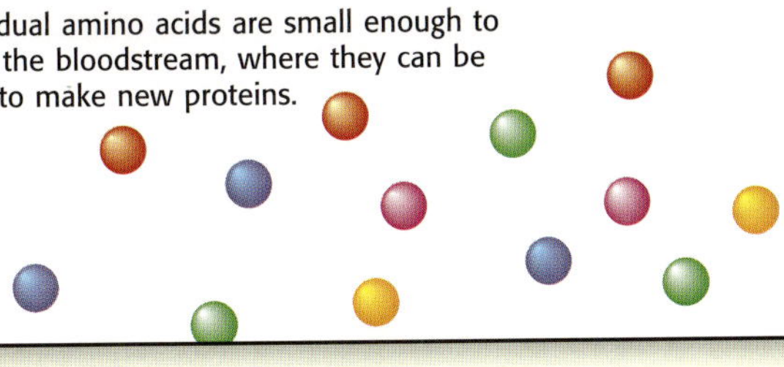

Quick Lab

Break It Up!

1. Drop **one piece of hard candy** into a **clear plastic cup of water.**
2. Wrap an **identical candy** in a **towel,** and crush the candy with a **hammer.** Drop the candy into a **second clear cup of water.**
3. The next day, examine both cups. What is different about the two candies?
4. What type of digestion is represented by breaking the hard candy?
5. How does chewing your food help the process of digestion?

Motivate

Demonstration — GENERAL

Measuring Saliva Show students a 2 L beaker or other transparent container. Pour 1.2 L (2.6 pt) of water into the container, and explain that this amount of water represents the amount of saliva an average person produces in one day. Ask students to discuss what roles saliva might play in the body, specifically in the process of digestion. (Answers might include lubricating the mouth and related organs, moistening food, and breaking down molecules into nutrients that cells can use.) **LS Logical/Visual**

Teach

Quick Lab

MATERIALS

FOR EACH STUDENT

- hammer
- hard candy, 2 identical pieces
- towel
- water, in 2 identical clear plastic cups

Safety Caution: Tell students not to ingest or taste any of the materials, including the candy, in this Quick Lab. Students should wear safety goggles while crushing the candy. Students should strike hammers away from their face and hands.

Answers

3. The crushed candy dissolved in the water. The whole candy did not dissolve as much.
4. Chewing, one form of mechanical digestion, is represented by breaking the candy.
5. Chewing breaks the large chunks of food into smaller pieces. Greater surface area is then available for enzymes to digest the food.

Homework — GENERAL

Kinds of Digestion Ask students to indicate which of the following activities represent mechanical digestion and which represent chemical digestion: The teeth cut and grind food, the stomach churns food, enzymes in saliva break down carbohydrates in food, and enzymes break down proteins in food. (mechanical—the teeth cutting and grinding food, the stomach churning food; chemical—enzymes in saliva breaking down carbohydrates in food, enzymes breaking down proteins in food) **LS Logical**

Answer to Reading Check

Enzymes cut nutrients into amino acids that the body can use.

Teach, *continued*

READING STRATEGY — BASIC

Activity Before students read the text on these pages, have them read the headings aloud. Then ask students to formulate one question that they expect the text under each heading to answer. Have students write their questions down. Possible questions include the following:

- "What happens in the mouth?"
- "Why is chewing important?"
- "What makes the stomach's environment harsh?"

LS Verbal

CONNECTION to Earth Science — GENERAL

Enamel Tooth enamel is made primarily of carbonated calcium hydroxyapatite. Fluoride compounds, taken internally while the teeth are growing, strengthen the enamel by forming the enamel out of fluoroapatite, an apatite material more acid resistant than natural enamel. Applying fluoride to the outside of teeth after the teeth have formed can also help strengthen the enamel by changing the natural hydroxyapatite in the tooth to fluoroapatite. Use the teaching transparency "Mohs' Hardness Scale" to illustrate that in geology, apatite is in the middle of the hardness scale for minerals. **LS Visual**

Digestion Begins in the Mouth

Chewing is important for two reasons. First, chewing creates small, slippery pieces of food that are easier to swallow than big, dry pieces are. Second, small pieces of food are easier to digest.

Teeth

Teeth are very important organs for mechanical digestion. With the help of strong jaw muscles, teeth break and grind food. The outermost layer of a tooth, the *enamel,* is the hardest material in the body. Enamel protects nerves and softer material inside the tooth. **Figure 3** shows a cross section of a tooth.

Have you ever noticed that your teeth have different shapes? Look at **Figure 4** to locate the different kinds of teeth. The molars are well suited for grinding food. The *premolars* are perfect for mashing food. The sharp teeth at the front of your mouth, the *incisors* and *canines,* are for shredding food.

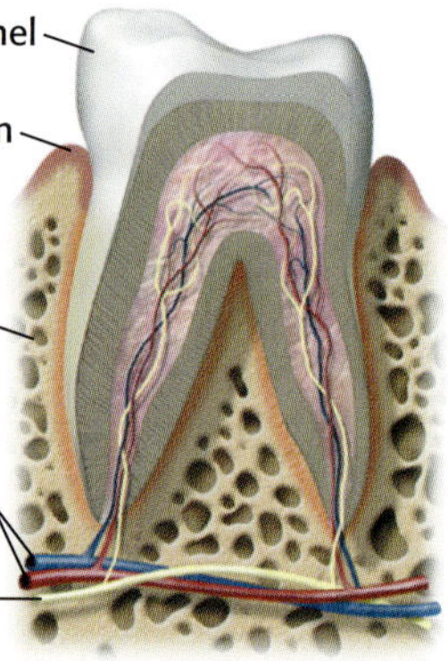

Figure 3 *A tooth, such as this molar, is made of many kinds of tissue.*

Saliva

As you chew, the food mixes with a liquid called *saliva.* Saliva is made in salivary glands located in the mouth. Saliva contains an enzyme that begins the chemical digestion of carbohydrates. Saliva changes complex carbohydrates into simple sugars.

Leaving the Mouth

Once the food has been reduced to a soft mush, the tongue pushes it into the throat, which leads to a long, straight tube called the **esophagus** (i SAHF uh guhs). The esophagus squeezes the mass of food with rhythmic muscle contractions called *peristalsis* (PER uh STAL sis). Peristalsis forces the food into the stomach.

esophagus a long, straight tube that connects the pharynx to the stomach

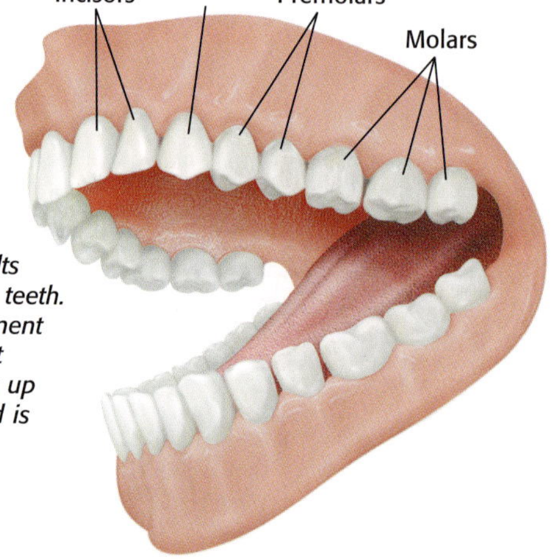

Figure 4 *Most adults have 32 permanent teeth. Each type of permanent tooth has a different function in breaking up food before the food is swallowed.*

CHAPTER RESOURCES

Technology

Transparencies
- LINK TO EARTH SCIENCE Mohs' Hardness Scale
- The Stomach, The Small Intestine and Villi

MISCONCEPTION ALERT

Teeth and Bones Students may mistakenly assume that teeth are made of bone. Point out that although enamel and dentin are similar to bone in appearance and calcium content, they are not bone. Bone contains blood vessels, but enamel and dentin do not.

Figure 5 The Stomach

The stomach squeezes and mixes food for hours before it releases the mixture into the small intestine.

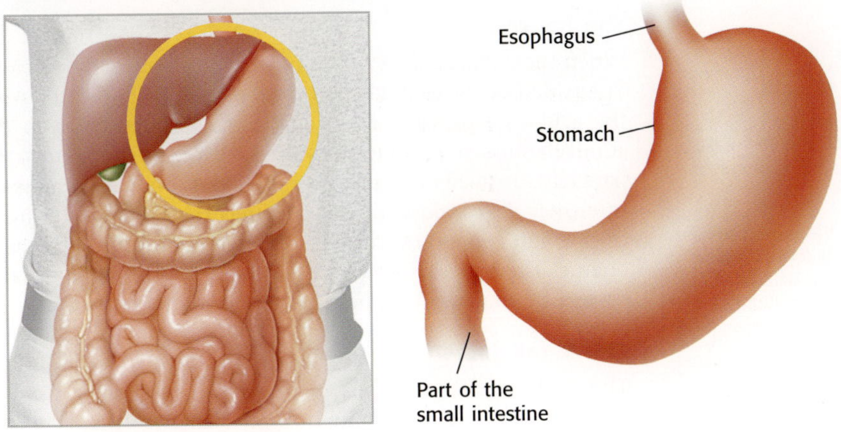

The Harsh Environment of the Stomach

The **stomach** is a muscular, saclike, digestive organ attached to the lower end of the esophagus. The stomach is shown in **Figure 5.** The stomach continues the mechanical digestion of your meal by squeezing the food with muscular contractions. While this squeezing is taking place, tiny glands in the stomach produce enzymes and acid. The enzymes and acid work together to break food into nutrients. Stomach acid also kills most bacteria that you might swallow with your food. After a few hours of combined mechanical and chemical digestion, your peanut butter and jelly sandwich has been reduced to a soupy mixture called *chyme* (KIEM).

stomach the saclike, digestive organ between the esophagus and the small intestine that breaks down food into a liquid by the action of muscles, enzymes, and acids

Reading Check What is chyme?

Leaving the Stomach

The stomach slowly releases the chyme into the small intestine through a small ring of muscle that works like a valve. This valve keeps food in the stomach until the food has been thoroughly mixed with digestive fluids. Each time the valve opens and closes, it lets a small amount of chyme into the small intestine. Because the stomach releases chyme slowly, the intestine has more time to mix the chyme with fluids from the liver and pancreas. These fluids help digest food and stop the harsh acids in chyme from hurting the small intestine.

MATH PRACTICE

Tooth Truth

Young children get a first set of 20 teeth called *baby teeth*. These teeth usually fall out and are replaced by 32 permanent teeth. How many more permanent teeth than baby teeth does a person have? What is the ratio of baby teeth to permanent teeth? Be sure to express the ratio in its most reduced form.

Demonstration — GENERAL

Peristalsis Help students visualize the process of peristalsis by performing the following demonstration. Fill a 75 cm section of rubber tubing with water, and have an assistant pinch and hold both ends of the tubing shut. Wrap your fists around the tubing near one of the assistant's hands, and slowly "walk" your hands to the other end of the tubing, being careful to squeeze the water from the starting end to the other end of the tubing. When your hands reach the other end of the tubing, point the end into a basin or bucket away from all students, and have the assistant release his or her grip on that end. Water should stream out of the tubing. Point out that this is what the esophagus does when it squeezes food into the stomach. English Language Learners

LS Kinesthetic/Visual

Answer to Reading Check

Chyme is a soupy mixture of partially digested food in the stomach.

Cultural Awareness — GENERAL

Lactose Intolerance Many adults of African and Asian descent have some form of lactose intolerance. They lack the enzyme lactase, which is needed to digest the sugar lactose in dairy products. Lactose intolerance is generally less common in people of European descent.

Answer to Math Practice

People have 12 more teeth as adults than they do as young children. The ratio of baby teeth to adult teeth is 20:32, which reduces to 5:8.

INCLUSION Strategies

- *Hearing Impaired*
- *Visually Impaired*
- *Developmentally Delayed*

Some students struggle with abstract concepts. Models that students can see and touch help make these ideas clearer. A model can help students understand how a long length of intestines fits within a small area in the human body. Organize the class into teams of three or four students. Have each team simulate intestinal folds using the following model:

1. Cut each of the three sheets of 8 1/2 × 11 paper into 8 strips (total of 24 strips per team).
2. Tape the 24 strips end to end.
3. Repeatedly bend the long strip into irregular lengths so that the folded paper also fits into about a 3 2/3 in. square.

English Language Learners

LS Visual

Teach, *continued*

CONNECTION ACTIVITY
Chemistry — GENERAL

Changes As food is digested, it undergoes both physical and chemical changes. Ask students to categorize the following activities as either physical changes or chemical changes:

- chewing food (physical change)
- enzymes in saliva breaking down carbohydrates (chemical change)
- squeezing and churning food in the stomach (physical change)
- breaking down food with pancreatic juice (chemical change)

LS Logical

ACTIVITY — ADVANCED

Examining Tissue Provide students with a compound light microscope and prepared slides of several different digestive organs. You might want to include cross sections of tissue from the stomach, the small intestine, the liver, and the pancreas. Review the proper use of microscopes before allowing students to operate them, and make sure that students do not work with broken or cracked slides. Have students observe the specimens, compare the different cross sections, and make sketches. LS Visual

pancreas the organ that lies behind the stomach and that makes digestive enzymes and hormones that regulate sugar levels

small intestine the organ between the stomach and the large intestine where most of the breakdown of food happens and most of the nutrients from food are absorbed

The Pancreas and Small Intestine

Most chemical digestion takes place after food leaves the stomach. Proteins, carbohydrates, and fats in the chyme are digested by the small intestine and fluids from the pancreas.

The Pancreas

When the chyme leaves the stomach, the chyme is very acidic. The pancreas makes fluids that protect the small intestine from the acid. The **pancreas** is an oval organ located between the stomach and small intestine. The chyme never enters the pancreas. Instead, the pancreatic fluid flows into the small intestine. This fluid contains enzymes that chemically digest chyme and contains bicarbonate, which neutralizes the acid in chyme. The pancreas also functions as a part of the endocrine system by making hormones that regulate blood sugar.

The Small Intestine

The **small intestine** is a muscular tube that is about 2.5 cm in diameter. Other than having a small diameter, it is really not that small. In fact, if you stretched the small intestine out, it would be longer than you are tall—about 6 m! If you flattened out the surface of the small intestine, it would be larger than a tennis court! How is this possible? The inside wall of the small intestine is covered with fingerlike projections called *villi,* shown in **Figure 6.** The surface area of the small intestine is very large because of the villi. The villi are covered with tiny, nutrient-absorbing cells. Once the nutrients are absorbed, they enter the bloodstream.

CONNECTION TO Social Studies

WRITING SKILL **Parasites** Intestinal parasites are organisms, such as roundworms and hookworms, that infect people and live in their digestive tract. Worldwide, intestinal parasites infect more than 1 billion people. Some parasites can be deadly. Research intestinal parasites in a library or on the Internet. Then, write a report on a parasite, including how it spreads, what problems it causes, how many people have it, and what can be done to stop it.

Figure 6 The Small Intestine and Villi

The highly folded lining of the small intestine has many fingerlike projections called *villi.*

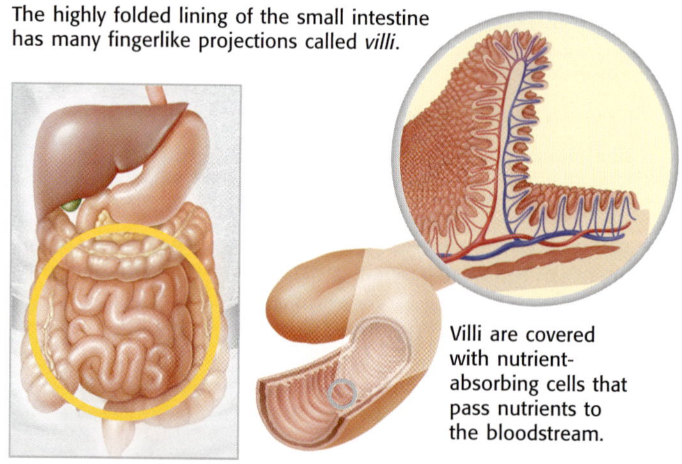

Villi are covered with nutrient-absorbing cells that pass nutrients to the bloodstream.

Answer to Connection to Social Studies

Sample answer: The World Health Organization reports that the intestinal parasite *Ascaris lumbricodes* (a kind of roundworm) kills 60,000 people every year. Symptoms include abdominal pain, intestinal blockage, and vomiting. The disease can be treated with medicines. Roundworms are spread by poor sanitary practices, such as using untreated human waste as fertilizer. Improved sanitary practices help control the spread of roundworms.

SCIENCE HUMOR

A man walks into a doctor's office with a stalk of celery in one ear and a carrot in the other. The man says, "Doc, I'm just not feeling well these days." The doctor replies, "I think that's because you haven't been eating right."

Figure 7 **The Liver and the Gallbladder**

Food does not move through the liver, gallbladder, and pancreas even though these organs are linked to the small intestine.

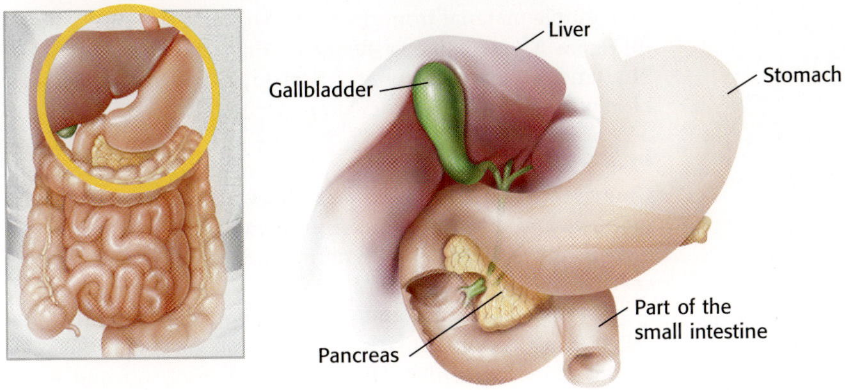

The Liver and Gallbladder

The **liver** is a large, reddish brown organ that helps with digestion. A human liver can be as large as a football. Your liver is located toward your right side, slightly higher than your stomach, as shown in **Figure 7.** The liver helps with digestion in the following ways:

- It makes bile to break up fat.
- It stores nutrients.
- It breaks down toxins.

liver the largest organ in the body; it makes bile, stores and filters blood, and stores excess sugars as glycogen

gallbladder a sac-shaped organ that stores bile produced by the liver

Breaking Up Fat

Although bile is made by the liver, bile is temporarily stored in a small, saclike organ called the **gallbladder,** shown in **Figure 7.** Bile is squeezed from the gallbladder into the small intestine, where the bile breaks large fat droplets into very small droplets. This mechanical process allows more fat molecules to be exposed to digestive enzymes.

Reading Check How does bile help digest fat?

Storing Nutrients and Protecting the Body

After nutrients are broken down, they are absorbed into the bloodstream and carried through the body. Nutrients that are not needed right away are stored in the liver. The liver then releases the stored nutrients into the bloodstream as needed. The liver also captures and detoxifies many chemicals in the body. For instance, the liver produces enzymes that break down alcohol and many other drugs.

SCHOOL to HOME

Bile Model

You can model the way bile breaks down fat and oil by using dish soap. At home with a parent, put a small amount of water in a small jar. Then, add a few drops of vegetable oil to the water. Notice that the two liquids separate. Draw a picture of the jar and its contents. Next, add a few drops of dishwashing soap to the water, tighten the lid securely onto the jar, and shake the jar. What happened to the three liquids in the jar? Draw another picture of the jar and its contents.

ACTIVITY

INTERNET ACTIVITY

Sequence Board — GENERAL

For an internet activity related to this chapter, have students go to **go.hrw.com** and type in the keyword **HL5BD3W.**

Answer to School-to-Home Activity

Sample answer: Before the dishwashing soap was added, the liquids separated. After the soap was added and the jar was shaken, the oil and water mixed.

ACTIVITY — BASIC

Digestive Tract Help students remember the route that food takes through the digestive tract by providing them with additional practice. Begin by showing students a model or illustration of the digestive tract. Then have students take turns naming and identifying, in order, the organs through which food travels through the digestive tract. (The proper sequence is as follows: mouth, pharynx, esophagus, stomach, small intestine, and large intestine.)
LS **Logical/Visual**

Homework — GENERAL

Concept Mapping List the following terms on the board:

digestive system, mouth, stomach, liver, pancreas, gallbladder, teeth, salivary glands, saliva, digestive tract, esophagus, throat, tongue, small intestine

Then have students copy the terms onto a piece of paper and construct a concept map using these terms and linking words between them. **LS** **Logical**

Answer to Reading Check

Bile breaks large fat droplets into very small droplets. This process allows more fat molecules to be exposed to digestive enzymes.

CONNECTION to *Math* — GENERAL

Liver Regeneration Tell students that as much as 75% of a person's liver can be removed or impaired before the liver stops functioning. Ask students to calculate how much of a 1 kg liver must remain for it to function. (25% × 1,000 g = 250 g or 0.25 kg)
LS **Logical**

Close

Reteaching — BASIC

Creating Graphs Provide students with the following average lengths of the digestive organs:

esophagus—25 cm, stomach—25 cm, small intestine—6 m, large intestine—1.5 m

Then have students prepare a bar graph to compare the average lengths of each digestive organ listed. English Language Learners

LS Logical

Quiz — GENERAL

Ask students whether each of the statements below is true or false.

1. Digestion begins when food reaches the stomach. (false)
2. Breaking, crushing, and mashing food is an example of chemical digestion. (false)
3. Saliva contains enzymes, which begin the chemical digestion of food. (true)
4. The esophagus connects the mouth with the small intestine directly. (false)

Alternative Assessment — ADVANCED

Writing **Owner's Guide** Have students develop an owner's guide for their digestive system. The guide should include information about the structures of the digestive system and a diagram of the location of these body structures.

LS Verbal

large intestine the wider and shorter portion of the intestine that removes water from mostly digested food and that turns the waste into semisolid feces, or stool

The End of the Line

Material that can't be absorbed into the blood is pushed into the large intestine. The **large intestine** is the organ of the digestive system that stores, compacts, and then eliminates indigestible material from the body. The large intestine, shown in **Figure 8,** has a larger diameter than the small intestine. The large intestine is about 1.5 m long, and has a diameter of about 7.5 cm.

In the Large Intestine

Undigested material enters the large intestine as a soupy mixture. The large intestine absorbs most of the water in the mixture and changes the liquid into semisolid waste materials called *feces,* or *stool.*

Whole grains, fruits, and vegetables contain a carbohydrate, called *cellulose,* that humans cannot digest. We commonly refer to this material as *fiber.* Fiber keeps the stool soft and keeps material moving through the large intestine.

Reading Check How does eating fiber help digestion?

Leaving the Body

The *rectum* is the last part of the large intestine. The rectum stores feces until they can be expelled. Feces pass to the outside of the body through an opening called the *anus.* It has taken your sandwich about 24 hours to make this journey through your digestive system.

CONNECTION TO Environmental Science

Waste Away Feces and other human wastes contain microorganisms and other substances that can contaminate drinking water. Every time you flush a toilet, the water and wastes go through the sewer to a wastewater treatment plant. At the wastewater treatment plant, the disease-causing microorganisms are removed, and the clean water is released back to rivers, lakes, and streams. Find out where the wastewater treatment plants are in your area. Report to your class where their wastewater goes.

ACTIVITY

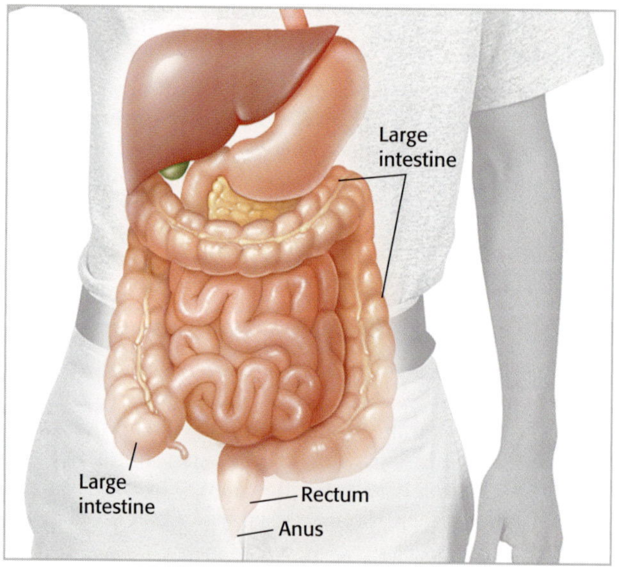

Figure 8 *The large intestine is the final organ of digestion.*

Is That a Fact!

Scientists once thought that only stess and heredity caused peptic ulcers. But in 1982, Australian scientists Robin Warren and Barry Marshall discovered a species of bacterium that was common to many ulcer patients. By infecting himself with the bacterium, succumbing to peptic ulcers, and then curing himself with antibiotics, Dr. Marshall proved that this species of bacterium was responsible for some ulcers.

Answer to Reading Check

Fiber keeps the stool soft and keeps material moving through the large intestine.

SECTION Review

Summary

- Your digestive system is a group of organs that work together to digest food so that the nutrients from food can be used by the body.
- The breaking and mashing of food is called *mechanical digestion*. Chemical digestion is the process that breaks large food molecules into simpler molecules.
- The stomach mixes food with acid and enzymes that break down nutrients. The mixture is called *chyme*.
- In the small intestine, pancreatic fluid and bile are mixed with chyme.
- From the small intestine, nutrients enter the bloodstream and are circulated to the body's cells.
- The liver makes bile, stores nutrients, and breaks down toxins.
- The large intestine absorbs water, changing liquid waste into semisolid stool, or feces.

Using Key Terms

1. Use each of the following terms in a separate sentence: *digestive system, large intestine,* and *small intestine.*

Understanding Key Ideas

2. Which of the following is NOT a function of the liver?
   - **a.** to secrete bile
   - **b.** to store nutrients
   - **c.** to detoxify chemicals
   - **d.** to compact wastes
3. What is the difference between mechanical digestion and chemical digestion?
4. What happens to the food that you eat when it gets to your stomach?
5. Describe the role of the liver, gallbladder, and pancreas in digestion.
6. Put the following steps of digestion in order.
   - **a.** Food is chewed by the teeth in the mouth.
   - **b.** Water is absorbed by the large intestine.
   - **c.** Food is reduced to chyme in the stomach.
   - **d.** Food moves down the esophagus.
   - **e.** Nutrients are absorbed by the small intestine.
   - **f.** The pancreas releases enzymes.

Critical Thinking

7. **Evaluating Conclusions** Explain the following statement: "Digestion begins in the mouth."
8. **Identifying Relationships** How would the inability to make saliva affect digestion?

Interpreting Graphics

9. Label and describe the function of each of the organs in the diagram below.

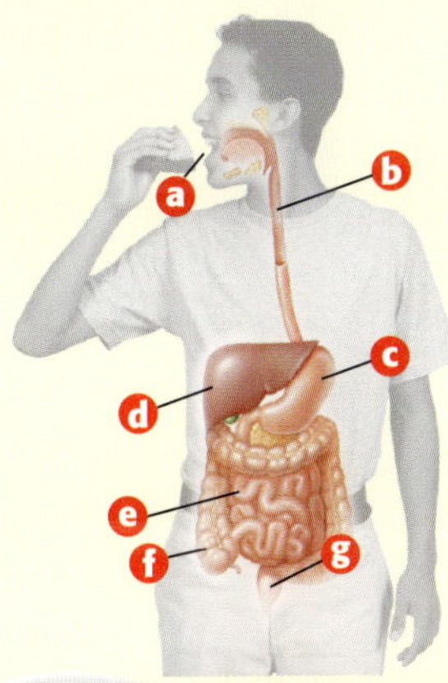

Answers to Section Review

1. Sample answer: The digestive system breaks down food so that cells can use the nutrients. The large intestine reabsorbs water from digested food. The small intestine absorbs nutrients from digested food.
2. d
3. Mechanical digestion usually involves the crushing and mashing of whole bites of food. Mechanical digestion prepares the food for chemical digestion, which is the process of changing large molecules of food into smaller molecules that can be absorbed into the bloodstream.
4. The contractions of the stomach break food into smaller pieces. Chemical digestion by enzymes, water, and acids breaks bonds in the food molecules.
5. The liver stores nutrients and makes bile, which is used in the digestion of fat. Bile is stored in a small, baglike organ called the gallbladder. The pancreas makes pancreatic juice, which contains enzymes for digesting chyme and contains bicarbonate for neutralizing the acid in chyme.
6. a, d, c, f, e, b
7. Mechanical and chemical digestion begin in the mouth. As you chew, you are physically breaking down food. Saliva contains an enzyme that begins the chemical digestion of carbohydrates.
8. Sample answer: Without saliva, swallowing would be more difficult and the chemical digestion of carbohydrates would not begin until the food reached the stomach.
9. Sample answer: a: The mouth and teeth chew food, beginning mechanical digestion. b: The esophagus squeezes food into the stomach. c: The stomach kneads food and adds acids and enzymes. d: The liver creates bile, which helps break down fat. e: The small intestine absorbs nutrients. f: The large intestine compacts wastes. g: The rectum passes food from the body.

CHAPTER RESOURCES

Chapter Resource File

- Section Quiz GENERAL
- Section Review GENERAL
- Vocabulary and Section Summary GENERAL
- Critical Thinking ADVANCED
- SciLinks Activity GENERAL
- Datasheet for Quick Lab

SECTION 2

Focus

Overview

This section introduces the structures and the functions of the urinary system. Students will learn how the kidneys filter blood. Students will also learn about some major problems of the urinary system and their causes.

Bellringer

Tell students that the blood must be cleaned regularly. Then ask students to speculate, without looking in the textbook, how the body cleans blood. Once students have written a description of the cleaning process, allow them to check their answers in their textbook.

Motivate

Discussion — GENERAL

Mongolian Gerbils Tell students that Mongolian gerbils are desert animals that never drink. Ask students how they think Mongolian gerbils obtain water. (They get water from the foods they eat.) Ask students how they think the gerbils have adapted to their habitat. (These animals have adapted to use water very efficiently, losing little or none from their lungs, skin, and urine.)

LS Verbal/Logical

SECTION 2

READING WARM-UP

Objectives

- Describe the parts and functions of the urinary system.
- Explain how the kidneys filter blood.
- Describe three disorders of the urinary system.

Terms to Learn

urinary system
kidney
nephron

READING STRATEGY

Reading Organizer As you read this section, create an outline of the section. Use the headings from the section in your outline.

The Urinary System

As blood travels through the tissues, it picks up waste produced by the body's cells. Your blood is like a train that comes to town to drop off supplies and take away garbage. If the waste is not removed, your body can actually be poisoned.

Excretion is the process of removing waste products from the body. Three of your body systems have a role in excretion. Your integumentary system releases waste products and water when you sweat. Your respiratory system releases carbon dioxide and water when you exhale. Finally, the **urinary system** contains the organs that remove waste products from your blood.

urinary system the organs that produce, store, and eliminate urine

Cleaning the Blood

As your body performs the chemical activities that keep you alive, waste products, such as carbon dioxide and ammonia, are made. Your body has to get rid of these waste products to stay healthy. The urinary system, shown in **Figure 1**, removes these waste products from the blood.

Figure 1 Urinary System

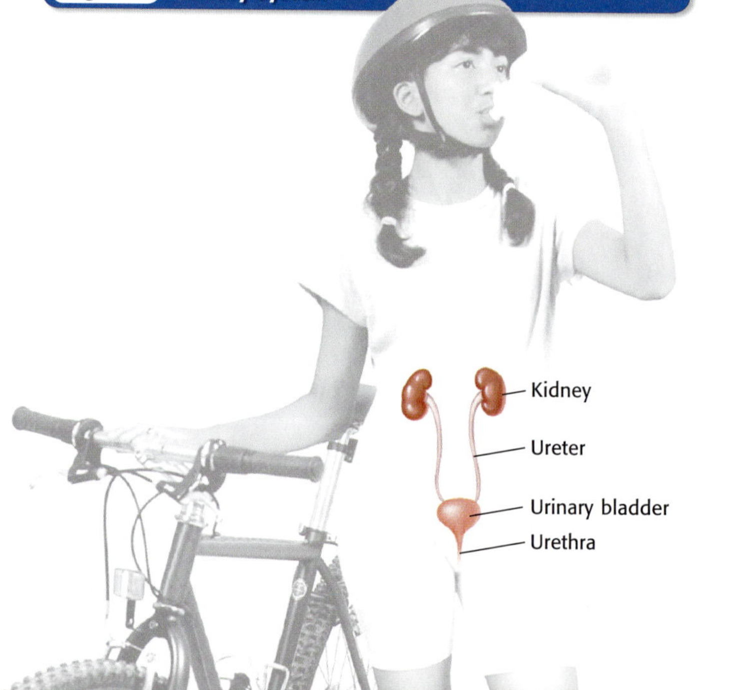

CHAPTER RESOURCES

Chapter Resource File

- Lesson Plan
- Directed Reading A BASIC
- Directed Reading B SPECIAL NEEDS

Technology

Transparencies
- Bellringer
- How the Kidneys Filter Blood

SCIENCE HUMOR

Q: What does the kidney say when it plays baseball with the other urinary organs?

A: Bladder up!

The Kidneys as Filters

The **kidneys** are a pair of organs that constantly clean the blood. Your kidneys filter about 2,000 L of blood each day. Your body holds only 5.6 L of blood, so your blood cycles through your kidneys about 350 times per day!

Inside each kidney, shown in **Figure 2,** are more than 1 million nephrons. **Nephrons** are microscopic filters in the kidney that remove wastes from the blood. Nephrons remove many harmful substances. One of the most important substances removed by nephrons is urea (yoo REE uh), which contains nitrogen and is formed when cells use protein for energy.

Reading Check How are nephrons related to kidneys? *(See the Appendix for answers to Reading Checks.)*

kidney one of the pair of organs that filter water and wastes from the blood and that excrete products as urine

nephron the unit in the kidney that filters blood

Figure 2 How the Kidneys Filter Blood

1. A large artery brings blood into each kidney.
2. Tiny blood vessels branch off the main artery and pass through part of each nephron.
3. Water and other small substances, such as glucose, salts, amino acids, and urea, are forced out of the blood vessels and into the nephrons.
4. As these substances flow through the nephrons, most of the water and some nutrients are moved back into blood vessels that wrap around the nephrons. A concentrated mixture of waste materials is left behind in the nephrons.
5. The cleaned blood, which has slightly less water and much less waste material, leaves each kidney in a large vein to recirculate in the body.
6. The yellow fluid that remains in the nephrons is called *urine.* Urine leaves each kidney through a slender tube called the *ureter* and flows into the *urinary bladder,* where urine is stored.
7. Urine leaves the body through another tube called the *urethra*. *Urination* is the process of expelling urine from the body.

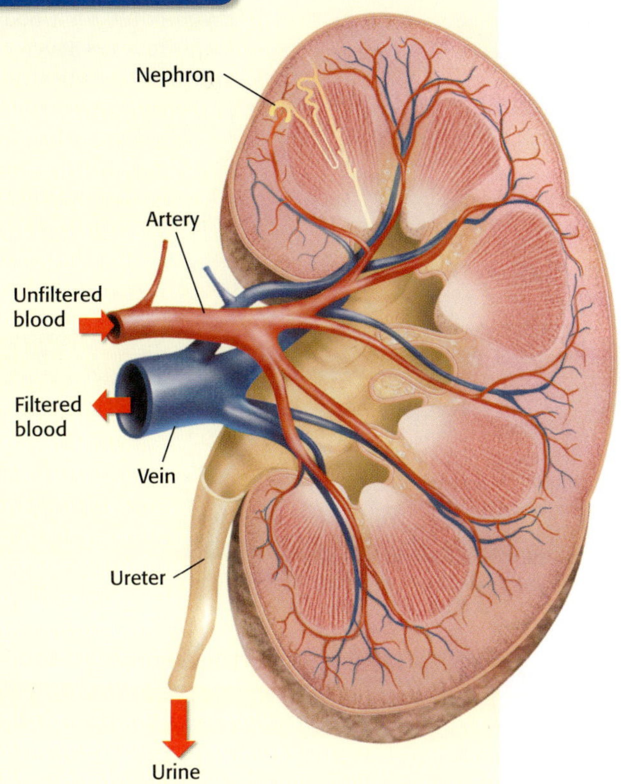

Teach

ACTIVITY — BASIC

Defining Terms Help students who have limited English proficiency with the new words on these pages. For example, write the terms *urine* and *urinary* on the board or on a transparency. As a class, define these terms and write working definitions of the terms beside each word. You may want to allow students to use their textbook for this exercise. Then point out that the term *urine* is a noun and that the term *urinary* is an adjective. English Language Learners
LS Verbal

Using the Figure — BASIC

Tracing a Path Have students trace the path of blood into, through, and out of the kidney illustrated in the figure. **LS Visual**

Demonstration — GENERAL

Kidney Structure Obtain a beef or sheep kidney from your local supermarket or butcher. Cut the kidney into two symmetrical halves. Allow students to observe and sketch the internal and external structures of the kidney and to compare the kidney with the kidney illustrated on this page. Have students wear goggles, protective gloves, and aprons during their examination of the kidney. Students should keep their hands away from their face and eyes and should wash their hands after examining the kidney. Dispose of the kidney as you would any other biohazard. Following the examination of the kidney, ask students why this kidney is so similar to a human kidney. (Both kidneys are mammalian.)
LS Visual/Logical

MISCONCEPTION ALERT

Water in Foods Students may think that they get water only from the water that they drink. But many foods, such as fruits and vegetables, supply some of the water that people need to carry on life processes.

Answer to Reading Check

Nephrons are microscopic filters inside the kidney.

Close

Reteaching — BASIC

Cooling You can demonstrate how sweat helps cool the body. Have students put a few drops of water on one arm. Tell students to blow air on their dry arm and then to blow air over the water on the other arm. Ask students: "Which arm was cooler?" (The arm with water was probably cooler. As the water evaporates from the arm, it takes thermal energy away from the skin on the arm.) English Language Learners
LS Kinesthetic

Quiz — GENERAL

Ask students whether these statements are true or false.

1. The main function of the urinary system is to remove wastes from the blood. (true)
2. The kidneys are made up of small filtering factories called *ureas*. (false)

Alternative Assessment — GENERAL

Concept Mapping Have students define the new terms in this section and use the terms to construct a concept map. Make sure the maps contain appropriate words and phrases linking the terms. LS Logical

Answer to Reading Check

Diuretics are chemicals that cause the kidneys to make more urine.

Figure 3 *Drinking water when you exercise helps replace the water you lose when you sweat.*

Water In, Water Out

You drink water every day. You lose water every day in sweat and urine. You need to get rid of as much water as you drink. If you don't, your body will swell up. So, how does your body keep the water levels in balance? The balance of fluids is controlled by chemical messengers in the body called *hormones*.

Sweat and Thirst

When you are too warm, as the boy in **Figure 3** is, you lose a lot of water in the form of sweat. The evaporation of water from your skin cools you down. As the water content of the blood drops, the salivary glands produce less saliva. This is one of the reasons you feel thirsty.

Antidiuretic Hormone

When you get thirsty, other parts of your body react to the water shortage, too. A hormone called *antidiuretic hormone* (AN tee DIE yoo RET ik HAWR MOHN), or ADH, is released. ADH signals the kidneys to take water from the nephrons. The nephrons return the water to the bloodstream. Thus, the kidneys make less urine. When your blood has too much water, small amounts of ADH are released. The kidneys react by allowing more water to stay in the nephrons and leave the body as urine.

Diuretics

Some beverages contain caffeine, which is a *diuretic* (DIE yoo RET ik). Diuretics cause the kidneys to make more urine, which decreases the amount of water in the blood. When you drink a beverage that contains water and caffeine, the caffeine increases fluid loss. So, your body gets to use less of the water from the caffeinated beverage than from a glass of water.

Reading Check What are diuretics?

CONNECTION TO Language Arts

WRITING SKILL **Beverage Ban** During football season, a football coach insists that all members of the team avoid caffeinated beverages. Many of the players are upset by the news. Pretend that you are the coach. Write a letter to the members of the team explaining why it is better for them to drink water than to drink beverages that contain caffeine. Read the letter aloud to members of your family. Ask them how you could make your letter more convincing.

INCLUSION Strategies

• Behavior Control Issues • Attention Deficit Disorder

Divide the class into teams of three. Assign these roles within each team: "record keeper," "balloon emptier," and "discharge watcher." Give each team a balloon and a very small pebble. Have students predict how many times out of 5 the "kidney stone" will pass. Tell students:

1. Place the pebble inside the balloon.
2. Have an adult fill the balloon with water.
3. Pinch the neck of the balloon so that the water can spray out in only a small spray.
4. While pinching the neck, empty the balloon into a bowl or bucket.
5. Record whether the pebble came out.
6. Repeat the above steps four times.
7. Compare the results with your predictions.

LS Kinesthetic English Language Learners

Urinary System Problems

The urinary system regulates body fluids and removes wastes from the blood. Any problems with water regulation can become dangerous for your body. Some common urinary system problems are described below.

- **Bacterial Infections** Bacteria can get into the bladder and ureters through the urethra and cause painful infections. Infections should be treated early, before they spread to the kidneys. Infections in the kidneys can permanently damage the nephrons.
- **Kidney Stones** Sometimes, salts and other wastes collect inside the kidneys and form kidney stones like the one in **Figure 4.** Some kidney stones interfere with urine flow and cause pain. Most kidney stones pass naturally from the body, but sometimes they must be removed by a doctor.
- **Kidney Disease** Damage to nephrons can prevent normal kidney functioning and can lead to kidney disease. If a person's kidneys do not function properly, a kidney machine can be used to filter waste from the blood.

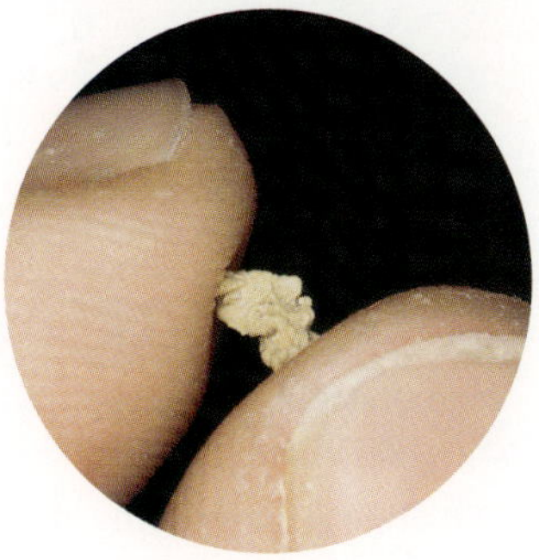

Figure 4 *This kidney stone had to be removed from a patient's urinary system.*

SECTION Review

Summary

- The urinary system removes liquid waste as urine. The filtering structures in the kidney are called *nephrons.*
- Most of the water in the blood is returned to the bloodstream. Urine passes through the ureter, into the bladder, and out of the body through the urethra.
- Disorders of the urinary system include infections, kidney stones, and kidney disease.

Using Key Terms

1. In your own words, write a definition for the term *urinary system.*

Understanding Key Ideas

2. Which event happens first?
   a. Water is absorbed into blood.
   b. A large artery brings blood into the kidney.
   c. Water enters the nephrons.
   d. The nephron separates water from wastes.
3. How do kidneys filter blood?
4. Describe three disorders of the urinary system.

Math Skills

5. A study has shown that 75% of teenage boys drink 34 oz of soda per day. How many 12 oz cans of soda would a boy drink in a week if he drank 34 oz per day?

Critical Thinking

6. **Applying Concepts** Which of the following contains more water: the blood going into the kidney or the blood leaving it?
7. **Predicting Consequences** When people have one kidney removed, their other kidney can often keep their blood clean. But the remaining kidney often changes. Predict how the remaining kidney may change to do the work of two kidneys.

SciLinks. NSTA Developed and maintained by the National Science Teachers Association

For a variety of links related to this chapter, go to www.scilinks.org

Topic: The Urinary System; Urinary System Ailments

SciLinks code: HSM1583; HSM1584

Answers to Section Review

1. Sample answer: The urinary system helps the body filter certain wastes out of the blood.
2. b
3. Blood enters the kidney. Tiny blood vessels bring blood to the nephrons. As the blood flows through the nephrons, water and nutrients are returned to the blood. Concentrated wastes form urine and are passed to the urinary bladder.
4. Bacterial infections happen when bacteria grow in the kidneys. Kidney stones happen when salts and other wastes form stones in the kidneys. Kidney disease happens when the nephrons are so damaged that the kidney does not function properly.
5. 19.8 (almost 20 cans)
6. The blood going into the kidney contains more water. Though the kidney returns most of the water back to the blood vessels, some of the water leaves the bloodstream when the kidneys produce urine.
7. Sample answer: The remaining kidney enlarges. By enlarging, the remaining kidney is able to clean much more blood.

CHAPTER RESOURCES

Chapter Resource File

- Section Quiz GENERAL
- Section Review GENERAL
- Vocabulary and Section Summary GENERAL
- Reinforcement Worksheet BASIC

Skills Practice Lab

As the Stomach Churns

Teacher's Notes

Time Required

One 45-minute class period and another 15 minutes after 24 hours

Lab Ratings

EASY → HARD

Teacher Prep 2

Student Set-Up 2

Concept Level 2

Clean Up 2

MATERIALS

The materials listed on the student page are enough for one student or a group of 2–4 students. The two tenderizers should be available at the grocery store. You will need to examine the different brands of powdered or liquid tenderizers. You may substitute fresh papaya or pineapple juices if they are available, if both types of the other tenderizers cannot be found.

Using Scientific Methods

Skills Practice Lab

As the Stomach Churns

OBJECTIVES

Demonstrate chemical digestion in the stomach.

Investigate three forms of chemical digestion.

MATERIALS

- beef stew meat, 1 cm cubes (3)
- eyedropper
- gloves, protective
- graduated cylinder, 25 mL
- hydrochloric acid, very dilute, 0.1 M
- measuring spoon, 1/4 tsp
- meat tenderizer, commercially prepared, containing bromelain
- meat tenderizer, commercially prepared, containing papain
- tape, masking
- test tubes (4)
- test-tube marker
- test-tube rack
- water

SAFETY

The stomach, as you know, performs not only mechanical digestion but also chemical digestion. As the stomach churns, which moves the food particles around, the digestive fluids—acid and enzymes—are added to begin protein digestion.

Commercially prepared meat tenderizers contain enzymes from plants that break down, or digest, proteins. Two types of meat tenderizer are commonly available at grocery stores. One type of tenderizer contains an enzyme called *papain,* from papaya. Another type of tenderizer contains an enzyme called *bromelain,* from pineapple. In this lab, you will test the effects of these two types of meat tenderizers on beef stew meat.

Ask a Question

1. Determine which question you will answer through your experiment. That question may be one of the following: Which meat tenderizer will work faster? Which one will make the meat more tender? Will the meat tenderizers change the color of the meat or water? What might these color changes, if any, indicate?

Form a Hypothesis

2. Form a hypothesis from the question you formed in step 1. **Caution:** Do not taste any of the materials in this activity.

Test the Hypothesis

3. Identify all variables and controls present in your experiment. In your notebook, make a data table that includes these variables and controls. Use this data table to record your observations and results.

4. Label one test tube with the name of one tenderizer, and label the other test tube with the name of the other tenderizer. Label the third test tube "Control." What will the test tube labeled "Control" contain?

Safety Caution

Remind students to review all safety cautions and icons before beginning this lab activity. Caution students that HCl is a chemical they should handle very carefully. You may choose to dispense the HCl yourself. Tell students that if any HCl comes in contact with their skin, they should wash the area with plenty of water immediately and they should notify you as they wash. Tell students not to taste any materials used in this or any lab.

CHAPTER RESOURCES

Chapter Resource File

- Datasheet for Chapter Lab
- Lab Notes and Answers

Technology

Classroom Videos

- Lab Video

LabBook

- Enzymes in Action

5. Pour 20 mL of water into each test tube.
6. Use the eyedropper to add four drops of very dilute hydrochloric acid to each test tube. **Caution:** Hydrochloric acid can burn your skin. If any acid touches your skin, rinse the area with running water and tell your teacher immediately.
7. Use the measuring spoon to add 1/4 tsp of each meat tenderizer to its corresponding test tube.
8. Add one cube of beef to each test tube.
9. Record your observations for each test tube immediately, after 5 min, after 15 min, after 30 min, and after 24 h.

Analyze the Results

1. **Describing Events** Did you immediately notice any differences in the beef in the three test tubes? At what time interval did you notice a significant difference in the appearance of the beef in the test tubes? Explain the differences.
2. **Examining Data** Did one meat tenderizer perform better than the other? Explain how you determined which performed better.

Draw Conclusions

3. **Evaluating Results** Was your hypothesis supported? Explain your answer.
4. **Applying Conclusions** Many animals that sting have venom composed of proteins. Explain how applying meat tenderizer to the wound helps relieve the pain of such a sting.

Test the Hypothesis

4. The "Control" tube will contain everything that goes into the other tubes except the meat tenderizer.

Analyze the Results

1. There should be very little difference in the three test tubes at first. Differences will be mild until after 24 hours, when a significant difference will be noticed between the "Control" tube and the experimental tubes.
2. Both tenderizers should work in a similar way. If one tenderizer is noted to be more efficient, students should be able to describe how they tested the stew meat to demonstrate the difference.

Draw Conclusions

3. Answers will vary, but students should be able to explain how their hypothesis was supported or how their experimental results did not support their hypothesis.
4. Students may conclude that certain digestive enzymes break down proteins. Breaking down the protein in venom often makes the venom less harmful and less painful. Explain to students that using meat tenderizer on an insect bite is only first aid and is not intended to substitute for medical attention. Emphasize that snakebites require immediate medical attention.

CHAPTER RESOURCES

Workbooks

Whiz-Bang Demonstrations
- Liver Let Live GENERAL

Long-Term Projects & Research Ideas
- Copying the Kidney ADVANCED

CLASSROOM TESTED & APPROVED

Yvonne Brannum
Hine Junior High School
Washington, D.C.

Chapter Review

Assignment Guide	
Section	**Questions**
1	1–2, 5–6, 8, 10–17, 19–26
2	3–4, 7, 9, 18

ANSWERS

Using Key Terms

1. pancreas
2. stomach
3. kidney
4. urinary system
5. digestive system
6. large intestine

Understanding Key Ideas

7. c
8. d
9. c
10. d
11. b
12. a
13. d
14. c
15. a

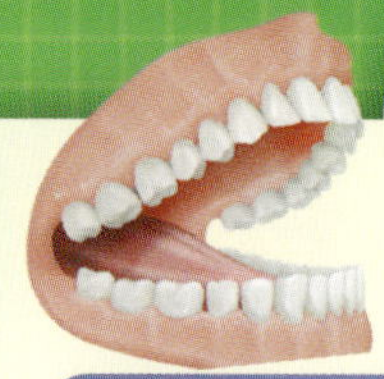

Chapter Review

USING KEY TERMS

Complete each of the following sentences by choosing the correct term from the word bank.

pancreas	digestive system
large intestine	stomach
kidney	small intestine
nephron	urinary system

1. The ____ secretes juices into the small intestine.
2. The saclike organ at the end of the esophagus is called the ____.
3. The ___ is an organ that contains millions of nephrons.
4. A group of organs that removes waste from the blood and excretes it from the body is called the ____.
5. The ____ is a group of organs that work together to break down food.
6. Indigestible material is formed into feces in the ____.

UNDERSTANDING KEY IDEAS

Multiple Choice

7. The hormone that signals the kidneys to make less urine is
   a. urea. **c.** ADH.
   b. caffeine. **d.** ATP.
8. Which of the following organs aids digestion by producing bile?
   a. stomach **c.** small intestine
   b. pancreas **d.** liver
9. The part of the kidney that filters the blood is the
   a. artery. **c.** nephron.
   b. ureter. **d.** urethra.
10. The fingerlike projections that line the small intestine are called
    a. emulsifiers.
    b. fats.
    c. amino acids.
    d. villi.
11. Which of the following is NOT part of the digestive tract?
    a. mouth **c.** stomach
    b. kidney **d.** rectum
12. The soupy mixture of food, enzymes, and acids in the stomach is called
    a. chyme. **c.** urea.
    b. villi. **d.** vitamins.
13. The stomach helps with
    a. storing food.
    b. chemical digestion.
    c. physical digestion.
    d. All of the above
14. The gall bladder stores
    a. food. **c.** bile.
    b. urine. **d.** villi.
15. The esophagus connects the
    a. pharynx to the stomach.
    b. stomach to the small intestine.
    c. kidneys to the nephrons.
    d. stomach to the large intestine.

Short Answer

16 Why is it important for the pancreas to release bicarbonate into the small intestine?

17 How does the structure of the small intestine help the small intestine absorb nutrients?

18 What is a kidney stone?

CRITICAL THINKING

19 **Concept Mapping** Use the following terms to create a concept map: *teeth, stomach, digestion, bile, saliva, mechanical digestion, gallbladder,* and *chemical digestion.*

20 **Predicting Consequences** How would digestion be affected if the liver were damaged?

21 **Analyzing Processes** When you put a piece of carbohydrate-rich food, such as bread, a potato, or a cracker, into your mouth, the food tastes bland. But if this food sits on your tongue for a while, the food will begin to taste sweet. What digestive process causes this change in taste?

22 **Making Comparisons** The recycling process for one kind of plastic begins with breaking the plastic into small pieces. Next, chemicals are used to break the small pieces of plastic down to its building blocks. Then, those building blocks are used to make new plastic. How is this process both like and unlike human digestion?

INTERPRETING GRAPHICS

The bar graph below shows how long the average meal spends in each portion of your digestive tract. Use the graph below to answer the questions that follow.

Length of Time in Digestive Organs

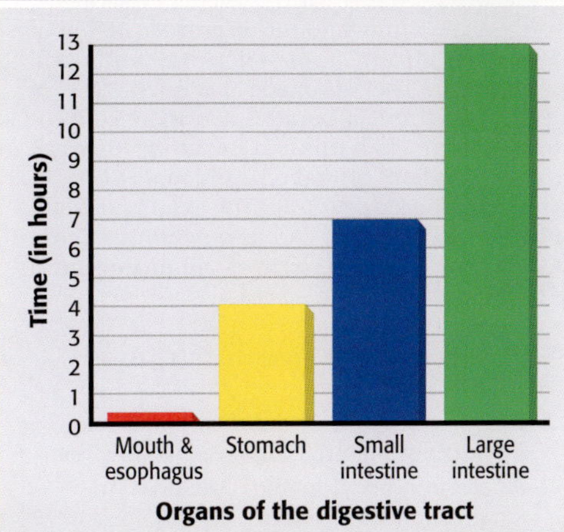

23 In which part of your digestive tract does the food spend the longest amount of time?

24 On average, how much longer does food stay in the small intestine than in the stomach?

25 Which organ mixes food with special substances to make chyme? Approximately how long does food remain in this organ?

26 Bile breaks large fat droplets into very small droplets. How long is the food in your body before it comes into contact with bile?

16. Bicarbonate neutralizes the acidic chyme coming in from the stomach. By neutralizing the acid, the bicarbonate protects the lining of the small intestine.

17. The long length, folds, and villi increase the surface area that comes into contact with the food that is being digested in the small intestine, allowing more nutrients to be absorbed.

18. a stone made of salts and the wastes that sometimes collect in the kidney

Critical Thinking

19. 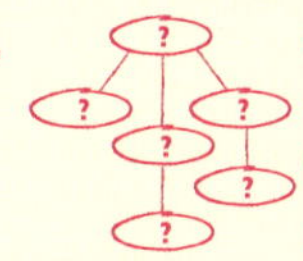

An answer to this exercise can be found at the end of this book.

20. Sample answer: Digestion would be poor. The liver makes bile, which aids in the digestion of fats. If the liver were damaged, it might not make bile, and so the body could not digest fats well.

21. Sample answer: Saliva causes chemical digestion of the carbohydrate-rich food, breaking it down into simple sugars, which taste sweet.

22. Sample answer: The process of breaking down the plastic mechanically and chemically so that its building blocks can be reused is similar to the way the body digests food. But the recycling process does not result in the growth or repair of a living organism, so recycling is different from digestion.

Interpreting Graphics

23. in the large intestine

24. 3 hours

25. The stomach mixes food with enzymes and acid to make a soupy mixture called chyme. Food stays in the stomach for approximately 4 hours.

26. just over 4 hours

CHAPTER RESOURCES

Chapter Resource File

- Chapter Review GENERAL
- Chapter Test A GENERAL
- Chapter Test B ADVANCED
- Chapter Test C SPECIAL NEEDS
- Vocabulary Activity GENERAL

Workbooks

Study Guide
- Assessment resources are also available in Spanish.

Standardized Test Preparation

Teacher's Note

To provide practice under more realistic testing conditions, give students 20 minutes to answer all of the questions in this Standardized Test Preparation.

MISCONCEPTION ALERT

Answers to the standardized test preparation can help you identify student misconceptions and misunderstandings.

READING

Passage 1

1. D
2. F

TEST DOCTOR

Question 1: Students may not understand that solutions have higher concentrations when the amount of dissolved material is the same in less liquid. Students may think the reverse is true: increasing the amount of liquid increases the concentration.

Question 2: Only answer F contains information contained in the passage. Though the other answers may be true, they are not facts from the passage.

Standardized Test Preparation

READING

Read the passage below. Then, read each question that follows the passage. Decide which is the best answer to each question.

Passage 1 When you lose water, your blood becomes more concentrated. Think about how you make a powdered drink, such as lemonade. If you use the same amount of powder in 1 L of water as you do in 2 L of water, the drinks will taste different. The lemonade made with 1 L of water will be stronger because it is more concentrated. Losing water through sweating increases the concentration of sodium and potassium in your blood. The kidneys force the extra potassium out of the blood stream and into nephrons. From the nephrons, the potassium is eliminated from the body in urine.

1. The words *more concentrated* in this passage refer to
   - **A** the same amount of water with different amounts of material dissolved in it.
   - **B** small amounts of material dissolved in small amounts of water.
   - **C** large amounts of material dissolved in large amounts of water.
   - **D** a given amount of material dissolved in a smaller amount of water.

2. Which of the following statements is a fact from the passage?
   - **F** Blood contains both potassium and sodium.
   - **G** Losing too much sodium is dangerous.
   - **H** Potassium and sodium can be replaced by drinking an exercise drink.
   - **I** Tears contain sodium.

Passage 2 Three major types of nutrients—carbohydrates, proteins, and fats—make up most of the food you eat. Chemical substances called *enzymes* break these nutrients into smaller particles for the body to use. For example, proteins, which are chains of smaller molecules called *amino acids,* are too large to be absorbed into the bloodstream. So, enzymes cut the chain of amino acids. These amino acids are small enough to pass into the bloodstream to be used by the body.

1. According to the passage, what is a carbohydrate?
   - **A** an enzyme
   - **B** a substance made of amino acids
   - **C** a nutrient
   - **D** the only substance in a healthy diet

2. Which of the following statements is a fact from the passage?
   - **F** Carbohydrates, fats, and proteins are three major types of nutrients.
   - **G** Proteins are made of fats and carbohydrates.
   - **H** Some enzymes create chains of proteins.
   - **I** Fats are difficult to digest.

3. Which of the following can be inferred from the passage?
   - **A** To be useful to the body, nutrients must be small enough to enter the bloodstream.
   - **B** Carbohydrates are made of amino acids.
   - **C** Amino acids are made of proteins.
   - **D** Without enough protein, the body cannot grow.

Passage 2

1. C
2. F
3. A

TEST DOCTOR

Question 3: In the passage, students are told that proteins need to be broken down because they are too large to be absorbed into the bloodstream. So, students may infer that, to be useful, nutrients have to be small enough to enter the bloodstream.

INTERPRETING GRAPHICS

Use the figure below to answer the questions that follow.

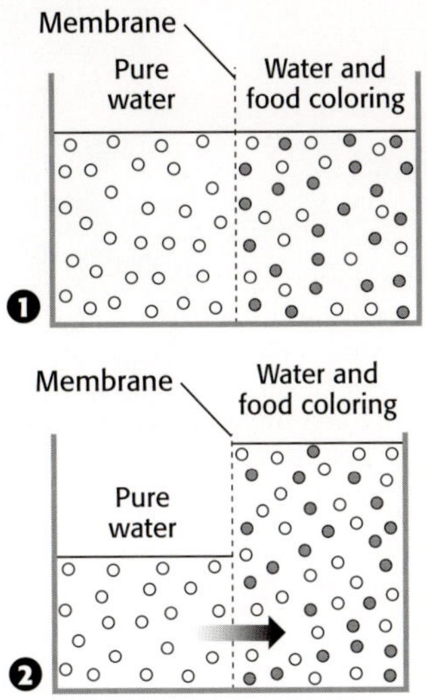

1. The container is divided by a barrier. What can you conclude from the diagram?

A Water molecules can pass through the membrane.

B Food-coloring molecules can pass through the membrane.

C Both water molecules and food-coloring molecules can pass through the membrane.

D Neither water molecules nor food-coloring molecules can pass through the membrane.

2. If the barrier has holes that separate molecules by size,

F food-coloring molecules are larger than water molecules.

G water molecules are larger than food-coloring molecules.

H water molecules and food-coloring molecules are the same size.

I the holes are smaller than both water molecules and food-coloring molecules.

3. The concentration of food-coloring molecules in the columns labeled "Water and food coloring"

A is greater in 2 than in 1.

B is greater in 1 than in 2.

C is the same in 1 and 2.

D cannot change.

MATH

Read each question below, and choose the best answer.

1. Cora is 1.5 m tall. Cora's small intestine is 6 m long. How many times longer is Cora's small intestine than her height?

A 3 times longer

B 4 times longer

C 5 times longer

D 6 times longer

2. During a water-balance study that was performed for one day, a woman drank 1,500 mL of water. The food she ate contained 750 mL of water, and her body produced 250 mL of water internally during normal body processes. She lost 900 mL of water in sweat, 1,500 mL in urine, and 100 mL in feces. Overall, how much water did she gain or lose during the day?

F She gained 1,500 mL of water.

G She lost 900 mL of water.

H She gained as much water as she lost.

I She lost twice as much water as she gained.

3. There are 6 blue marbles, 2 red marbles, and 4 green marbles in a bag. If someone selects 1 marble at random from the bag, what is the probability that the marble will be blue?

A 1/5

B 1/4

C 1/3

D 1/2

Standardized Test Preparation

INTERPRETING GRAPHICS

1. A

2. F

3. B

TEST DOCTOR

Question 1: Students interpreting the graphic may not understand that the arrow indicates that water molecules pass from one side of the membrane to the other. Students also may not realize that some membranes have holes small enough to separate molecules.

Question 3: Students may not realize that as water passes through the membrane, the water is diluting the food coloring. Increasing the number of water molecules without increasing the number of food-coloring molecules decreases the concentration of food-coloring.

MATH

1. B

2. H

3. D

TEST DOCTOR

Question 2: Students struggling with this type of question may benefit from practicing organizing information. This type of gain-and-loss question is best answered by setting up a table with gains in one column and losses in another. After adding up all the gains and losses separately, students can subtract the sum of the losses column from the sum of the gains column. This final number is the net loss or gain. A positive answer represents a gain; a negative answer represents a loss.

CHAPTER RESOURCES

Chapter Resource File

- Standardized Test Preparation GENERAL

State Resources

For specific resources for your state, visit **go.hrw.com** and type in the keyword **HSMSTR.**

Science in Action

Weird Science

Background

In many parts of the world, dangerous parasitic infections are common health problems. For example, onchocerciasis, or river blindness, is a parasitic disease spread by flies. Worldwide, almost 18 million people suffer from this disease. And dracunculiasis, or Guinea worm disease, is a painful parasitic illness caused by a long, thin worm that infests many tissues and ulcerates the soles of the feet as the worm leaves the body.

Science, Technology, and Society

Teaching Strategy-GENERAL

Invite students to brainstorm the ways that doctors diagnose diseases. List their ideas on the board. Have students identify diseases with each diagnostic method. Then have students read the feature and compare this method of diagnosis with the others on their list. Have them discuss the advantages and disadvantages of the pill camera and other diagnostic tools with which they are familiar.

Science in Action

Weird Science

Tapeworms

What if you found out that you had a constant mealtime companion who didn't want just a bite but wanted it all? And what if that companion never asked for your permission? This mealtime companion might be a tapeworm. Tapeworms are invertebrate flatworms. These flatworms are parasites. A parasite is an organism that obtains its food by living in or on another organism. A tapeworm doesn't have a digestive tract of its own. Instead, a tapeworm absorbs the nutrients digested by the host. Some tape worms can grow to be over 10 m long. Cooking beef, pork, and fish properly can help prevent people from getting tapeworms. People or animals who get tapeworms can be treated with medicines.

Social Studies ACTIVITY

WRITING SKILL The World Health Organization and the Pan American Health Organization have made fighting intestinal parasites in children a high priority. Conduct library or Internet research on Worm Busters, which is a program for fighting parasites. Write a brief report of your findings.

Science, Technology, and Society

Pill Cameras

Open wide and say "Ahhhh." When you have a problem with your mouth or teeth, doctors can examine you pretty easily. But when people have problems that are further down their digestive tract, examination becomes more difficult. So, some doctors have recently created a tiny, disposable camera that patients can swallow. As the camera travels down the digestive tract, the camera takes pictures and sends them to a tiny recorder that patients wear on their belt. The camera takes about 57,000 images during its trip. Later, doctors can review the pictures and see the pictures of the patient's entire digestive tract.

Math ACTIVITY

If a pill camera takes 57,000 images while it travels through the digestive system and takes about two pictures per second, how many hours is the camera in the body?

Answer to Social Studies Activity

Sample answer: The Worm Busters program helps eliminate parasites in Latin America. The program provides medicine to millions of children who suffer from intestinal parasites.

Answer to Math Activity

57,000 images at 2 images per second means the camera is in the body for 28,500 seconds, or 475 minutes, or almost 8 hours.

Careers

Christy Krames

Medical Illustrator Christy Krames is a medical illustrator. For 19 years, she has created detailed illustrations of the inner workings of the human body. Medical illustrations allow doctors and surgeons to share concepts, theories, and techniques with colleagues and allow students to learn about the human body.

Medical illustrators often draw tiny structures or body processes that would be difficult or impossible to photograph. For example, a photograph of a small intestine can show the entire organ. But a medical illustrator can add to the photograph an enlarged drawing of the tiny villi inside the intestine. Adding details helps to better explain how small parts of organs work together so that the organs can function.

Medical illustration requires knowledge of both art and science. So, Christy Krames studed both art and medicine in college. Often, Krames must do research before she draws a subject. Her research may include reading books, observing surgical procedures, or even dissecting a pig's heart. This research results in accurate and educational drawings of the inner body.

Language Arts Activity

WRITING SKILL Pretend you are going to publish an atlas of the human body. Write a classified advertisement to hire medical illustrators. Describe the job, and describe the qualities that the best candidates will have. As you write the ad, remember you are trying to persuade the best illustrators to contact you.

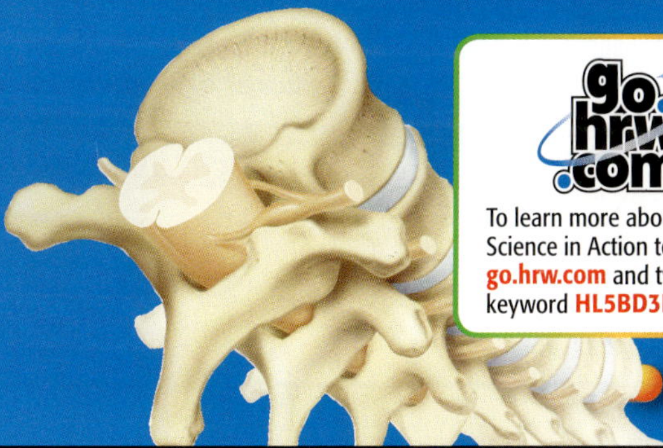

go.hrw.com

To learn more about these Science in Action topics, visit go.hrw.com and type in the keyword HL5BD3F.

Current Science

Check out Current Science® articles related to this chapter by visiting go.hrw.com. Just type in the keyword HL5CS24.

Careers

Teaching Strategy-GENERAL

Students can practice medical illustration by drawing one of their own hands. Have students use colored pencils to draw one hand as accurately as possible. They can illustrate and label the nails, pores, hairs, and wrinkles. Ask students what the most difficult part of illustrating their hand is. Student volunteers can display their drawings around the room.

Activity — GENERAL

Have interested students conduct Internet research to find out more about the Association of Medical Illustrators. Have students investigate what the group does and what it suggests for people who want to become medical illustrators.

Answer to Language Arts Activity

Student advertisements may emphasize that medical illustrators need to have a strong background in biology, artistic skills, and problem-solving ability.

Communication and Control

Chapter Planning Guide

Compression guide: To shorten instruction because of time limitations, omit Section 2.

OBJECTIVES	LABS, DEMONSTRATIONS, AND ACTIVITIES	TECHNOLOGY RESOURCES
PACING • 90 min pp. 78–87 **Chapter Opener**	SE **Start-up Activity,** p. 79 GENERAL	OSP **Parent Letter** ■ GENERAL CD **Student Edition on CD-ROM** CD **Guided Reading Audio CD** ■ TR **Chapter Starter Transparency*** VID **Brain Food Video Quiz**
Section 1 The Nervous System • Describe the relationship between the central nervous system and the peripheral nervous system. • Compare the somatic nervous system with the autonomic nervous system. • List one function of each part of the brain.	TE **Activity** Nerve Cells and Other Cells, p. 81 ADVANCED TE **Group Activity** Simulating Neuronal Impulses, p. 82 BASIC TE **Group Activity** Model the PNS, p. 83 ADVANCED TE **Group Activity** Reteaching the Brain, p. 84 BASIC TE **Connection Activity** Language Arts, p. 85 GENERAL TE **Connection Activity** Real World, p. 85 GENERAL SE **Quick Lab** Building a Neuron, p. 86 ◆ GENERAL CRF **Datasheet for Quick Lab***	CRF **Lesson Plans*** TR **Bellringer Transparency*** TR A Typical Neuron* TR What Is a Nerve?* TR Regions of the Brain* TR The Spinal Cord* CRF **SciLinks Activity*** GENERAL
PACING • 90 min pp. 88–93 **Section 2 Responding to the Environment** • List five sensations that are detected by receptors in the skin. • Describe how a feedback mechanism works. • Describe how light relates to sight. • Describe how the senses of hearing, taste, and smell work.	TE **Group Activity** Losing Your Senses, p. 88 GENERAL TE **Activity** "Ol' Three Eyes," p. 89 BASIC TE **Activity** Do You Hear Colors?, p. 89 ADVANCED TE **Activity** Pupil Action, p. 90 ◆ GENERAL SE **Quick Lab** Where's the Dot?, p. 91 GENERAL CRF **Datasheet for Quick Lab*** SE **Connection to Physics,** p. 92 GENERAL SE **Skills Practice Lab** You've Gotta Lotta Nerve, p. 98 ◆ GENERAL CRF **Datasheet for Chapter Lab*** LB **Whiz-Bang Demonstrations** Now You See It, Now You Don't* ◆ GENERAL LB **Labs You Can Eat** A Salty Sweet Experiment* ◆ GENERAL	CRF **Lesson Plans*** TR **Bellringer Transparency*** TR ***LINK TO PHYSICAL SCIENCE*** Measuring Wavelengths; Measuring Frequency* SE **Internet Activity,** p. 90 GENERAL VID **Lab Videos for Life Science**
PACING • 45 min pp. 94–97 **Section 3 The Endocrine System** • Explain why the endocrine system is important to the body. • Identify five glands of the endocrine system, and describe what their hormones do. • Describe how feedback mechanisms stop and start hormone release. • Name two hormone imbalances.	SE **Connection to Language Arts** Fight or Flight?, p. 95 GENERAL LB **Long-Term Projects & Research Ideas** Man Versus Machine* ◆ ADVANCED SE **Science in Action** Math, Social Studies, and Language Arts Activities, pp. 104–105 GENERAL	CRF **Lesson Plans*** TR **Bellringer Transparency***

PACING • 90 min

CHAPTER REVIEW, ASSESSMENT, AND STANDARDIZED TEST PREPARATION

CRF **Vocabulary Activity*** GENERAL
SE **Chapter Review,** pp. 100–101 GENERAL
CRF **Chapter Review*** ■ GENERAL
CRF **Chapter Tests A*** ■ GENERAL, **B*** ADVANCED, **C*** SPECIAL NEEDS
SE **Standardized Test Preparation,** pp. 102–103 GENERAL
CRF **Standardized Test Preparation*** GENERAL
CRF **Performance-Based Assessment*** GENERAL
OSP **Test Generator** GENERAL
CRF **Test Item Listing*** GENERAL

Online and Technology Resources

go.hrw.com

Visit **go.hrw.com** for a variety of free resources related to this textbook. Enter the keyword **HL5BD4.**

Holt Online Learning

Students can access interactive problem-solving help and active visual concept development with the *Holt Science and Technology* Online Edition available at **www.hrw.com.**

Guided Reading Audio CD

A direct reading of each chapter using instructional visuals as guideposts. For auditory learners, reluctant readers, and Spanish-speaking students. Available in English and Spanish.

KEY

- **SE** Student Edition
- **TE** Teacher Edition
- **CRF** Chapter Resource File
- **OSP** One-Stop Planner
- **LB** Lab Bank
- **TR** Transparencies
- **SS** Science Skills Worksheets
- **MS** Math Skills for Science Worksheets
- **CD** CD or CD-ROM
- **VID** Classroom Video/DVD
- * Also on One-Stop Planner
- ◆ Requires advance prep
- ■ Also available in Spanish

SKILLS DEVELOPMENT RESOURCES	SECTION REVIEW AND ASSESSMENT	STANDARDS CORRELATIONS
SE Pre-Reading Activity, p. 78 GENERAL **OSP Science Puzzlers, Twisters & Teasers** GENERAL		National Science Education Standards SAI 1; HNS 3; LS 1e
CRF Directed Reading A* ■ BASIC, **B*** SPECIAL NEEDS **CRF Vocabulary and Section Summary*** ■ GENERAL **SE Reading Strategy** Discussion, p. 80 GENERAL **SE Math Practice** Time to Travel, p. 81 GENERAL **TE Reading Strategy** Formulating Questions, p. 81 BASIC **TE Inclusion Strategies,** p. 81 **SE Connection to Chemistry** Keeping Your Balance, p. 83 GENERAL **TE Reading Strategy** Prediction Guide, p. 84 GENERAL **CRF Reinforcement Worksheet** This System is Just "Two" Nervous!* BASIC	**SE Reading Checks,** pp. 80, 81, 82, 83, 84, 85, 86 GENERAL **TE Homework,** p. 84 GENERAL **TE Homework,** p. 85 BASIC **TE Reteaching,** p. 86 BASIC **TE Quiz,** p. 86 GENERAL **TE Alternative Assessment,** p. 86 GENERAL **SE Section Review,*** p. 87 ■ GENERAL **CRF Section Quiz*** ■ GENERAL	UCP 1, 2, 3, 4, 5; SAI 1, 2; SPSP 1; LS 1a, 1d, 1e, 3a, 3b, 3c
CRF Directed Reading A* ■ BASIC, **B*** SPECIAL NEEDS **CRF Vocabulary and Section Summary*** ■ GENERAL **SE Reading Strategy** Reading Organizer, p. 88 GENERAL **MS Math Skills for Science** Multiplying and Dividing in Scientific Notation* ADVANCED **CRF Reinforcement Worksheet** The Eyes Have It* BASIC **CRF Critical Thinking** There's a Microchip in My Eye!* ADVANCED	**SE Reading Checks,** pp. 88, 89, 90, 91, 92 GENERAL **TE Homework,** p. 89 ADVANCED **TE Reteaching,** p. 92 BASIC **TE Quiz,** p. 92 GENERAL **TE Alternative Assessment,** p. 92 GENERAL **SE Section Review,*** p. 93 ■ GENERAL **CRF Section Quiz*** ■ GENERAL	UCP 2, 4, 5; SAI 1; LS 1a, 1d, 1e, 3a; *Chapter Lab:* UCP 2; SAI 1; LS 3a
CRF Directed Reading A* ■ BASIC, **B*** SPECIAL NEEDS **CRF Vocabulary and Section Summary*** ■ GENERAL **SE Reading Strategy** Discussion, p. 94 GENERAL **TE Inclusion Strategies,** p. 96 **CRF Reinforcement Worksheet** Every Gland Lends a Hand* BASIC	**SE Reading Checks,** pp. 95, 96 GENERAL **TE Reteaching,** p. 96 BASIC **TE Quiz,** p. 96 GENERAL **TE Alternative Assessment,** p. 96 GENERAL **SE Section Review,*** p. 97 ■ GENERAL **TE Homework,** p. 97 GENERAL **CRF Section Quiz*** ■ GENERAL	UCP 1; LS 1e, 1f, 3a, 3b, 3c

One-Stop Planner® CD-ROM

This convenient CD-ROM includes:
- **Lab Materials QuickList Software**
- **Holt Calendar Planner**
- **Customizable Lesson Plans**
- **Printable Worksheets**
- **ExamView® Test Generator**

CNN Student News

cnnstudentnews.com

Find the latest news, lesson plans, and activities related to important scientific events.

SciLinks NSTA

www.scilinks.org

Maintained by the **National Science Teachers Association.** See Chapter Enrichment pages for a complete list of topics.

Current Science®

Check out ***Current Science*** articles and activities by visiting the HRW Web site at **go.hrw.com.** Just type in the keyword **HL5CS25T.**

Classroom Videos

- **Lab Videos** demonstrate the chapter lab.
- **Brain Food Video Quizzes** help students review the chapter material.

4 Chapter Resources

Visual Resources

CHAPTER STARTER TRANSPARENCY

BELLRINGER TRANSPARENCIES

TEACHING TRANSPARENCIES

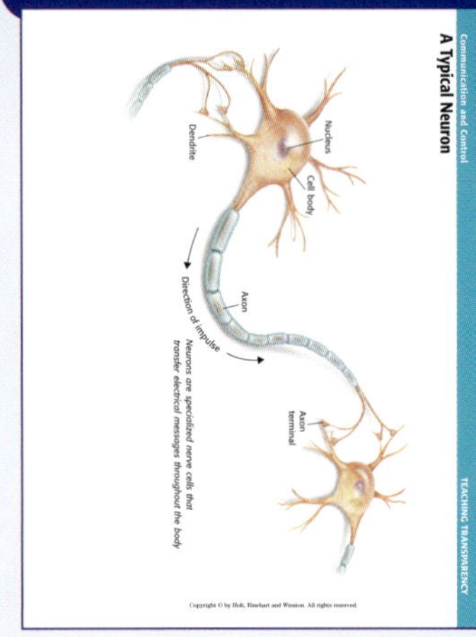

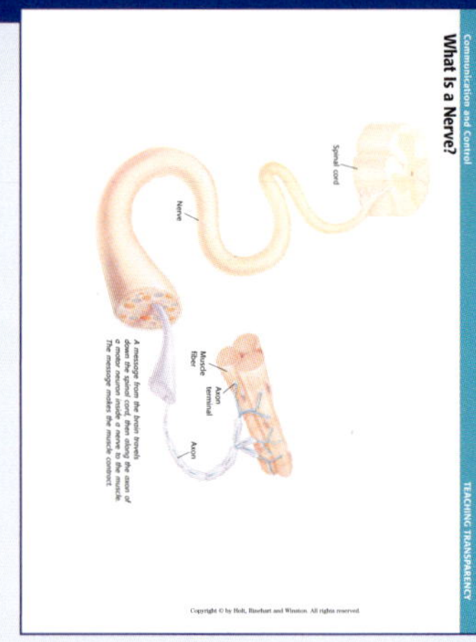

TEACHING TRANSPARENCIES

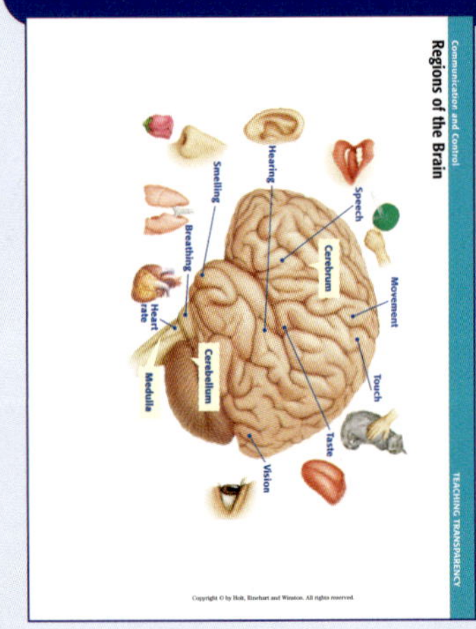

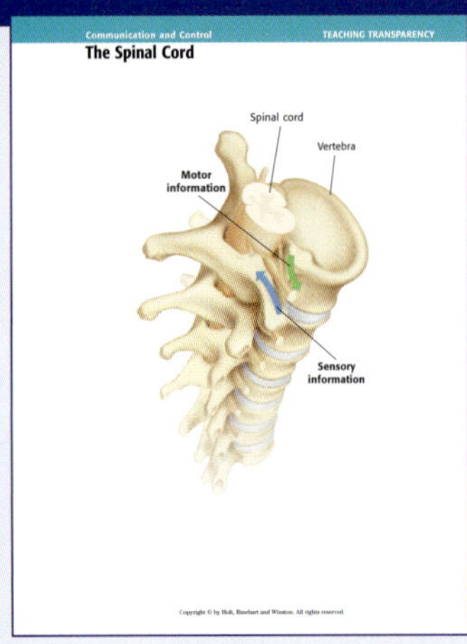

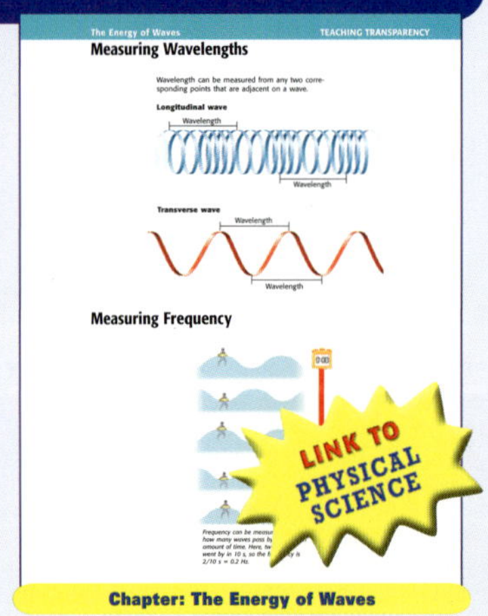

CONCEPT MAPPING TRANSPARENCY

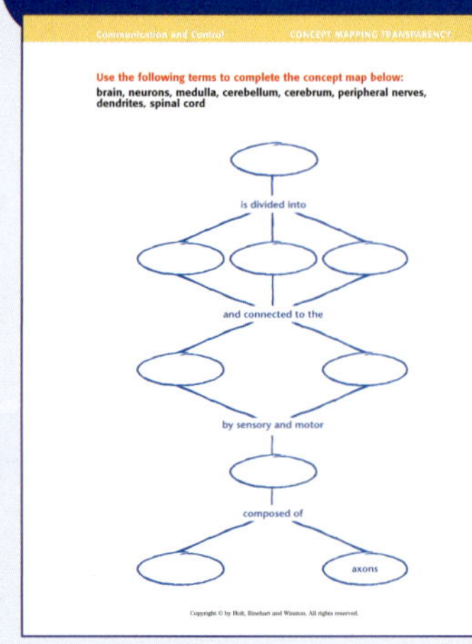

Planning Resources

LESSON PLANS

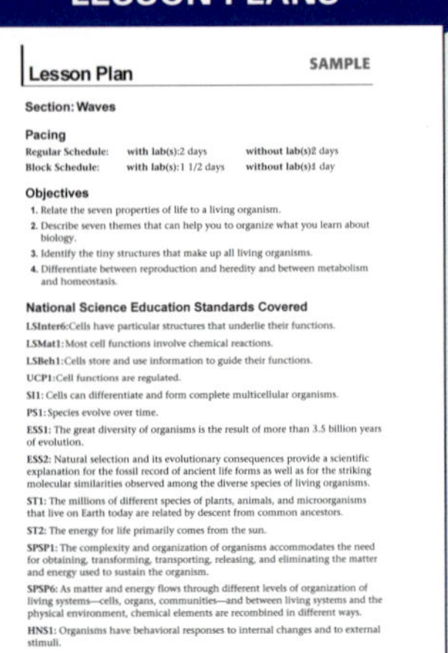

PARENT LETTER

TEST ITEM LISTING

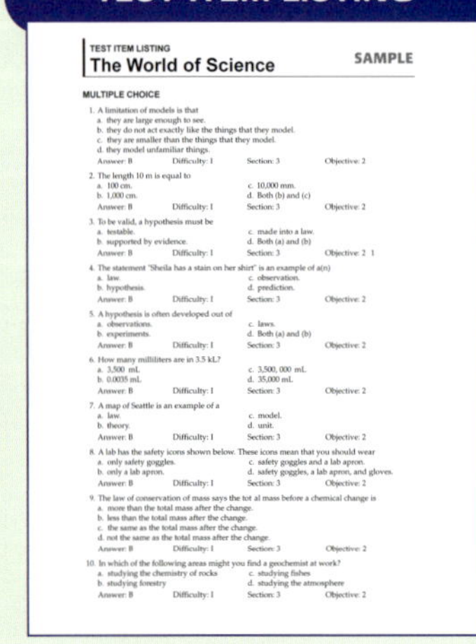

One-Stop Planner® CD-ROM

This CD-ROM includes all of the resources shown here and the following time-saving tools:

- *Lab Materials QuickList Software*
- *Customizable lesson plans*
- *Holt Calendar Planner*
- *The powerful ExamView® Test Generator*

For a preview of available worksheets covering math and science skills, see pages T12–T19. All of these resources are also on the One-Stop Planner®.

Meeting Individual Needs

DIRECTED READING A
Skills Worksheet — Directed Reading A — SAMPLE

BASIC — ALSO IN SPANISH

VOCABULARY ACTIVITY
Activity — Vocabulary Activity — SAMPLE

Getting the Dirt on the Soil

GENERAL

REINFORCEMENT
Skills Worksheet — Reinforcement — SAMPLE

The Plane Truth

BASIC

SCILINKS ACTIVITY
Activity — SciLinks Activity — SAMPLE

MARINE ECOSYSTEMS

GENERAL

DIRECTED READING B
Skills Worksheet — Directed Reading B — SAMPLE

SPECIAL NEEDS

VOCABULARY AND SECTION SUMMARY
Skills Worksheet — Vocabulary & Notes — SAMPLE

GENERAL — ALSO IN SPANISH

CRITICAL THINKING
Skills Worksheet — Critical Thinking — SAMPLE

A Solar Solution

ADVANCED

SCIENCE PUZZLERS, TWISTERS & TEASERS
Chapter 25 — Science Puzzlers, Twisters & Teasers — *Communication and Control*

GENERAL

Labs and Activities

LONG-TERM PROJECTS & RESEARCH IDEAS
Project 25 — Student Worksheet — *Man Versus Machine*

ADVANCED

WHIZ-BANG DEMONSTRATIONS
Demo 15 — Teacher-Led Demonstration — *Now You See It, Now You Don't*

GENERAL

LABS YOU CAN EAT
Lab 9 — Student Worksheet — *A Salty Sweet Experiment*

GENERAL

DATASHEETS FOR QUICK LABS
Quick Lab — Reaction to Stress — SAMPLE

DATASHEETS FOR CHAPTER LABS
Skills Practice Lab — Using Scientific Methods — SAMPLE

DATASHEETS FOR LABBOOK
Skills Practice Lab — Does It All Add Up? — SAMPLE

Review and Assessments

SECTION QUIZ
Assessment — Section Quiz — SAMPLE

GENERAL — ALSO IN SPANISH

CHAPTER REVIEW
Skills Worksheet — Chapter Review — SAMPLE

GENERAL — ALSO IN SPANISH

CHAPTER TEST B
Assessment — Chapter Test B — SAMPLE

ADVANCED

STANDARDIZED TEST PREPARATION
Assessment — Standardized Test Preparation — SAMPLE

GENERAL

SECTION REVIEW
Skills Worksheet — Section Review — SAMPLE

GENERAL — ALSO IN SPANISH

CHAPTER TEST A
Assessment — Chapter Test A — SAMPLE

GENERAL — ALSO IN SPANISH

CHAPTER TEST C
Assessment — Chapter Test C — SAMPLE

SPECIAL NEEDS

PERFORMANCE-BASED ASSESSMENT
Assessment — Performance-Based Assessment — SKILL BUILDER SAMPLE

GENERAL

4 Chapter Enrichment

This Chapter Enrichment provides relevant and interesting information to expand and enhance your presentation of the chapter material.

Section 1

The Nervous System

The Brain

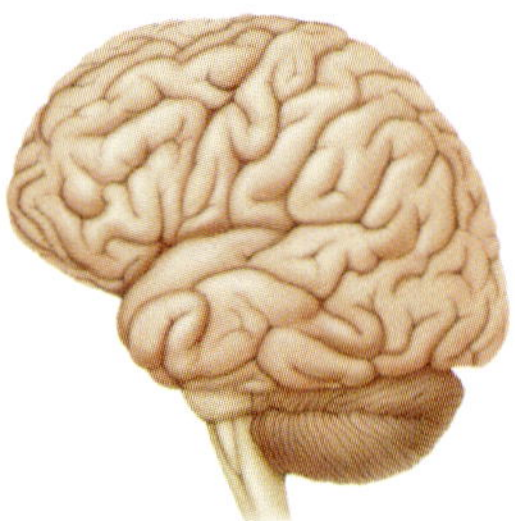

- The cerebral cortex is the main part of the exterior surface of the brain. The surface area of the brain is about 1,500 cm^2 to 2,000 cm^2, or about the size of 1 to 2 pages of a newspaper. To fit within the skull, the cortex is folded, forming *gyri* (folds) and *sulci* (grooves). Several large sulci divide the cortex into lobes: the frontal lobe, parietal lobe, occipital lobe, and temporal lobe. Each lobe has a different function.

- Computerized scanning techniques allow physicians to observe the brain in action and to detect brain abnormalities. Scanning techniques include CAT scanning, MRI scanning, radionucleotide scanning, ultrasound scanning, and PET scanning.

Is That a Fact!

- ◆ The cerebral cortex makes up more than 80% of the total human brain mass.

The Spinal Cord

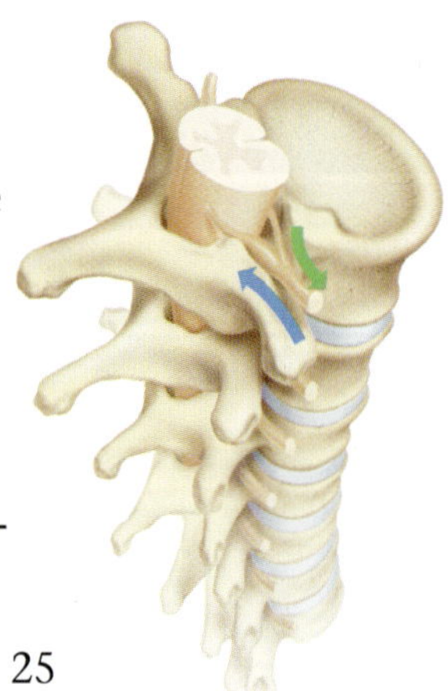

- Like the brain, the spinal cord contains both gray matter and white matter. The center of the spinal cord is made up of neuron cell bodies and is called *gray matter.* The outer layer of the spinal cord is made up of axons that traverse the spinal cord. This outer part of the spinal cord is called *white matter.*

- The spinal cord is protected by 25 bones, including the vertebrae and the sacrum. These bones are connected by joints and are separated by cartilaginous disks.

Section 2

Responding to the Environment

Hearing Loss

- There are two principal kinds of deafness: conductive deafness and sensorineural deafness.

- Conductive deafness results when transmission of sound from the outer ear to the inner ear fails. It may occur as a result of earwax buildup or damage to the middle ear.

- Sensorineural deafness results when sounds reach the inner ear but are not transmitted to the brain because of damaged inner ear structures or damaged nerves that carry information from the ear to the brain.

Is That a Fact!

- ◆ Sensorineural deafness occurs in 1 out of every 1,000 babies.

The Eye Doctor

- A variety of healthcare professionals have qualifications to treat different aspects of eye disorders and to correct vision problems.

- Ophthalmologists are physicians who specialize in the eyes. An ophthalmologist can examine eyes, prescribe corrective lenses, treat eye disorders, and perform eye surgery.

- An optometrist can examine and test eyes and can prescribe corrective lenses in the form of glasses or contact lenses.

- An optician may only fit and adjust glasses and contact lenses.

Is That a Fact!

- ◆ Using contact lenses to correct poor vision was first recorded in 1508 by Leonardo da Vinci (1452–1519).
- ◆ The first contact lens was made of glass. It covered the entire frontal surface of the eyeball. This first contact lens was made by Adolf Fick in 1887.

The Sense of Taste

- The water and enzymes in saliva break down the food and drink that we consume into small molecules and ions. After passing through pores in the taste buds, these molecules and ions stimulate small nerve endings, which send messages to the brain. These messages form our sense of taste.

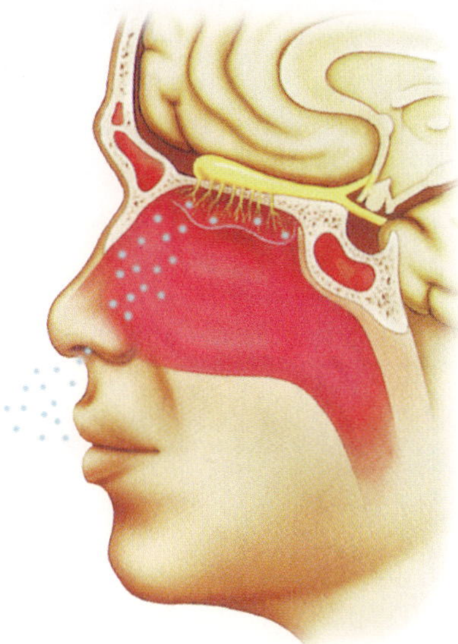

- People often lose their sense of taste when they lose their sense of smell. This loss occurs when olfactory bulbs are damaged or when the person has a stuffy nose. It is rare for a person to maintain the sense of smell and to lose the sense of taste.

Section 3

The Endocrine System

Exocrine and Endocrine Glands

- There are two main types of glands in the body: exocrine glands and endocrine glands. Exocrine glands, such as sweat glands and salivary glands, secrete substances through ducts to a local area.
- Unlike exocrine glands, endocrine glands secrete substances directly into the bloodstream (no ducts are involved). The substances secreted by endocrine glands—called hormones—are carried by the bloodstream to all parts of the body. A hormone may affect just a few cells or may have an effect on several organs or tissues.

The Pituitary Gland

- The pituitary gland is often called the *master gland* because its secretions regulate several other endocrine glands.
- About 10% of brain tumors affect the pituitary gland. Although usually benign, these tumors can have a great effect on the body because they can affect the production of the pituitary hormones.
- Because of the pituitary gland's location in the brain, enlargement of the gland can cause vision disorders by creating pressure on the optic nerve.

Diabetes

- The term *diabetes* refers to more than one disorder. The three kinds of diabetes include *diabetes insipidus* (a rare condition) and two types of *diabetes mellitus* (called *type 1* and *type 2 diabetes*). Type 1 diabetes requires regular injections of insulin. Type 2 diabetes can often be controlled by changes in diet and plenty of exercise, although insulin injections may be necessary.

Is That a Fact!

- ◆ Approximately one in every 400 to 500 children and adolescents in the United States has type 1 diabetes.
- ◆ About 16.9 million people age 20 years or older have diabetes. Of these 16.9 million people, 90% to 95% have type 2 diabetes.

SciLinks

NSTA Developed and maintained by the National Science Teachers Association

SciLinks is maintained by the National Science Teachers Association to provide you and your students with interesting, up-to-date links that will enrich your classroom presentation of the chapter.

Visit www.scilinks.org and enter the SciLinks code for more information about the topic listed.

Topic: Nervous System
SciLinks code: HSM1023

Topic: The Senses
SciLinks code: HSM1378

Topic: The Eye
SciLinks code: HSM0560

Topic: Sensory Receptors
SciLinks code: HSM1379

Topic: Hormones
SciLinks code: HSM0758

Topic: Endocrine System
SciLinks code: HSM0504

Overview

In this chapter, students will learn how the body receives, processes, and responds to information. The chapter describes the nervous system and how senses respond to stimuli. The chapter also describes how the endocrine system is part of the body's communication and control system.

Assessing Prior Knowledge

Students should be familiar with the following topics:

- cells and the activities of cells
- sensory interactions with the environment
- body structure and organization

Identifying Misconceptions

Some students may be confused about the different levels of the nervous system. A large diagram or other visual aid may help you teach this chapter by showing the central and peripheral nervous systems and the parts of the peripheral nervous system. Also, many students associate a stimulus with its source (i.e., light comes from the bulb) and do not understand that some stimuli travel as energy that from one place to another. So, students may have difficulty understanding how vision works.

4 Communication and Control

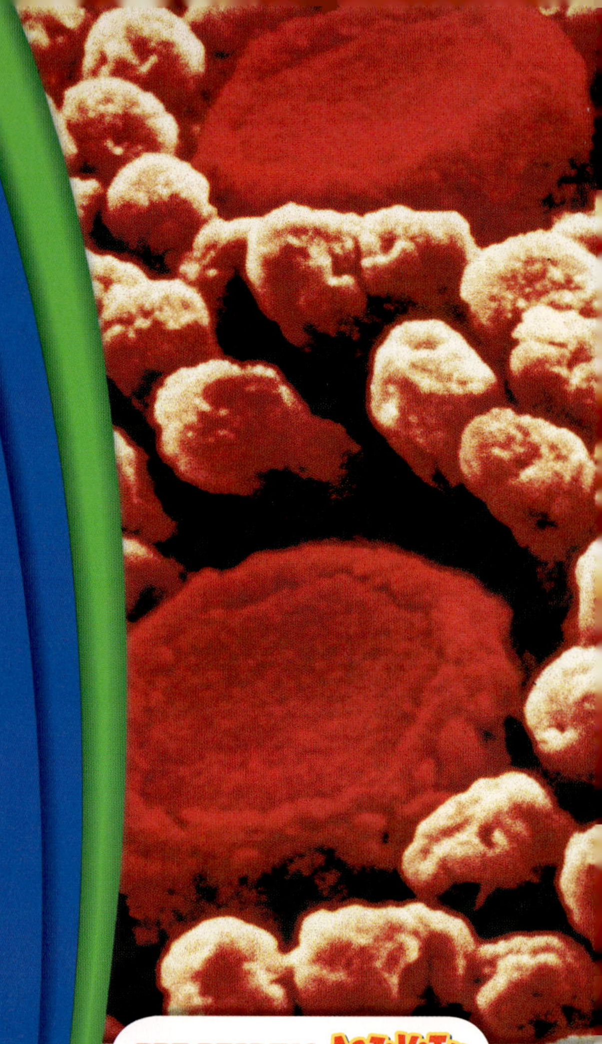

About the PHOTO

This picture may look like it shows a flower garden or a coral reef. But it really shows something much closer to home. It shows the human tongue (magnified thousands of times, of course). Those round bumps are taste buds. You use taste and other senses to gather information about your surroundings.

PRE-READING ACTIVITY

Graphic Organizer

Concept Map Before you read the chapter, create the graphic organizer entitled "Concept Map" described in the **Study Skills** section of the Appendix. As you read the chapter, fill in the concept map with details about each part or division of the nervous system. Include details about what each part or division does.

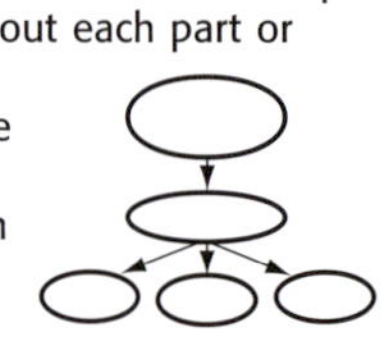

Standards Correlations

National Science Education Standards

The following codes indicate the National Science Education Standards that correlate to this chapter. The full text of the standards is at the front of the book.

Chapter Opener
SAI 1; HNS 3; LS 1e

Section 1 The Nervous System
UCP 1, 2, 3, 4, 5; SAI 1, 2; SPSP 1; LS 1a, 1d, 1e, 3a, 3b, 3c

Section 2 Responding to the Environment
UCP 2, 4, 5; SAI 1; LS 1a, 1d, 1e, 3a

Section 3 The Endocrine System
UCP 1; LS 1e, 1f, 3a, 3b, 3c

Chapter Lab
UCP 2; SAI 1; LS 3a

Chapter Review
UCP 1; LS 1a, 1d, 1e, 3b

Science in Action
UCP 2; SAI 2; ST 2; SPSP 4, 5; HNS 1, 2; LS 1a, 1d, 1e, 3a, 3b, 3c

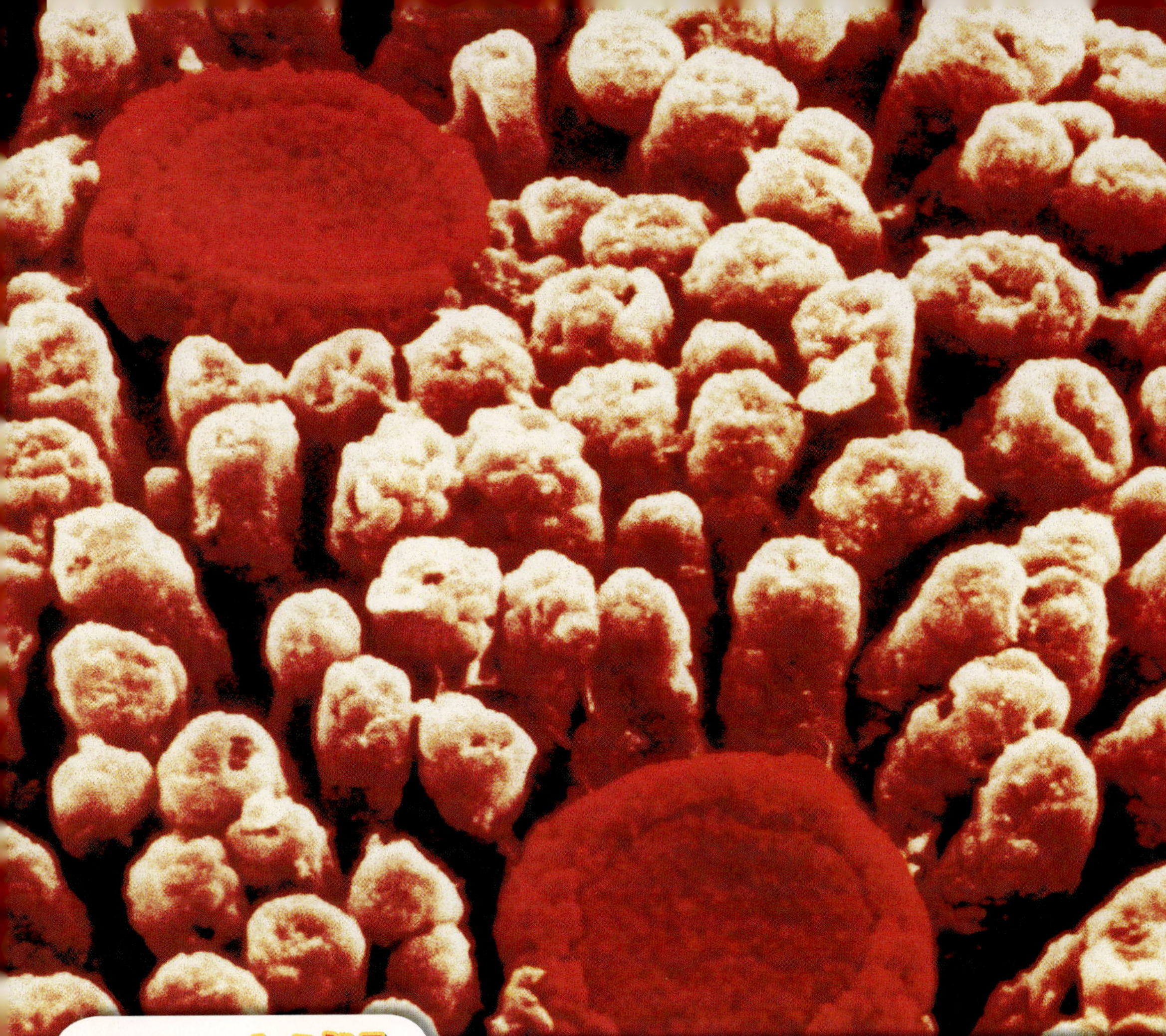

START-UP ACTIVITY

Act Fast!

If you want to catch an object, your brain sends a message to the muscles in your arm. In this exercise, you will see how long sending that message takes.

Procedure

1. Sit in a **chair** with one arm in a "handshake" position. Your partner should stand facing you, holding a **meterstick** vertically. The stick should be positioned so that it will fall between your thumb and fingers.
2. Tell your partner to let go of the meterstick without warning you. Catch the stick between your thumb and fingers. Your partner should catch the meterstick if it tips over.
3. Record the number of centimeters that the stick dropped before you caught it. That distance represents your reaction time.
4. Repeat steps 1–3 three times. Calculate the average distance.
5. Repeat steps 1–4 with your other hand.
6. Trade places with your partner, and repeat steps 1–5.

Analysis

1. Compare the reaction times of your own hands. Why might one hand react more quickly than the other?
2. Compare your results with your partner's. Why might one person react more quickly than another?

START-UP ACTIVITY

MATERIALS

For Each Group
- chair
- meterstick

Safety Caution: Remind students to handle the meterstick carefully and to keep it away from their face and far away from the faces and eyes of their classmates.

Teacher's Notes: Allow students to have a practice trial. Instruct students to look only at the ruler and not at their partner. Point out that looking at their partner could distort the results of the investigation because the partner might give a clue about when he or she will drop the meterstick.

Answers

1. Sample answer: I am right handed, and I use each hand differently. Each hand has different abilities.
2. Sample answer: One person might respond more quickly than another depending on how rested a person is, how well a person can concentrate on the activity, and how strong a person's hands are.

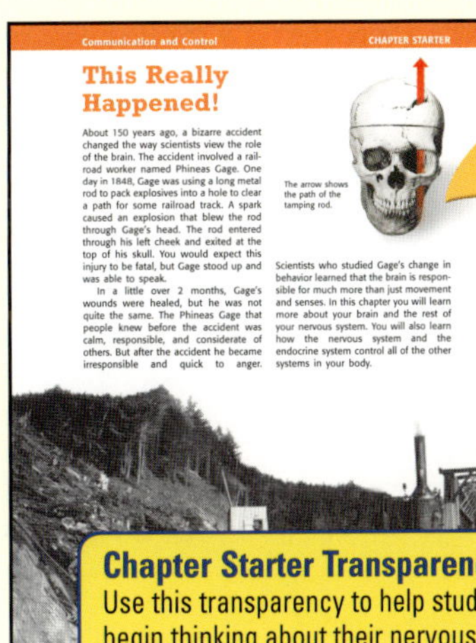

Communication and Control — CHAPTER STARTER

This Really Happened!

About 150 years ago, a bizarre accident changed the way scientists view the role of the brain. The accident involved a railroad worker named Phineas Gage. One day in 1848, Gage was using a long metal rod to pack explosives into a hole to clear a path for some railroad track. A spark caused an explosion that blew the rod through Gage's head. The rod entered through his left cheek and exited at the top of his skull. You would expect this injury to be fatal, but Gage stood up and was able to speak.

In a little over 2 months, Gage's wounds were healed, but he was not quite the same. The Phineas Gage that people knew before the accident was calm, responsible, and considerate of others. But after the accident he became irresponsible and quick to anger. Scientists who studied Gage's change in behavior learned that the brain is responsible for much more than just movement and senses. In this chapter you will learn more about your brain and the rest of your nervous system. You will also learn how the nervous system and the endocrine system control all of the other systems in your body.

The arrow shows the path of the tamping rod.

Chapter Starter Transparency
Use this transparency to help students begin thinking about their nervous system and their environment.

CHAPTER RESOURCES

Technology

Transparencies (READING SKILLS)
- Chapter Starter Transparency

Student Edition on CD-ROM

Guided Reading Audio CD
- English or Spanish

Classroom Videos
- Brain Food Video Quiz

Workbooks

Science Puzzlers, Twisters & Teasers
- Communication and Control (GENERAL)

SECTION 1

Focus

Overview

This section introduces the structures and functions of the nervous system. Students will learn the differences between the central nervous system and the peripheral nervous system, and will learn about the parts of the peripheral nervous system.

Bellringer

Have students list as many different functions of the brain as they can. Have students predict how the brain coordinates all the different activities.

Motivate

Discussion — GENERAL

Reacting to Stimuli Ask students to describe a time when they reacted quickly to something. Have them describe not only what happened but also how quickly they were able to react and what they were thinking about as they reacted. Sample experiences include jerking a hand away from a hot object, quickly catching a falling object, and extending one's hand out to brace for a fall. Based on students' experiences, lead a discussion about how quickly the nervous system is able to respond to a stimulus.
LS Intrapersonal

SECTION 1

READING WARM-UP

Objectives

- Describe the relationship between the central nervous system and the peripheral nervous system.
- Compare the somatic nervous system with the autonomic nervous system.
- List one function of each part of the brain.

Terms to Learn

central nervous system
peripheral nervous system
neuron
nerve
brain

READING STRATEGY

Discussion Read this section silently. Write down questions that you have about this section. Discuss your questions in a small group.

The Nervous System

Which of the following activities do NOT involve your nervous system: eating, playing a musical instrument, reading a book, running, or sleeping?

This is a trick question. All of these activities involve your nervous system. In fact, your nervous system controls almost everything you do.

Two Systems Within a System

The nervous system acts as the body's central command post. It has two basic functions. First, it gathers and interprets information. This information comes from inside your body and from the world outside your body. Then, the nervous system responds to that information as needed.

The nervous system has two parts: the central nervous system and the peripheral (puh RIF uhr uhl) nervous system. The **central nervous system** (CNS) is your brain and spinal cord. The CNS processes and responds to all messages coming from the peripheral nervous system. The **peripheral nervous system** (PNS) is all of the parts of the nervous system except for the brain and the spinal cord. The PNS connects all parts of the body to the CNS. The PNS uses specialized structures, called *nerves,* to carry information between your body and your CNS. **Figure 1** shows the major divisions of the nervous system.

Reading Check **Explain the difference between the CNS and the PNS.** (*See the Appendix for answers to Reading Checks.*)

central nervous system (CNS) the brain and the spinal cord

peripheral nervous system (PNS) all of the parts of the nervous system except for the brain and the spinal cord

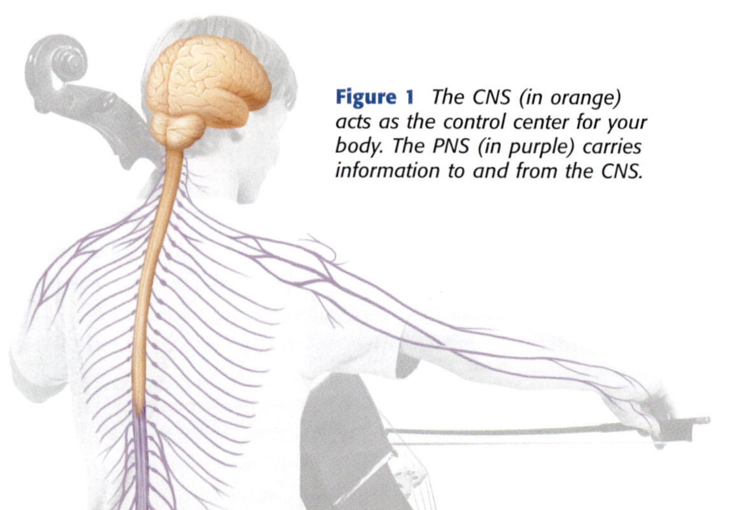

Figure 1 *The CNS (in orange) acts as the control center for your body. The PNS (in purple) carries information to and from the CNS.*

CHAPTER RESOURCES

Chapter Resource File

- Lesson Plan
- Directed Reading A BASIC
- Directed Reading B SPECIAL NEEDS

Technology

Transparencies
- Bellringer
- A Typical Neuron

Answer to Reading Check

The CNS is the brain and the spinal cord. The PNS is all of the parts of the nervous system except the brain and the spinal cord.

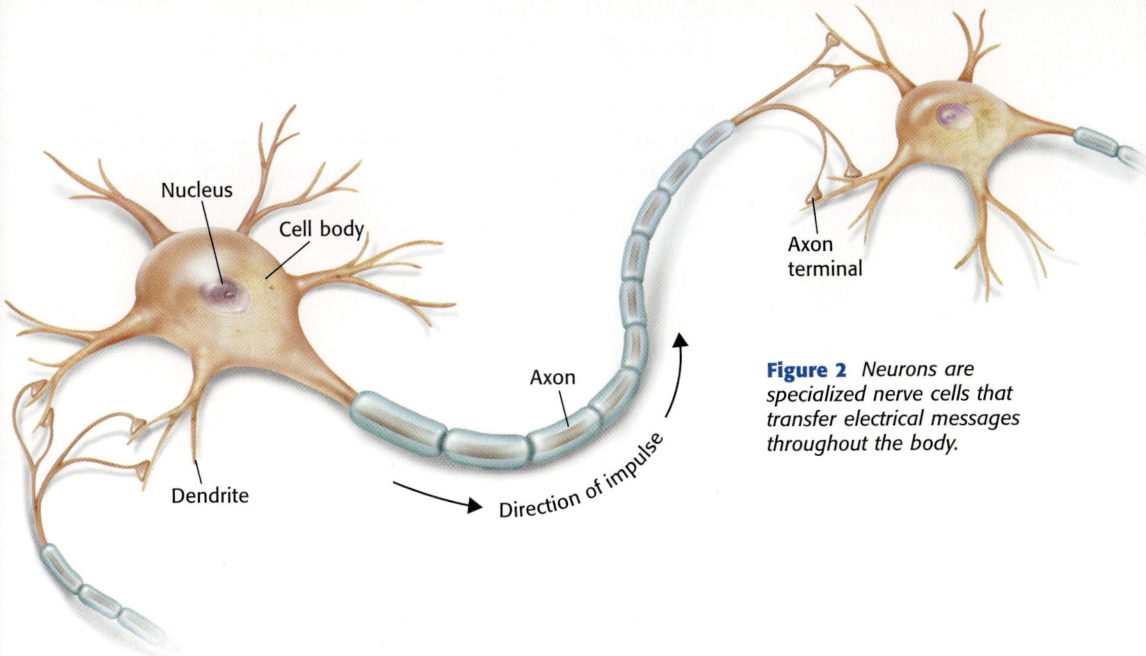

Figure 2 *Neurons are specialized nerve cells that transfer electrical messages throughout the body.*

The Peripheral Nervous System

Messages about your environment travel through the nervous system along neurons. A **neuron** (NOO RAHN) is a nerve cell that is specialized to transfer messages in the form of fast-moving electrical energy. These electrical messages are called *impulses*. Impulses may travel as fast as 150 m/s or as slow as 0.2 m/s. **Figure 2** shows a typical neuron transferring an impulse.

neuron a nerve cell that is specialized to receive and conduct electrical impulses

Neuron Structure

In many ways, a neuron is similar to other cells. A neuron has a large region in its center called the *cell body*. The cell body has a nucleus and cell organelles. But neurons also have special structures called dendrites and axons. *Dendrites* are usually short, branched extensions of the cell. Neurons receive information from other cells through their dendrites. A neuron may have many dendrites, which allows it to receive impulses from thousands of other cells.

Impulses are carried away from the cell body by axons. *Axons* are elongated extensions of a neuron. They can be very short or quite long. Some long axons extend almost 1 m from your lower back to your toes. The end of an axon often has branches that allow information to pass to other cells. The tip of each branch is called an *axon terminal*.

Reading Check In your own words, describe a neuron.

MATH PRACTICE

Time to Travel

To calculate how long an impulse takes to travel a certain distance, you can use the following equation:

$$time = \frac{distance}{speed}$$

If an impulse travels 100 m/s, about how long would it take an impulse to travel 10 m?

Teach

READING STRATEGY — BASIC

Writing **Formulating Questions** Before students read these pages, have them identify and read aloud the headings. Ask students to formulate one question that they expect the text under each heading to answer. Have students write the questions in their **science journal.** Students can answer the questions as a homework assignment. LS Logical

INCLUSION Strategies

• ***Visually Impaired***

Students with visual impairments can benefit from physical explanations of diagrams that are shown in the textbook. Assign students to partners. Pair all students who have visual impairments with full-sighted students. Have full-sighted students look at **Figure 1**, and use rulers to follow lightly the path of the central nervous system on their partners' backs. Ask them to trace both the spinal cord and the peripheral nervous system radiating out from the spinal cord. Ask partners to reverse positions and repeat the process. LS Kinesthetic

Is That a Fact!

Male canaries sing a new song every year. Male canaries replace old brain cells related to song production with new neurons each spring. In the spring, the brain-cell clusters associated with vocalization grow larger. So, males compose their new melodies and females learn to recognize the males by their new songs. In the fall, when the birds stop singing, the brain clusters shrink.

Answer to Reading Check

A neuron is a cell that has a cell body and a nucleus. A neuron also has dendrites that receive signals from other neurons and axons that send signals to other neurons.

Answer to Math Practice

T = 10 m ÷ 100 m/s = 1/10 s or 0.1 s

ACTIVITY — ADVANCED

Nerve Cells and Other Cells

Provide students with a compound light microscope and prepared slides of different kinds of cells, such as blood cells, liver cells, skin cells, and neurons. Ask students to sketch what they see on each slide. Have students describe differences they observe. LS Visual

Teach, *continued*

Group Activity — Basic

Simulating Neuronal Impulses Ask students to form a circle and hold hands. Explain that each person in the circle represents a neuron. Every left hand represents a dendrite, every body represents a cell body, and every right hand represents an axon. Join the circle, and initiate a nerve impulse by gently squeezing the hand of the student to your right. Instruct students to pass the nerve impulse to the person to their right by gently squeezing his or her hand. Once students understand the mechanics of the activity, have them call out *dendrite, cell body,* and *axon* as the impulse is passed along the circle.
LS Kinesthetic/Interpersonal English Language Learners

Discussion — General

Following an Impulse Ask students to imagine that an electrical impulse is sent from the brain along neurons, similar to the one shown in **Figure 2,** and along a nerve, as shown in **Figure 3.** Discuss with students the path of the impulse from the brain, along the neuron to the spinal cord, and then along the nerve to the muscle. Draw a diagram on the board. Then, ask students to predict what will happen in the muscle next. (The muscle will move and send a message back to the brain to tell the brain of its new position.)
LS Visual/Verbal

nerve a collection of nerve fibers (axons) through which impulses travel between the central nervous system and other parts of the body

Information Collection

Remember that neurons are a type of nerve cell that carries impulses. Some neurons are *sensory neurons.* These neurons gather information about what is happening in and around your body. They have specialized nerve endings called *receptors.* Receptors detect changes inside and outside the body. For example, receptors in your eyes detect light. Sensory neurons then send this information to the CNS for processing.

Delivering Orders

Neurons that send impulses from the brain and spinal cord to other systems are called *motor neurons.* When muscles get impulses from motor neurons, they respond by contracting. For example, motor neurons cause muscles around your eyes to contract when you are in bright light. These muscles make you squint. Squinting lets less light enter the eyes. Motor neurons also send messages to your glands, such as sweat glands. These messages tell sweat glands to start or stop making sweat.

Nerves

The central nervous system is connected to the rest of your body by nerves. A **nerve** is a collection of axons bundled together with blood vessels and connective tissue. Nerves are everywhere in your body. Most nerves have axons of both sensory neurons and motor neurons. Axons are parts of nerves, but nerves are more than just axons. **Figure 3** shows the structure of a nerve. The axon in this nerve transmits information from the spinal cord to muscle fibers.

Reading Check What is a nerve?

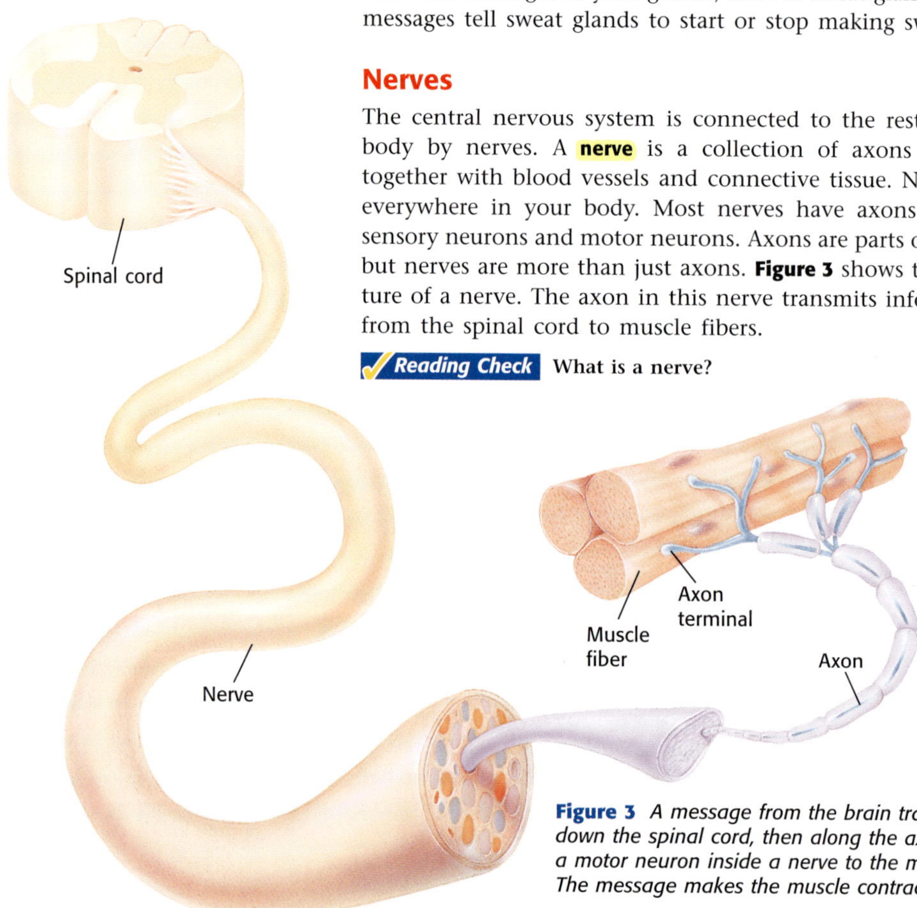

Figure 3 *A message from the brain travels down the spinal cord, then along the axon of a motor neuron inside a nerve to the muscle. The message makes the muscle contract.*

CHAPTER RESOURCES

Technology

Transparencies
- What Is a Nerve?

Answer to Reading Check

A nerve is a collection of nerve fibers, or axons, bundled together with blood vessels through which impulses travel between the central nervous system and other parts of the body.

Somatic and Autonomic Nervous Systems

Remember, the PNS connects your CNS to the rest of your body. And the PNS has two main parts—the sensory part (sensory neurons) and the motor part (motor neurons). You know that sensory nerves collect information from your senses and send that information to the CNS. You also know that motor nerves carry out the CNS's responses to that sensory information. To carry those responses, the motor part of the PNS has two kinds of nerves: somatic nerves and autonomic nerves.

Somatic Nervous System

Most of the neurons that are part of the *somatic nervous system* are under your conscious control. These are the neurons that stimulate skeletal muscles. They control voluntary movements, such as writing, talking, smiling, or jumping.

Autonomic Nervous System

Autonomic nerves do not need your conscious control. These neurons are part of the autonomic nervous system. The *autonomic nervous system* controls body functions that you don't think about, such as digestion and heart rate (the number of times your heart beats per minute).

The main job of the autonomic nervous system is to keep all the body's functions in balance. Depending on the situation, the autonomic nervous system can speed up or slow down these functions. The autonomic nervous system has two divisions: the *sympathetic nervous system* and the *parasympathetic nervous system.* These two divisions work together to keep your internal environment stable. This is called *homeostasis.* Some of these functions are shown in **Table 1.**

Reading Check Describe three functions of the PNS.

CONNECTION TO Chemistry

Keeping Your Balance The autonomic nervous system has two parts—the sympathetic division and the parasympathetic division. These parts of your nervous system help keep all of your body systems in balance. Research these two parts of the nervous system, and make a poster showing how they keep your body healthy.

ACTIVITY

Table 1 Effects of the Autonomic Nervous System on the Body

Organ	Effect of sympathetic division	Effect of parasympathetic division
Eyes	pupils dilate (grow larger; makes it easier to see objects)	pupils constrict (vision normal)
Heart	heart rate increases (increases blood flow)	heart rate slows (blood flow slows)
Lungs	bronchioles dilate (grow larger; increases oxygen in blood)	bronchioles constrict
Blood vessels	blood vessels dilate (increases blood flow except to digestion)	little or no effect
Intestines	digestion slows (reduces blood flow to stomach and intestines)	digestion returns to normal

Answer to Connection to Chemistry

The sympathetic division and the parasympathetic division of the nervous system help maintain homeostasis. Students' research should show how these parts of the nervous system work with other body systems to maintain homeostasis. Students' posters should be clear and properly labeled.

Group ACTIVITY—ADVANCED

Model the PNS The peripheral nervous system comprises the 12 pairs of cranial nerves and the 31 pairs of spinal nerves. Organize the class into four groups. Assign one of the following topics to the first three groups: somatic nervous system, sympathetic nervous system, and parasympathetic nervous system. Have groups research the parts and functions of their assigned part of the nervous system. Each group should present its findings on a poster or in a model. Ask the fourth group to take the results from the first three groups and put the results together in a poster or model of the peripheral nervous system. LS Visual

Answer to Reading Check

The PNS connects your CNS to the rest of your body, controls voluntary movements, and keeps your body's functions in balance.

CONNECTION ACTIVITY Language Arts—GENERAL

Writing **Treating Spinal Injuries** Have students conduct Internet or library research on current treatments for spinal injuries and write a story about someone who has a spinal injury and receives treatment for it. (By using the search term "spinal injury current research" or by going to the Web site of the American Spinal Injury Association, students can find information about recent developments in treating spinal cord injuries.) LS Visual

CONNECTION to Real World—GENERAL

Writing **Nerve Block** Sometimes, it isn't possible to inject anesthetic into a part of the body that needs to be anesthetized. In these cases, a *nerve block* is performed. In this procedure, anesthetic is injected into or around a nerve that serves the body part that needs to be anesthetized. Have students research nerve block anesthesia and create a pamphlet or brochure that explains the procedure to a patient who may need it. LS Verbal

SCIENCE HUMOR

Q: How do nerves shop?

A: They buy only on impulse.

Teach, continued

READING STRATEGY — GENERAL

Prediction Guide Before students read about the central nervous system, ask them whether the following statements are true or false.

1. The brain is the body's largest organ. (false)
2. The largest part of the brain is the cerebrum. (true)
3. The medulla is responsible for speech and balance. (false)
4. The spinal cord is about as big around as your thumb. (true)

Have students evaluate their answers after they have read this section. LS Verbal

Group ACTIVITY — BASIC

Reteaching the Brain If students have difficulty distinguishing the structures of the brain, have them make a life-size model of the brain. Organize the class into groups of two or three. Give each group modeling clay of different colors. Each group should have enough clay to make a life-size model of the brain. Instruct students to make labels for the parts of the brain. They should include the cerebrum, cerebellum, medulla, and the top of the spinal cord. Also, have students label the hemispheres. Allow students to use their models as you finish teaching about the brain. English Language Learners LS Kinesthetic

The Central Nervous System

The central nervous system receives information from the sensory neurons. Then it responds by sending messages to the body through motor neurons in the PNS.

The Control Center

brain the mass of nerve tissue that is the main control center of the nervous system

The largest organ in the nervous system is the brain. The **brain** is the main control center of the nervous system. Many processes that the brain controls happen automatically. These processes are called *involuntary.* For example, you couldn't stop digesting food even if you tried. On the other hand, some actions controlled by your brain are *voluntary.* When you want to move your arm, your brain sends signals along motor neurons to muscles in your arm. Then, the muscles contract, and your arm moves. The brain has three main parts—the cerebrum (suh REE bruhm), the cerebellum (SER uh BEL uhm), and the medulla (mi DUHL uh). Each part has its own job.

Reading Check What is the difference between a voluntary action and an involuntary action?

The Cerebrum

The largest part of your brain is called the *cerebrum.* It looks like a mushroom cap. This dome-shaped area is where you think and where most memories are stored. It controls voluntary movements and allows you to sense touch, light, sound, odors, taste, pain, heat, and cold.

The cerebrum has two halves, called *hemispheres.* The left hemisphere directs the right side of the body, and the right hemisphere directs the left side of the body. **Figure 4** shows some of the activities that each hemisphere controls. However, most brain activities use both hemispheres.

Figure 4 The Cerebral Hemispheres

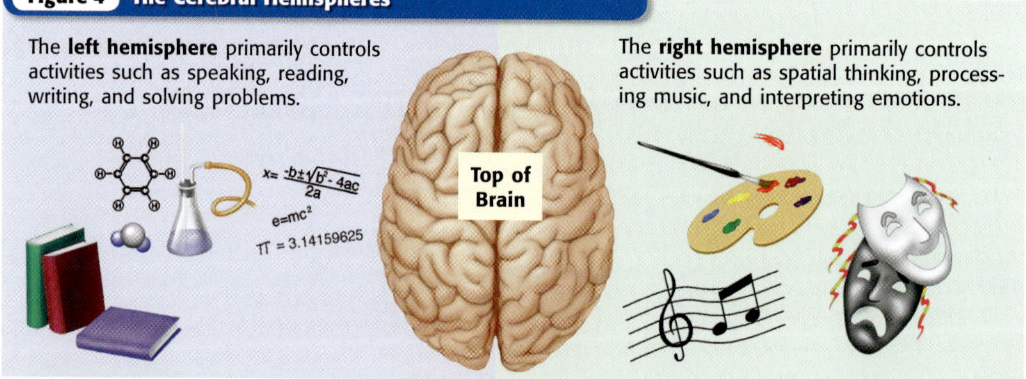

Homework — GENERAL

Writing **Dream Research** Dreaming, sleepwalking, and daydreaming are all phenomena of the brain. Have students research one of these topics and give group presentations to the class. They may use posters, signs, songs, skits, oral reports, and other techniques for their presentations. LS Verbal/Visual

Answer to Reading Check

A voluntary action is an action over which you have conscious control. Voluntary activities include throwing a ball, playing a video game, talking to your friends, taking a bite of food, and raising your hand to answer a question in class. An involuntary action is an action that happens automatically. It is an action or process over which you have no conscious control.

The Cerebellum

The second-largest part of your brain is the *cerebellum.* It lies beneath the back of the cerebrum. The cerebellum processes sensory information from your body, such as from skeletal muscles and joints. This allows the brain to keep track of your body's position. If you begin to lose your balance, the cerebellum sends impulses telling different skeletal muscles to contract. Those muscles shift a person's weight and keep a person, such as the girl in **Figure 5,** from losing her balance.

Figure 5 *Your cerebellum causes skeletal muscles to make adjustments so that you will stay upright.*

The Medulla

The *medulla* is the part of your brain that connects to your spinal cord. The medulla is about 3 cm long, and you can't live without it. The medulla controls involuntary processes, such as blood pressure, body temperature, heart rate, and involuntary breathing.

Your medulla constantly receives sensory impulses from receptors in your blood vessels. It uses this information to regulate your blood pressure. If your blood pressure gets too low, the medulla sends out impulses that tell blood vessels to tighten up. As a result, blood pressure rises. The medulla also sends impulses to the heart to make the heart beat faster or slower. **Figure 6** shows the location of the parts of the brain and some of the functions of each part.

Reading Check Explain why the medulla is important.

Figure 6 Areas of the Brain at Work

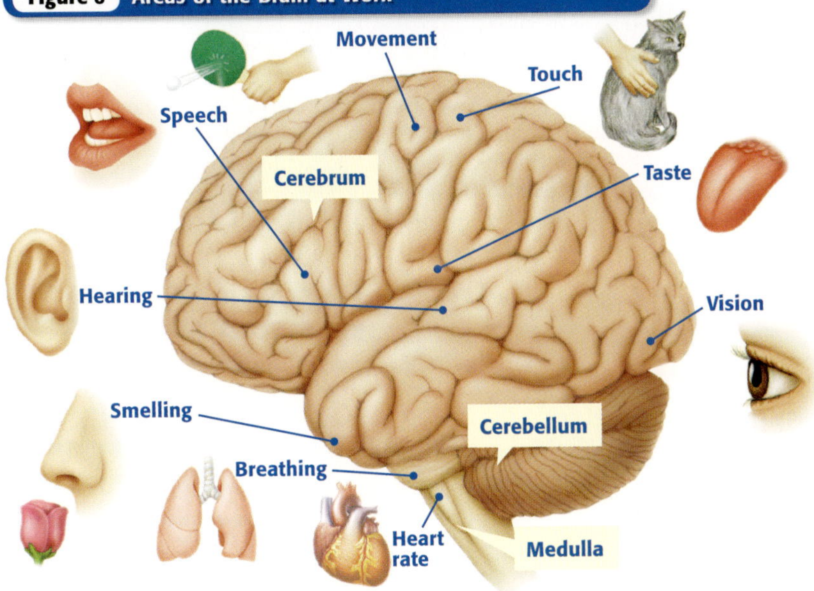

Is That a Fact!

Synapses form in a human baby's brain at the rate of 3 billion a second. At 8 months old, a baby's brain has about 1,000 trillion connections. After that, the number begins to decline. Half the connections die by the time the child reaches age 10, leaving about 500 trillion connections.

CHAPTER RESOURCES

Technology

Transparencies
- Regions of the Brain

Homework — Basic

Concept Mapping List the following terms on the board:

brain, hemispheres, cerebrum, cerebellum, medulla, spinal cord, central nervous system, peripheral nervous system, sensory neurons, and *motor neurons.*

Discuss each term with students. Then, for homework, have students use the terms to construct a concept map. **LS Verbal**

Connection Activity — Language Arts — General

Thanks for the Memories Long-term memory enables us to recall events that happened to us long ago. These memories can be triggered by a stimulus, such as a song or a smell, or they can be deliberately recalled. Recalling memories helps us to refresh them and makes them last a lifetime. Ask students to write down their earliest memory, their happiest recollection, or their most embarrassing moment. Interested students may share their favorites. **LS Verbal**

Answer to Reading Check

The medulla is important because it controls your heart rate, blood pressure, and ordinary breathing.

Connection Activity — Real World — General

Fainting Students have probably seen people faint in movies or television shows. Fainting and other forms of unconsciousness, except sleeping, are usually the result of some problem in the brain. Have students research fainting and write a short explanation about what causes fainting. (Fainting is often caused by suddenly low blood pressure and insufficient blood flow to the cerebrum.) **LS Verbal**

Close

Reteaching — BASIC

Nervous Systems Have students work in pairs to trace the outline of one student's body on butcher paper. Next, have each pair fill in the outline, using different colors for each of the parts of the nervous systems.
LS Kinesthetic/Visual English Language Learners

Quiz — GENERAL

Ask students whether each of the statements below is true or false. Have students correct false statements.

1. An axon transmits an electrical impulse to a dendrite. (true)
2. Most neurons are made up of either axons or dendrites; few neurons have both. (false)
3. The central nervous system is made up of the brain and the spinal cord. (true)

Alternative Assessment — GENERAL

Writing **Owner's Guide** Have students develop an owner's guide for their central nervous system. The guide should include information about the various components of the central nervous system and a diagram showing their location in the body. Encourage students to share their guide with the class in the form of a presentation or a poster.
LS Verbal

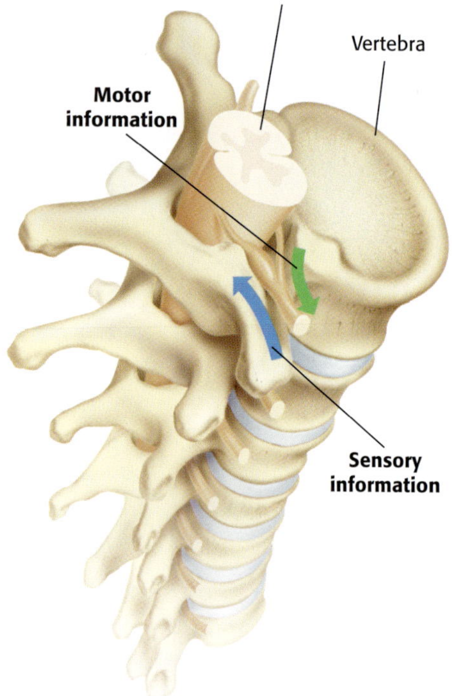

Figure 7 *The spinal cord carries information to and from the brain. Vertebrae protect the spinal cord.*

The Spinal Cord

Your spinal cord, which is part of your central nervous system, is about as big around as your thumb. The spinal cord is made of neurons and bundles of axons that pass impulses to and from the brain. As shown in **Figure 7,** the spinal cord is surrounded by protective bones called *vertebrae* (VUHR tuh BRAY).

The nerve fibers in your spinal cord allow your brain to communicate with your peripheral nervous system. Sensory neurons in your skin and muscles send impulses along their axons to your spinal cord. The spinal cord carries impulses to your brain. The brain interprets these impulses as pain, temperature, or other sensations. The brain then responds to the situation. Impulses moving from the brain down the spinal cord are relayed to motor neurons. Motor neurons carry the impulses along their axons to muscles and glands all over your body.

Reading Check **Describe the path of an impulse from the skin to the brain and the path of the response.**

Spinal Cord Injury

A spinal cord injury may block all information to and from the brain. Sensory information coming from below the injury may not get to the brain. For example, a spinal cord injury may block all sensory impulses from the feet and legs. People with such an injury would not be able to sense pain, touch, or temperature with their feet. And motor commands from the brain to the injured area may not reach the peripheral nerves. So, the person would not be able to move his or her legs.

Each year, thousands of people are paralyzed by spinal cord injuries. Many of these injuries happen in car accidents and could be avoided by wearing a seat belt. Among young people, spinal cord injuries are sometimes related to sports or other activities. These injuries might be prevented by wearing proper safety equipment.

Quick Lab

Building a Neuron

1. Your teacher will provide at least four different colors of **modeling clay.** Build a model of a neuron by using different-colored clay for the various parts of the neuron.
2. Use **tape** to attach your model neuron to a **piece of plain white paper.**
3. On the paper, label each part of the neuron. Draw an arrow from the label to the part.
4. Using a **colored pencil, marker,** or **crayon,** draw arrows showing the path of an impulse traveling in your neuron. Tell whether the impulse is a sensory impulse or a motor impulse. Then, describe what will happen when the impulse reaches its destination.

Quick Lab

MATERIALS

For Each Group

- clay, modeling
- colored pencil, marker, or crayon
- paper, white (1 piece)
- tape

Safety Caution: Students should wear a safety apron while doing this Quick Lab.

Answer to Reading Check

When someone touches your skin, an impulse that travels along a sensory neuron to your spinal cord and then to your brain is created. The response travels back from your brain to your spinal cord and then along a motor neuron to a muscle.

SECTION
Review

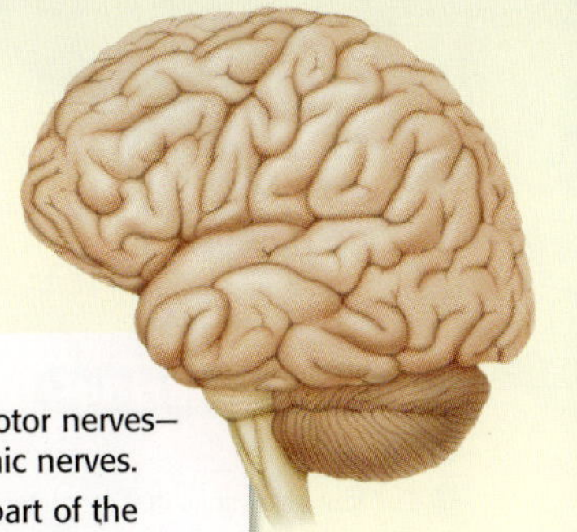

Summary

- The central nervous system (CNS) includes the brain and the spinal cord.
- The peripheral nervous system (PNS) is all the parts of the nervous system except the brain and spinal cord.
- The peripheral nervous system has nerves made up of axons of neurons.
- Sensory neurons have receptors that detect information about the body and its environment. Motor neurons carry messages from the brain and spinal cord to other parts of the body.
- The PNS has two types of motor nerves—somatic nerves and autonomic nerves.
- The cerebrum is the largest part of the brain and controls thinking, sensing, and voluntary movement.
- The cerebellum is the part of the brain that keeps track of the body's position and helps maintain balance.
- The medulla controls involuntary processes, such as heart rate, blood pressure, body temperature, and breathing.

Using Key Terms

1. In your own words, write a definition for each of the following terms: *neuron* and *nerve*.
2. Use the following terms in the same sentence: *brain* and *peripheral nervous system*.

Understanding Key Ideas

3. Someone touches your shoulder and you turn around. Which sequence do your impulses follow?
   a. motor neuron, sensory neuron, CNS response
   b. motor neuron, CNS response, sensory neuron
   c. sensory neuron, motor neuron, CNS response
   d. sensory neuron, CNS response, motor neuron
4. Describe one function of each part of the brain.
5. Compare the somatic nervous system with the autonomic nervous system.
6. Explain how a severe injury to the spinal cord can affect other parts of the body.

Critical Thinking

7. **Applying Concepts** Some medications slow a person's nervous system. These drugs are often labeled "May cause drowsiness." Explain why a person needs to know about this side effect.
8. **Predicting Consequences** Explain how your life would change if your autonomic nervous system suddenly stopped working.

Interpreting Graphics

Use the figure below to answer the questions that follow.

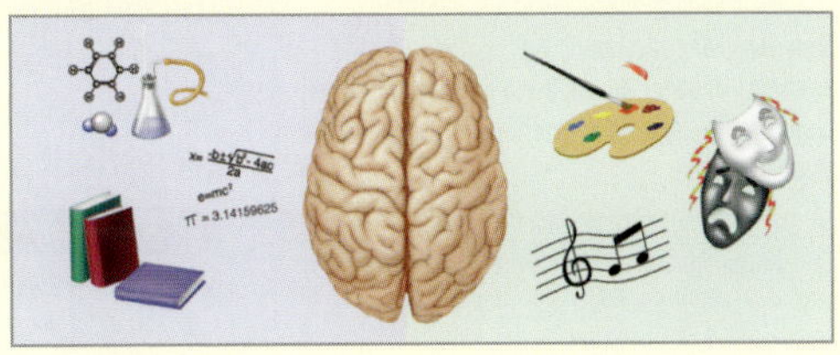

9. Which hemisphere of the brain recognizes and processes words, numbers, and letters? faces, places, and objects?
10. For a person whose left hemisphere is primarily in control, would it be easier to learn to play a new computer game by reading the rules and following instructions or by watching a friend play and imitating his actions?

SCILINKS. NSTA Developed and maintained by the National Science Teachers Association

For a variety of links related to this chapter, go to www.scilinks.org

Topic: Nervous System
SciLinks code: HSM1023

Answers to Section Review

1. Sample answer: A neuron is a special nerve cell that can receive and conduct electrical impulses. A nerve is a bundle of axons and other tissues connecting the CNS and PNS.
2. Sample answer: The peripheral nervous system consists of nerves that connect all parts of your body to your brain and spinal cord.
3. d
4. Sample answer: The cerebrum controls speech and touch. The cerebellum receives sensory input and helps you balance. The medulla controls some involuntary processes, such as heart rate and blood pressure.
5. Sample answer: The somatic nervous system controls actions that are under conscious control, such as singing or shaking hands. The autonomic nervous system controls body processes and functions that are not under your conscious control, such as pupil dilation and digestion.
6. Sample answer: A spinal injury may damage nerves in the spinal cord that send or receive messages between the brain and the body. Damage to the nerves may stop those messages and keep the person from either receiving sensory inputs or making any movements that would be controlled by the damaged nerves.
7. Sample answer: A drug that causes drowsiness may affect the way that impulses travel in neurons. Affecting the impulses may increase the time it takes a message to get to the brain or the time it takes a message to get from the brain to muscles. It is important to be aware of this side effect so that activities that require a fast reaction time, such as driving, can be avoided.
8. Sample answer: If my autonomic nervous system suddenly stopped working, I would not be able to digest my food, my eyes would not adjust to light or darkness, my heart rate and blood pressure couldn't change as they needed to, and my lungs wouldn't work properly.
9. left; right
10. Sample answer: It would be easier to learn the new game by reading the rules and following the instructions.

CHAPTER RESOURCES

Chapter Resource File

- Section Quiz GENERAL
- Section Review GENERAL
- Vocabulary and Section Summary GENERAL
- Reinforcement Worksheet BASIC
- SciLinks Activity GENERAL
- Datasheet for Quick Lab

Technology

Transparencies
- The Spinal Cord

SECTION 2

Focus

Overview

This section introduces the five senses: touch, sight, hearing, taste, and smell. Students will learn about the receptor cells that are unique to each sense.

Bellringer

Ask students to list the five senses and draw an organ associated with each sense as well as an object detected by each sense. For example, students may draw an ear and a bell to represent the sense of hearing.

Motivate

Group Activity—GENERAL

Losing Your Senses

Organize the class into groups of three or four students, and assign each group a sense. Each group should imagine what it would be like to live without that sense. Allow students 10 to 15 minutes to develop a skit or example of life without the assigned sense. LS Intrapersonal/Kinesthetic

Co-op Learning English Language Learners

Answer to Reading Check

Skin can detect pressure, temperature, pain, and vibration.

SECTION 2

Responding to the Environment

READING WARM-UP

Objectives

- List four sensations that are detected by receptors in the skin.
- Describe how a feedback mechanism works.
- Describe how light relates to sight.
- Describe how the senses of hearing, taste, and smell work.

Terms to Learn

integumentary system
reflex
feedback mechanism
retina
cochlea

READING STRATEGY

Reading Organizer As you read this section, create an outline of the section. Use the headings from the section in your outline.

You feel a tap on your shoulder. Who tapped you? You turn to look, hoping to see a friend. Your senses are on the job!

The tap produces impulses in sensory receptors on your shoulder. These impulses travel to your brain. Once the impulses reach your brain, they create an awareness called a *sensation*. In this case, the sensation is of your shoulder being touched. But you still do not know who tapped you. So, you turn around. The sensory receptors in your eyes send impulses to your brain. Now, your brain recognizes your best friend.

Sense of Touch

Touch is what you feel when sensory receptors in the skin are stimulated. It is the sensation you feel when you shake hands or feel a breeze. As shown in **Figure 1,** skin has different kinds of receptors. Each kind of receptor responds mainly to one kind of stimulus. For example, *thermoreceptors* respond to temperature change. Each kind of receptor produces a specific sensation of touch, such as pressure, temperature, pain, or vibration. Skin is part of the integumentary (in TEG yoo MEN tuhr ee) system. The **integumentary system** protects the body from damage. It includes hair, skin, and nails.

Reading Check List four sensations that your skin can detect. (*See the Appendix for answers to Reading Checks.*)

integumentary system the organ system that forms a protective covering on the outside of the body

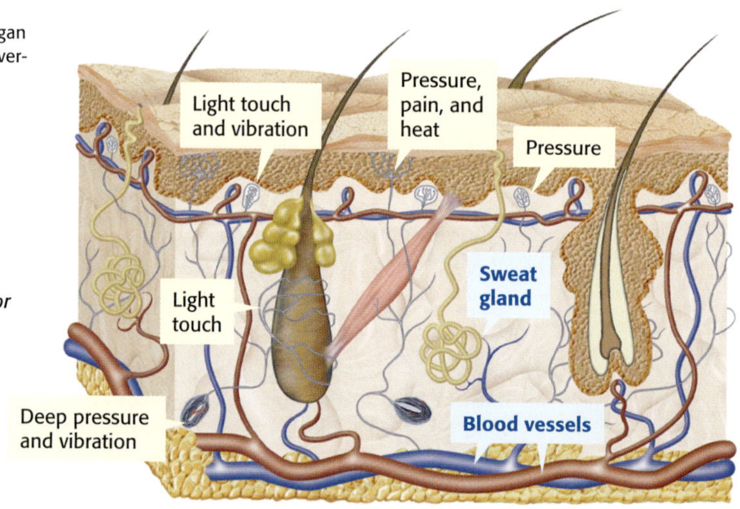

Figure 1 *Each type of receptor in your skin has its own structure and function.*

CHAPTER RESOURCES

Chapter Resource File

- Lesson Plan
- Directed Reading A BASIC
- Directed Reading B SPECIAL NEEDS

Technology

Transparencies
- Bellringer

Workbooks

Study Guide
- Multilplying and Dividing in Scientific Notation GENERAL

Responding to Sensory Messages

When you step on something sharp, as the man in **Figure 2** did, pain receptors in your foot or toe send impulses to your spinal cord. Almost immediately, a message to move your foot travels back to the muscles in your leg and foot. Without thinking, you quickly lift your foot. This immediate, involuntary action is called a **reflex.** Your brain isn't telling your leg to move. In fact, by the time the message reaches your brain, your leg and foot have already moved. If you had to wait for your brain to act, you toes might be seriously hurt!

Reading Check Why are reflexes important?

reflex an involuntary and almost immediate movement in response to a stimulus

feedback mechanism a cycle of events in which information from one step controls or affects a previous step

Feedback Mechanisms

Most of the time, the brain processes information from skin receptors. For example, on a hot day, heat receptors in your skin detect an increase in your temperature. The receptors send impulses to the brain. Your brain responds by sending messages to your sweat glands to make sweat. As sweat evaporates, it cools your body. Your brain also tells the blood vessels in your skin to dilate (open wider). Blood flow increases. Thermal energy from the blood in your skin moves to your surroundings. This also cools your body. As your body cools, it sends messages to your brain. The brain responds by sending messages to sweat glands and blood vessels to reduce their activity.

This cooling process is one of your body's feedback mechanisms. A **feedback mechanism** is a cycle of events in which information from one step controls or affects a previous step. The temperature-regulating feedback mechanism helps keep your body temperature within safe limits. This cooling mechanism works like a thermostat on an air conditioner. Once a room reaches the right temperature, the thermostat sends a message to the air conditioner to stop blowing cold air.

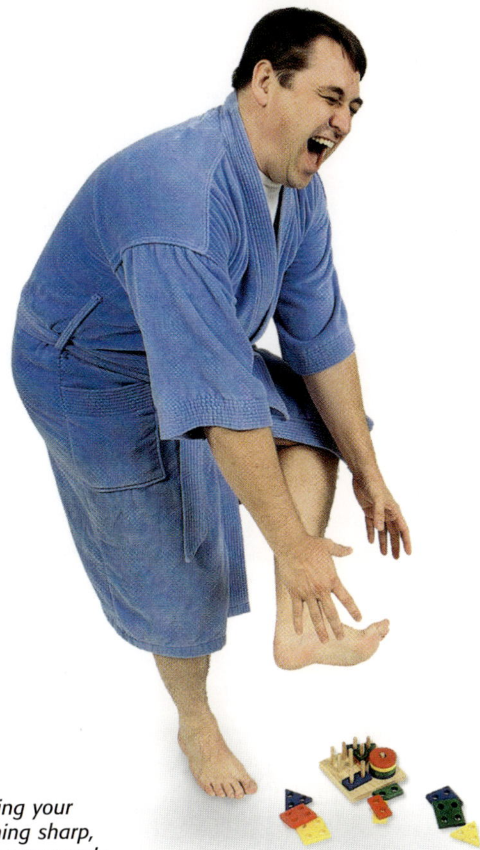

Figure 2 *A reflex, such as lifting your foot when you step on something sharp, is one way your nervous system responds to your environment.*

Teach

Homework ADVANCED

Other Reflexes Have students use the Internet or the library to research different reflexes the body has. They are probably familiar with the knee-jerk reflex, but they may not be aware of reflexes associated with eye blinks and loud noises. Have students create a visual or other presentation to show what they have learned. English Language Learners
LS Visual

CONNECTION to Environmental Science ADVANCED

Writing **Climate and Feedback** Scientists have learned that there are feedback mechanisms in the environment. Changes in the environment can produce positive or negative feedback that affects Earth's climate. Have students research feedback mechanisms and the environment and write a report explaining how the feedback mechanisms may affect humans. LS Verbal

ACTIVITY BASIC

Writing **"Ol' Three Eyes"** *Tuatara* are an ancient species of reptile. Their closest relatives became extinct 65 million years ago. One unusual feature of tuatara is that they have three eyes. The third eye is really a light-sensitive gland. Have students research tuatara and write a report on these unusual "living fossils." Encourage students to illustrate their reports. LS Verbal

ACTIVITY ADVANCED

Writing **Do You Hear Colors?** About one out of every 25,000 people has *synesthesia*. These people experience a blend of senses. For example, one person sees blue when a particular note is played on a piano. Another person gets a very bitter taste in his mouth when he makes hamburger patties with his hands. Have students conduct Internet or library research and write a report on synesthesia and how it affects people who have it.
LS Verbal

Answer to Reading Check

Reflexes are important because they can protect you from injury.

Teach, *continued*

ACTIVITY — GENERAL

Writing **Pupil Action** Ask students to write a paragraph that explains what happens to their eyes and vision when they first leave a dark movie theater on a sunny day. Students should discuss the change from dim to bright light and its effects on the pupils as well as on their ability to see. **LS Verbal**

Discussion — GENERAL

Eye Strain For us to see objects within a distance of 6 m, the muscles in the eye must work constantly to focus. Long periods of focusing on very near objects, such as a book, can tire the eye muscles and cause eyestrain. Ask students to suggest ways to avoid straining the eye muscles during these activities. (Answers may include taking breaks from the activity once an hour and looking up and allowing the eyes to relax occasionally.) **LS Logical**

Answer to Reading Check

Light strikes cells on the retina and triggers impulses in those cells. The impulses are carried to the brain, which interprets the impulses as images that you "see."

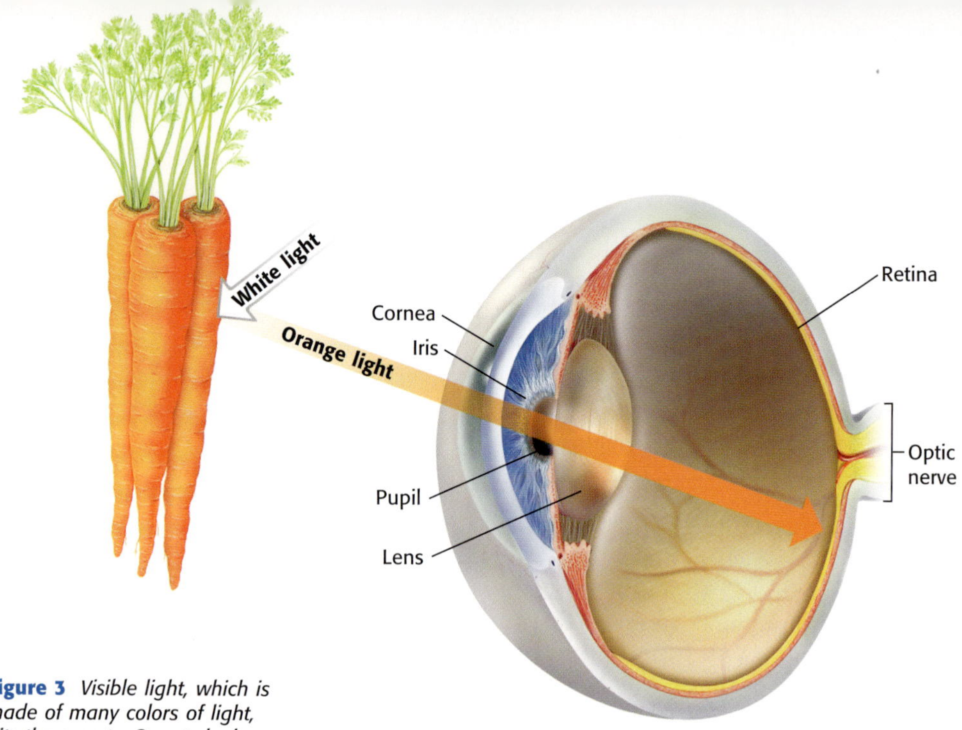

Figure 3 *Visible light, which is made of many colors of light, hits the carrots. Carrots look orange because they reflect orange light to your eyes.*

Sense of Sight

Sight is the sense that allows you to see the size, shape, motion, and color of objects around you. You see an object when it sends or reflects visible light toward your eyes. Your eyes detect this light, which enables your brain to form visual images.

Your eyes are complex sensory organs, as you can see in **Figure 3.** The front of the eye is covered by a clear membrane called the *cornea.* The cornea protects the eye but allows light to enter. Light from an object enters the front of your eye through an opening called the *pupil.* The light then travels through the lens to the back of the eye. There, the light strikes the **retina,** a layer of light-sensitive cells.

The retina is packed with photoreceptors. A *photoreceptor* is a special neuron that changes light into electrical impulses. The retina has two kinds of photoreceptors: rods and cones. Rods are very sensitive to dim light. They are important for night vision. Impulses from rods are interpreted as black-and-white images. Cones are very sensitive to bright light. Impulses from cones allow you to see fine details and colors.

Impulses from the rods and cones travel along axons. The impulses leave the back of each eye through an optic nerve. The optic nerve carries the impulses to your brain, where the impulses are interpreted as the images that you see.

Reading Check Describe how light and sight are related.

retina the light-sensitive inner layer of the eye; it receives images formed by the lens and transmits them through the optic nerve to the brain

INTERNET ACTIVITY

For another activity related to this chapter, go to **go.hrw.com** and type in the keyword **HL5BD4W.**

CHAPTER RESOURCES

Technology

Transparencies
- **LINK TO PHYSICAL SCIENCE** Measuring Wavelengths

MISCONCEPTION ALERT

Colorblindness Students may think that people who are colorblind see in black and white. People who are colorblind can usually perceive colors, but certain colors may appear very similar to one another. This similarity is caused by a lack of at least one of the three cones in the eye. Many people do not know that they are colorblind because they have learned to distinguish other differences in their perception of colors.

Reacting to Light

Your pupil looks like a black dot in the center of your eye. In fact, it is an opening that lets light enter the eye. The pupil is surrounded by the *iris,* a ring of muscle. The iris controls the amount of light that enters the eye and gives the eye its color. In bright light, the iris contracts, which makes the pupil smaller. A smaller pupil reduces the amount of light entering the eye and passing onto the retina. In dim light, the iris opens the pupil and lets in more light.

Reading Check How does your iris react to bright light?

Focusing the Light

Light travels in straight lines until it passes through the cornea and the lens. The *lens* is an oval-shaped piece of clear, curved material behind the iris. Muscles in the eye change the shape of the lens in order to focus light onto the retina. When you look at objects close to the eye, the lens becomes more curved. When you look at objects far away, the lens gets flatter.

Figure 4 shows some common vision problems. In some eyes, the lens focuses the light in front of the retina, which results in nearsightedness. If the lens focuses the light just behind the retina, the result is farsightedness. Glasses, contact lenses, or surgery can usually correct these vision problems.

Quick Lab

Where's the Dot?

1. Hold your **book** at arm's length, and close your right eye. Focus your left eye on the black dot below.

○ ●

2. Slowly move the book toward your face until the white dot disappears. You may need to try a few times to get this result. The white dot doesn't always disappear for every person.
3. Describe your observations.
4. Use the library or the Internet to research the optic nerve and to find out why the white dot disappears.

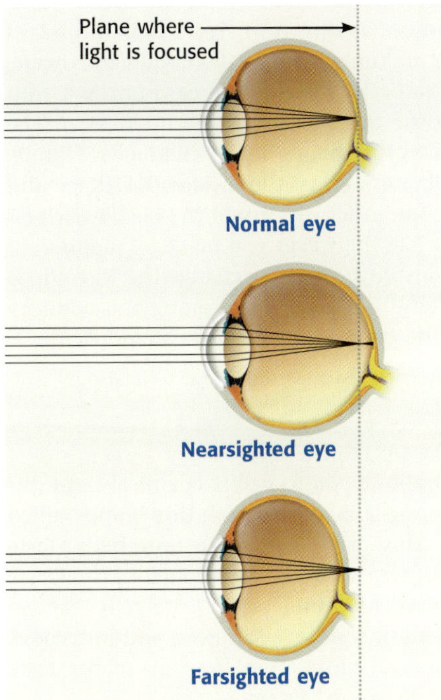

Figure 4 *A concave lens bends light rays outward to correct nearsightedness. A convex lens bends light rays inward to correct farsightedness.*

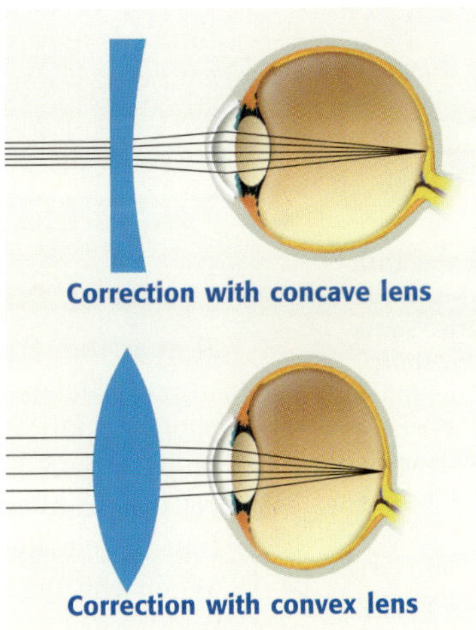

Answers to Quick Lab

3. Students should observe that the white dot disappears when the image is held about 10 cm from their face. The exact point where this happens will vary from person to person. Have students try this activity several times until the dot disappears.
4. Students should discover that the area where the optic nerve leaves the back of the eyeball does not contain any photoreceptors and is called the blind spot. So, when the white dot is focused on the eye's blind spot, the dot disappears.

CONNECTION to Physical Science—GENERAL

Wavelength and Color Vision

In the retina, there are three types of cones that respond maximally to red, blue, or green light. When red cones and green cones are stimulated at the same time, the color yellow is perceived. White is perceived when all three kinds of cones are equally stimulated. Different colors of visible light have different wavelengths. Red has the longest wavelength, and violet has the shortest. Use the teaching transparency entitled "Measuring Wavelengths" to illustrate how wavelengths are measured. **LS Visual**

Is That a Fact!

Young people often hold a book close to their face when reading, while adults hold a book farther away. The closest point on which a person can focus, called the *near point of vision,* changes with age. A typical child can focus on an object about 10 cm from his or her eyes (objects closer than the near point of vision can be seen but appear fuzzy). The change in the near point of vision is caused by the lens's decreasing elasticity over time.

Answer to Reading Check

In bright light, your iris contracts and reduces the amount of light entering the eye.

Close

Reteaching — Basic

Model of a Sense Organ Have students choose one of the sense organs and make a model that shows how the organ interacts with the brain. Allow students to choose from an assortment of materials for their model. Ask students to explain their model to the class. English Language Learners

LS Kinesthetic

Quiz — General

Ask students whether each of the statements below is true or false. Have students correct false statements.

1. Rods help you see color and detail in bright light. (false)
2. Cones provide a colorful view of the world. (true)
3. Your brain combines signals from the senses of smell and hearing to give you a sensation of flavor. (false)

Alternative Assessment — General

Writing **Making Sense of It** Have students write and illustrate a pamphlet or brochure that explains each of the senses. The pamphlet or brochure should be written for somone who does not know much about science. English Language Learners

LS Verbal/Visual

Answer to Reading Check

Neurons in the cochlea convert waves into electrical impulses that the brain interprets as sound.

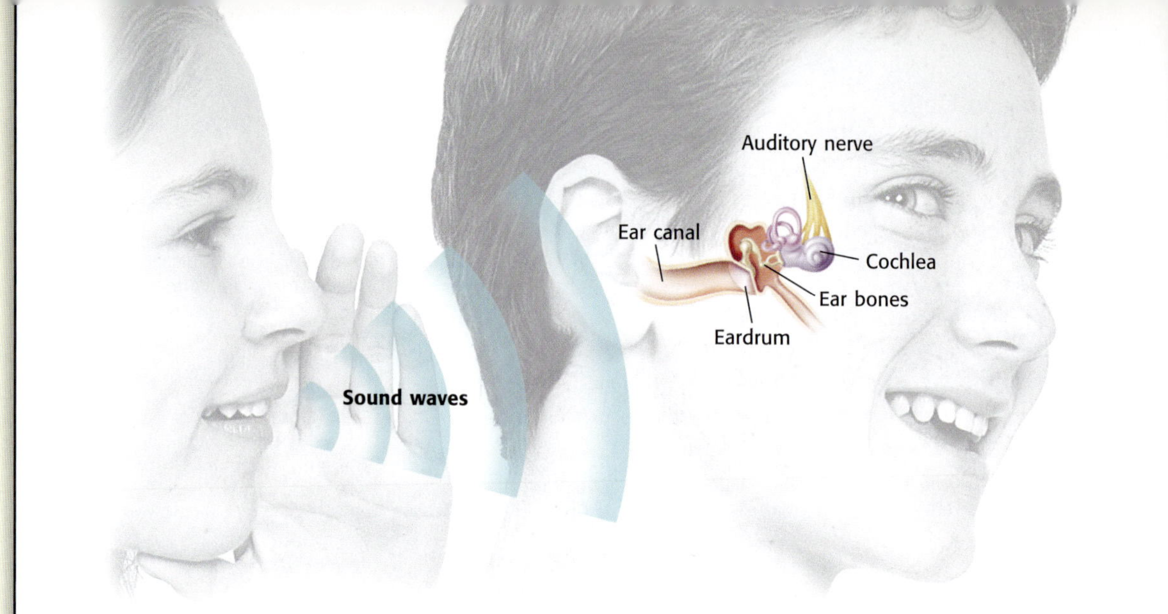

Figure 5 *A sound wave travels into the outer ear. It is converted into bone vibrations in the middle ear, then into liquid vibrations in the inner ear, and finally, into nerve impulses that travel to the brain.*

Sense of Hearing

Sound is produced when something, such as a drum, vibrates. Vibrations push on nearby air particles, which push on other air particles. The vibrations create waves of sound energy. Hearing is the sense that allows you to experience sound energy.

Ears are organs specialized for hearing. Each ear has an outer, middle, and inner portion, as shown in **Figure 5.** Sound waves reaching the outer ear are funneled into the middle ear. There, the waves make the eardrum vibrate. The eardrum is a thin membrane separating the outer ear from the middle ear. The vibrating eardrum makes tiny bones in the middle ear vibrate. One of these bones vibrates against the **cochlea** (KAHK lee uh), a fluid-filled organ of the inner ear. Inside the cochlea, vibrations make waves just like the waves you make by tapping on a glass of water. Neurons in the cochlea change the waves into electrical impulses. These impulses travel along the auditory nerve to the area of the brain that interprets sound.

cochlea a coiled tube that is found in the inner ear and that is essential to hearing

Reading Check Why is the cochlea important to hearing?

Sense of Taste

Taste is the sense that allows you to detect chemicals and distinguish flavors. Your tongue is covered with tiny bumps called *papillae* (puh PIL ee). Most papillae contain taste buds. Taste buds contain clusters of *taste cells,* the receptors for taste. Taste cells respond to dissolved food molecules. Taste cells react to four basic tastes: sweetness, sourness, saltiness, and bitterness. When the brain combines information from all of the taste buds, you taste a "combination" flavor.

CONNECTION TO Physics

WRITING SKILL **Elephant Talk** Sound is produced by vibrating objects. Some sounds, called *infrasonic sounds,* are too low for human ears to detect. Research how elephants use infrasonic sounds to communicate with each other, and write a report about what you learn.

CONNECTION to Physical Science — Advanced

Owls Hunting by Hearing Students are probably familiar with the idea that owls see well in the dark. Students may not know that some owls actually hunt by listening for mice tunneling under the snow. Have students research the many adaptations that help owls hear well. Students can present their findings on a poster or by making a model of an owl's head that shows how an owl uses its hearing to hunt. LS Verbal

WEIRD SCIENCE

Some spicy foods, such as chile peppers, contain a chemical compound called *capsaicin.* Capsaicin triggers the pain receptors in the mouth. These receptors react to capsaicin exactly as they would to heat. This is why spicy foods, such as jalapeño peppers, feel like they are burning your mouth. In fact, no damage from heat actually happens—the burning is all in your mind!

Sense of Smell

As you can see in **Figure 6,** receptors for smell are located on *olfactory cells* in the upper part of your nasal cavity. An olfactory cell is a nerve cell that responds to chemical molecules in the air. You smell something when the receptors react to molecules that have been inhaled. The molecules dissolve in the moist lining of the nasal cavity and trigger an impulse. Olfactory cells send those impulses to the brain, which interprets the impulses as odors.

Taste buds and olfactory cells both detect dissolved molecules. Your brain combines information from both senses to give you sensations of flavor.

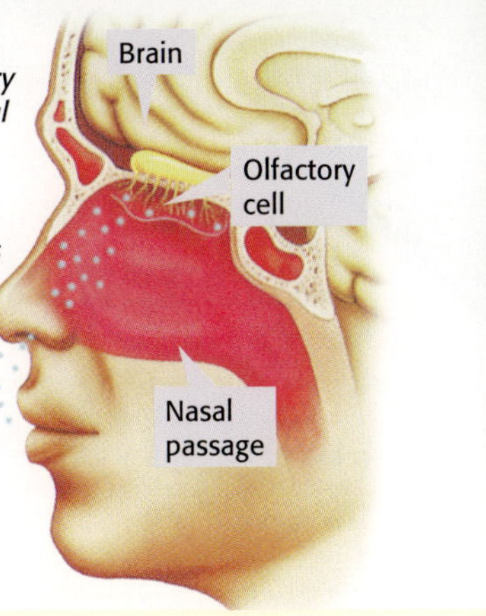

Figure 6 *Olfactory cells line the nasal cavity. These cells are sensory receptors that react to chemicals in the air.*

SECTION Review

Summary

- Touch allows you to respond to temperature, pressure, pain, and vibration on the skin.
- Reflexes and feedback mechanisms help you respond to your environment.
- Sight allows you to respond to light energy.
- Hearing allows you to respond to sound energy.
- Taste allows you to distinguish flavors.
- Smell allows you to perceive different odors.

Using Key Terms

1. In your own words, write a definition for each of the following terms: *reflex* and *feedback mechanism.*
2. Use each of the following terms in a separate sentence: *retina* and *cochlea.*

Understanding Key Ideas

3. Three sensations that receptors in the skin detect are
   a. light, smell, and sound.
   b. touch, pain, and odors.
   c. temperature, pressure, and pain.
   d. pressure, sound, and touch.
4. Explain how light and sight are related.
5. Describe how your senses of hearing, taste, and smell work.
6. Explain why you might have trouble seeing bright colors at a candlelit dinner.
7. How is your sense of taste similar to your sense of smell, and how do these senses work together?
8. Describe how the feedback mechanism that regulates body temperature works.

Math Skills

9. Suppose a nerve impulse must travel 0.90 m from your toe to your central nervous system. If the impulse travels at 150 m/s, calculate how long it will take the impulse to arrive. If the impulse travels at 0.2 m/s, how long will it take the impulse to arrive?

Critical Thinking

10. **Making Inferences** Why is it important for the human body to have reflexes?
11. **Applying Concepts** Rods help you detect objects and shapes in dim light. Explain why it is important for human eyes to have both rods and cones.

SCI LINKS. NSTA Developed and maintained by the National Science Teachers Association

For a variety of links related to this chapter, go to www.scilinks.org

Topic: The Senses; The Eye
SciLinks code: HSM1378; HSM0560

Answers to Section Review

1. Sample answer: A reflex is an unconscious, immediate response to a stimulus. A feedback mechanism is a series of steps in which one step affects a previous step.
2. Sample answer: The retina converts light to electrical impulses. The cochlea converts sound to electrical impulses.
3. c
4. Sample answer: Light enters the eye and strikes the retina. Cells in the retina are triggered by light and produce electrical impulses that are sent to the brain, where they are interpreted as images you see.
5. Sample answer: The ear converts sound waves into electrical impulses. The brain interprets these impulses as sounds. Taste buds in the papillae respond to four basic tastes (sweetness, bitterness, sourness, and saltiness) and send messages to the brain, which produces a "flavor." Smell receptors on olfactory cells respond to chemicals in the air and send impulses to the brain, which interprets them as odors.
6. Sample answer: Cones in the retina, which produce impulses that are interpreted as colors, do not work as well in dim light.
7. Sample answer: Receptors for smell detect chemicals in air. Receptors for taste detect chemicals in your mouth. The chemicals trigger impulses that the brain interprets as smells or tastes. The brain combines information from the mouth and the nose to give you a sensation of flavor.
8. Sample answer: The brain receives information about body temperature. If body temperature is too high or too low, the brain tells the body to take steps to correct its temperature. The brain receives information about the adjustments. When body temperature is correct, the brain tells the body to stop what it was doing to raise or lower the temperature.
9. 0.90 m ÷ 150 m/s = 0.006 s; 0.90 m ÷ 0.2 m/s = 4.5 s
10. Sample answer: Reflexes are important becaus they protect us by acting faster than the body would act if the brain had to process information and send a message.
11. In bright light, cones trigger impulses that are interpreted as images that provide information about color, shape, and motion. At night, rods allow you to detect something in the dark, even if you can't tell much about its color or shape. Together, cones and rods give you information about your environment.

CHAPTER RESOURCES

Chapter Resource File

- Section Quiz GENERAL
- Section Review GENERAL
- Vocabulary and Section Summary GENERAL
- Reinforcement Worksheet BASIC
- Critical Thinking ADVANCED

SECTION 3

Focus

Overview

In this section, students will learn about the endocrine system and how endocrine glands control the body's slower, long-term processes via hormones. They will also learn the location and function of specific endocrine glands.

Bellringer

Write the following on the board:

Unscramble the following words, and write them on a piece of paper:

nalgd (gland)
meornoh (hormone)
noclotr (control)

Motivate

Discussion — GENERAL

Endocrine System Ask students to think of a time when they were suddenly frightened. Discuss with students how their pulse rate and breathing rate were different before and after being scared. (Both rates should be elevated after the scare.) Tell students that the endocrine system, along with the autonomic nervous system, is responsible for the changes that occurred in their pulse rate and breathing rate. **LS Verbal**

SECTION 3

The Endocrine System

Have you ever heard of an epinephrine (EP uh NEPH rin) rush? You might have had one without realizing it. Exciting situations, such as riding a roller coaster or watching a scary movie, can cause your body to release epinephrine.

Epinephrine is one of the body's chemical messengers made by the endocrine system. Your endocrine system regulates body processes, such as fluid balance, growth, and development.

READING WARM-UP

Objectives

- Explain why the endocrine system is important to the body.
- Identify five glands of the endocrine system, and describe what their hormones do.
- Describe how feedback mechanisms stop and start hormone release.
- Name two hormone imbalances.

Terms to Learn

endocrine system
gland
hormone

READING STRATEGY

Discussion Read this section silently. Write down questions that you have about this section. Discuss your questions in a small group.

Hormones as Chemical Messengers

The **endocrine system** controls body functions by using chemicals that are made by the endocrine glands. A **gland** is a group of cells that make special chemicals for your body. Chemical messengers made by the endocrine glands are called hormones. A **hormone** is a chemical messenger made in one cell or tissue that causes a change in another cell or tissue in another part of the body. Hormones flow through the bloodstream to all parts of the body. Thus, an endocrine gland near your brain can control an organ that is somewhere else in your body.

Endocrine glands may affect many organs at one time. For example, in the situation shown in **Figure 1,** the adrenal glands release the hormone *epinephrine*, which is sometimes called *adrenaline*. Epinephrine increases your heartbeat and breathing rate. This response is called the "fight-or-flight" response. When you are frightened, angry, or excited, the "fight-or-flight" response prepares you to fight the danger or to run from it.

endocrine system a collection of glands and groups of cells that secrete hormones that regulate growth, development, and homeostasis

gland a group of cells that make special chemicals for the body

Figure 1 *When you have to move quickly to avoid danger, your adrenal glands make more blood glucose available for energy.*

CHAPTER RESOURCES

Chapter Resource File
- Lesson Plan
- Directed Reading A BASIC
- Directed Reading B SPECIAL NEEDS

Technology

Transparencies
- Bellringer

Is That a Fact!

Epinephrine occurs naturally in the human body, but it is also administered as a drug by doctors. It can be injected into the heart to help revive a person who has suffered a heart attack. Epinephrine is also sometimes used to dilate the bronchioles in the lungs of people who have asthma or who have severe allergic reactions, such as reactions to bee stings.

More Endocrine Glands

Your body has several other endocrine glands. Some of these glands have many functions. For example, your pituitary gland stimulates skeletal growth and helps the thyroid gland work properly. It also regulates the amount of water in the blood. And the pituitary gland stimulates the birth process in women.

Your thyroid gland is very important during infancy and childhood. Thyroid hormones control the secretion of growth hormones for normal body growth. Thyroid hormones also control the development of the central nervous system. And they control your metabolism. *Metabolism* is the sum of all the chemical processes that take place in an organism.

Your thymus gland is important to your immune system. Cells called *killer T cells* grow and mature in the thymus gland. These T cells help destroy or neutralize cells or substances that invade your body. The names and some of the functions of endocrine glands are shown in **Figure 2.**

Reading Check Name two endocrine glands, and explain why they are important to your body. *(See the Appendix for answers to Reading Checks.)*

hormone a substance that is made in one cell or tissue and that causes a change in another cell or tissue in a different part of the body

CONNECTION TO Language Arts

WRITING SKILL **Fight or Flight?** Write a paragraph describing a time when you had a fight-or-flight experience. Include in your description the following terms: *hormones, fight-or-flight,* and *epinephrine.* If you cannot think of a personal experience, write a short story describing someone else's fight-or-flight experience.

Figure 2 Endocrine Glands and Their Functions

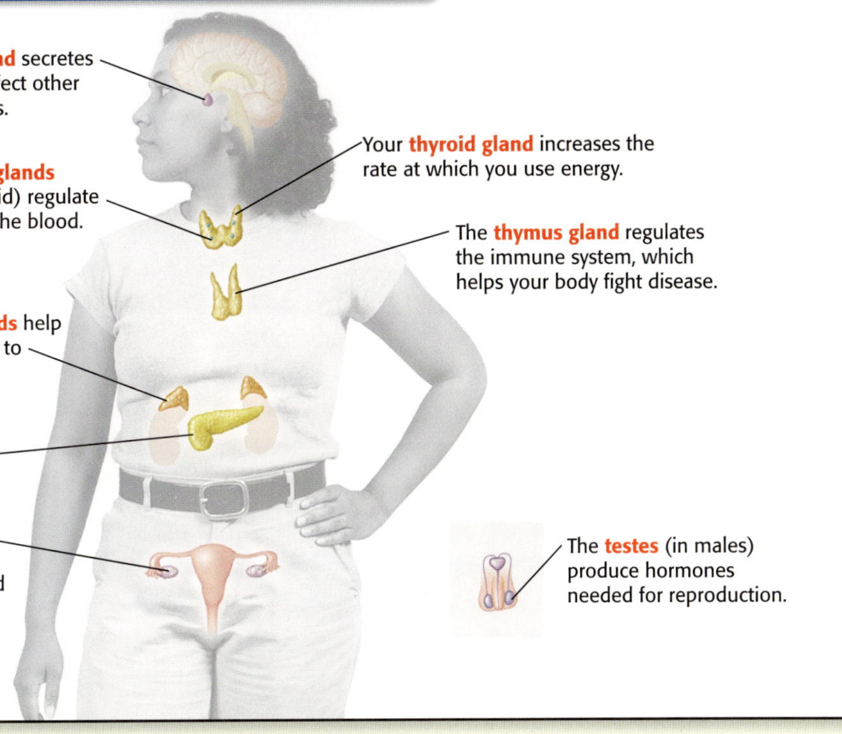

Teach

Answer to Connection to Language Arts

Paragraphs may vary based on students' lives and experiences. Students should write about a time when they experienced a fight-or-flight response to something. They should include the terms *hormone, fight-or-flight,* and *epinephrine.*

Using the Figure—BASIC

Endocrine Glands Write the following terms on the board: *endocrine, pituitary, thyroid, parathyroid, thymus, adrenals, pancreas, ovaries,* and *testes.*

As a class, pronounce each term. Then, locate each gland, as shown in **Figure 2.**

LS Visual/Auditory

Discussion—ADVANCED

Cancer Cells and Hormones Some cells found in tumors secrete hormones that are identical to those secreted by endocrine glands. Typically, tumor cells secrete too much of the hormone and do not respond to feedback control. Ask students to predict the effects on the body of these tumor cells and the hormones they produce. (Students should recognize that when hormones are produced in improper amounts or in an uncontrolled way, the body's homeostasis will be disrupted. The specific effects will depend on the body's internal environment and on which hormones the tumor cells secrete.)

LS Auditory

Is That a Fact!

Since the mid-1980s research at Rutgers University shows that some potent hormones produced in the last trimester of pregnancy prepare and motivate mothers to care for their young. The most important of these hormones is oxytocin. Oxytocin causes the uterus to contract, so oxytocin is in the mother's brain at the time the mother meets her newborn. Some scientists believe oxytocin helps the mother to bond with her new baby.

Answer to Reading Check

Sample answer: The thyroid gland increases the rate at which the body uses energy. The thymus gland regulates the immune system, which helps your body fight disease.

Close

Reteaching — BASIC

Fight-or-Flight Have students work in pairs. Ask students to tell each other about a time when they were frightened and to describe the physical responses they had. Make a list of all responses. Discuss with students how one hormone causes all these changes. Ask students, "How are the changes related?" (They all get the body ready to fight or run from danger.)
LS Interpersonal/Intrapersonal

Quiz — GENERAL

Ask students whether each of the statements below is true or false. Have students correct false statements.

1. Hormones are chemicals secreted into the bloodstream. (true)
2. All endocrine glands come in pairs. (false)
3. Hormones are regulated by feedback mecanisms. (true)

Alternative Assessment — GENERAL

Writing **New Hormone** Have students write a story about a new hormone that controls a function, such as the ability to tell jokes, not discussed in this lesson. Students should describe a feedback control of this hormone. **LS Verbal**

Answer to Reading Check

Insulin helps regulate the amount of glucose in the blood.

Controlling the Endocrine Glands

Do you remember the feedback mechanisms at work in the nervous system? Endocrine glands control similar feedback mechanisms. For example, the pancreas has specialized cells that make two different hormones, *insulin* and *glucagon*. As shown in **Figure 3**, these two hormones control the level of glucose in the blood. Insulin lowers blood-glucose levels by telling the liver to convert glucose into glycogen and to store glycogen for future use. Glucagon has the opposite effect. It tells the liver to convert glycogen into glucose and to release the glucose into the blood.

Reading Check What does insulin do?

Figure 3 Blood-Glucose Feedback Control

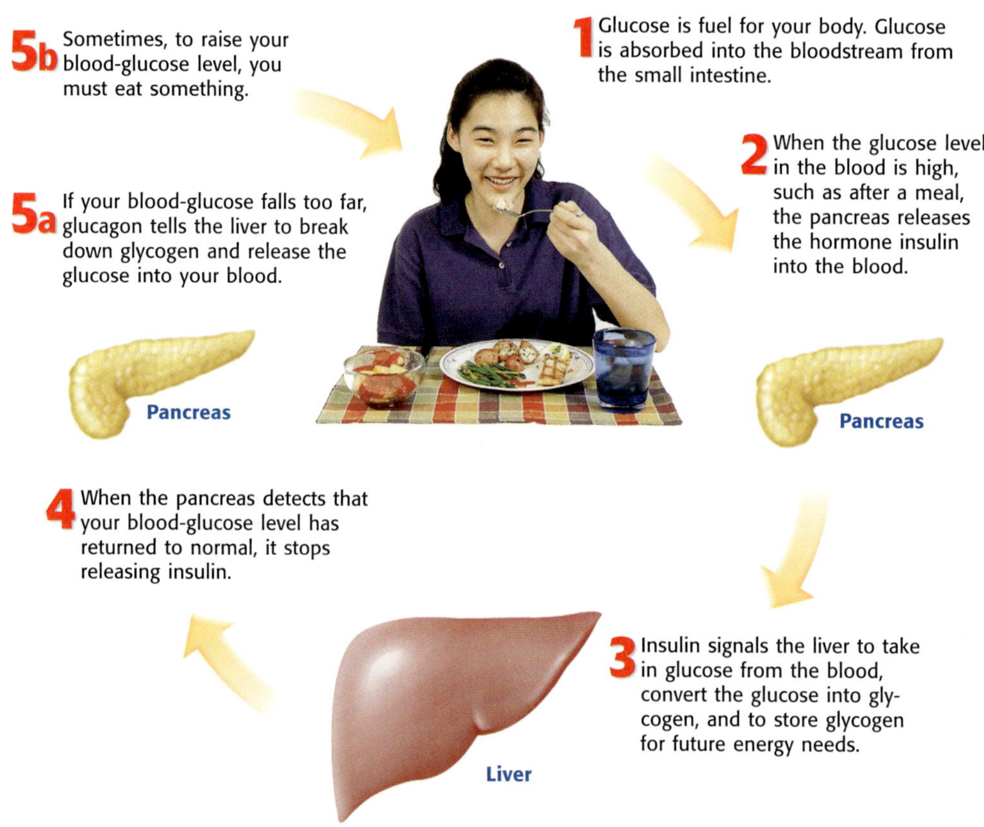

INCLUSION Strategies

• Gifted and Talented • Hearing Impaired

Some students benefit from exploring topics in greater depths. Ask students to divide a large piece of poster board in half. On one half, have them show a healthy pancreas with healthy insulin production and usage. On the other half, have them show a pancreas that could belong to a person with *diabetes mellitus*. English Language Learners
LS Visual

Hormone Imbalances

Occasionally, an endocrine gland makes too much or not enough of a hormone. For example, when a person's blood-glucose level rises, the pancreas secretes insulin. Insulin sends a message to the liver to convert glucose into glycogen. The liver stores glycogen for future use. But a person whose body does not use insulin properly or whose pancreas does not make enough insulin has a condition called *diabetes mellitus* (DIE uh BEET EEZ muh LIET uhs). A person who has diabetes may need daily injections of insulin to keep his or her blood-glucose levels within safe limits. Some patients, such as the woman in **Figure 4,** receive their insulin automatically from a small machine worn next to the body.

Another hormone imbalance is when a child's pituitary gland doesn't make enough growth hormone. As a result, the child's growth is stunted. Fortunately, if the problem is detected early, a doctor can prescribe growth hormone and monitor the child's growth. If the pituitary makes too much growth hormone, a child may grow taller than expected.

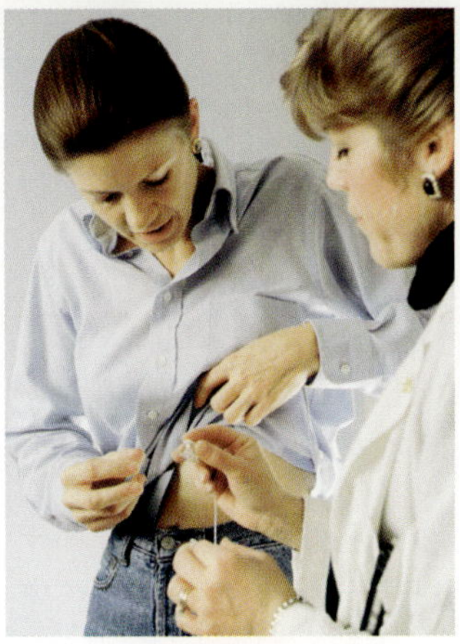

Figure 4 *This woman has diabetes and receives insulin from a device that monitors her blood-glucose level.*

SECTION Review

Summary

- Glands in the endocrine system use chemical messengers called *hormones*.
- Hormones regulate body functions by causing changes in cells or tissues.
- Feedback mechanisms tell endocrine glands when to turn hormones on and off.
- A hormone imbalance is when a gland releases too much or too little of a hormone.

Using Key Terms

1. Use the following terms in the same sentence: *endocrine system, glands,* and *hormone.*

Understanding Key Ideas

2. Identify five endocrine glands, and explain why their hormones are important to your body.
3. Hormone imbalances may cause
   a. feedback and insulin.
   b. diabetes and stunted growth.
   c. thyroid and pituitary.
   d. glucose and glycogen.
4. How do feedback mechanisms control hormone production?

Math Skills

5. One's bedtime blood-glucose level is normally 140 mg/dL. Ty's blood-glucose level is 189 mg/dL at bedtime. What percentage above 140 mg/dL is Ty's level?

Critical Thinking

6. **Making Inferences** Glucose is a source of energy. Epinephrine quickly increases the blood-glucose level. Why is epinephrine important in times of stress?
7. **Applying Concepts** The hormone glucagon is released when glucose levels fall below normal. Explain how the hormones glucagon and insulin work together to control blood-glucose levels.

SCILINKS. NSTA Developed and maintained by the National Science Teachers Association

For a variety of links related to this chapter, go to www.scilinks.org

Topic: Hormones
SciLinks code: HSM0758

Answers to Section Review

1. Sample answer: Your endocrine system controls body processes such as fluid balance, growth, and sexual development with the use of chemicals, called hormones, produced in glands.
2. b
3. Answers may vary depending on which five glands a student chooses.
4. Feedback mechanisms stop and start hormone release in response to conditions in the body. Students may illustrate this answer by drawing a diagram demonstrating how a thermostat keeps water or air within a certain temperature range.
5. $[(189 - 140) \div 140] = 35\%$
6. Sample answer: During stressful situations, your body functions need to speed up. These changes require more energy. Epinephrine instructs the body to release more glucose, a source of energy, into the bloodstream.
7. Sample answer: When your blood-glucose level falls below the normal level, glucagon is released. Glucagon tells your body to convert glycogen into glucose, which raises your blood-glucose level to a normal level. When your blood-glucose level is normal, insulin is released to tell the body to store any extra glucose as glycogen.

Homework — GENERAL

Writing **Researching Steroids** Tell students that hormones are used as medicines to treat endocrine disorders. However, other hormones, such as anabolic steroids, are abused to increase athletic prowess. Have students research the effects and dangers of abusing anabolic steroids. Have them write a brief report or prepare an oral presentation to share their findings. LS **Verbal**

CHAPTER RESOURCES

Chapter Resource File

- Section Quiz GENERAL
- Section Review GENERAL
- Vocabulary and Section Summary GENERAL
- Reinforcement Worksheet BASIC

Skills Practice Lab

You've Gotta Lotta Nerve

Teacher's Notes

Time Required

One 45-minute class period

Lab Ratings

EASY → HARD

Teacher Prep 1 flask

Student Set-Up 1 flask

Concept Level 2 flasks

Clean Up 1 flask

MATERIALS

The materials listed on the student page are enough for a group of 3 students. If more appropriate for your class, substitute fine point ball point pens for the dissecting pins.

Safety Caution

Remind students to review all safety cautions and icons before beginning this lab activity. Remind students to be safe and gentle with each other in this exercise and to respect the sensitivity and comfort of their peers. Tell students that they will not be testing for pain. The protective cover on the sharp end of the dissecting pin must remain in place at all times.

Skills Practice Lab

You've Gotta Lotta Nerve

OBJECTIVES

Locate areas on the skin that respond to certain stimuli.

Determine which areas on the skin are more sensitive to certain kinds of stimuli.

MATERIALS

- dissecting pin with a small piece of cork or a small rubber stopper covering the sharp end
- eyedropper, plastic
- paper, graphing
- pens or markers, washable, fine point
- ruler, metric
- tap water, hot
- water, very cold

SAFETY

Your skin has thousands of nerve receptors that detect sensations, such as temperature, pain, and pressure. Your brain is designed to filter out or ignore most of the input it receives from these skin receptors. If the brain did not filter input, simply wearing clothes would trigger so many responses that you couldn't function.

Some areas of the skin, such as the back of your hand, are more sensitive than others. In this activity, you will map the skin receptors for heat, cold, and pressure on the back of your hand.

Procedure

1. Form a group of three. One of you will volunteer the back of your hand for testing, one will do the testing, and the third will record the results.
2. Use a fine-point, washable marker or pen and a metric ruler to mark a 3 cm × 3 cm square on the back of one person's hand. Draw a grid within the area. Space the lines approximately 0.5 cm apart. You will have 36 squares in the grid when you are finished, as shown in the photograph below.
3. Mark off three 3 cm × 3 cm areas on a piece of graph paper. Make a grid in each area exactly as you did on the back of your partner's hand. Label one grid "Cold," another grid "Hot," and the third grid "Pressure."

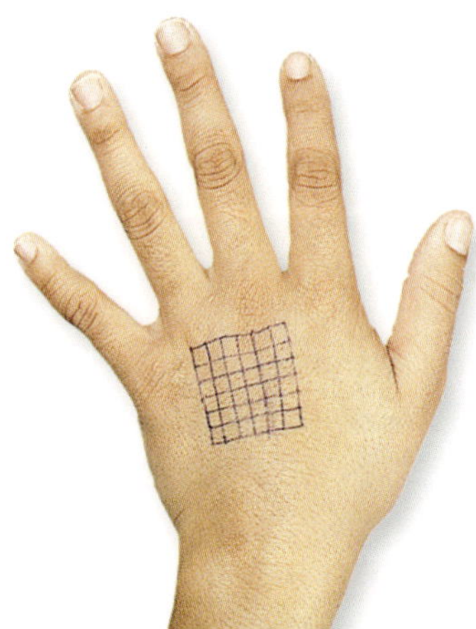

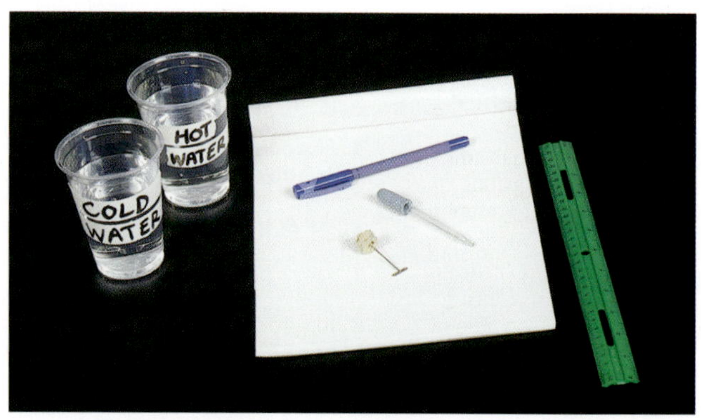

Lab Notes

This activity works best if the student whose hand is being tested looks away or is loosely blindfolded while his or her hand is being tested. Often, students will say they feel something when they think they should feel something. Students should be given the choice of being blindfolded or looking away.

CHAPTER RESOURCES

Chapter Resource File

- Datasheet for Chapter Lab
- Lab Notes and Answers

Technology

Classroom Videos

- Lab Video

4. Use the eyedropper to apply one small droplet of cold water on each square in the grid on your partner's hand. Your partner should turn away while being tested. On your graph paper, mark an X on the "Cold" grid to show where your partner felt the cold droplet. Carefully blot the water off your partner's hand after several drops.

5. Repeat the test using hot-water droplets. The hot water should not be hot enough to hurt your partner. Mark an X on the "Hot" grid to indicate where your partner felt the hot droplet.

6. Repeat the test by using the head (not the point!) of the pin. Touch the skin to detect pressure receptors. Use a very light touch. On the graph paper, mark an X on the "Pressure" grid to indicate where your partner felt the pressure.

Analyze the Results

1. **Organizing Data** Count the number of Xs in each grid. How many heat receptor responses are there per 3 cm^2? How many cold receptor responses are there? How many pressure receptor responses are there?

2. **Explaining Events** Do you have areas on the back of your hand where the receptors overlap? Explain your answer.

3. **Recognizing Patterns** How do you think the results of this experiment would be similar or different if you mapped an area of your forearm? of the back of your neck? of the palm of your hand?

Draw Conclusions

4. **Interpreting Information** Prepare a written report that includes a description of your investigation and a discussion of your answers to items 1–3. What conclusions can you draw from your results?

Applying Your Data

Use the library or the Internet to research what happens if a receptor is continuously stimulated. Does the kind of receptor make a difference? Does the intensity or strength of the stimulus make a difference? Explain your answers.

Disposal Information

Provide receptacles for used dissecting pins and eyedroppers. Be sure students know the location of the receptacles and deposit their sharp items in the designated containers. Dispose of sharp items properly to eliminate the risk of injury.

Analyze the Results

1. Answers may vary. Some groups will have more receptors in each category than others.
2. Students may notice that receptors in the same square on the grid can sense heat, cold, and pressure. Students may explain this reaction by noting that they have different kinds of receptors in the same spot. Students may also explain this by noting that the same receptor responds to more than one stimulus. Some students may interpret hot and cold as variations of a single sensation—temperature. They may also determine that the blunt head of the pin might have been cold and that they felt its temperature instead of the pressure.
3. Answers may vary. Different areas of the body are more sensitive than others.

Draw Conclusions

4. Answers may vary. Students should provide logical, well-reasoned conclusions that are supported by the details and results of the experiment. Students should discuss the answers to items 1–3.

Applying Your Data

Students will find that responses to a constant stimulus differ depending on the type and intensity of the stimulus. A person may become insensitive to an odor or even mild pain over time. Intense, sudden, or continued stimuli, such as heat, pain, or noise, may do much damage to a person over time.

CHAPTER RESOURCES

Workbooks

Whiz-Bang Demonstrations
- Now You See It, Now You Don't GENERAL

Labs You Can Eat
- A Salty Sweet Experiment GENERAL

Long-Term Projects & Research Ideas
- Man Versus Machine ADVANCED

Christopher Wood
Western Rockingham Middle School
Madison, North Carolina

Chapter Review

Assignment Guide

Section	Questions
1	1–2, 6, 10–11, 13, 15, 18, 22
2	4, 7, 8, 16, 19–21
3	3, 5, 9, 12, 14, 17, 23–26

ANSWERS

Using Key Terms

1. central nervous system
2. retina
3. hormone
4. reflex
5. feedback mechanism
6. neuron

Understanding Key Ideas

7. c
8. a
9. d
10. a
11. d
12. d
13. Sample answer: The main difference is that the somatic nervous system controls voluntary movements and activities, and the autonomic nervous system controls body functions that you do not need to think about. Both systems are important because together they keep your body active and alive. You are able to do all of your voluntary activities while your body functions are controlled by the autonomic nervous system.

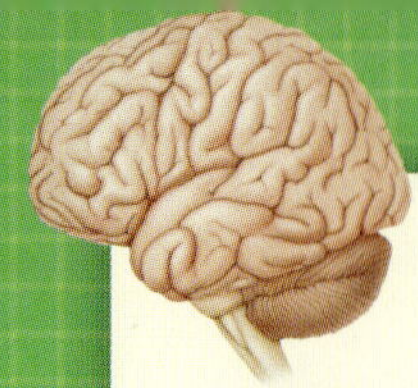

Chapter Review

USING KEY TERMS

Complete each of the following sentences by choosing the correct term from the word bank.

insulin	axon
hormone	nerve
retina	central nervous system
neuron	
reflex	

1. The two parts of your _____ are your brain and spinal cord.
2. Sensory receptors in the _____ detect light.
3. Epinephrine is a(n) _____ that triggers the fight-or-flight response.
4. A(n) _____ is an involuntary and almost immediate movement in response to a stimulus.
5. One hormone that helps to regulate blood-glucose levels is _____ .
6. A(n) _____ is a specialized cell that receives and conducts electrical impulses.

UNDERSTANDING KEY IDEAS

Multiple Choice

7. Which of the following has receptors for smelling?
   a. cochlea cells
   b. thermoreceptors
   c. olfactory cells
   d. optic nerve

8. Which of the following allow you to see the world in color?
   a. cones
   b. rods
   c. lenses
   d. retinas
9. Which of the following glands makes insulin?
   a. adrenal gland
   b. pituitary gland
   c. thyroid gland
   d. pancreas
10. The peripheral nervous system does NOT include
   a. the spinal cord.
   b. axons.
   c. sensory receptors.
   d. motor neurons.
11. Which part of the brain regulates blood pressure?
   a. right cerebral hemisphere
   b. left cerebral hemisphere
   c. cerebellum
   d. medulla
12. The process in which the endocrine system, the digestive system, and the circulatory system control the level of blood glucose is an example of
   a. a reflex.
   b. an endocrine gland.
   c. the fight-or-flight response.
   d. a feedback mechanism.

14. Sample answer: The endocrine system is important because it produces chemical messengers called *hormones* that regulate body processes such as growth, fluid balance, and development.
15. Sample answer: The PNS receives stimuli from inside and outside the body and sends messages to the CNS. The CNS processes the messages and sends responses. The PNS carries the responses to the part of the body that will respond to a stimulus.

Short Answer

13 What is the difference between the somatic nervous system and the autonomic nervous system? Why are both systems important to the body?

14 Why is the endocrine system important to your body?

15 What is the relationship between the CNS and the PNS?

16 What is the function of the bones in the middle ear?

17 Describe two interactions between the endocrine system and the body that happen when a person is frightened.

CRITICAL THINKING

18 **Concept Mapping** Use the following terms to create a concept map: *nervous system, spinal cord, medulla, peripheral nervous system, brain, cerebrum, central nervous system,* and *cerebellum.*

19 **Making Comparisons** Compare a feedback mechanism with a reflex.

20 **Analyzing Ideas** Why is it important to have a lens that can change shape inside the eye?

21 **Applying Concepts** Why it is important that reflexes happen without thinking about them?

22 **Predicting Consequences** What would happen if your autonomic nervous system stopped working?

23 **Making Comparisons** How are the nervous system and the endocrine system similar? How are they different?

INTERPRETING GRAPHICS

Use the diagram below to answer the questions that follow.

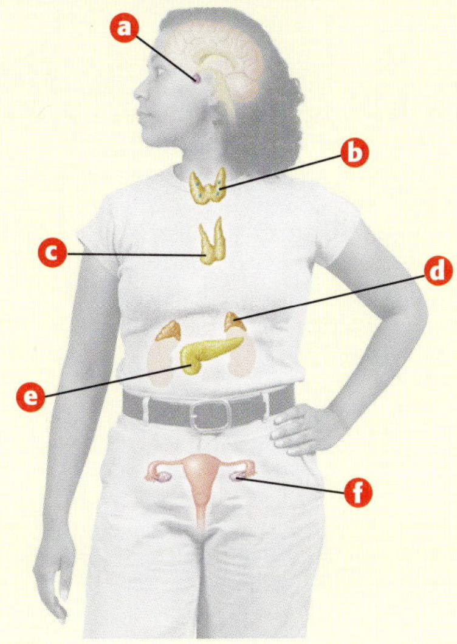

24 Which letter identifies the gland that regulates blood-glucose level?

25 Which letter identifies the gland that releases a hormone that stimulates the birth process?

26 Which letter identifies the gland that helps the body fight disease?

16. Sample answer: The bones of the middle ear transmit vibrations from the eardrum to the cochlea, which then converts the vibrations into electrical impulses that are interpreted by the brain as sound.

17. Sample answer: Fright makes the endocrine system release the hormone epinephrine, which increases the heart rate and the rate of breathing.

CHAPTER RESOURCES

Chapter Resource File

- Chapter Review GENERAL
- Chapter Test A GENERAL
- Chapter Test B ADVANCED
- Chapter Test C SPECIAL NEEDS
- Vocabulary Activity GENERAL

Workbooks

Study Guide

- Assessment resources are also available in Spanish.

Critical Thinking

18. 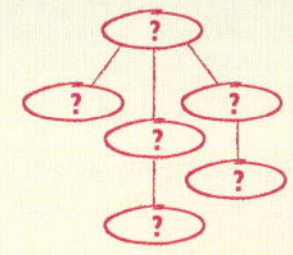 An answer to this exercise can be found at the end of this book.

19. Sample answer: A feedback mechanism is a series of events in which one step affects an earlier step in the process. Usually, a feedback mechanism operates over a period of time. A reflex is an immediate and unconscious response to a stimulus. It is not a process.

20. Sample answer: The lens of your eye changes shape so that light rays are focused on the retina and so that you are able to see images from a wide range of distances.

21. Sample answer: The speed of a reflex response can prevent serious harm to the body. Thinking about taking an action might take too long, and an injury could happen while you were thinking about what to do.

22. Sample answer: You would eventually die if your autonomic nervous system shut down. Your body could not maintain homeostasis, and functions such as digestion, breathing, and heart rate would soon all stop working properly.

23. Sample answer: The endocrine system is similar to the nervous system because both systems send messages to control body functions and to respond to internal and external stimuli. The two systems are different because electrical messages travel along nerves faster than chemicals can travel in the bloodstream.

Interpreting Graphics

24. e (pancreas)

25. a (pituitary)

26. c (thymus)

Standardized Test Preparation

Teacher's Note

To provide practice under more realistic testing conditions, give students 20 minutes to answer all of the questions in this Standardized Test Preparation.

MISCONCEPTION ALERT

Answers to the standardized test preparation can help you identify student misconceptions and misunderstandings.

READING

Passage 1

1. B
2. H
3. D

TEST DOCTOR

Question 1: The correct answer, B, is not explicitly stated in the passage. Students must read the passage to find out that a synapse is the gap between an axon and another cell. Students may select answer A if they confuse the neurotransmitters with the gap across which neurotransmitters work.

Question 3: Students may select incorrect answers A or C if they are not clear about what a synapse is. A synapse is not part of a cell; a synapse is a space between nerve cells. Impulses cannot just a synapse; they are transmitted by neurotransmitters.

Standardized Test Preparation

READING

Read each of the passages below. Then, answer the questions that follow each passage.

Passage 1 The axon terminals of neurons usually do not touch the other cells. There is a small gap between an axon terminal and another cell. This space where a neuron meets another cell is called a *synapse*. When a nerve impulse arrives at an axon terminal, the impulse cannot cross the gap. Instead, the impulse triggers the release of chemicals called *neurotransmitters*. These neurotransmitters cross the synapse between the axon terminal and the cell. When neurotransmitters reach the next cell, they signal the cell to react in a certain way. There are many kinds of neurotransmitters. Some neurotransmitters tell cells to start an action. Other neurotransmitters tell cells to stop an action.

1. What is the space between a neuron terminal and a receiving cell called?
   - **A** a neurotransmitter
   - **B** a synapse
   - **C** an axon
   - **D** a nerve

2. Why are neurotransmitters necessary?
   - **F** They tell muscle cells to contract or relax.
   - **G** They create a gap that axons must cross.
   - **H** They carry messages across the synapse.
   - **I** They release chemical signals called *impulses*.

3. Which of the following statements is a fact in the passage?
   - **A** A synapse is an extension of a nerve cell.
   - **B** The space between an axon terminal and another cell is filled with neurons.
   - **C** Nerve impulses jump from an axon to another cell.
   - **D** There are many kinds of neurotransmitters.

Passage 2 Hormones are chemical messengers released by cells that <u>regulate</u> other cells in the body. Hormones regulate many body processes. Hormones control growth, direct the production and use of energy, keep body temperature within normal limits, and direct responses to stimuli outside the body. Hormones carry chemical messages that tell cells to change their activities. For example, one hormone tells the heart to beat faster. Another hormone tells certain cells to make proteins and stimulates bone and muscle growth. Each hormone communicates with specific cells. Each hormone is like a key that opens only one kind of lock. A hormone's message can be received only by cells that have the right kind of lock. Hormones control many important body functions, so their messages must be delivered properly.

1. According to the passage, which of the following statements about hormones is true?
   - **A** Hormones tell cells to change their activities.
   - **B** Hormones are electrical messengers.
   - **C** Hormones are like locks.
   - **D** Hormones are not important to your body.

2. What does the word *regulate* mean?
   - **F** to control or direct
   - **G** to beat faster
   - **H** to raise your temperature
   - **I** to reverse

3. According to the passage, what are two ways that one particular hormone affects the body?
   - **A** controls your temperature and heart rate
   - **B** responds to stimuli and makes proteins
   - **C** stimulates bone growth and makes proteins
   - **D** coordinates energy production and use and decreases temperature

Passage 2

1. A
2. F
3. C

TEST DOCTOR

Question 2: This question asks students to determine the meaning of the word *regulate* from the context of the passage. Students may select incorrect answers G, H, or I if they confuse a specific type of command or regulation with the more general definition of the word.

Question 3: Students may select incorrect answers A, B, or D if they read only the general list of hormone effects. Only one sentence in the passage refers to a particular hormone that has two effects, so the correct answer is C.

INTERPRETING GRAPHICS

The diagram below shows a typical neuron. Use the diagram below to answer the questions that follow.

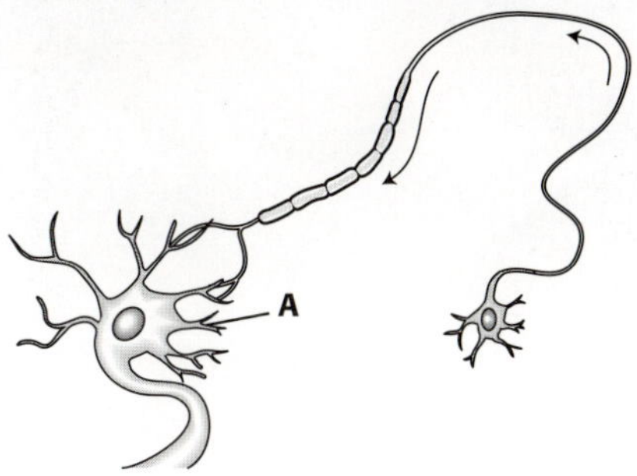

1. What does A represent?

A a cell body
B an axon
C a dendrite
D an axon terminal

2. Which of the following represents the path that an impulse in a neuron travels?

F dendrite, cell body, axon, axon terminal
G axon, axon terminal, cell body, dendrite
H dendrite, nucleus, cell body, axon
I nucleus, cell body, nucleus, axon

3. To where is an impulse that reaches an axon terminal transmitted?

A another axon terminal
B the brain
C a reflex
D dendrites of another neuron

4. What does having many dendrites allow a neuron to do?

F to be locked into place in the body
G to receive impulses from many other cells
H to send impulses to surrounding cells
I to get necessary nutrition

5. Which of the following statements about an axon is true?

A An axon is part of a gland.
B An axon connects the cell body to the axon terminal.
C An axon detects sights and sounds.
D An axon carries chemical messages.

MATH

Read each question below, and choose the best answer.

1. Sound travels about 335 m/s. How many kilometers would a sound travel in 1 min? (One kilometer is equal to 1,000 meters.)

A 335,000 km
B 20,100 km
C 20.1 km
D 0.335 km

2. Some axons send one impulse every 2.5 milliseconds. How many impulses could one of these axons send every second? (One second is equal to 1,000 milliseconds.)

F 4 impulses
G 40 impulses
H 400 impulses
I 4,000 impulses

3. The table below shows the results of Miguel's blood-glucose tests. Miguel ate lunch at 12:00 noon. His blood glucose was measured every hour after that time. What was the average hourly decrease in blood-glucose level?

Blood Glucose

Time tested	Blood-glucose level (mg/1,000 mL)
1:00 P.M.	178
2:00 P.M.	112
3:00 P.M.	100
4:00 P.M.	89

A approximately 160 mg/1,000 mL
B approximately 120 mg/1,000 mL
C approximately 30 mg/1,000 mL
D approximately 22 mg/1,000 mL

4. Your brain has about 1 billion neurons. How is 1 billion expressed in scientific notation?

F 1×10^3
G 1×10^6
H 1×10^9
I 1×10^{12}

Standardized Test Preparation

INTERPRETING GRAPHICS

1. C
2. F
3. D
4. G
5. B

TEST DOCTOR

Question 3: Answer D is correct because when an impulse reaches an axon terminal, it is carried across the synapse by neurotransmitters to the dendrites of another neuron. Students may select incorrect answer A if they forget the difference between a dendrite and an axon terminal.

Question 5: Answer B is correct because an axon is the part of the nerve cell that runs from the nerve cell body and ends in an axon terminal. Students may select incorrect answer D if they confuse the chemical messages carried across a synapse by neurotransmitters with the electrical impulses carried by axons.

MATH

1. D
2. H
3. C
4. H

TEST DOCTOR

Question 1: Students may select incorrect answers A or C if they simply multiply or divide 335 by 1,000. They will have neglected to include the factor of 60 s/min. Students may select incorrect answer B if they forget to convert meters to kilometers.

Question 4: Students may select incorrect answers F, G, or I if they count the wrong number of decimal places when figuring the exponent.

CHAPTER RESOURCES

Chapter Resource File

• Standardized Test Preparation GENERAL

State Resources

For specific resources for your state, visit **go.hrw.com** and type in the keyword **HSMSTR.**

Science in Action

Scientific Discoveries

Discussion — GENERAL

Encourage a class discussion on the placebo effect by asking students the following questions:

- Do you think other therapies, such as acupuncture, homeopathy, or chiropractic care could act as placebos? (Answers may vary.)
- Do you think it would ever be ethical for a doctor to give a patient a placebo? (Answers may vary.)
- Do you think a person could become addicted to a placebo? (Answers may vary.)

Science, Technology, and Society

Discussion — GENERAL

If a disabled person could be hooked up to a computer, then so could anyone. Discuss the possibilities of having computers directly connected to the human brain. Ask students the following questions: What are some things that computers can do that humans can't? (Answers may vary.) What are some of the possibilities for memory, information retrieval, and new sorts of senses? (Answers may vary.) What can humans do that computers can't? (Answers may vary.)

Science in Action

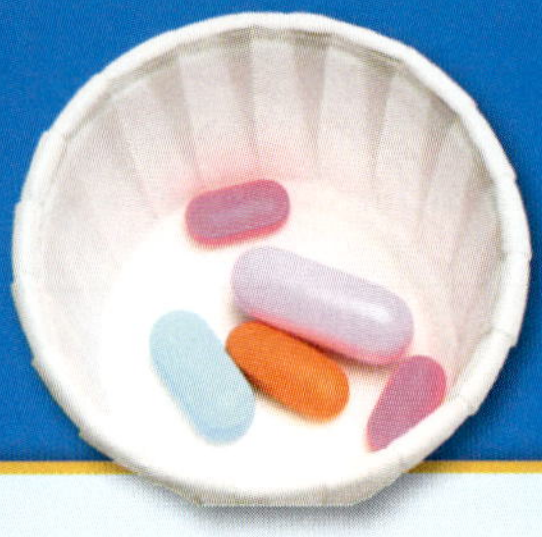

Scientific Discoveries

The Placebo Effect

A placebo (pluh SEE boh) is an inactive substance, such as a sugar pill, used in experimental drug trials. Some of the people who are test subjects are given a placebo as if it were the drug being tested. Usually, neither the doctor conducting the trial nor the test subjects know whether a person is taking a placebo or the test drug. In theory, any change in a subject's condition should be the result of the test drug. But for many years, scientists have known about the *placebo effect,* the effect of feeling better after taking the placebo pill. What makes someone who takes the placebo feel better? By studying brain activity, scientists are beginning to understand the placebo effect.

Social Studies ACTIVITY

Research the differences and similarities between ancient Chinese medical practices and traditional Western medical treatment. Both types of treatment rely in part on a patient's mental and emotional response to treatment. How might the placebo effect be part of both medical traditions? Create a poster showing the results of your research.

Science, Technology, and Society

Robotic Limbs

Cyborgs, or people that are part human and part robot, have been part of science fiction for many years and usually have super-human strength and X-ray vision. Meanwhile there are ordinary people on Earth who have lost the use of their arms and legs and could use some robot power. However, until recently, they have had to settle for clumsy mechanical limbs that were not a very good substitute for a real arm or hand. Today, thanks to advances in technology, scientists are developing artificial limbs—and eyes and ears—that can be wired directly into the nervous system and can be controlled by the brain. In the near future, artificial limbs and some artificial organs will be much more like the real thing.

Language Arts ACTIVITY

WRITING SKILL At the library or on the Internet, find examples of optical or visual illusions. Research how the brain processes visual information and how the brain "sees" and interprets these illusions. Write a report about why the brain seems to be fooled by visual tricks. How can understanding the brain's response to illusions help scientists create artificial vision?

Answer to Social Studies Activity

Students' posters may include a variety of ancient medical practices, including acupuncture, acupressure, and herbal medicines. The student should indicate that, in addition to the physical effect of the treatment, any medical treatment always includes a mental or emotional component that may be strengthened or weakened by the treatment being employed.

Answer to Language Arts Activity

Current brain research is still trying to find the answers to why the brain is fooled by optical or visual illusions. Students may find information in a variety of places, including science magazines such as *Scientific American* and a number of Internet Web sites.

People in Science

Bertha Madras

Studying Brain Activity The brain is an amazing organ. Sometimes, though, drugs or disease keep the brain from working properly. Bertha Madras is a biochemist who studies drug addiction. Dr. Madras studies brain activity to see how substances, such as cocaine, target cells or areas in the brain. Using a variety of brain scanning techniques, Dr. Madras can observe a brain on drugs. She can see how a drug affects the normal activity of the brain. During her research, Dr. Madras realized that some of her results could be applied to Parkinson's disease and to attention deficit hyperactivity disorder (ADHD) in adults. Her research has led to new treatments for both problems.

Math ACTIVITY

Using a search engine on a computer connected to the Internet, search the Internet for "reaction time experiment." Go to one of the Web sites and take the response-time experiment. Record the time that it took you to respond. Repeat the test nine more times, and record your response time for each trial. Then, make a line graph or a bar graph of your response times. Did your response times change? In what way did they change?

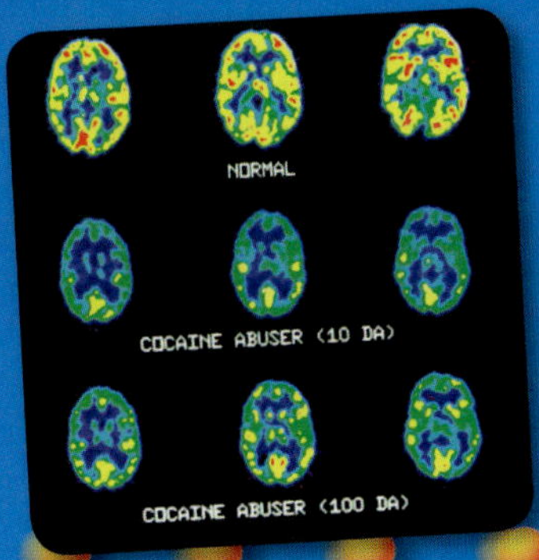

go.hrw.com

To learn more about these Science in Action topics, visit go.hrw.com and type in the keyword HL5BD4F.

Current Science

Check out Current Science® articles related to this chapter by visiting go.hrw.com. Just type in the keyword HL5CS25.

People in Science

Background

Bertha Madras, Ph.D., is a Professor of Psychobiology in the Department of Psychiatry at Harvard Medical School. She also serves as Chair of the Division of Neurochemistry at the New England Regional Primate Research Center (part of Harvard Medical School).

Madras has published more than 100 articles and served on a number of committees dedicated to brain and drug research. Madras also volunteers her time as a science teacher and regularly speaks to groups about the impact of drugs on the brain. To help people understand how drugs affect the brain, Madras has developed an exhibit and a CD-ROM called "Changing Your Mind: Drugs in the Brain."

Answer to Math Activity

Students will find several response-time experiments on the Internet. Students may find that their response times improved as they repeated the experiment, perhaps indicating that they were improving their ability to anticipate the cue and to respond to it. Other students may find that their response times declined as they played, indicating that their brains and muscles had become fatigued. Students' graphs should indicate what kind of experiment they took and should reflect their response times.

5 Reproduction and Development

Chapter Planning Guide

Compression guide: To shorten instruction because of time limitations, omit the Chapter Lab.

OBJECTIVES	LABS, DEMONSTRATIONS, AND ACTIVITIES	TECHNOLOGY RESOURCES
PACING • 90 min pp. 106–111 **Chapter Opener**	**SE Start-up Activity,** p. 107 ◆ GENERAL	**OSP Parent Letter** ■ GENERAL **CD Student Edition on CD-ROM** **CD Guided Reading Audio CD** ■ **TR Chapter Starter Transparency*** **VID Brain Food Video Quiz**
Section 1 Animal Reproduction • Distinguish between asexual and sexual reproduction. • Explain the difference between external and internal fertilization. • Identify the three different types of mammalian reproduction.	**TE Activity** Asexual or Sexual Reproduction?, p. 109 BASIC **TE Demonstration** The Amazing Egg, p. 109 ◆ GENERAL **TE Connection Activity** Chemistry, p. 109 ADVANCED **SE Science in Action** Math, Social Studies, and Language Arts Activities, pp. 128–129 ◆ GENERAL	**CRF Lesson Plans*** **TR Bellringer Transparency*** **TE Internet Activity,** p. 111 GENERAL
PACING • 45 min pp. 112–115 **Section 2 Human Reproduction** • Identify the structures and functions of the male and female reproductive systems. • Describe two reproductive system problems.	**TE Connection Activity** Language Arts, p. 113 GENERAL **SE School-to-Home Activity** Twins and More, p. 114 GENERAL **TE Group Activity** Reproductive System Cancer, p. 114 GENERAL **SE Connection to Social Studies** Understanding STDs, p. 115 GENERAL **TE Connection Activity** Language Arts, p. 114 ADVANCED	**CRF Lesson Plans*** **TR Bellringer Transparency*** **TR** The Male Reproductive System* **TR** The Female Reproductive System*
PACING • 90 min pp. 116–121 **Section 3 Growth and Development** • Summarize the processes of fertilization and implantation. • Describe the development of the embryo and the fetus. • Identify the stages of human development from birth to death.	**SE School-to-Home Activity** Growing Up, p. 117 GENERAL **TE Activity** Making Models, p. 117 ◆ GENERAL **SE Connection to Physics** Using Ultrasound, p. 118 GENERAL **TE Connection Activity** Language Arts, p. 119 GENERAL **TE Connection Activity** Earth Science, p. 119 GENERAL **SE Quick Lab** Life Grows On, p. 120 GENERAL **CRF Datasheet for Quick Lab*** **SE Skills Practice Lab** It's a Comfy, Safe World!, p. 122 ◆ GENERAL **CRF Datasheet for Chapter Lab*** **SE Skills Practice Lab** My, How You've Grown!, p. 184 ◆ GENERAL **CRF Datasheet for LabBook*** **LB Long-Term Projects & Research Ideas** Get a Whiff of This!* ◆ ADVANCED	**CRF Lesson Plans*** **TR Bellringer Transparency*** **TR** Stages of Human Development* **TR** *LINK TO PHYSICAL SCIENCE* How Sonar Works* **CRF SciLinks Activity*** GENERAL **VID Lab Videos for Life Science**

PACING • 90 min

CHAPTER REVIEW, ASSESSMENT, AND STANDARDIZED TEST PREPARATION

CRF Vocabulary Activity* GENERAL
SE Chapter Review, pp. 124–125 GENERAL
CRF Chapter Review* ■ GENERAL
CRF Chapter Tests A* ■ GENERAL, **B*** ADVANCED, **C*** SPECIAL NEEDS
SE Standardized Test Preparation, pp. 126–127 GENERAL
CRF Standardized Test Preparation* GENERAL
CRF Performance-Based Assessment* GENERAL
OSP Test Generator GENERAL
CRF Test Item Listing* GENERAL

Online and Technology Resources

go.hrw.com — Visit **go.hrw.com** for a variety of free resources related to this textbook. Enter the keyword **HL5BD5.**

Holt Online Learning — Students can access interactive problem-solving help and active visual concept development with the *Holt Science and Technology* Online Edition available at **www.hrw.com.**

Guided Reading Audio CD — A direct reading of each chapter using instructional visuals as guideposts. For auditory learners, reluctant readers, and Spanish-speaking students. Available in English and Spanish.

KEY

SE Student Edition **TE** Teacher Edition	**CRF** Chapter Resource File **OSP** One-Stop Planner **LB** Lab Bank **TR** Transparencies	**SS** Science Skills Worksheets **MS** Math Skills for Science Worksheets **CD** CD or CD-ROM **VID** Classroom Video/DVD	* Also on One-Stop Planner ◆ Requires advance prep ■ Also available in Spanish

SKILLS DEVELOPMENT RESOURCES	SECTION REVIEW AND ASSESSMENT	STANDARDS CORRELATIONS
SE Pre-Reading Activity, p. 106 GENERAL **OSP Science Puzzlers, Twisters & Teasers** GENERAL		National Science Education Standards UCP 3; SAI 1; LS 2a, 3a
CRF Directed Reading A* ■ BASIC, **B*** SPECIAL NEEDS **CRF Vocabulary and Section Summary*** ■ GENERAL **SE Reading Strategy** Prediction Guide, p. 108 GENERAL **SE Connection to Language Arts** Nature or Nurture, p. 109 GENERAL **TE Inclusion Strategies,** p. 110 GENERAL **SS Science Skills** Organizing Your Research BASIC	**TE Homework,** p. 108 GENERAL **SE Reading Checks,** pp. 109, 110, 111 GENERAL **TE Reteaching,** p. 110 BASIC **TE Quiz,** p. 110 GENERAL **TE Alternative Assessment,** p. 110 GENERAL **TE Homework,** p. 110 GENERAL **SE Section Review,*** p. 111 ■ GENERAL **CRF Section Quiz*** ■ GENERAL	UCP 5; SAI 2; LS 2a, 2b, 2c, 2d, 2e
CRF Directed Reading A* ■ BASIC, **B*** SPECIAL NEEDS **CRF Vocabulary and Section Summary*** ■ GENERAL **SE Reading Strategy** Reading Organizer, p. 112 GENERAL **SE Math Practice** Counting Eggs, p. 113 GENERAL **TE Reading Strategy** Activity, p. 113 GENERAL **MS Math Skills for Science** Multiplying Whole Numbers* GENERAL **MS Math Skills for Science** Dividing Whole Numbers with Long Division* GENERAL **MS Math Skills for Science** Parts of 100: Calculating Percentages* GENERAL	**SE Reading Checks,** pp. 112, 114 GENERAL **TE Reteaching,** p. 114 BASIC **TE Quiz,** p. 114 GENERAL **TE Alternative Assessment,** p. 114 GENERAL **SE Section Review,*** p. 115 ■ GENERAL **CRF Section Quiz*** ■ GENERAL	UCP 1, 5; SAI 1; SPSP 1, 4; LS 1d, 1e, 1f, 2b
CRF Directed Reading A* ■ BASIC, **B*** SPECIAL NEEDS **CRF Vocabulary and Section Summary*** ■ GENERAL **SE Reading Strategy** Discussion, p. 116 GENERAL **TE Reading Strategy** Answering Questions, p. 117 GENERAL **TE Inclusion Strategies,** p. 117 GENERAL **TE Connection to Math** Factor of Increase, p. 118 GENERAL **CRF Reinforcement Worksheet** The Beginning of a Life* GENERAL **CRF Critical Thinking** One to Grow On!* ADVANCED	**SE Reading Checks,** pp. 116, 117, 118, 120 GENERAL **TE Reteaching,** p. 120 BASIC **TE Quiz,** p. 120 GENERAL **TE Alternative Assessment,** p. 120 ADVANCED **SE Section Review,*** p. 121 ■ GENERAL **CRF Section Quiz*** ■ GENERAL	UCP 1, 2, 3; SAI 1; ST 2; LS 1a, 1b, 1c, 1e, 2b; *Chapter Lab:* SAI 1, 2; ST 2; SPSP 4; HNS 1; LS 1a, 2b; *LabBook:* UCP 2, 3; SAI 1; ST 1

One-Stop Planner® CD-ROM

This convenient CD-ROM includes:
- **Lab Materials QuickList Software**
- **Holt Calendar Planner**
- **Customizable Lesson Plans**
- **Printable Worksheets**
- **ExamView® Test Generator**

CNN Student News

cnnstudentnews.com

Find the latest news, lesson plans, and activities related to important scientific events.

SciLinks NSTA

www.scilinks.org

Maintained by the **National Science Teachers Association.** See Chapter Enrichment pages for a complete list of topics.

Current Science®

Check out ***Current Science*** articles and activities by visiting the HRW Web site at **go.hrw.com.** Just type in the keyword **HL5CS26T.**

Classroom Videos
- **Lab Videos** demonstrate the chapter lab.
- **Brain Food Video Quizzes** help students review the chapter material.

5 Chapter Resources

Visual Resources

CHAPTER STARTER TRANSPARENCY

Reproduction and Development CHAPTER STARTER

Strange but True!

A tiny animal is resting peacefully in a warm, dark space. Suddenly it finds itself driven out into the cold and light. The chilly temperature is almost unbearable for its hairless body. The glaring light is disorienting to its blurry vision. Driven by instinct, the animal starts crawling up and up to the promise of warmth and security. If it slips and falls to the ground, it will die. Almost 30 minutes pass before it reaches its destination—a pouch in its mother's abdomen. Inside the pouch, it finds a nipple that will supply milk for several months as the animal grows and develops.

Can you guess what animal has just been born? That's right, a baby kangaroo, called a joey. Joeys are born in a very early stage of development. They continue to develop inside their mother's pouch until they are able to eat solid food such as grasses and other plants.

Like all living things, kangaroos eventually grow old and die. They must reproduce to pass on their genetic heritage. There are many ways of reproducing, as you will discover in this chapter. But reproduction is only part of the story. Development, or the way organisms grow, is also an important part of the cycle of life.

BELLRINGER TRANSPARENCIES

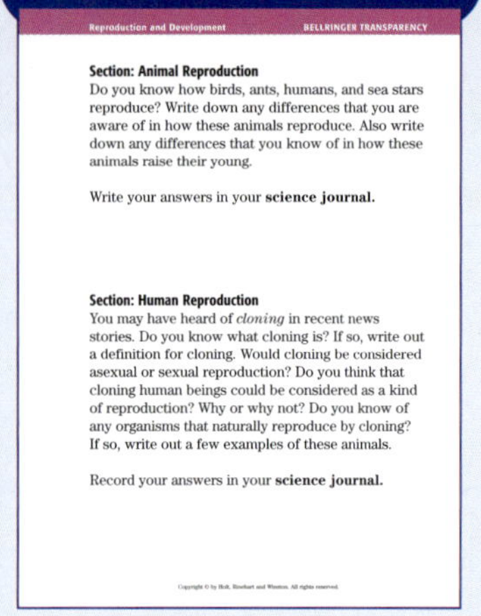
Reproduction and Development BELLRINGER TRANSPARENCY

Section: Animal Reproduction

Do you know how birds, ants, humans, and sea stars reproduce? Write down any differences that you are aware of in how these animals reproduce. Also write down any differences that you know of in how these animals raise their young.

Write your answers in your **science journal.**

Section: Human Reproduction

You may have heard of *cloning* in recent news stories. Do you know what cloning is? If so, write out a definition for cloning. Would cloning be considered asexual or sexual reproduction? Do you think that cloning human beings could be considered as a kind of reproduction? Why or why not? Do you know of any organisms that naturally reproduce by cloning? If so, write out a few examples of these animals.

Record your answers in your **science journal.**

TEACHING TRANSPARENCIES

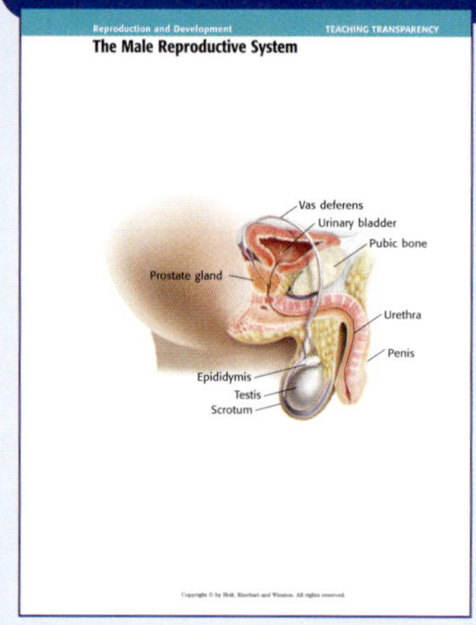
Reproduction and Development TEACHING TRANSPARENCY

The Male Reproductive System

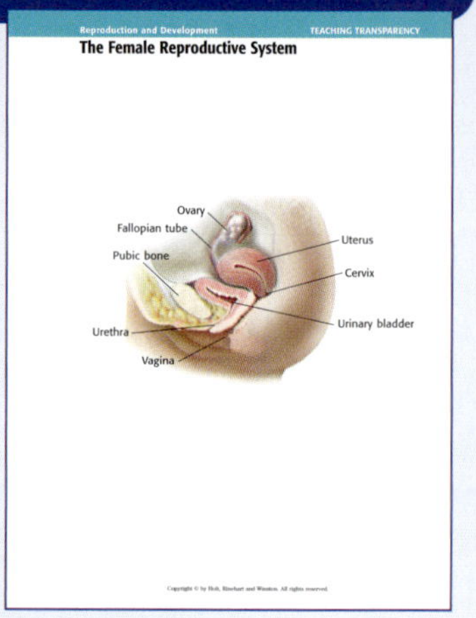
Reproduction and Development TEACHING TRANSPARENCY

The Female Reproductive System

TEACHING TRANSPARENCIES

Reproduction and Development

Stages of Human Development

The Nature of Sound TEACHING TRANSPARENCY

How Sonar Works

A fish finder sends ultrasonic waves down into the water. The time it takes for the echo to return helps determine the location of the fish.

LINK TO PHYSICAL SCIENCE

Chapter: The Nature of Sound

CONCEPT MAPPING TRANSPARENCY

Reproduction and Development CONCEPT MAPPING

Use the following terms to complete the concept map below:

fragmentation, reproduction, sexual reproduction, internal fertilization, an embryo, zygote, egg, budding, asexual reproduction

Planning Resources

LESSON PLANS

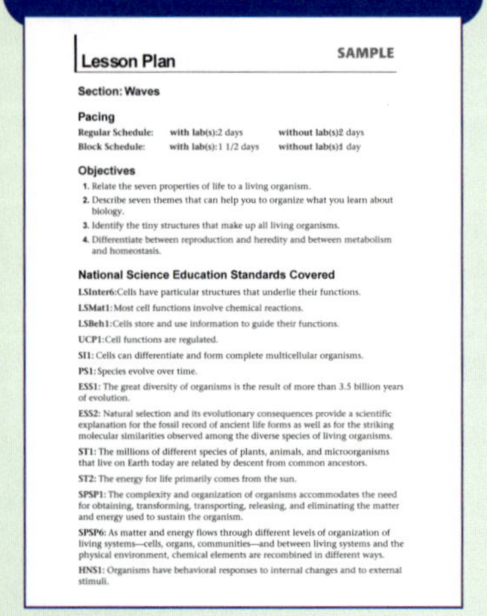
SAMPLE

Lesson Plan

Section: Waves

Pacing

Objectives

National Science Education Standards Covered

PARENT LETTER

SAMPLE

Dear Parent,

Questions to Ask Along the Way

ALSO IN SPANISH

TEST ITEM LISTING

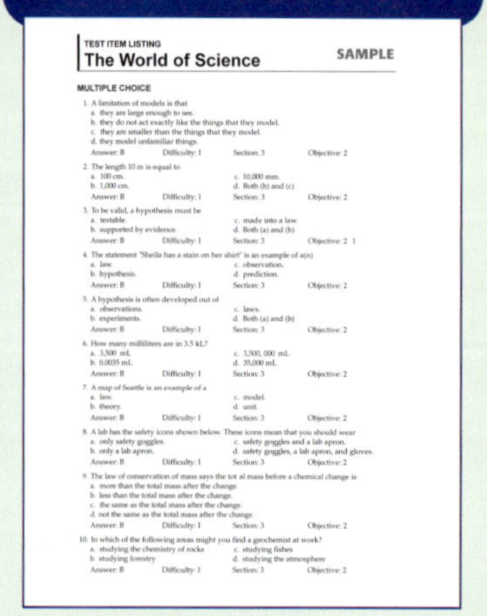
TEST ITEM LISTING SAMPLE

The World of Science

MULTIPLE CHOICE

One-Stop Planner® CD-ROM

This CD-ROM includes all of the resources shown here and the following time-saving tools:

- *Lab Materials QuickList Software*
- *Customizable lesson plans*
- *Holt Calendar Planner*
- *The powerful ExamView® Test Generator*

For a preview of available worksheets covering math and science skills, see pages T12–T19. All of these resources are also on the One-Stop Planner®.

Meeting Individual Needs

- DIRECTED READING A — Basic — Also in Spanish
- VOCABULARY ACTIVITY — General
- REINFORCEMENT — Basic
- SCILINKS ACTIVITY — General
- DIRECTED READING B — Special Needs
- VOCABULARY AND SECTION SUMMARY — General — Also in Spanish
- CRITICAL THINKING — Advanced
- SCIENCE PUZZLERS, TWISTERS & TEASERS — General

Labs and Activities

- LONG-TERM PROJECTS & RESEARCH IDEAS — Advanced
- DATASHEETS FOR QUICK LABS
- DATASHEETS FOR CHAPTER LABS
- DATASHEETS FOR LABBOOK

Review and Assessments

- SECTION QUIZ — General — Also in Spanish
- CHAPTER REVIEW — General — Also in Spanish
- CHAPTER TEST B — Advanced
- STANDARDIZED TEST PREPARATION — General
- SECTION REVIEW — General — Also in Spanish
- CHAPTER TEST A — General — Also in Spanish
- CHAPTER TEST C — Special Needs
- PERFORMANCE-BASED ASSESSMENT — General

5 Chapter Enrichment

This Chapter Enrichment provides relevant and interesting information to expand and enhance your presentation of the chapter material.

Section 1

Animal Reproduction

Asexual Reproduction

- The most common forms of asexual reproduction are binary fission, budding, and fragmentation. Binary fission is used mainly by bacteria.

- A major advantage of asexual reproduction is that it does not require a mate. Asexual reproduction also allows animals to produce many offspring in a short period of time. Many animals that do not move around, such as sea sponges, reproduce asexually.

Sexual Reproduction

- Because sexual reproduction brings together genetic material from two parents, there is greater variation among animals that reproduce sexually versus those that reproduce asexually.

Fertilization

- In external fertilization, eggs can be fertilized without physical contact between the parents. Instead, chemical signals coordinate the fertilization process, ensuring that the parents release their sex cells at the appropriate time.

- Internal fertilization requires a more sophisticated reproductive system, including organs for delivering and storing sperm. Fertilized eggs can develop externally, as with birds, or internally, as with placental mammals. Internally protected embryos are more likely to survive, but placental females do not usually produce as many offspring as do egg-laying females.

Is That a Fact!

◆ Some animal species that reproduce sexually don't have separate sexes. Instead, every individual contains male and female sexual characteristics. This situation is known as *hermaphrodism*. Hermaphrodites, such as earthworms, usually exchange sex cells with one another. Some hermaphrodites can reproduce by themselves as well.

Section 2

Human Reproduction

The Male Reproductive System

- Male reproductive functions mainly concern sperm production. The head of a sperm contains DNA, and the tail region contains mitochondria. The mitochondria are "engines" for the sperm, providing the sperm's tail with the energy to whip back and forth.

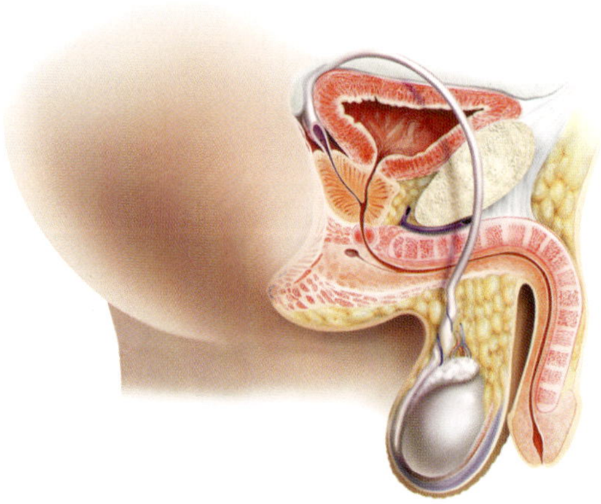

- The process of maturation of sperm from germ cells to spermatozoa takes about 74 days. Even then, they cannot yet penetrate an ovum. First they must "ripen" in the epididymis, a process that takes about 10 days. Though the maturation process is lengthy, once sperm are fully mature, they can remain viable for about 6 weeks.

Is That a Fact!

◆ If the seminiferous tubules—the bundle of tubes that makes up each testicle—were joined together and extended, they would be more than 200 m long!

The Female Reproductive System

- The ovaries are the primary female reproductive organs. About the size of large almonds, the ovaries are located on either side of the uterus, each anchored by an ovarian ligament. These tiny organs secrete the hormones largely responsible for development during puberty. They are also responsible for releasing eggs.

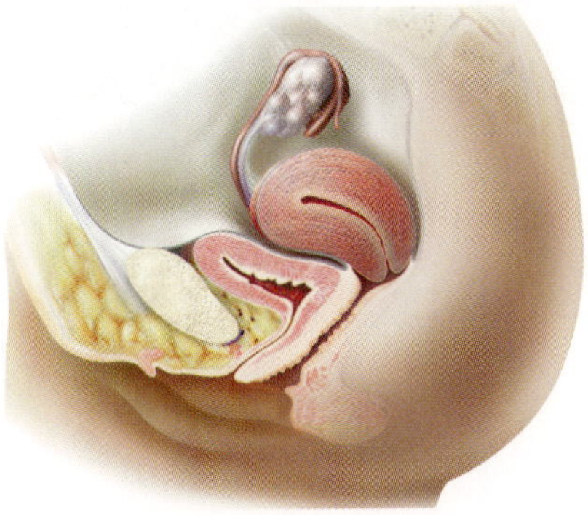

- Every menstrual cycle, several ova begin to ripen. In most cases, however, only one egg reaches maturity at a time. This mature, ripened ovum, encased in a Graafian follicle, travels to the surface of the ovary, where it remains until midcycle, when ovulation occurs. Then, the Graafian follicle, distended with fluid, ruptures, sending the egg into the abdominal cavity. The fallopian tube then captures the ovum, and the ovum begins its descent to the uterus.

Sexually Transmitted Diseases

- Chlamydia is the most prevalent sexually transmitted disease (STD) in North America. Caused by an organism called *Chlamydia trachomatis,* its symptoms include a frequent desire to urinate, pain with urination, and penile or vaginal discharge. In women, there are often few symptoms in the early stages. Troublesome as the symptoms are, the consequences are worse: chlamydia is a major cause of infertility. If left untreated, it can cause *pelvic inflammatory disease* and, in women, the subsequent inability to conceive. In men, infection that reaches the testes results in infertility.

- Also known as *salpingitis,* pelvic inflammatory disease (PID) is an infection of the fallopian tubes, uterus, and cervix. While a number of bacteria can cause PID, the usual culprits are chlamydia and gonorrhea. Although these infections can be completely cured with antibiotics, PID often leaves the fallopian tubes—the conduits between the ovaries and the uterus—scarred, making conception difficult or impossible. Other potential consequences of PID include ectopic pregnancy, peritonitis, and death.

Section 3

Growth and Development

Sex Determination

- One pair of human chromosomes determines the sex of a baby. There are two types of these sex chromosomes: X and Y. Because the egg contains only the X chromosome, the gender of the baby is determined by the father's sperm, which may contain either an X or a Y chromosome. If a sperm containing an X chromosome joins with an egg, the baby will be a girl. If a sperm containing a Y chromosome fertilizes the egg, the baby will be a boy.

Fetal Development

- The development of a baby from a single cell progresses at an astounding rate. Early in development, the embryo resembles a tiny tadpole, with a rounded body and tail. About 4 weeks after fertilization (the 6th week of pregnancy), however, limb buds—with knee and elbow joints evident—form, and facial features are recognizable. By the ninth week, nerves and muscles have developed enough that the fetus can move independently. By 12 weeks, the fetus is about 7.6 cm long and has a mass of about 28 g.

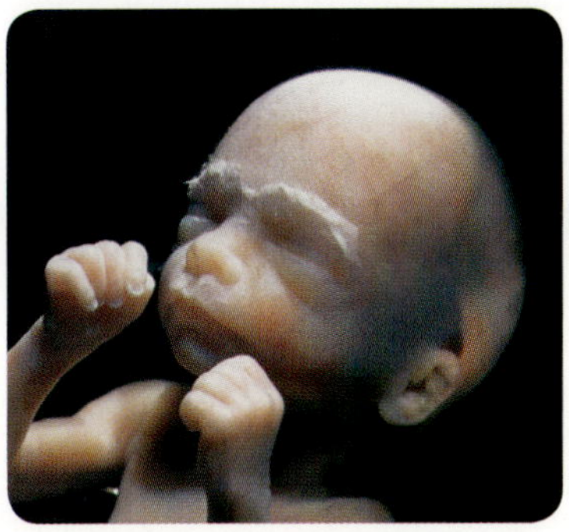

SCILINKS

NSTA Developed and maintained by the National Science Teachers Association

SciLinks is maintained by the National Science Teachers Association to provide you and your students with interesting, up-to-date links that will enrich your classroom presentation of the chapter.

Visit www.scilinks.org and enter the SciLinks code for more information about the topic listed.

Topic: Animal Repoduction
SciLinks code: HSM0075

Topic: Repoduction
SciLinks code: HSM1293

Topic: Reproductive System Irregularities and Disorders
SciLinks code: HSM1298

Topic: Repoduction
SciLinks code: HSM1293

Topic: Before Birth
SciLinks code: HSM0140

Topic: Growth and Development
SciLinks code: HSM0700

5

Overview

Tell students that this chapter will help them learn about how animals, including people, reproduce. The chapter describes asexual and sexual reproduction in animals. The chapter also describes human reproduction and development.

Assessing Prior Knowledge

Students should be familiar with the following topics:

- cells and cellular activities
- mammals
- human body organization and structure

Identifying Misconceptions

Students may have some confusion about forms of reproduction that do not involve "mating," such as asexual reproduction or external fertilization. Students may also be confused that offspring inherit traits from both parents equally. Girls may believe they get more of their traits from their mother, and boys may believe they share more with their father. Students may or may not be familiar with the words *gene* and *chromosome*. Very few students have a good grasp of the chemical basis for inheritance.

5 Reproduction and Development

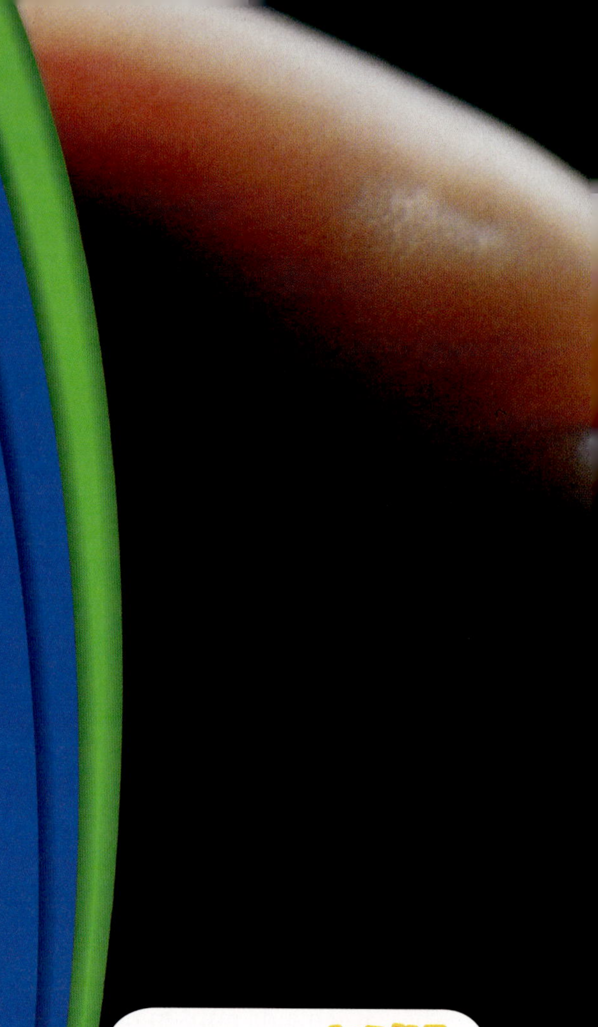

About the PHOTO

If someone had taken your picture when your mother was about 13 weeks pregnant with you, that picture would have looked much like this photograph. You have changed a lot since then, haven't you? You started out as a single cell, and you became a complete person. And you haven't stopped growing and changing yet. In fact, you will continue to change for the rest of your life.

PRE-READING ACTIVITY

Graphic Organizer

Spider Map Before you read the chapter, create the graphic organizer entitled "Spider Map" described in the **Study Skills** section of the Appendix. Label the circle "Reproduction and Development." Create a leg for each section title. As you read the chapter, fill in the map with details about reproduction and development from each section.

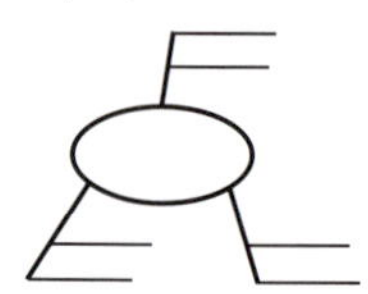

Standards Correlations

National Science Education Standards

The following codes indicate the National Science Education Standards that correlate to this chapter. The full text of the standards is at the front of the book.

Chapter Opener
UCP 3; SAI 1; LS 2a, 3a

Section 1 Animal Reproduction
UCP 5; SAI 2; LS 2a, 2b, 2c, 2d, 2e

Section 2 Human Reproduction
UCP 1, 5; SAI 1; SPSP 1, 4; LS 1d-1f, 2b

Section 3 Growth and Development
UCP 1, 2, 3; SAI 1; ST 2; LS 1a, 1c, 1d, 1e, 2b; *Labbook:* UCP2, 3; SAI 1; ST 1

Chapter Lab
SAI 1, 2; ST 2; SPSP 3, 4; HNS 1; LS 1a, 2b

Chapter Review
LS 2a, 2b

Science in Action
SAI 1; ST 2; SPSP 2, 5; HNS 1; LS 1e, 1f

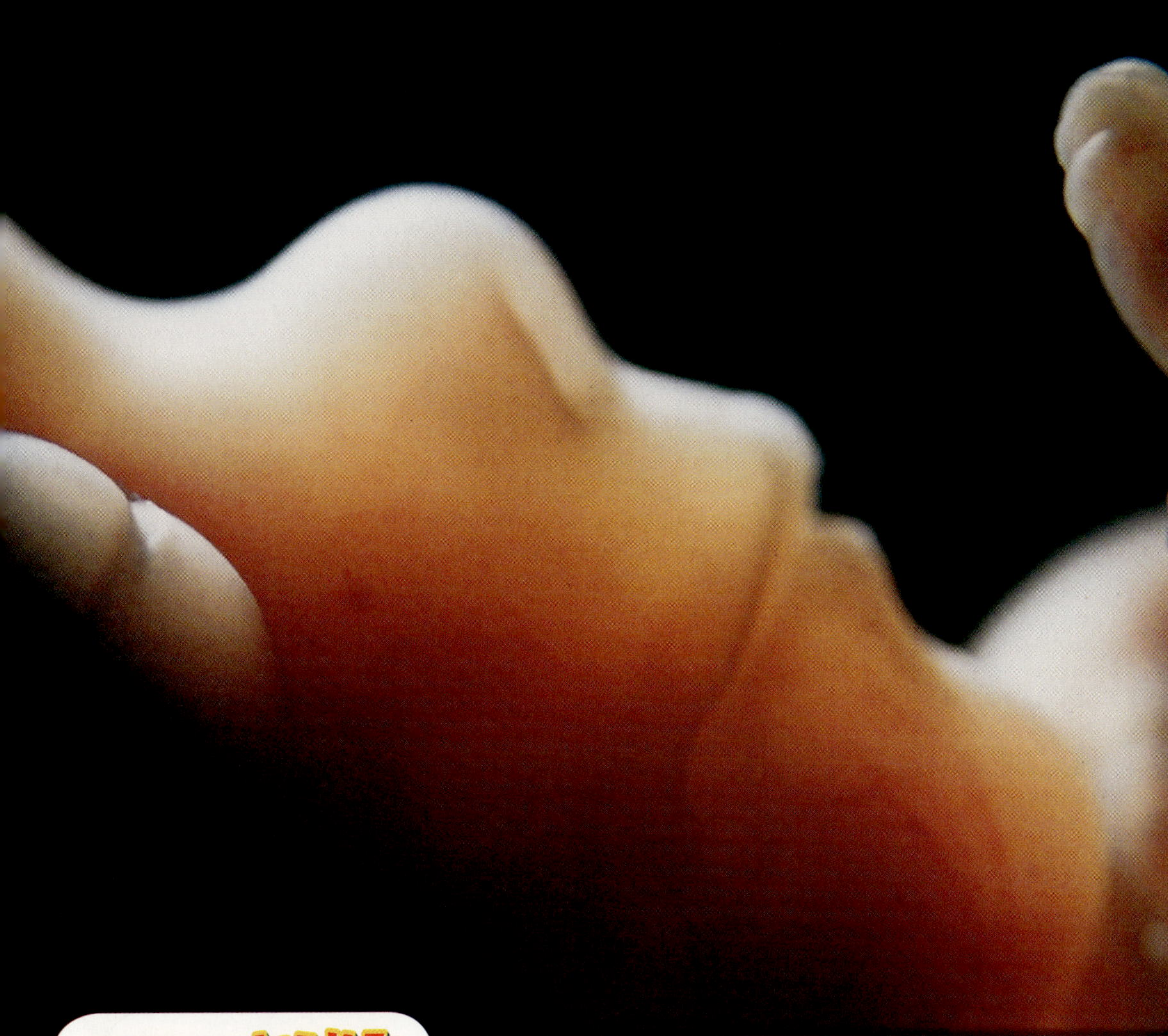

START-UP ACTIVITY

How Grows It?

As you read this paragraph, you are slowly aging. Your body is growing into the body of an adult. But does your body have the same proportions that an adult's body has? Complete this activity to find out.

Procedure

1. Have a classmate use a **tape measure** and **meterstick** to measure your total height, head height, and leg length. Your teacher will tell you how to take these measurements.
2. Use the following equations to calculate your head height–to–total body height proportion and your leg length–to–total body height proportion.

$$\text{head proportion} = \frac{\text{head height}}{\text{body height}} \times 100$$

$$\text{leg proportion} = \frac{\text{leg length}}{\text{body height}} \times 100$$

3. Your teacher will give you the head, body, and leg measurements of three adults. Calculate the head-body and leg-body proportions of each of the three adults. Record all of the measurements and calculations.

Analysis

1. Compare your proportions with the proportions of the three adults.

START-UP ACTIVITY

MATERIALS

FOR EACH PAIR
- tape measure and meterstick

Teacher's Notes: Instructions for taking measurements (demonstrate these techniques for your students): Measure head height by standing next to the board with one ear against it. On the board, mark the top of the head and the bottom of the chin. Measure the distance between the marks. Find the total body height by measuring the distance from the top-of-the-head mark to the ground. Leg length should be measured from the hip where the leg bends when you sit down. Hold a meterstick parallel to the floor at your hips. Have someone then use the tape measure to measure the distance from the meterstick to the floor. Prior to class, write the measurements of at least three adults, such as yourself and two other teachers, on the board.

Answer

1. The student's head height will probably be a greater proportion of his or her overall height than the adults' head height. The student's leg length should be about 50% of his or her overall height, which should match the leg-length proportion of adults.

Reproduction and Development — CHAPTER STARTER

Strange but True!

A tiny animal is resting peacefully in a warm, dark space. Suddenly it finds itself driven out into the cold and light. The chilly temperature is almost unbearable for its hairless body. The glaring light is disorienting to its blurry vision. Driven by instinct, the animal starts crawling up and up to the promise of warmth and security. If it slips and falls to the ground, it will die. Almost 30 minutes pass before it reaches its destination—a pouch in its mother's abdomen. Inside the pouch, it finds a nipple that will supply milk for several months as the animal grows and develops.

Can you guess what animal has just been born? That's right, a baby kangaroo, called a joey. Joeys are born in a very early stage of development. They continue to develop inside their mother's pouch until they are able to eat solid food such as grasses and other plants.

Like all living things, kangaroos eventually grow old and die. They must reproduce to pass on their genetic heritage. There are many ways of reproducing, as you will discover in this chapter. But reproduction is only part of the story. Development, or the way organisms grow, is also an important part of the cycle of life.

Chapter Starter Transparency
Use this transparency to help students begin thinking about reproduction and human growth and development.

CHAPTER RESOURCES

Technology

- **Transparencies** (READING SKILLS)
  - Chapter Starter Transparency
- **Student Edition on CD-ROM**
- **Guided Reading Audio CD**
  - English or Spanish
- **Classroom Videos**
  - Brain Food Video Quiz

Workbooks

- **Science Puzzlers, Twisters & Teasers**
  - Reproduction and Development (GENERAL)

SECTION 1

Focus

Overview

In this section, students learn about asexual and sexual reproduction. Students also learn about internal and external fertilization. Finally, students learn about differences in mammalian reproduction.

Bellringer

Write the following list on the board:

a. bird **c.** ants

b. human **d.** sea stars

Ask students to write a paragraph explaining how they think reproduction differs between these four animals.

Motivate

Discussion — GENERAL

Reproduction If appropriate for your students, lead a discussion and ask students to think about the similarities and differences between the ways animals, such as the ones listed above, reproduce. Birds and ants lay eggs, but humans and sea stars don't. Females and males mate to reproduce in humans, ants, and birds, but not in sea stars. Help them understand that the end result of reproduction is the same for all animal species, but the means differ widely.
LS Verbal

SECTION 1

READING WARM-UP

Objectives

- Distinguish between asexual and sexual reproduction.
- Explain the difference between external and internal fertilization.
- Identify the three different types of mammalian reproduction.

Terms to Learn

asexual reproduction
sexual reproduction
egg
sperm
external fertilization
internal fertilization

READING STRATEGY

Prediction Guide Before reading this section, write the title of each heading in this section. Next, under each heading, write what you think you will learn.

Animal Reproduction

The life span of some living things is short compared with ours. For example, a fruit fly lives only about 40 days. Other organisms live much longer than we do. Some bristlecone pine trees, for example, are nearly 5,000 years old.

But all living things eventually die. If a species is to survive, its members must reproduce.

Asexual Reproduction

Some animals, particularly simpler ones, reproduce asexually. In **asexual reproduction,** a single parent has offspring that are genetically identical to the parent.

One kind of asexual reproduction is called budding. *Budding* happens when a part of the parent organism pinches off and forms a new organism. The new organism separates from the parent and lives independently. The hydra, shown in **Figure 1,** reproduces by budding. The new hydra is genetically identical to its parent.

Fragmentation is a second kind of asexual reproduction. In *fragmentation,* parts of an organism break off and then develop into a new individual that is identical to the original one. Certain organisms, such as flatworms called *planaria,* reproduce by fragmentation. A third type of asexual reproduction, similar to fragmentation, is *regeneration.* When an organism capable of regeneration, such as the sea star in **Figure 2,** loses a body part, that part may develop into an entirely new organism.

asexual reproduction reproduction that does not involve the union of sex cells and in which a single parent produces offspring that are genetically identical to the parent

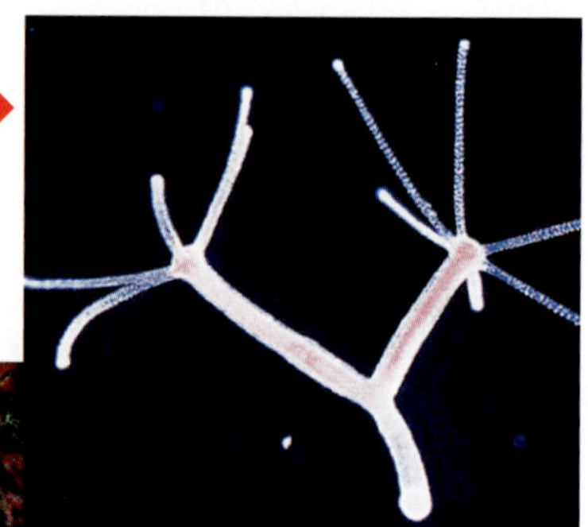

Figure 1 *The hydra bud will separate from its parent. Buds from other organisms, such as certain corals, remain attached to the parent.*

Figure 2 *The largest arm on this sea star was a fragment, from which a new sea star will regenerate. In time, all of the sea star's arms will grow to the same size.*

CHAPTER RESOURCES

Chapter Resource File

- Lesson Plan
- Directed Reading A BASIC
- Directed Reading B SPECIAL NEEDS

Technology

Transparencies
- Bellringer

Workbooks

Science Skills
- Organizing Your Research BASIC

Homework — GENERAL

Making Tables Ask students to make a table in their **science journal.** In one column, they will list 10 mammals. In the next column, they will indicate how each mammal produces young. From the information in this section, students should be able to indicate whether the animal is a monotreme, a marsupial, or a placental. Students should fill in the table to the best of their knowledge, research the correct answers, and put them in an additional column. **LS Logical**

Sexual Reproduction

Most animals reproduce sexually. In **sexual reproduction,** offspring are formed when genetic information from more than one parent combines. Sexual reproduction in animals usually requires two parents—a male and a female. The female parent produces sex cells called **eggs.** The male parent produces sex cells called **sperm.** When an egg's nucleus and a sperm's nucleus join, a fertilized egg, called a *zygote* (ZIE GOHT), is created. This joining of an egg and sperm is known as *fertilization.*

Human cells—except eggs and sperm and mature red blood cells—contain 46 chromosomes. Eggs and sperm are formed by a process called *meiosis.* In humans, meiosis is the division of one cell that has 46 chromosomes into four cells that have 23 chromosomes each. When an egg and a sperm join to form a zygote, the original number of 46 chromosomes is restored.

Genetic information is found in *genes.* Genes are located on *chromosomes* (KROH muh SOHMZ) made of the cell's DNA. During fertilization, the egg and sperm each contribute chromosomes to the zygote. The combination of genes from the two parents results in a zygote that grows into a unique individual. **Figure 3** shows how genes mix through three generations.

Reading Check What is sexual reproduction? (*See the Appendix for answers to Reading Checks.*)

sexual reproduction reproduction in which sex cells from two parents unite to produce offspring that share traits from both parents

egg a sex cell produced by a female

sperm the male sex cell

Figure 3 Inheriting Genes

Eggs and sperm contain chromosomes. You inherit chromosomes—and the genes on them—from both of your parents. Your parents each inherited chromosomes from their parents.

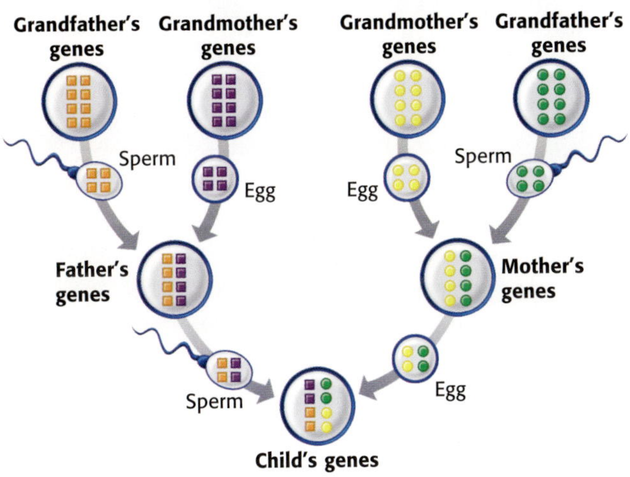

CONNECTION TO Language Arts

WRITING SKILL **Nature or Nurture?** Scientists debate whether genetics or upbringing is more important in shaping people. Use the Internet or library to research the issue of "nature versus nurture." Find information about identical twins who were raised apart. When you finish your research, write a persuasive essay supporting one side of the debate. Include evidence to support your argument.

Teach

Demonstration — GENERAL

The Amazing Egg Bird eggs are fertilized internally but develop externally, which requires good packaging. The egg shell gives the developing embryo a stable, safe environment. Using an ordinary chicken egg, demonstrate that the shape of the egg keeps it from rolling away from the mother. Next, wrap your hand around the egg. Demonstrate that the shape of the egg is protective against squeezing pressure from the outside. Tell students that the egg is shaped to withstand the pressures of passing through the chicken and of incubation. It is designed to be easy to break from within so that even a young chick can peck its way out. English Language Learners
LS Visual

Answer to Reading Check

Sexual reproduction is reproduction in which the sex cells (egg and sperm) of two parents unite to form a new individual.

Answer to Connection to Language Arts

Opinions may vary. Make sure students have provided information to support their answers. Sources of information may include journals such as *Scientific American* and reputable studies of twins.

ACTIVITY — BASIC

Asexual or Sexual Reproduction? Help students differentiate between sexual and asexual reproduction by having them make a poster that includes all the pertinent information. Students should use a different primary color for sexual and asexual reproduction and include artwork depicting the cells involved. The diagrams should state whether or not the resulting offspring are genetically identical to their parent(s). English Language Learners
LS Visual

CONNECTION ACTIVITY
Chemistry — ADVANCED

PORTFOLIO Encourage interested students to use the library or the Internet to research the role that incubation temperature plays in the reproductive process of some reptile and amphibian species. Have students draw diagrams or make a poster illustrating the effect of temperature on reproduction and development. Give students time to present their findings to the class. LS Visual

Close

Answer to Reading Check

External fertilization is when sex cells unite outside of the female's body. Internal fertilization is when sex cells unite inside the female's body.

Reteaching — BASIC

Outline Discuss with students how to make an outline of this section. Ask students to volunteer to share their outline to reteach the class. **LS Logical**

Quiz — GENERAL

1. How do the offspring created by asexual reproduction differ from those created by sexual reproduction? (Offspring from asexual reproduction are genetically identical to the parent. Offspring from sexual reproduction have genetic material from two parents and are genetically different from their parents.)
2. Why does external fertilization usually take place in moist environments? (External fertilization usually takes place in moist environments so that the delicate zygotes do not dry out.)

Alternative Assessment — GENERAL

Concept Mapping Have students create a concept map using the new terms in this section and the section title. **LS Verbal**

Figure 4 *Some fish, such as these clownfish, fertilize their eggs externally. The eggs are the orange mass on the rock.*

Internal and External Fertilization

Fertilization can happen either outside or inside the female's body. When the sperm fertilizes the eggs outside the female's body, the process is called **external fertilization.** External fertilization must take place in a moist environment so that the delicate zygotes won't dry out. Some fishes, such as those in **Figure 4,** reproduce by external fertilization.

Many amphibians, such as frogs, use external fertilization. For example, the female frog releases her eggs. At the same time, the male frog releases his sperm over the eggs to fertilize them. Frogs usually leave the zygotes to develop on their own. In about two weeks, the fertilized eggs hatch into tadpoles.

Internal Fertilization

When the egg and sperm join inside the female's body, the process is called **internal fertilization**. Internal fertilization allows the female animal to protect the developing egg inside her body. Reptiles, birds, mammals, and some fishes reproduce by internal fertilization. Many animals that use internal fertilization can lay fertilized eggs. Female chickens, for example, usually lay one or two eggs after internal fertilization has taken place.

In most mammals, one or more fertilized eggs develop inside the mother's body. Many mammals give birth to young that are well developed. Young zebras, such as the one in **Figure 5,** can stand up and nurse almost immediately after birth.

external fertilization the union of sex cells outside the bodies of the parents

internal fertilization fertilization of an egg by sperm that occurs inside the body of a female

✓ ***Reading Check*** **What is the difference between external and internal fertilization?**

Figure 5 *This zebra has just been born, but he is already able to stand. Within an hour, he will be able to run.*

Homework — GENERAL

Writing **Montremes** Encourage students to research an egg-laying mammal. Then, have them write and illustrate a short story about how the egg develops into a new individual. **LS Verbal**

INCLUSION *Strategies*

- ***Behavior Control Issues***
- ***Visually Impaired***
- ***Attention Deficit Disorder***

Organize the students into teams of two or three. Have each team find an animal that has a pouch for its babies. Ask the teams to create a presentation giving what the name of the animal is, where the animal lives, how the baby gets into the pouch, and how long the baby stays in the pouch. Make sure no animal choices are repeated. **LS Verbal**

Mammals

All mammals reproduce sexually. All mammals nurture their young with milk. And all mammals reproduce in one of the following three ways:

- **Monotreme** *Monotremes* (MAHN oh TREEMZ) are mammals that lay eggs. After the eggs are incubated and hatch, the young are nourished by milk that oozes from pores on the mother's belly. Echidnas and platypuses are monotremes.
- **Marsupial** Mammals that give birth to partially developed live young, such as the kangaroo in **Figure 6,** are *marsupials* (mahr SOO pee uhlz). Most marsupials have pouches where their young continue to develop after birth. Opossums, koalas, wombats, and Tasmanian devils are marsupials.
- **Placental Mammal** There are more than 4,000 species of placental mammals, including armadillos, humans, and bats. Placental mammals are nourished inside their mother's body before birth. Newborn placental mammals are more developed than newborn monotremes or marsupials are.

Figure 6 *The red kangaroo is a marsupial. A young kangaroo, such as this one in its mother's pouch, is called a* joey.

Reading Check Name two ways that all mammals are alike.

SECTION Review

Summary

- In asexual reproduction, a single parent produces offspring that are genetically identical to the parent.
- In sexual reproduction, an egg from one parent combines with a sperm from the other parent.
- Fertilization can be external or internal.
- All mammals reproduce sexually and nurture their young with milk.

Using Key Terms

For each pair of terms, explain how the meanings of the terms differ.

1. *internal fertilization* and *external fertilization*
2. *asexual reproduction* and *sexual reproduction*

Understanding Key Ideas

3. In humans, each egg and each sperm contain
   a. 23 chromosomes.
   b. 46 chromosomes.
   c. 69 chromosomes.
   d. 529 chromosomes.
4. List three types of asexual reproduction.
5. How do monotremes differ from marsupials?
6. Describe the process of meiosis.
7. Are humans placental mammals, monotremes, or marsupials? Explain.

Math Skills

8. Some bristlecone pine needles last 40 years. If a tree lives for 3,920 years, how many sets of needles might it grow?

Critical Thinking

9. **Making Inferences** Why is reproduction as important to a bristlecone pine as it is to a fruit fly?
10. **Applying Concepts** Describe one advantage of internal fertilization over external fertilization.

SCILINKS. NSTA Developed and maintained by the National Science Teachers Association

For a variety of links related to this chapter, go to www.scilinks.org

Topic: Reproduction

SciLinks code: HSM1293

Answers to Section Review

1. Internal fertilization occurs inside the female's body, while external fertilization occurs outside the female's body.
2. Asexual reproduction is when a single parent produces offspring that are genetically identical to the parent. Sexual repro duction requires sex cells from two parents to unite.
3. a
4. budding, fragmentation, and regeneration
5. Monotremes lay eggs; marsupials give birth to live young.
6. Meiosis is when a male or a female sex cell that has 46 chromosomes divides into 4 eggs or 4 sperm that have 23 chromosomes each.
7. Humans are placental mammals because they nourish their young internally before giving birth to live young.
8. $3,920 \div 40 = 98$ sets of needles
9. Every living organism is a member of a species. Every living organism eventually dies. In order for a species to survive, individual organisms must reproduce.
10. Internal fertilizaton allows the female to protect the developing egg inside her body.

Answer to Reading Check

All mammals reproduce sexually and nurture their young with milk.

INTERNET ACTIVITY

Sequence Board—GENERAL

For an internet activity related to this chapter, have students go to **go.hrw.com** and type in the keyword **HL5BD5W.**

CHAPTER RESOURCES

Chapter Resource File

- Section Quiz GENERAL
- Section Review GENERAL
- Vocabulary and Section Summary GENERAL
- Reinforcement Worksheet BASIC

SECTION 2

Focus

Overview

In this section, students learn about the male and female reproductive systems. Students also learn about multiple births and some problems of the reproductive system.

Bellringer

Have students write answers to these questions: Do you think that cloning human beings could be considered reproduction? Why or why not? What kind of reproduction is it?

Motivate

Discussion — GENERAL

Bird and Human Ask students to compare reproduction in birds with reproduction in humans. (Both birds and humans fertilize their eggs internally. Birds lay eggs and protect them. Birds keep the eggs warm while obtaining food for themselves. Human mothers carry a baby inside their body, so the baby is always protected. After eggs hatch or the baby is born, parents protect and care for the baby. Humans take care of their offspring for much longer than birds care for theirs.) **LS Verbal**

Answer to Reading Check

testes, epididymis, vas deferens, urethra, penis

SECTION 2

Human Reproduction

About nine months after a human sperm and egg combine, a mother gives birth to her baby. But how do humans make eggs and sperm?

READING WARM-UP

Objectives

- Identify the structures and functions of the male and female reproductive systems.
- Describe two reproductive system problems.

Terms to Learn

testes, penis, ovary, uterus, vagina

READING STRATEGY

Reading Organizer As you read this section, create an outline of the section. Use the headings from the section in your outline.

The Male Reproductive System

The male reproductive system, shown in **Figure 1,** produces sperm and delivers it to the female reproductive system. The **testes** (singular, *testis*) are a pair of organs that make sperm and testosterone (tes TAHS tuhr OHN). Testosterone is the main male sex hormone. It helps regulate the production of sperm and the development of male characteristics.

As sperm leave a testis, they are stored in a tube called an *epididymis* (EP uh DID i mis). Sperm mature in the epididymis. Another tube, called a *vas deferens* (vas DEF uh RENZ), passes from the epididymis into the body and through the *prostate gland*. The prostate gland surrounds the neck of the bladder. As sperm move through the vas deferens, they mix with fluids from several glands, including the prostate gland. This mixture of sperm and fluids is called *semen.*

To leave the body, semen passes through the vas deferens into the *urethra* (yoo REE thruh). The urethra is the tube that runs through the penis. The **penis** is the external organ that transfers semen into the female's body.

Reading Check Describe the path that sperm take from the testes to the penis. (*See the Appendix for answers to Reading Checks.*)

testes the primary male reproductive organs, which produce sperm and testosterone (singular, *testis*)

penis the male organ that transfers sperm to a female and that carries urine out of the body

Figure 1 The Male Reproductive System

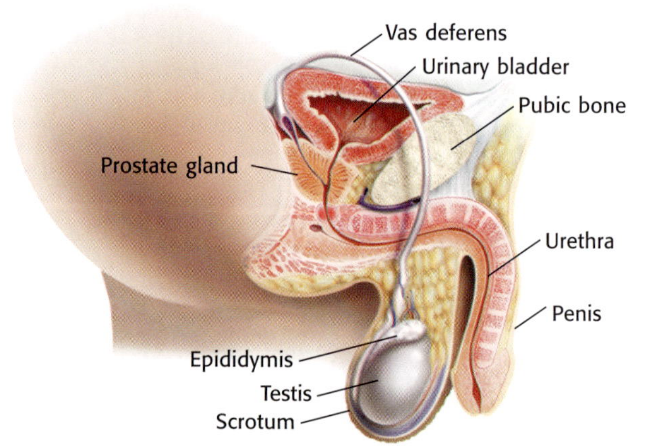

CHAPTER RESOURCES

Chapter Resource File

- Lesson Plan
- Directed Reading A BASIC
- Directed Reading B SPECIAL NEEDS

Technology

Transparencies

- Bellringer
- The Male Reproductive System

Is That a Fact!

Mumps, a common childhood disease, poses a risk to males who contract it during puberty or adulthood. When mumps occurs after childhood, it can cause inflammation of the testes, known as *acute orchitis,* which, in rare cases, can result in sterility.

Figure 2 The Female Reproductive System

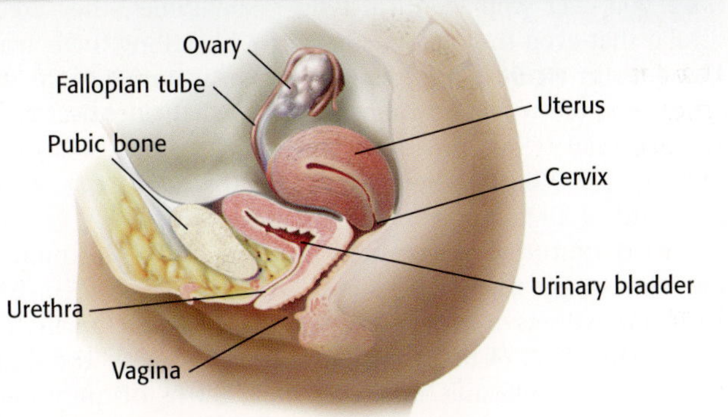

The Female Reproductive System

The female reproductive system, shown in **Figure 2,** produces eggs, nurtures fertilized eggs (zygotes), and gives birth. The two **ovaries** are the organs that make eggs. Ovaries also release estrogen (ES truh juhn) and progesterone (proh JES tuhr OHN), the main female sex hormones. These hormones regulate the release of eggs and development of female characteristics.

The Egg's Journey

During *ovulation* (AHV yoo LAY shuhn), an egg is released from an ovary and passes into a *fallopian* (fuh LOH pee uhn) *tube.* A fallopian tube leads from each ovary to the uterus. The egg passes through the fallopian tube into the uterus. Fertilization usually happens in the fallopian tube. If the egg is fertilized, the resulting zygote enters the uterus. The zygote may become embedded in the thickened lining of the uterus. The **uterus** is the organ in which a zygote develops into a baby.

When a baby is born, he or she passes from the uterus through the vagina and emerges outside the body. The **vagina** is the canal between the outside of the body and the uterus.

Menstrual Cycle

From puberty through her late 40s or early 50s, a woman's reproductive system goes through monthly changes. These changes prepare the body for pregnancy and are called the *menstrual cycle* (MEN struhl SIE kuhl). The first day of *menstruation* (MEN STRAY shuhn), the monthly discharge of blood and tissue from the uterus, is counted as the first day of the cycle. Menstruation lasts about 5 days. When menstruation ends, the lining of the uterus thickens. Ovulation occurs on about the 14th day of the cycle. If the egg is not fertilized within a few days, menstruation begins and flushes the egg away. The cycle—which usually takes about 28 days—starts again.

ovary in the female reproductive system of animals, an organ that produces eggs

uterus in female mammals, the hollow, muscular organ in which a fertilized egg is embedded and in which the embryo and fetus develop

vagina the female reproductive organ that connects the outside of the body to the uterus

MATH PRACTICE

Counting Eggs

1. The average woman ovulates each month from about age 12 to about age 50. How many mature eggs could she produce from age 18 to age 50?
2. A female's ovaries typically contain 2 million immature eggs. If she ovulates regularly from age 12 to age 50, what percentage of her eggs will mature?

CONNECTION ACTIVITY

Language Arts — GENERAL

Word Origins The words *male* and *female* actually originate from two unrelated Latin words. *Female* comes from *femella,* meaning "girl," while *male* comes from *masculus,* meaning "male." Have students select the names of any five parts of the male and female reproductive systems. Then, have students find the origins of those terms and write a brief history of each one.
LS Verbal

CHAPTER RESOURCES

Technology

Transparencies
- The Female Reproductive System

Workbooks

Math Skills for Science
- Multiplying Whole Numbers GENERAL
- Dividing Whole Numbers with Long Division GENERAL
- Parts of 100: Calculating Percentages GENERAL

Teach

READING STRATEGY — GENERAL

Activity Draw students' attention to **Figures 1** and **2.** Have them write the new terms in their **science journal,** leaving a few lines between each term. As students read the section, have them fill in definitions that explain the new terms in their own words. LS Verbal

Discussion — GENERAL

Multiple Births A Russian woman in the 18th century gave birth to 69 children, with 67 living to reach adulthood. Ask students: "If this woman began ovulation at age 12 and stopped ovulating at age 50, and pregnancy lasts 9 months, how could she have had so many children?" Write the following answer on the board to demonstrate how it is possible that she had 69 children: She had 27 pregnancies, all of them producing 2 or more children. She had 16 pairs of twins, 7 sets of triplets, 4 sets of quadruplets.
$(16 \times 2) + (7 \times 3) + (4 \times 4) = 69$
LS Logical

Answers to Math Practice

1. 50 y − 12 y = 38 y
   38 y × 12 eggs/year = 456 eggs
   She can produce about 456 mature eggs in her lifetime.
2. 456 mature eggs/2,000,000 possible eggs = 0.000228 × 100 = 0.0228% of her eggs may possibly mature.

Close

Answer to Reading Check

Twins happen about 30 times in every 1,000 births.

Reteaching —— BASIC

Puzzle Making Organize the class into small groups. Challenge each group to create a crossword puzzle using the Terms to Learn and italicized terms from the section. Have each group write appropriate clues and construct a puzzle. Then, have groups exchange puzzles. Allow time for students to solve the puzzles. LS **Logical/Interpersonal**

Quiz —— GENERAL

1. What is puberty? (the stage of development when sex organs become mature)
2. What purpose does the epididymis serve? (the site where sperm mature and are stored before sperm leave the testes)
3. What is menstruation? (the monthly discharge of blood and tissue from the uterus)

Alternative Assessment —— GENERAL

Tracing the Path If appropriate for your class, ask students to make diagrams that illustrate the path an egg or sperm must travel before fertilization. Have them label anatomical structures and indicate, with arrows, the direction in which the sex cell travels. English Language Learners LS **Visual**

Figure 3 *Identical twins have genes that are exactly the same. Many identical twins who are raised apart have similar personalities and interests.*

Multiple Births

Have you ever seen identical twins? Sometimes, they are so similar that even their parents have trouble telling them apart. The boys in **Figure 3** are identical twins. Fraternal twins, the other type of twins, are more common than identical twins are. Fraternal twins can look very different from each other. In every 1,000 births, there are about 30 sets of twins. About one-third of all twin births are identical twins.

Twins are the most common multiple births. But humans sometimes have triplets (3 babies). In the United States, there are about two sets of triplets in every 1,000 births. Humans also have quadruplets (4 babies), quintuplets (5 babies), and more. These types of multiple births are rare. Births of quintuplets or more happen only once in about 53,000 births.

Reading Check What is the frequency of twin births?

Reproductive System Problems

In most cases, the reproductive system functions flawlessly. But like any body system, the reproductive system sometimes has problems. These problems include disease and infertility.

STDs

Chlamydia, herpes, and hepatitis B are common sexually transmitted diseases. A *sexually transmitted disease,* or STD, is a disease that can pass from a person who is infected with the STD to an uninfected person during sexual contact. STDs are also called *sexually transmitted infections,* or STIs. These diseases affect many people each year, as shown in **Table 1.**

An STD you may have heard of is *acquired immune deficiency syndrome* (AIDS). AIDS is caused by *human immunodeficiency virus* (HIV). But you may not have heard of the STD *hepatitis B,* a liver disease also caused by a virus. This virus is spread in several ways, including sexual contact. In the United States, about 140,000 new cases of hepatitis B happen each year.

SCHOOL to HOME

Twins and More

With a parent, discuss some challenges that are created by the birth of twins, triplets, quadruplets, or other multiples. Include financial, mental, emotional, and physical challenges.

Create a poster that shows these challenges and ways to meet them.

If twins or other multiples are in your family, discuss how the individuals differ and how they are alike.

ACTIVITY

Table 1 The Spread of STDs in the United States

STD	Approximate number of new cases each year
Chlamydia	3 to 10 million
Genital HPV (human papillomavirus)	5.5 million
Genital herpes	1 million
Gonorrhea	650,000
Syphilis	70,000
HIV/AIDS	40,000 to 50,000

CONNECTION ACTIVITY
Language Arts —— ADVANCED

Writing **Fertility Drugs and Multiple Births** Have students conduct Internet or library research and write a report on how fertility drugs may have affected the number and frequency of multiple births in the last 15 years. LS **Verbal**

Group ACTIVITY —— GENERAL

Reproductive System Cancer Have students work in small groups to research breast, testicular, ovarian, prostate, or cervical cancer. Students should focus on the incidence of the disease, its risk factors, and the importance of early detection. Have students use the information they gather to design and create a public service brochure—complete with artwork—designed to educate the public about the disease. Allow students time to present and discuss their brochures in class. LS **Visual**

Cancer

Sometimes, cancer happens in reproductive organs. *Cancer* is a disease in which cells grow at an uncontrolled rate. Cancer cells start out as normal cells. Then, something triggers uncontrolled cell growth. Different kinds of cancer have different triggers.

In men, the two most common reproductive system cancers are cancer of the testes and cancer of the prostate gland. In women, the two most common reproductive system cancers are breast cancer and cancer of the cervix. The *cervix* is the lower part, or neck, of the uterus. The cervix opens to the vagina.

Infertility

In the United States, about 15% of married couples have difficulty producing offspring. Many of these couples are *infertile,* or unable to have children. Men may be infertile if they do not produce enough healthy sperm. Women may be infertile if they do not ovulate normally.

Sexually transmitted diseases, such as gonorrhea and chlamydia, can lead to infertility in women. STD-related infertility occurs in men, but not as commonly as it does in women.

CONNECTION TO Social Studies

Understanding STDs Select one of the STDs in **Table 1.** Make a poster or brochure that identifies the cause of the disease, describes its symptoms, explains how it affects the body, and tells how it can be treated. Include a bar graph that shows the number of cases in different age groups.

ACTIVITY

SECTION Review

Summary

- The male reproductive system produces sperm and delivers it to the female reproductive system.
- The female reproductive system produces eggs, nurtures zygotes, and gives birth.
- Humans usually have one child per birth, but multiple births, such as those of twins or triplets, are possible.
- Human reproduction can be affected by cancer, infertility, and disease.

Using Key Terms

1. Use the following terms in the same sentence: *uterus* and *vagina.*

Understanding Key Ideas

2. Describe two problems of the reproductive system.
3. Identify the structures and functions of the male and female reproductive systems.
4. Identical twins happen once in 250 births. How many pairs of these twins might be at a school with 2,750 students?
   a. 1
   b. 11
   c. 22
   d. 250

Math Skills

5. In one country, 7 out of 1,000 infants die before their first birthday. Convert this figure to a percentage. Is your answer greater than or less than 1%?

Critical Thinking

6. **Making Inferences** What is the purpose of the menstrual cycle?
7. **Applying Concepts** Twins can happen when a zygote splits in two or when two eggs are fertilized. How can these two ways of twin formation explain how identical twins differ from fraternal twins?
8. **Predicting Consequences** How might cancer of the testes affect a man's ability to make sperm?

SCI LINKS. NSTA Developed and maintained by the National Science Teachers Association

For a variety of links related to this chapter, go to www.scilinks.org

Topic: Reproduction System Irregularities or Disorders

SciLinks code: HSM1298

CHAPTER RESOURCES

Chapter Resource File

- Section Quiz GENERAL
- Section Review GENERAL
- Vocabulary and Section Summary GENERAL

Answers to Section Review

1. Sample answer: A woman's vagina is the canal that connects the uterus to the outside world.
2. One reproductive system problem is infertility. Infertility is when a man or a woman is unable to have children. Sexually transmitted diseases (STDs) are a second reproductive system problem. STDs can infect the reproductive system and may cause infertility.
3. testes (males): make sperm and produce testosterone (helps regulate sperm production and the development of male characteristics); penis (males): the external male organ that transfers semen into the female's body; ovaries (females): organs that make eggs and release estrogen and progesterone (which regulate the release of eggs and the development of female characteristics); uterus (females): the organ in which the fertilized egg develops; vagina (females): the canal that connects the uterus to the outside world, through which a baby passes when it is born
4. b
5. 7/1,000 = 0.007; Then, convert to a percentage by moving the decimal point two places to the right: 0.7%; This is less than 1%.
6. The menstrual cycle is the series of changes through which a female's reproductive system goes in order to prepare the female for pregnancy. The cycle usually takes about 28 days.
7. When a single fertilized egg splits in two, each cell contains identical genes. If each cell develops into a separate baby, they will be identical twins. When two separate eggs are fertilized, their genes are different. This results in fraternal twins.
8. Sample answer: Cancer is uncontrolled cell growth. In the testes, this uncontrolled growth could destroy cells and damage the production of sperm and the production of testosterone.

SECTION 3

Focus

Overview

In this section, students learn about fertilization and implantation. Students are also introduced to the different stages of growth of a fetus in utero, culminating in the birth of a baby. Finally, students learn about stages of human development, from birth through adulthood.

Bellringer

Write the following on the board:

Name the stages of physical development you have passed through thus far in your life.

Have students list the stages in their **science journal.** Remind students that their growth and development began while they were still in the uterus. (Students may list stages such as the following: crawling, walking, talking, growing taller, and puberty.)

Motivate

Discussion — GENERAL

Life Stages Ask students to list as many characteristics of each of the following as they can: infancy, childhood, adolescence, and adulthood. Tell students that while there are individual differences, all people go through these stages. LS Logical

SECTION 3

Growth and Development

READING WARM-UP

Objectives

- Summarize the processes of fertilization and implantation.
- Describe the development of the embryo and the fetus.
- Identify the stages of human development from birth to death.

Terms to Learn

embryo
placenta
umbilical cord
fetus

READING STRATEGY

Discussion Read this section silently. Write down questions that you have about this section. Discuss your questions in a small group.

Every one of us started out as a single cell. How did that cell become a person made of trillions of cells?

A single cell divides many times and develops into a baby. But the development of a baby from a single cell is only the first stage of human development. Think about how you will change between now and when you become a grandparent!

From Fertilization to Embryo

Ordinarily, the process of human development starts when a man deposits millions of sperm into a woman's vagina. A few hundred sperm make it through the uterus into a fallopian tube. There, a few sperm cover the egg. Usually, only one sperm gets through the outer coating of the egg. When this happens, it triggers a response—a membrane forms around the egg to keep other sperm from entering. When the sperm's nucleus joins with the nucleus of the egg, the egg becomes fertilized.

The fertilized egg (zygote) travels down the fallopian tube toward the uterus. This journey takes 5 to 6 days. During the trip, the zygote undergoes cell division many times. Eleven to 12 days after fertilization, the zygote has become a tiny ball of cells called an **embryo.** The embryo implants itself in the uterus. *Implantation* happens when the zygote embeds itself in the thick, nutrient-rich lining of the uterus. Fertilization and implantation are outlined in **Figure 1.**

Reading Check Describe the process of fertilization and implantation. (*See the Appendix for answers to Reading Checks.*)

embryo a developing human, from fertilization through the first 8 weeks of development (the 10th week of pregnancy)

placenta the partly fetal and partly maternal organ by which materials are exchanged between fetus and mother

Figure 1 Fertilization and Implantation

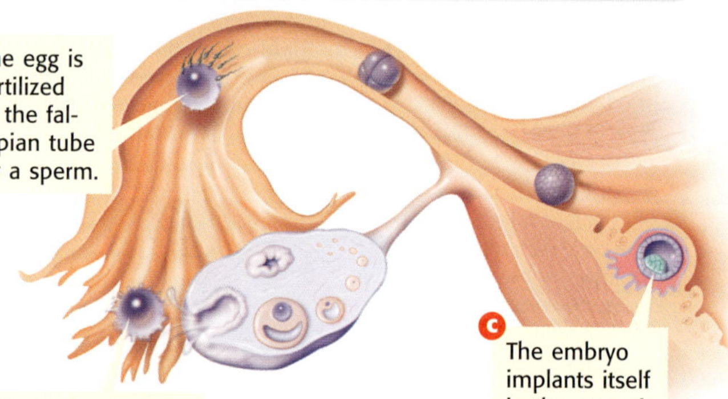

b The egg is fertilized in the fallopian tube by a sperm.

a The egg is released from the ovary.

c The embryo implants itself in the uterus's wall.

CHAPTER RESOURCES

Chapter Resource File

- Lesson Plan
- Directed Reading A BASIC
- Directed Reading B SPECIAL NEEDS

Technology

Transparencies
- Bellringer

Answer to Reading Check

Fertilization happens when the nucleus of a sperm unites with the nucleus of an egg. Implantaton happens after the fertilized egg travels down the fallopian tube to the uterus and embeds itself in the wall of the uterus.

From Embryo to Fetus

After implantation, the placenta (pluh SEN tuh) begins to grow. The **placenta** is a special two-way exchange organ. It has a network of blood vessels that provides the embryo with oxygen and nutrients from the mother's blood. Wastes produced by the embryo are removed in the placenta. They are carried by the mother's blood so that her body can excrete them. The embryo's blood and the mother's blood flow very near each other in the placenta, but they normally do not mix.

Reading Check Why is the placenta important?

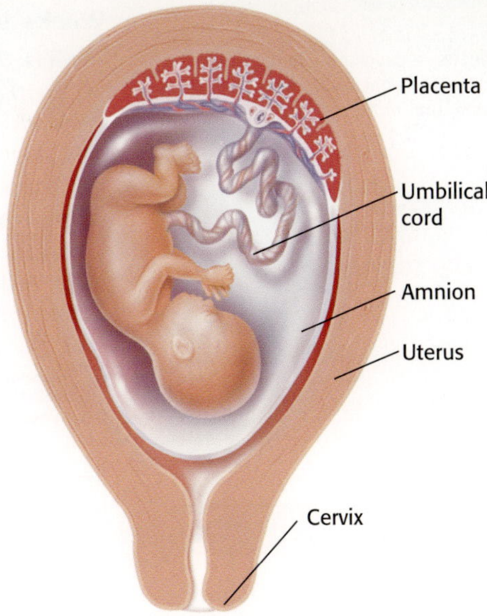

Figure 2 *The placenta, amnion, and umbilical cord are the life support system for the fetus. This fetus is about 20 to 22 weeks old.*

Weeks 1 and 2

Doctors commonly count the time of a woman's pregnancy as starting from the first day of her last menstrual period. Even though fertilization has not yet taken place, that day is a convenient date from which to start counting. A normal pregnancy lasts about 280 days, or 40 weeks, from that day.

Weeks 3 and 4

Fertilization takes place at about the end of week 2. In week 3, after fertilization, the zygote moves to the uterus. As the zygote travels, it divides many times. It becomes a ball of cells that implants itself in the wall of the uterus. The zygote is now called an *embryo*. At the end of week 4, implantation is complete and the woman is pregnant. The embryo's blood cells begin to form. At this point, the embryo is about 0.2 mm long.

umbilical cord the structure that connects the fetus to the placenta

Weeks 5 to 8

Weeks 5 to 8 of pregnancy are weeks 3 to 6 of embryonic development. In this stage, the embryo becomes surrounded by a thin membrane called the *amnion* (AM nee AHN). The amnion is filled with amniotic fluid and protects the growing embryo from bumps and injury. During week 5, the umbilical cord forms. The **umbilical cord** (uhm BIL i kuhl KAWRD) is a cord that connects the embryo to the placenta. **Figure 2** shows the umbilical cord, amnion, and placenta.

In this stage, the heart, brain, other organs, and blood vessels start to form. They grow quickly. In weeks 5 and 6, eyes and ears take shape. The spinal cord begins to develop. In week 6, tiny limb buds appear. These buds will become arms and legs. In week 8, muscles start developing. Nerves grow into the shoulders and upper arms. Fingers and toes start to form. The embryo, now about 16 mm long, can swallow and blink.

School to Home

Growing Up
With a parent, discuss the physical and mental changes that you went through between your birth and your first day of school. Make a poster illustrating those changes.

Activity

Teach

Activity — General

Making Models Students may have difficulty understanding how the placenta functions. To help them understand, have groups of students make a clay model of the diagram in **Figure 2.** Encourage groups to learn more about the placenta in order to add more detail to their diagram. English Language Learners
LS Kinesthetic

Answer to Reading Check

The placenta is important because it provides the embryo with oxygen and nutrients from the mother's blood. Wastes from the embryo also travel to the placenta, where they are carried to the mother so the mother can excrete them.

Reading Strategy — General

Answering Questions Ask students to write down the following question and fill in the correct answers as they read the section: At what stage of development does each of the following events take place? The fetus is felt kicking (weeks 15–16), the brain and spinal cord begin to form (weeks 5–6), and the fetus practices breathing by moving its lungs up and down (weeks 25–26). LS Verbal

Inclusion Strategies

- *Behavior Control Issues* • *Visually Impaired*
- *Attention Deficit Disorder*

Divide the class into six teams. Give each team several pages of newspapers and an empty plastic grocery bag. Choose a "stuffer" in each group. Make sure students with behavior control issues are "stuffers." Assign each team one of the pregnancy months three through nine. Have teams research the actual size of the baby during their assigned month, and have them stuff their bags to simulate the baby's size. Have teams line their "babies" up in order. Add the following to the lineup for months one through three:

- small pebble or piece of rice (month 1)
- small rock or a peanut (month 2)
- grocery bag that has a very small filled area about 2 inches in diameter and whose unneeded ends are cut off (month 3)

LS Kinesthetic

Teach, *continued*

BRAIN FOOD

Writing **Seahorses** Seahorses are one of very few kinds of organisms in which the male carries the fertilized eggs. Ask students to research the reproductive and parenting behaviors of seahorses. Have them make a poster showing their findings. (After courtship, the female deposits eggs into the pouch of her male mate. The male then fertilizes the eggs and carries them until they develop. After a gestation period of between 10 and 60 days, depending on the species, tiny, fully formed sea horses emerge from an opening in the pouch.)

Answer to Reading Check

The embryo is now called a *fetus;* the fetus's face begins to look more human and it can swallow, the fetus grows rapidly (it triples in size), and the fetus begins to make movements that the mother can feel.

Research — GENERAL

Writing **FAS** Drinking alcohol during the first few weeks of pregnancy can lead to birth defects and miscarriages. Have students research the causes and consequences of fetal alcohol syndrome (FAS) and present their findings in a short report. LS **Verbal**

fetus a developing human from seven or eight weeks after fertilization until birth

Weeks 9 to 16

At week 9, the fetus may begin to make tiny movements. After week 10, the embryo is called a **fetus** (FEET uhs). In about week 13, the fetus's face begins to look more human. During this stage, fetal muscles grow stronger. As a result, the fetus can make a fist and begins to move. The fetus grows rapidly during this stage. It doubles, and then triples, its size within a month. For example, in week 10, the fetus is about 36 mm long. A little later, at week 16, the fetus is about 108 mm to 116 mm long. Use **Figure 3** to follow some of the changes that take place in the fetus as it develops.

Reading Check Describe three changes the fetus undergoes during weeks 9 to 16.

CONNECTION TO Physics

Using Ultrasound Doctors often use ultrasound to view a fetus in the uterus. Research how an ultrasound machine works, and make a poster explaining how sound waves can show what is happening inside a human body.

ACTIVITY

Weeks 17 to 24

By week 17, the fetus can make faces. Usually, in week 18, the fetus starts to make movements that the mother can feel. By week 18, the fetus can hear sounds through the mother's uterus. It may even jump at loud noises. By week 23, the fetus's movements may be quite vigorous! If the fetus were born after week 24, it might survive. But babies born at 24 weeks require a lot of help. In weeks 17 to 24, the fetus grows to between 25 cm and 30 cm in length.

Weeks 25 to 36

At about 25 or 26 weeks, the fetus's lungs are well developed. But the fetus still gets oxygen from its mother through the placenta. The fetus will not take its first breath of air until it is born. By the 32nd week, the fetus's eyes are open and perceive light through the mother's abdominal wall. At this point, sunlight affects the fetus's sleeping pattern. Some scientists have observed brain activity and eye movements in sleeping fetuses that resemble those activities in sleeping children or adults. These scientists think that a sleeping fetus may dream. After 36 weeks, the fetus is almost ready to be born.

Birth

At 37 to 38 weeks, the fetus is fully developed. A full-term pregnancy usually lasts about 40 to 42 weeks. Typically, as birth begins, the mother's uterus begins a series of muscular contractions called *labor.* Usually, these contractions push the fetus through the mother's vagina, and the baby is born. The newborn is still connected to the placenta by its umbilical cord, which is tied and cut. All that will remain of the point where the umbilical cord was attached is the baby's navel. Soon, the mother expels the placenta, and labor is complete.

CONNECTION to Math — GENERAL

Factor of Increase Have students calculate the factor of increase in a fetus's body length from the 8th week, when it is about 2.5 cm long, to the 5th month, when it is about 25 cm long, to birth, when it is about 50 cm long. (2.5 cm to 25 cm is an increase by a factor of 10, 25 cm to 50 cm is an increase by a factor of 2, and 2.5 cm to 50 cm is an increase by a factor of 20) LS **Logical**

Is That a Fact!

Between conception and birth, the developing fetus increases in size from a single cell to 6 trillion cells! After birth, infants and children continue to grow rapidly. For example, girls attain three-quarters of their adult height by the age of 7 1/2. Boys attain three-quarters of their adult height by the age of 9.

Figure 3 Pregnancy Timeline

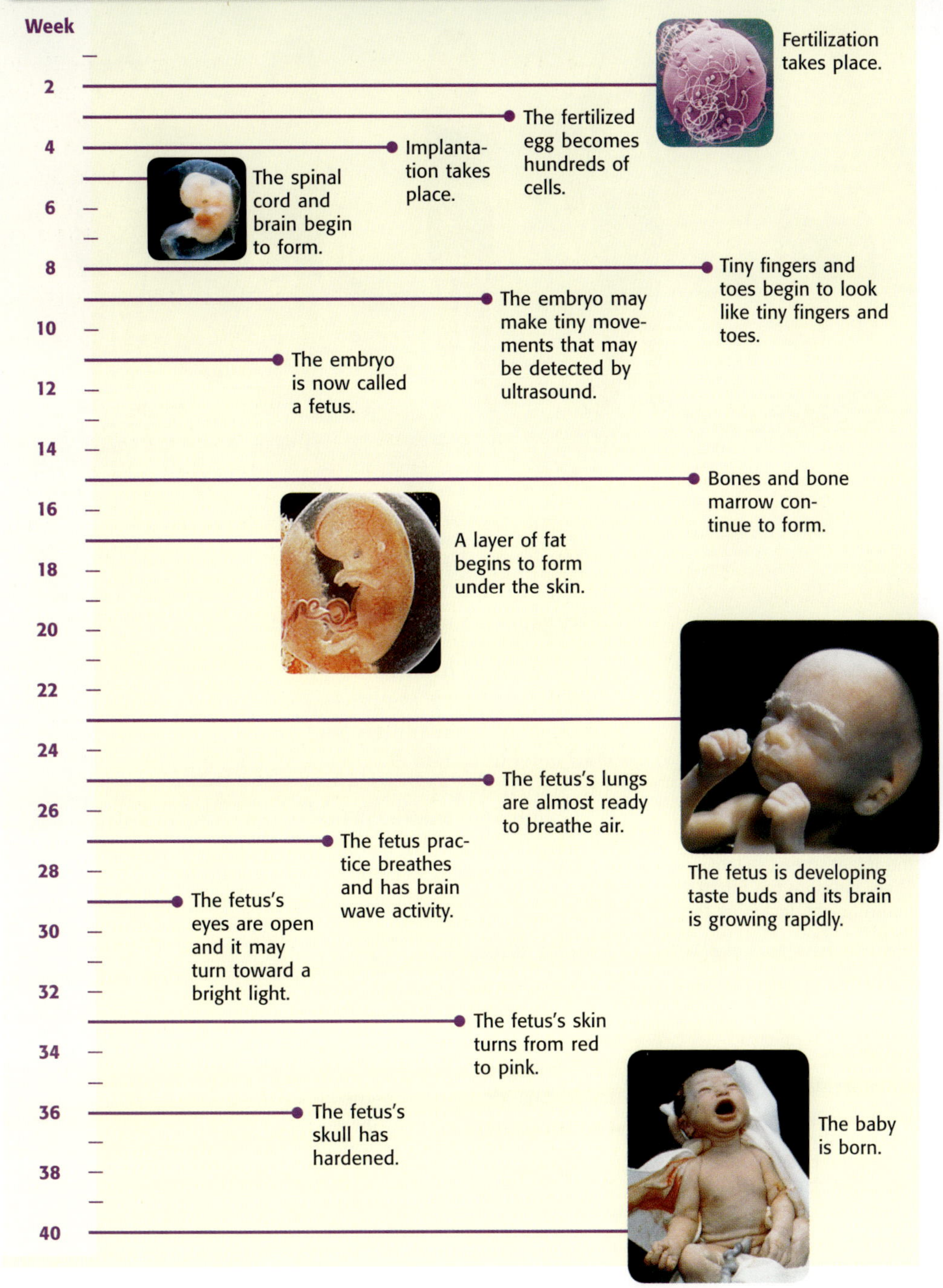

Using the Figure—GENERAL

Pregnancy Timeline Ask students to identify in **Figure 3** the weeks during which the major organs and body structures form. (weeks 5–12)

Ask them to describe the events that occur during the remainder of the pregnancy. (All the organs and body structures formed earlier in pregnancy grow and mature during weeks 13–40. As all the structures grow and develop, the fetus is more and more prepared to be born and to survive outside the uterus.) **LS Visual**

CONNECTION ACTIVITY Earth Science—GENERAL

Sound Waves and Sonograms Sound waves can be used to make images. *Sonograms* are images obtained by ultrasound (bouncing high-frequency sound waves off of an object). Sonar uses sound waves to make pictures of things underwater. Have students research how sonar is used to navigate ships and locate objects underwater. Use the teaching transparency "How Sonar Works" to illustrate how sound is used to make images. **LS Visual**

CONNECTION ACTIVITY Language Arts—GENERAL

Navel History After the umbilical cord is cut and falls off, all that remains is the *navel*. Have students research the origin of the word *navel*, and have them write a short explanation of the word's history. (*Navel* comes from the Anglo-Saxon word *nafu*, which means "the hub of the wheel." Early Anglo-Saxons believed that the navel was the center of the body.) **LS Verbal/Logical**

CHAPTER RESOURCES

Technology

Transparencies
- **LINK TO PHYSICAL SCIENCE** How Sonar Works

Close

Reteaching — BASIC

Writing **From Birth to Adulthood** Have students work in pairs to design, create, and illustrate a pamphlet or brochure that briefly describes human growth and development from birth to old age.
LS Interpersonal

Quiz — GENERAL

1. What is implantation? (It is the process by which an embryo embeds itself in the uterus.)
2. What functions does the placenta serve? (It is a two-way exchange organ that allows oxygen and nutrients to travel to the fetus from the mother and allows wastes to travel from the fetus to the mother.)

Alternative Assessment — ADVANCED

Writing **Life as a Fetus** Ask students to imagine that they have not yet been born. Have them write first-person stories describing their time in utero. Encourage creativity, but direct students to include the stages of development they went through as a fetus. Allow time for students to share their stories with the class.

Figure 4 Stages of Human Development

From Birth to Death

After birth, the human body goes through several stages of development. Some of those stages are shown in **Figure 4.**

Infancy and Childhood

Generally, infancy is the stage from birth to age 2. During infancy, you grew quickly and your baby teeth appeared. As your nervous system developed, you became more coordinated and started to walk.

Childhood—another period of fast growth—lasts from age 2 to puberty. Your baby teeth were replaced by permanent teeth. And your muscles became more coordinated, which allowed you to ride a bicycle, jump rope, and do other activities.

Adolescence

The stage from puberty to adulthood is adolescence. During puberty, a person's reproductive system becomes mature. In most boys, puberty takes place between the ages of 11 and 16. During this time, the young male's body becomes more muscular, his voice becomes deeper, and body and facial hair appear. In most girls, puberty takes place between the ages of 9 and 14. During puberty in females, the amount of fat in the hips and thighs increases, the breasts enlarge, body hair appears, and menstruation begins.

Reading Check Name an important change that takes place during adolescence.

Quick Lab

Life Grows On

Use **Figure 4** to complete this activity.

1. Use a **ruler** to measure the infant's head height. Then, measure the infant's entire height, including the head.
2. Calculate the ratio of the infant's head height to the infant's total height.
3. Repeat these measurements and calculations for the other stages.
4. Does a baby's head grow faster or slower than the rest of the body? Why do you think this is so?

Answer to Reading Check

A person's reproductive system becomes mature.

Quick Lab

MATERIALS

FOR EACH STUDENT
- calculator, if available
- ruler

Answers

Slower; Newborns have large heads to hold the large brains that enable them to learn quickly. In infancy, their bodies begin to catch up in size.

Adulthood

From about age 20 to age 40, you will be a young adult. You will be at the peak of your physical development. Beginning around age 30, changes associated with aging begin. These changes are gradual and different for everyone. Some early signs of aging include loss of flexibility in muscles, deterioration of eyesight, increase in body fat, and some loss of hair.

The aging process continues in middle age (between 40 and 65 years old). During this time, hair may turn gray, athletic abilities will decline, and skin may wrinkle. A person who is more than 65 years old is considered an older adult. Although the aging process continues, many older adults lead very active lives, as is shown in **Figure 5.**

Figure 5 *Older adults can still enjoy activities that they enjoyed when they were younger.*

SECTION Review

Summary

- Fertilization occurs when a sperm from the male joins with an egg from the female.
- The embryo and fetus undergo many changes between implantation and birth.
- The first stage of human development lasts from fertilization to birth.
- After birth, a human goes through four more stages of growth and development.

Using Key Terms

1. In your own words, write a definition for the term *umbilical cord.*
2. Use the following terms in the same sentence: *embryo* and *fetus.*

Understanding Key Ideas

3. After birth, the two periods of most rapid growth are
   a. infancy and adolescence.
   b. childhood and adulthood.
   c. infancy and childhood.
   d. adolescence and adulthood.
4. After birth, which stage of human development is the longest?
   a. infancy
   b. childhood
   c. adolescence
   d. adulthood
5. Describe the development of the embryo and the fetus.
6. What is the function of the placenta?
7. Summarize the processes of fertilization and implantation.
8. What are five stages of human development?

Math Skills

9. Suppose a person is 80 years old and that puberty took place when he or she was 12 years old.
   a. Calculate the percentage of the person's life that he or she spent in each of the four stages of development that follow birth.
   b. Make a bar graph showing the percentage for each stage.

Critical Thinking

10. **Applying Concepts** Why does the egg's covering change after a sperm has entered the egg?
11. **Analyzing Ideas** Do you think any one stage of development is more important than other stages? Explain your answer.

SCILINKS. NSTA Developed and maintained by the National Science Teachers Association

For a variety of links related to this chapter, go to www.scilinks.org

Topic: Before Birth; Growth and Development

SciLinks code: HSM0140; HSM0700

Answers to Section Review

1. The umbilical cord is a cord that forms during the 5th week of pregnancy and connects the embryo or fetus to the placenta.
2. An embryo is less developed than a fetus.
3. c
4. d
5. Sample answer: An embryo begins as a single fertilized egg cell, then develops into a ball of cells that embeds itself in the wall of the uterus. Soon, the amnion, placenta, and umbilical cord form to protect and nourish the embryo. The embryo develops and, after week 10, it is called a fetus. By week 12, all the major organ systems have begun to form and grow. By week 16, the fetus is moving strongly enough for the mother to feel the movements. From week 12 to week 40, all body systems continue to grow and strengthen. At week 40 or so, the fetus is ready to be born.
6. The placenta is the organ that allows oxygen and nutrients to get from the mother's blood to the fetus and allows wastes from the fetus to get to the mother's blood.
7. Fertilization happens when a sperm's nucleus unites an egg's nucleus to form a zygote or embryo. Implantation takes place when an embryo embeds itself in the wall of the uterus.
8. fertilization to birth; infancy, childhood, adolescence, and adulthood
9. a. • infancy: 2 y/80 y = 0.025, or 2.5%
   • childhood (10 years, from age 2 to age 12): 10 y/80 y = 12.5%
   • adolescence (age 12 to age 20): 8 y/80 y = 10%
   • adulthood (age 20 to age 80): 60 y/80 y = 75%
   b. Length of bars on bar graph should reflect percentages shown above.
10. The membrane keeps other sperm from entering the egg and uniting with the nucleus. This is important because it helps maintain the characteristic chromosome number.
11. Sample answer: Adolescence is the most important stage because that is when humans reach sexual maturity. Without sexual maturity, the species could not survive.

CHAPTER RESOURCES

Chapter Resource File

- Section Quiz GENERAL
- Section Review GENERAL
- Vocabulary and Section Summary GENERAL
- Reinforcement Worksheet BASIC
- Critical Thinking ADVANCED
- SciLinks Activity GENERAL

Technology

Transparencies
- The Stages of Human Development

Skills Practice Lab

It's a Comfy, Safe World!

Teacher's Notes

Time Required

Two 45-minute class periods

Lab Ratings

EASY → HARD

Teacher Prep 2
Student Set-Up 1
Concept Level 1
Clean Up 3

MATERIALS

This lab may require some large plastic bags, a meterstick, and various other materials depending on the students' designs. Soft-boiled eggs will simplify the cleanup. Students may wear gloves.

Safety Caution

Remind students to review all safety cautions and icons before beginning this lab activity. Students should wash their hands after handling the eggs.

Lab Notes

Students will be dropping their models, so this lab should be done over a large plastic sheet. You may want to do this lab outside because it is may be quite messy. Students can modify their models in any way they feel will improve the protection.

Using Scientific Methods

Skills Practice Lab

It's a Comfy, Safe World!

Before birth, baby birds live inside a hard, protective shell until the baby has used up all the food supply. Most mammal babies develop within their mother's uterus, in which they are surrounded by fluid and connected to a placenta, before they are born. Before human babies are born, they lead a comfy life. By the seventh month, they lie around sucking their thumb, blinking their eyes, and perhaps even dreaming.

OBJECTIVES

Construct a model of a human uterus protecting a fetus.

Compare the protection that a bird's egg gives a developing baby bird with the protection that a human uterus gives a fetus.

MATERIALS

- computer (optional)
- cotton, soft fabric, or other soft materials
- eggs, soft-boiled and in the shell (2 to 4)
- eggs, soft-boiled and peeled (3 or 4)
- gloves, protective
- mineral oil, cooking oil, syrup, or other thick liquid
- plastic bags, sealable
- water

SAFETY

Ask a Question

1. Inside which structure is a developing organism better protected from bumps and blows: the uterus of a placental mammal or the egg of a bird?

Form a Hypothesis

2. A placental mammal's uterus protects a developing organism from bumps and blows better than a bird's egg does.

Test the Hypothesis

3. Brainstorm several ideas about how you will construct and test your model of a mammalian uterus. Then, use the materials provided by your teacher to build your model. A peeled, soft-boiled egg will represent the fetus inside your model uterus.
4. Make a data table similar to **Table 1** below. Test your model, examine the egg for damage, and record your results.

***Table 1* First Model Test**

Original model	Modified model
DO NOT	WRITE
IN B	OOK

5. Modify your model as necessary; test this modified model using another peeled, soft-boiled egg; and record your results.

Randy Christian
Stovall Junior High School
Houston, Texas

CHAPTER RESOURCES

Chapter Resource File

- Datasheet for Chapter Lab
- Lab Notes and Answers

Technology

Classroom Videos
- Lab Video

LabBook
- My, How You've Grown!

6 When you are satisfied with the design of your model, obtain another peeled, soft-boiled egg and an egg in the shell. The egg in the shell represents the baby bird inside the egg.

7 Make a data table similar to **Table 2** below. Test your new eggs, examine them for damage, and record your results in your data table.

Table 2 Final Model Test	
	Test Results
Model	DO NOT WRITE IN BOOK
Egg in shell	

Analyze the Results

1 **Explaining Events** Explain any differences in the test results for the model and the egg in a shell.

2 **Analyzing Results** What modification to your model was the most effective in protecting the fetus?

Draw Conclusions

3 **Evaluating Data** Review your hypothesis. Did your data support your hypothesis? Why or why not?

4 **Evaluating Models** What modifications to your model might make it more like a uterus?

Applying Your Data

Use the Internet or the library to find information about the development of monotremes, such as the echidna or the platypus, and marsupials, such as the koala or the kangaroo. Then, using what you have learned in this lab, compare the development of placental mammals with that of marsupials and monotremes.

CHAPTER RESOURCES

Workbooks

Long-Term Projects & Research Ideas
• Get a Whiff of This! ADVANCED

Analyze the Results

1. Answers may vary, but an egg inside a viscous liquid in a plastic bag, wrapped in soft cotton and placed inside another bag, should not be damaged when dropped from a height of 1 m. An egg protected only by a shell should break. In general, the more protected the fetus or egg, the more resistant to damage it is and the less damage it will suffer when dropped from a height of 1 m.
2. Answers may vary according to students' modifications. Most students will observe that the more soft wrapping and/or viscous fluid protecting the egg, the better the egg survived the drop from 1 m.

Draw Conclusions

3. Students' answers may vary depending on their hypothesis and the data collected. Accept all reasonable answers.
4. Students' answers may vary depending on their model and the data collected. Accept all reasonable answers.

Applying Your Data

Students should recognize that all mammals have internal fertilization of eggs. Students should include the ideas that monotreme mammals lay eggs with leathery shells and, after the eggs hatch, nurture their young with milk that oozes from pores in the mother's skin. Marusupial mammals give birth to live young that are very underdeveloped; the young move to a pouch and continue their development in the pouch. Placental mammals also give birth to live young, but their young are more developed than those of marsupial mammals. Even so, placental mammals may have to care for their young for several years.

Chapter Review

Assignment Guide	
Section	**Questions**
1	1, 3, 6, 10, 18–19, 21
2	2, 4–5, 7–8, 11, 12, 13 16–17, 20, 21
3	9, 12, 14–15
2 and 3	22–25

ANSWERS

Using Key Terms

1. Internal fertilization means that the eggs are fertilized inside the female's body. External fertilization is when the eggs are fertilized outside the female's body.
2. Testes are male organs that produce sperm. Ovaries are female organs that produce eggs.
3. Asexual reproduction is when a single parent has offspring that are genetically identical to the parent. Sexual reproduction happens when sex cells from two parents unite.
4. Fertilization happens when the nucleus of a sperm unites with the nucleus of an egg. Implantation is when the embryo embeds itself in the wall of the uterus.
5. The umbilical cord links the fetus with the placenta. The placenta is the organ that helps the fetus get oxygen and nutrients and get rid of wastes.

Chapter Review

USING KEY TERMS

For each pair of terms, explain how the meanings of the terms differ.

1. *internal fertilization* and *external fertilization*
2. *testes* and *ovaries*
3. *asexual reproduction* and *sexual reproduction*
4. *fertilization* and *implantation*
5. *umbilical cord* and *placenta*

UNDERSTANDING KEY IDEAS

Multiple Choice

6. The sea star reproduces asexually by
   - **a.** fragmentation.
   - **b.** budding.
   - **c.** external fertilization.
   - **d.** internal fertilization.
7. Which list shows in order sperm's path through the male reproductive system?
   - **a.** testes, epididymis, urethra, vas deferens
   - **b.** epididymis, urethra, testes, vas deferens
   - **c.** testes, vas deferens, epididymis, urethra
   - **d.** testes, epididymis, vas deferens, urethra
8. Identical twins are the result of
   - **a.** a fertilized egg splitting in two.
   - **b.** two separate eggs being fertilized.
   - **c.** budding in the uterus.
   - **d.** external fertilization.
9. If the onset of menstruation is counted as the first day of the menstrual cycle, on what day of the cycle does ovulation typically occur?
   - **a.** 2nd day
   - **b.** 5th day
   - **c.** 14th day
   - **d.** 28th day
10. How do monotremes differ from placental mammals?
    - **a.** Monotremes are not mammals.
    - **b.** Monotremes have hair.
    - **c.** Monotremes nurture their young with milk.
    - **d.** Monotremes lay eggs.
11. All of the following are sexually transmitted diseases EXCEPT
    - **a.** chlamydia.
    - **b.** AIDS.
    - **c.** infertility.
    - **d.** genital herpes.
12. Where do fertilization and implantation, respectively, take place?
    - **a.** uterus, fallopian tube
    - **b.** fallopian tube, vagina
    - **c.** uterus, vagina
    - **d.** fallopian tube, uterus

Short Answer

13. Which human reproductive organs produce sperm? produce eggs?
14. Explain how the fetus gets oxygen and nutrients and how it gets rid of waste.
15. What are four stages of human life following birth?

Understanding Key Ideas

6. a
7. d
8. a
9. c
10. d
11. c
12. d
13. The testes produce sperm, and the ovaries produce eggs.
14. The placenta is a specialized organ that allows the fetus to get oxygen and get rid of wastes.
15. infancy, childhood, adolescence, adulthood
16. Sample answer: Infertility is a problem because it may prevent people from having babies. STDs are a problem because they can damage a person's reproductive system, and they may also cause infertility. Cancer is a problem because it can damage a person's reproductive system.
17. Drawings should resemble the images in Section 2, Figure 1, The Male Reproductive System, and Figure 2, The Female Reproductive System.

16 Name three problems that can affect the human reproductive system, and explain why each is a problem.

17 Draw a diagram showing the structures of the male and female reproductive systems. Label each structure, and explain how each structure contributes to fertilization and implantation.

CRITICAL THINKING

18 **Concept Mapping** Use the following terms to create a concept map: *asexual reproduction, budding, external fertilization, fragmentation, reproduction, internal fertilization,* and *sexual reproduction.*

19 **Identifying Relationships** The environment in which organisms live may change over time. For example, a wet, swampy area may gradually become a grassy area with a small pond. Explain how sexual reproduction may give species that live in a changing environment a survival advantage.

20 **Applying Concepts** What is the function of the uterus? How is this function related to the menstrual cycle?

21 **Making Inferences** In most human body cells, the 46 chromosomes are duplicated during cell division so that each new cell receives 46 chromosomes. Cells that make eggs and sperm also split and duplicate their 46 chromosomes. But then, in the process of meiosis, the two cells split again to form four cells (egg or sperm) that each have 23 chromosomes. Why is meiosis important to human reproduction and to the human species?

INTERPRETING GRAPHICS

The following graph illustrates the cycles of the female hormone estrogen and the male hormone testosterone. The blue line shows the estrogen level in a female over 28 days. The red line shows the testosterone level in a male over the same amount of time. Use the graph below to answer the questions that follow.

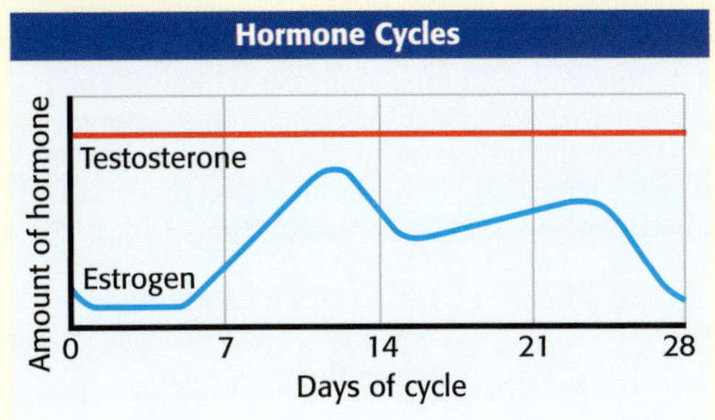

22 What is the major difference between the levels of the two hormones over the 28 days?

23 What cycle do you think estrogen affects?

24 Why might the level of testosterone stay the same?

25 Do you think that the above estrogen cycle would change in a pregnant woman? Explain your answer.

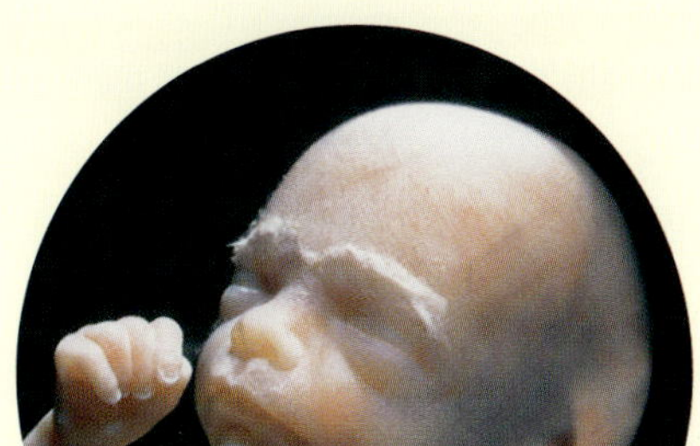

CHAPTER RESOURCES

Chapter Resource File

- Chapter Review GENERAL
- Chapter Test A GENERAL
- Chapter Test B ADVANCED
- Chapter Test C SPECIAL NEEDS
- Vocabulary Activity GENERAL

Workbooks

Study Guide

- Assessment resources are also available in Spanish.

Critical Thinking

18. 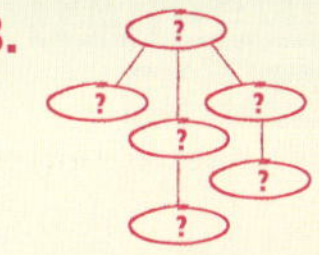
An answer to this exercise can be found at the end of this book.

19. Sample answer: Sexual reproduction produces offspring that have genetic information from two different parents. Some of that information may enable the offspring to do better in a changing environment than other members of the species. As a result, there is a greater chance that the beneficial information will be passed on to the next generation.

20. The uterus, an organ in the female reproductive system, is where an embryo develops into a fetus. Every month, it builds up tissue that can help nourish a developing embryo. If no embryo implants, the tissue will discharge in menstruation.

21. Meiosis ensures that the characteristic human chromosome number remains the same. Because sex cells from two separate parents combine to form a zygote, each sex cell needs only half the 46 chromosomes that all other human body cells have. Meiosis also ensures that offspring will have a mix of genetic information.

Interpreting Graphics

22. Estrogen levels fluctuate, but testosterone stays at the same level throughout the month.

23. Estrogen affects the menstrual cycle.

24. Testosterone levels stay the same because men continually produce sperm and do not prepare for a possible pregnancy.

25. Sample answer: Yes, I think it might change because estrogen affects the menstrual cycle. During pregnancy, the menstrual cycle stops, so the level of estrogen will level off until the pregnancy is over.

Standardized Test Preparation

READING

Read each of the passages below. Then, answer the questions that follow each passage.

Passage 1 The male reproductive system is made up of internal and external organs. The external organs of this system are the penis and the scrotum. The scrotum is a skin-covered sac that hangs outside the body. Normal human body temperature is about 37°C. Normal sperm production and development cannot take place at that temperature. Normal sperm production and development takes place at lower temperatures. That is why the testes rest in the scrotum, outside the body. The scrotum is about 2°C cooler than the body. Inside each testis are masses of tightly coiled tubes, called *seminiferous tubules,* in which sperm are produced when conditions are right.

1. In this passage, what does the word *external* mean?

- **A** not part of the body
- **B** outside the body
- **C** inside the body
- **D** lasting a long time

2. Which of the following statements is a fact according to the passage?

- **F** The temperature in the scrotum is higher than body temperature.
- **G** Testes are internal organs of the male reproductive system.
- **H** Normal sperm production cannot take place at normal body temperature.
- **I** Normal human body temperature is about 37°F.

3. What are the tubes in which sperm are made called?

- **A** testes
- **B** scrotum
- **C** seminiferous tubules
- **D** external organs

Passage 2 In a normal pregnancy, the fertilized egg travels to the uterus and implants itself in the uterus's wall. But, in about 7 out of 1,000 pregnancies in the United States, a woman has an ectopic pregnancy. The term *ectopic* is from two Greek words meaning "out of place." In an ectopic pregnancy, the fertilized egg implants itself in an ovary, a fallopian tube, or another area of the female reproductive system that is not the lining of the uterus. Because the zygote cannot develop properly outside of the uterus, an ectopic pregnancy can be very dangerous for both the mother and fetus. As the fetus grows, it causes the mother pain and bleeding. For example, an ectopic pregnancy in a fallopian tube can rupture the tube and cause abdominal bleeding. If an ectopic pregnancy is not treated quickly enough, both the mother and the fetus may die.

1. In the passage, what does the term *ectopic pregnancy* probably mean?

- **A** a pregnancy that takes place at the wrong time
- **B** a type of pregnancy that happens about 7 out of 100 times in the United States
- **C** a type of pregnancy caused by a problem with a fallopian tube
- **D** a pregnancy in which the zygote implants itself in the wrong place

2. Which of the following statements is a fact according to the passage?

- **F** Ectopic pregnancies take place in about 7% of all pregnancies.
- **G** The ectopic pregnancy rate in the United States is less than 1%.
- **H** Ectopic pregnancies take place in the uterus.
- **I** An ectopic pregnancy is harmless.

Teacher's Note

To provide practice under more realistic testing conditions, give students 20 minutes to answer all of the questions in this Standardized Test Preparation.

MISCONCEPTION ALERT

Answers to the standardized test preparation can help you identify student misconceptions and misunderstandings.

READING

Passage 1

1. B
2. H
3. C

TEST DOCTOR

Question 2: This fact is mentioned in the following sentences: "Normal human body temperature is about 37°C. Normal sperm production and development cannot take place at that high temperature." Students may not put the two sentences together to extract the fact. Some students may choose answer I because they have mistakenly read "37°F" for the correct "37°C."

Question 3: Students may skip over the word "tubes" in the question and answer the more general question "Where are sperm made?" by choosing answer A.

Passage 2

1. D
2. G

TEST DOCTOR

Question 2: The passage states that ectopic pregnancies happen in about 7 out of 1,000 pregnancies, which is 0.7%. The correct answer is therefore answer G. Students may read "7 out of 1000" as "7 out of 100," and would pick incorrect answer F.

INTERPRETING GRAPHICS

Use the diagrams below to answer the questions that follow.

A.

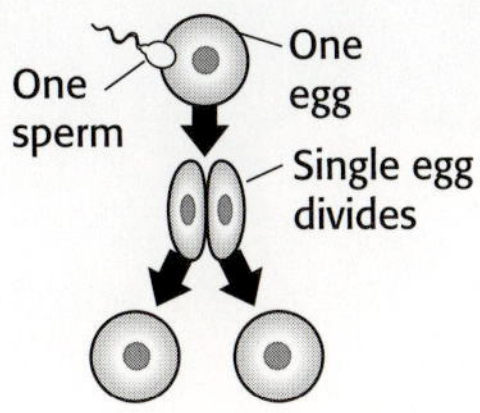

B.

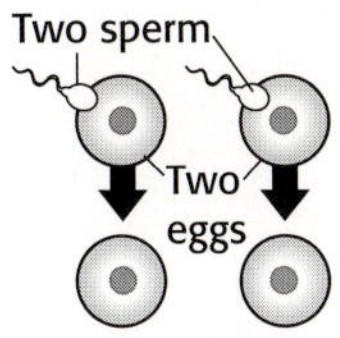

1. Which diagram of cell division would produce identical twins: A or B?

A diagram B, because each egg is fertilized by a separate sperm cell
B both diagram A and diagram B, because twins result in both cases
C diagram A, because a single fertilized egg separates into two halves
D diagram B, because two eggs are released by an ovary

2. Which of the following could describe fraternal twins?

F both boys
G both girls
H one girl and one boy
I any of these combinations

3. Which diagram of cell division could explain triplets, two of whom are identical and one of whom is fraternal?

A diagram A
B diagram B
C either diagram A or diagram B
D neither diagram A or diagram B

MATH

Read each question below, and choose the best answer.

1. Identify the group that contains equivalent fractions, decimals, and percents.

A 7/10, 0.7, 7%
B 1/2, 0.5, 50%
C 3/8, 0.38, 38%
D 3/100, 0.3, 33%

2. A geologist was exploring a cave. She spent 2.7 h exploring on Saturday and twice as many hours exploring on Sunday. Which equation could be used to find n, the total number of hours the geologist spent exploring the cave on those 2 days?

F $n = 2 \div 2.7$
G $n = 2.7 + (2 \times 2.7)$
H $n = 2.7 + 2.7 + 2$
I $n = 2 \times 2.7$

3. Which of the following story problems can be solved by the equation below?

$$(60 + 70 + 68 + 80 + x) \div 5 = 70$$

A The heights of four buildings in South Braintree are 60 ft, 70 ft, 68 ft, and 80 ft. Find x, the average height of the buildings.
B The weights of four dogs Jason is raising are 60 lb, 70 lb, 68 lb, and 80 lb. Find x, the sum of the weights of the four dogs.
C Kayla's first four handmade bracelets sold for \$60, \$70, \$68, and \$80. Find x, the amount for which Kayla needs to sell her fifth bracelet to have an average selling price of \$70.
D The times it took Taylor to complete each of four 100 m practice swims were 60 s, 70 s, 68 s, and 80 s. Find x, the average time it took Taylor to complete his practice swims.

CHAPTER RESOURCES

Chapter Resource File

• Standardized Test Preparation GENERAL

State Resources

For specific resources for your state, visit **go.hrw.com** and type in the keyword **HSMSTR.**

INTERPRETING GRAPHICS

1. C
2. I
3. D

TEST DOCTOR

Question 1: Students may select answer B because twins do result in both cases. However, identical twins happen only when a single fertilized egg splits in two.

Question 2: Students may think that fraternal twins, even if they are not identical, must still be of the same sex. The correct answer is I; fraternal twins may be two boys, two girls, or one of each, because fraternal twins develop from two separate fertilized eggs, not from the splitting of one egg.

MATH

1. B
2. G
3. C

TEST DOCTOR

Question 2: Students often have trouble converting a word problem into a number statement. Students may select answer H or I because those answers are similar to the correct answer. Students who have problems writing number statements often overlook an important clue in the word problem, and their statement is incomplete.

Question 3: This is another word problem, but it is a little more complex than question 2. Students must realize that the average has already been calculated and that what they are seeking is the quantity that gives the average. Answer C is the only answer in which the average is given; the other three answers ask to solve for the average.

Science in Action

Science, Technology, and Society

Discussion — ADVANCED

Lead students in a discussion about the implications of fetal-surgery technology for treating fetal disorders. What sorts of disorders might be successfully treated? (anatomical disorders and discrete tumors) What types of disorders would be more difficult to treat? (genetic disorders and disorders affecting the entire fetus)

Scientific Discoveries

Background

The word *laser* is actually an abbreviation for **l**ight **a**mplification by **s**timulated **e**mission of **r**adiation. Lasers can pump energy into a solid, liquid, gas, or semiconductor to produce a beam. Carbon dioxide and helium-neon are common gas lasers. Ruby is a common solid-state laser. Each medium produces a laser with a different wavelength. Semiconductor (or diode) lasers emit a wavelength of 630 nm. They are so weak that they can be used as pointers. Carbon dioxide lasers, emitting a wavelength of 10,600 nm, are so strong that they can cut through steel!

Science in Action

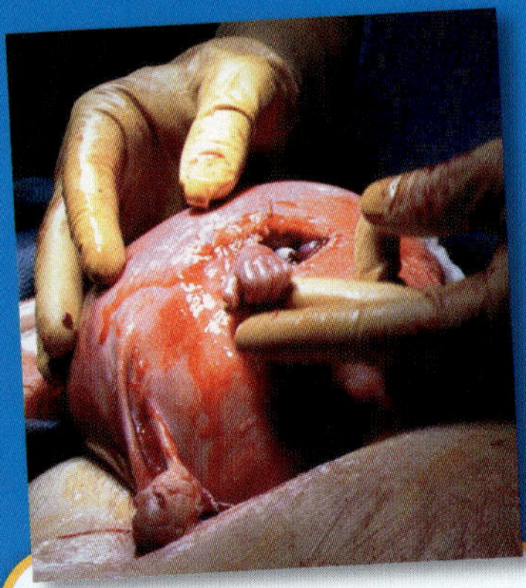

Doctors operated on a fetus, whose hand is visible in this photo, to correct spina bifida.

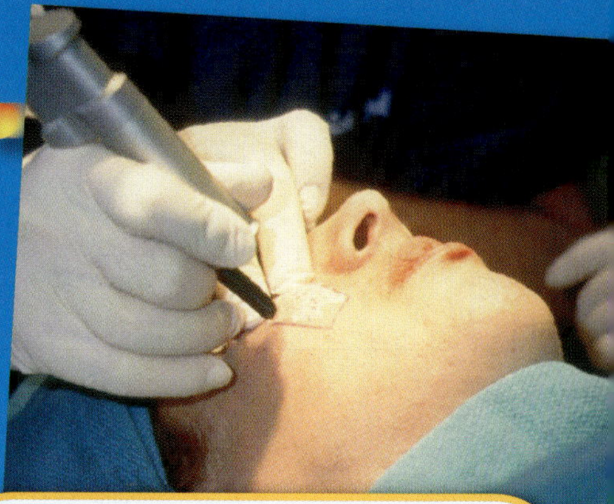

Science, Technology, and Society

Fetal Surgery

Sometimes, a developing fetus has a serious medical problem. In many cases, surgery after birth can correct the problem. But some problems can be treated while the fetus is still in the uterus. For example, fetal surgery is used to correct spina bifida (part of the spinal cord is exposed because the backbone doesn't form properly). Doctors now can fix several types of problems before a baby is born.

Social Studies ACTIVITY

WRITING SKILL Research the causes of spina bifida. Write a brochure telling expectant mothers what precautions they can take to prevent spina bifida.

Scientific Discoveries

Lasers and Acne

Many people think that acne affects only teenagers, but acne can strike at any age. Some acne is mild, but some is severe. Now, for some severe cases of acne, lasers may provide relief. That's right—lasers can be used to treat acne! Surgeons who specialize in the health and diseases of the skin use laser light to treat the skin disease known as *acne*.

In addition, laser treatments may stimulate the skin cells that produce collagen. Collagen is a protein found in connective tissue, such as skin. Increased production of collagen in the skin improves the skin's texture and helps smooth out acne scars.

Language Arts ACTIVITY

WRITING SKILL Write a story about how severe acne affects a teen's life. Tell what happens when a doctor refers the teen to a specialist for laser treatment and how the successful treatment changes the teen's life.

Answer to Social Studies Activity

Spina bifida is a Latin term meaning *split spine.* It is a group of congenital birth defects that affect the development of the central nervous system—the brain and the spinal cord—and nerve tissues. Spina bifida is relatively common, occurring about 10 to 20 times per 1,000 births. With severe spina bifida, a person's legs and feet are paralyzed. There are problems with bowel and bladder control. Learning disabilities are common, and mental retardation sometimes occurs. Up to 75% of the cases of spina bifida could be prevented by taking folic acid daily before pregnancy and during the first trimester.

Answer to Language Arts Activity

Students' stories may reflect the emotional, social, and psychological impact that severe acne can have and the positive effects laser treatments may have for the acne sufferer. Improved self-confidence, a better self-image, and more-successful social interactions may be some of the benefits.

Careers

Reva Curry

Diagnostic Medical Sonographer Sounds are everywhere in our world. But only some of those sounds—such as your favorite music playing on the stereo or the dog barking next door—are sounds that we can hear. There are sound waves whose frequency is too high for us to hear. These high-pitched sounds are called *ultrasound*. Some animals, such as bats, use ultrasound to hunt and to avoid midair collisions.

Humans use ultrasound, too. Ultrasound machines can peer inside the human body to look at hearts, blood vessels, and fetuses. Diagnostic medical sonographers are people who use sonography equipment to diagnose medical problems and to follow the growth and development of a fetus before it is born. One of the leading professionals in the field of diagnostic medical sonography is Dr. Reva Curry. Dr. Curry spent many years as a sonographer. Her primary job was to use high-tech instruments to create ultrasound images of parts of the body and interpret the results for other medical professionals. Today, Dr. Curry works with students as the dean of a community college.

Math Activity

At 20°C, the speed of sound in water is 1,482 m/s and in steel is 5,200 m/s. How long would it take a sound to travel 815.1 m in water? In that same length of time, how far would a sound travel in a steel beam?

To learn more about these Science in Action topics, visit go.hrw.com and type in the keyword HL5BD5F.

Current Science

Check out Current Science® articles related to this chapter by visiting go.hrw.com. Just type in the keyword HL5CS26.

Careers

Discussion — General

Generate interest in diagnostic medical sonography by inviting a diagnostic medical sonographer from a local hospital or clinic. Before the visit, discuss with the class Dr. Curry's story and any experiences the students may have had with sonography or other diagnostic imaging techniques. Have students prepare questions for the guest speaker prior to the visit. Your guest may show a video of the procedure itself and interpret some actual sonograms. The students should have the opportunity for an open discussion with the practitioner and thereby get first-hand knowledge about the profession.

Answers to Math Activity

1. 0.55 s
2. 2,860 m

Body Defenses and Disease

Chapter Planning Guide

Compression guide: To shorten instruction because of time limitations, omit the Chapter Lab.

OBJECTIVES	LABS, DEMONSTRATIONS, AND ACTIVITIES	TECHNOLOGY RESOURCES
PACING • 90 min pp. 130–135 **Chapter Opener**	SE **Start-up Activity,** p. 131 GENERAL	OSP **Parent Letter** ■ GENERAL CD **Student Edition on CD-ROM** CD **Guided Reading Audio CD** ■ TR **Chapter Starter Transparency*** VID **Brain Food Video Quiz**
Section 1 Disease • Explain the difference between infectious diseases and noninfectious diseases. • Identify five ways that you might come into contact with a pathogen. • Discuss four methods that have helped reduce the spread of disease.	SE **Connection to Social Studies** Disease and History, p. 133 GENERAL TE **Activity** Using Microscopes, p. 133 BASIC SE **School-to-Home Activity** Label Check, p. 134 GENERAL TE **Connection to Physical Science** Pasteurization, p. 134 SE **Skills Practice Lab** Passing the Cold, p. 144 ◆ GENERAL CRF **Datasheet for Chapter Lab*** SE **Science In Action** Math, Social Studies and Language Arts Activites, pp. 150–151	CRF **Lesson Plans*** TR **Bellringer Transparency*** TR **LINK TO PHYSICAL SCIENCE** Thermal Energy in Water* VID **Lab Videos for Life Science**
PACING • 45 min pp. 136–143 **Section 2 Your Body's Defenses** • Describe how your body keeps out pathogens. • Explain how the immune system fights infections. • Describe four challenges to the immune system.	TE **Activity** Reactions to Illness, p. 136 GENERAL SE **Quick Lab** Only Skin Deep, p. 137 GENERAL CRF **Datasheet for Quick Lab*** TE **Activity** Follow the Path, p. 138 BASIC TE **Group Activity** Immunity Skit, p. 139 ADVANCED SE **Connection to Chemistry** Bent out of Shape, p. 140 GENERAL TE **Connection Activity** Real World, p. 140 GENERAL TE **Activity** Concept Mapping, p. 142 GENERAL SE **Model-Making Lab** Antibodies to the Rescue, p. 185 GENERAL CRF **Datasheet for LabBook*** LB **Long-Term Projects & Research Ideas** A Chuckle a Day Keeps the Doctor Away* ADVANCED	CRF **Lesson Plans*** TR **Bellringer Transparency*** TR An Antibody's Shape Fits an Antigen* TR Immune Response: A* TR Immune Response: B* SE **Internet Activity,** p. 138 GENERAL CRF **SciLinks Activity*** GENERAL

PACING • 90 min

CHAPTER REVIEW, ASSESSMENT, AND STANDARDIZED TEST PREPARATION

CRF **Vocabulary Activity*** GENERAL
SE **Chapter Review,** pp. 146–147 GENERAL
CRF **Chapter Review*** ■ GENERAL
CRF **Chapter Tests A*** ■ GENERAL, **B*** ADVANCED, **C*** SPECIAL NEEDS
SE **Standardized Test Preparation,** pp. 148–149 GENERAL
CRF **Standardized Test Preparation*** GENERAL
CRF **Performance-Based Assessment*** GENERAL
OSP **Test Generator** GENERAL
CRF **Test Item Listing*** GENERAL

Online and Technology Resources

go.hrw.com

Visit **go.hrw.com** for a variety of free resources related to this textbook. Enter the keyword **HL5BD6.**

Holt Online Learning

Students can access interactive problem-solving help and active visual concept development with the *Holt Science and Technology* Online Edition available at **www.hrw.com.**

Guided Reading Audio CD

A direct reading of each chapter using instructional visuals as guideposts. For auditory learners, reluctant readers, and Spanish-speaking students. Available in English and Spanish.

KEY

SE Student Edition
TE Teacher Edition
CRF Chapter Resource File
OSP One-Stop Planner
LB Lab Bank
TR Transparencies
SS Science Skills Worksheets
MS Math Skills for Science Worksheets
CD CD or CD-ROM
VID Classroom Video/DVD
* Also on One-Stop Planner
◆ Requires advance prep
■ Also available in Spanish

SKILLS DEVELOPMENT RESOURCES	SECTION REVIEW AND ASSESSMENT	STANDARDS CORRELATIONS
SE **Pre-Reading Activity,** p. 130 GENERAL OSP **Science Puzzlers, Twisters & Teasers*** GENERAL		National Science Education Standards SAI 1, 2; LS 1f
CRF **Directed Reading A*** ■ BASIC, **B*** SPECIAL NEEDS CRF **Vocabulary and Section Summary*** ■ GENERAL SE **Reading Strategy** Paired Summarizing, p. 132 GENERAL TE **Inclusion Strategies,** p. 133 SE **Math Practice** Epidemic!, p. 135 GENERAL CRF **Critical Thinking** Vaccine for Super Bug Found!* ADVANCED	SE **Reading Checks,** pp. 133, 135 GENERAL TE **Reteaching,** p. 134 BASIC TE **Quiz,** p. 134 GENERAL TE **Alternative Assessment,** p. 134 GENERAL SE **Section Review,*** p. 135 ■ GENERAL CRF **Section Quiz*** ■ GENERAL	SAI 1, 2; ST 2; SPSP 1, 5; HNS 1, 3; LS 1f; *Chapter Lab:* UCP 2, SAI 1
CRF **Directed Reading A*** ■ BASIC, **B*** SPECIAL NEEDS CRF **Vocabulary and Section Summary*** ■ GENERAL SE **Reading Strategy** Reading Organizer, p. 136 GENERAL TE **Inclusion Strategies,** p. 140 CRF **Reinforcement Worksheet** Immunity Teamwork* GENERAL	SE **Reading Checks,** pp. 137, 138, 141, 142 GENERAL TE **Homework,** p. 141 GENERAL TE **Reteaching,** p. 142 BASIC TE **Quiz,** p. 142 GENERAL TE **Alternative Assessment,** p. 142 ADVANCED SE **Section Review,*** p. 143 ■ GENERAL CRF **Section Quiz*** ■ GENERAL	UCP 1, 2, 3, 4, 5; SAI 1; HNS 3; LS 1b, 1d, 1e, 1f, 3a, 3b; *LabBook:* UCP 2; SAI 1

One-Stop Planner® CD-ROM

This convenient CD-ROM includes:
- **Lab Materials QuickList Software**
- **Holt Calendar Planner**
- **Customizable Lesson Plans**
- **Printable Worksheets**
- **ExamView® Test Generator**

CNN Student News

cnnstudentnews.com

Find the latest news, lesson plans, and activities related to important scientific events.

SciLinks NSTA

www.scilinks.org

Maintained by the **National Science Teachers Association.** See Chapter Enrichment pages for a complete list of topics.

Current Science®

Check out ***Current Science*** articles and activities by visiting the HRW Web site at **go.hrw.com.** Just type in the keyword **HL5CS27T.**

Classroom Videos

- **Lab Videos** demonstrate the chapter lab.
- **Brain Food Video Quizzes** help students review the chapter material.

Chapter Resources

Visual Resources

CHAPTER STARTER TRANSPARENCY

BELLRINGER TRANSPARENCIES

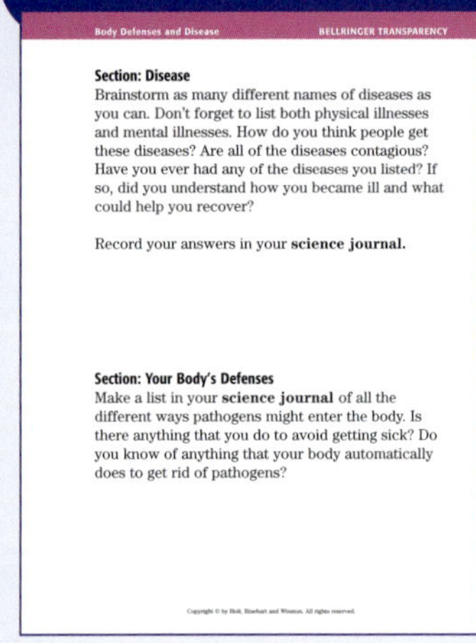
Body Defenses and Disease — BELLRINGER TRANSPARENCY

Section: Disease
Brainstorm as many different names of diseases as you can. Don't forget to list both physical illnesses and mental illnesses. How do you think people get these diseases? Are all of the diseases contagious? Have you ever had any of the diseases you listed? If so, did you understand how you became ill and what could help you recover?

Record your answers in your **science journal.**

Section: Your Body's Defenses
Make a list in your **science journal** of all the different ways pathogens might enter the body. Is there anything that you do to avoid getting sick? Do you know of anything that your body automatically does to get rid of pathogens?

TEACHING TRANSPARENCIES

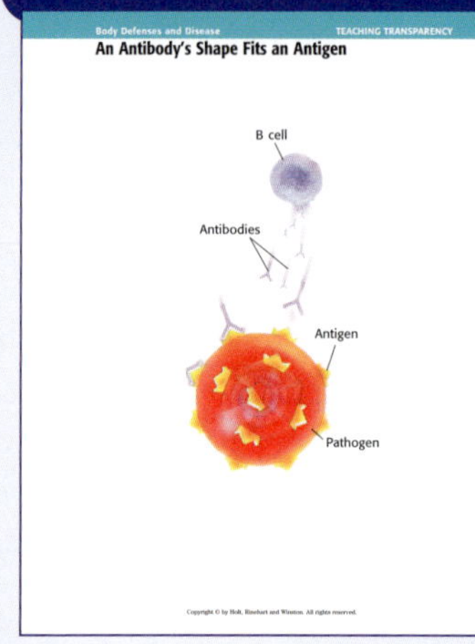

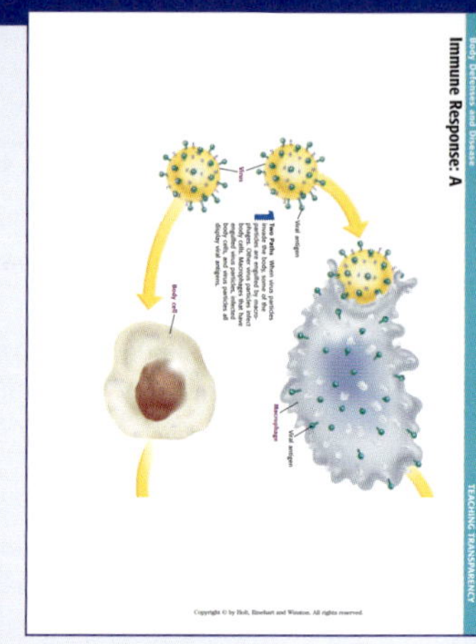

TEACHING TRANSPARENCIES

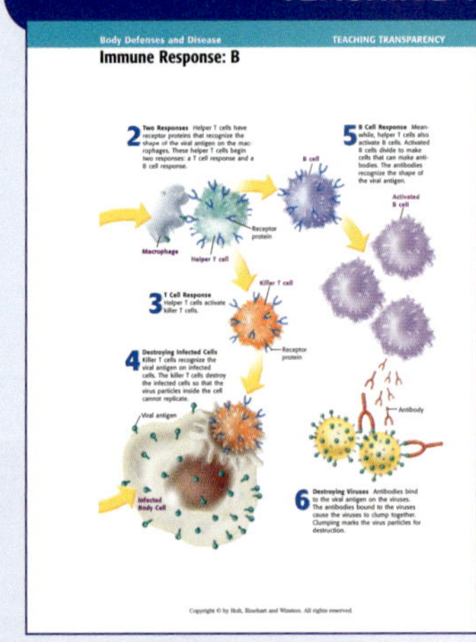

CONCEPT MAPPING TRANSPARENCY

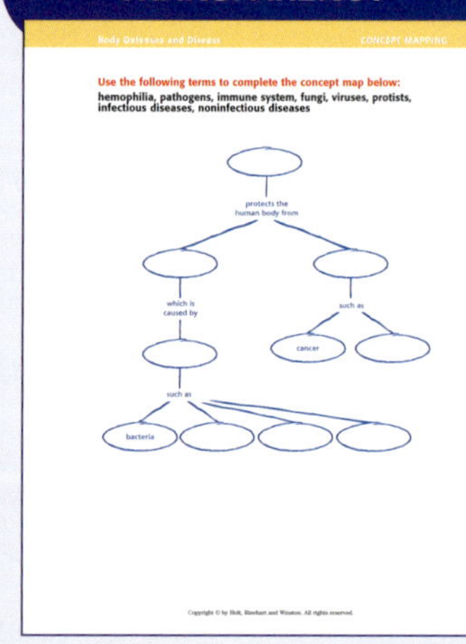

Planning Resources

LESSON PLANS

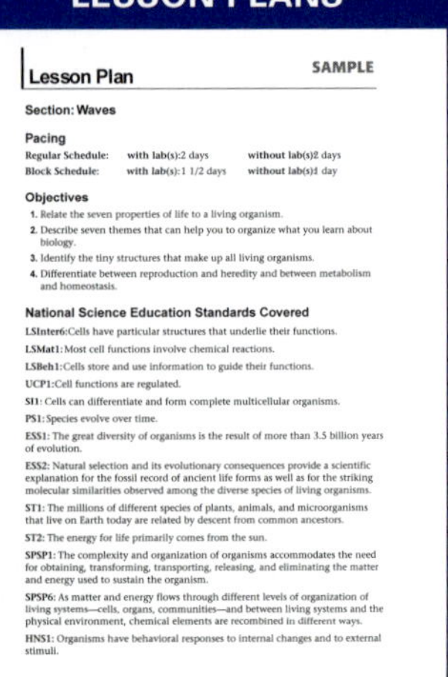

PARENT LETTER

TEST ITEM LISTING

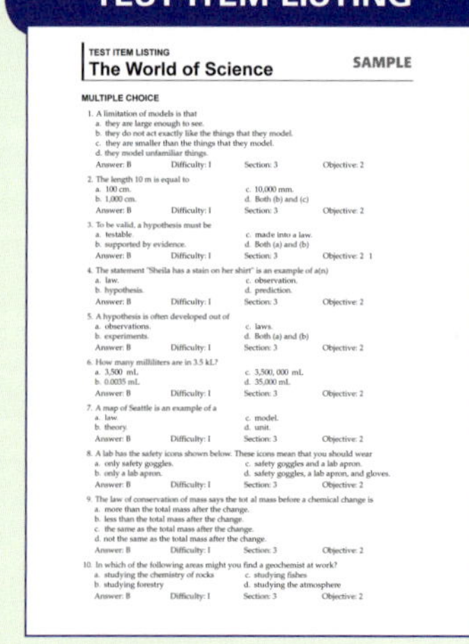

One-Stop Planner® CD-ROM

This CD-ROM includes all of the resources shown here and the following time-saving tools:

- *Lab Materials QuickList Software*
- *Customizable lesson plans*
- *Holt Calendar Planner*
- *The powerful ExamView® Test Generator*

For a preview of available worksheets covering math and science skills, see pages T12–T19. All of these resources are also on the One-Stop Planner®.

Meeting Individual Needs

DIRECTED READING A
Directed Reading A — SAMPLE
BASIC · ALSO IN SPANISH

DIRECTED READING B
Directed Reading B — SAMPLE
SPECIAL NEEDS

VOCABULARY ACTIVITY
Vocabulary Activity — SAMPLE
GENERAL

VOCABULARY AND SECTION SUMMARY
Vocabulary & Notes — SAMPLE
GENERAL · ALSO IN SPANISH

REINFORCEMENT
Reinforcement — SAMPLE
BASIC

CRITICAL THINKING
Critical Thinking — SAMPLE
ADVANCED

SCILINKS ACTIVITY
SciLinks Activity — SAMPLE
GENERAL

SCIENCE PUZZLERS, TWISTERS & TEASERS
Body Defenses and Disease
GENERAL

Labs and Activities

LONG-TERM PROJECTS & RESEARCH IDEAS
A Chuckle a Day Keeps the Doctor Away
ADVANCED

DATASHEETS FOR QUICK LABS
Reaction to Stress — SAMPLE

DATASHEETS FOR CHAPTER LABS
Using Scientific Methods — SAMPLE

DATASHEETS FOR LABBOOK
Does It All Add Up? — SAMPLE

Review and Assessments

SECTION QUIZ
Section Quiz — SAMPLE
GENERAL · ALSO IN SPANISH

SECTION REVIEW
Section Review — SAMPLE
GENERAL · ALSO IN SPANISH

CHAPTER REVIEW
Chapter Review — SAMPLE
GENERAL · ALSO IN SPANISH

CHAPTER TEST A
Chapter Test A — SAMPLE
GENERAL · ALSO IN SPANISH

CHAPTER TEST B
Chapter Test B — SAMPLE
ADVANCED

CHAPTER TEST C
Chapter Test C — SAMPLE
SPECIAL NEEDS

STANDARDIZED TEST PREPARATION
Standardized Test Preparation — SAMPLE
GENERAL

PERFORMANCE-BASED ASSESSMENT
Performanced-Based Assessment — SAMPLE
GENERAL

Chapter Enrichment

This Chapter Enrichment provides relevant and interesting information to expand and enhance your presentation of the chapter material.

Section 1

Disease

Pathogens and the Diseases That They Cause

- Pathogens are agents that cause disease. Pathogens can be living or nonliving. Living pathogens include bacteria, protists, fungi, worms, and insects. Nonliving pathogens include viruses, viroids, and prions.
- Bacterial diseases include bubonic plague, cholera, dental caries, Lyme disease, pneumonia, and typhoid fever.
- Viral diseases include colds, influenza, chickenpox, measles, rubella, mumps, smallpox, infectious hepatitis, polio, and AIDS.

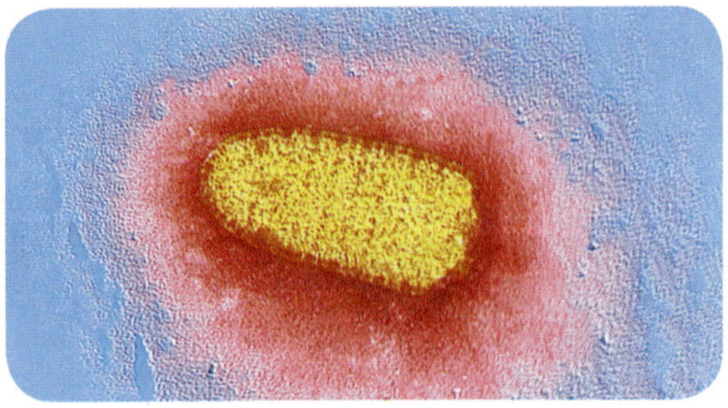

- Protists cause malaria, Chagas' disease, toxoplasmosis, and giardiasis. Common fungal diseases include ringworm, athlete's foot, vaginal yeast infections, jock itch, and histoplasmosis.

Is That a Fact!

- Bacteria were the first organisms to inhabit Earth. They are thought to have been the only organisms on Earth for about 1 billion years.
- All other life-forms are thought to have evolved from ancestral forms of bacteria.
- Some species of bacteria can reproduce every 20 min. Given unlimited resources, one bacterium could produce 1 million kilograms of bacteria in a day!

Emerging Viruses

- The news is full of instances in which new viruses have suddenly begun to infect people. Examples include HIV, which became common in the 1980s, hantavirus, which caused an outbreak in 1993, new strains of Ebola, which were announced in the 1990s, and SARS, which became common in 2003. Such viruses are called *emerging viruses.*
- Emerging viruses can evolve or mutate from an existing virus that was not infectious. They can also spread from one species to another. Another way that a virus can suddenly emerge is through globalization. When people build roads through a previously isolated area or travel to new places, viruses can spread.

Types of Vaccines

- Vaccines are generally harmless forms of a pathogen or parts of a pathogen that are introduced into the body to help the immune system develop antibodies against the pathogen should the disease-causing form enter the body in the future.
- There are several ways that vaccines are prepared today. One way to make a vaccine is to inactivate the viruses or to kill the bacteria that cause the disease. Another way is to use a similar strain of pathogen or an attenuated or weakened version of the pathogen. The latter kind of vaccine is sometimes called a *live attenuated vaccine.*

Section 2

Your Body's Defenses

The Body's Defenses Against Pathogens

- The body has two main kinds of defenses against disease—specific and nonspecific defenses.

- The body's so-called first and second lines of defense are its nonspecific defense mechanisms. The first line of defense consists of the skin, the mucous membranes, and the secretions of the mucous membranes. Oil, sweat, tears, mucus, and saliva wash pathogens away and contain enzymes that digest the cell walls of microbes.

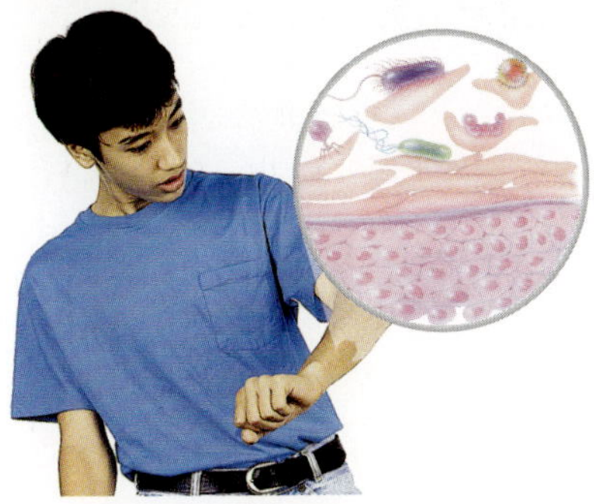

- The body's second line of defense includes white blood cells that ingest or destroy foreign agents detected in the body. The second line of defense also includes proteins that help cells avoid infection by foreign agents. Antimicrobial proteins include complement proteins and interferons. Also part of the body's second line of defense is its inflammatory response, which occurs in response to cuts or other incisions through the skin.

Fever

- A fever occurs when the body's temperature rises above 98.6°F (37°C) when measured orally.
- Fevers, which commonly accompany infectious diseases, help the body thwart invading pathogens. However, fevers may also accompany noninfectious conditions, such as dehydration and heart attack.

Is That a Fact!

- A very high fever can cause a coma, seizures, or brain damage.

AIDS

- AIDS is the final stage in an HIV infection. Upon infection, the viruses first multiply quickly and are quickly fought off by immune system cells. Some viruses remain in the body and replicate slowly over time. Immune system cells continue the fight, sometimes for many years. HIV attacks T4 helper cells, and when the number of helper T cells falls below a certain level, a person is said to have AIDS.
- HIV is spread through the exchange of blood and other body fluids. It is not spread through casual contact.
- There is no cure for AIDS at this time. There are only drug regimens that may prolong the life of some patients.
- The best way to avoid contracting AIDS is to avoid the behaviors known to put one at risk for acquiring HIV. These behaviors include sharing needles and engaging in unprotected sexual contact.

Allergic Reactions

- The first time a person is exposed to an allergen, such as pollen, he or she usually shows no allergic response. As when a vaccine is administered, the body produces antibodies against the allergen. Thus prepared, the body develops the allergic response upon subsequent exposures to the allergen.

Is That a Fact!

- Allergic responses may be a defense mechanism left over from the body's defenses against parasitic worms. This hypothesis comes from the observation that the body's method of combating parasitic worms is very similar to the allergic response seen in hay fever and asthma.

SCILINKS

NSTA Developed and maintained by the National Science Teachers Association

SciLinks is maintained by the National Science Teachers Association to provide you and your students with interesting, up-to-date links that will enrich your classroom presentation of the chapter.

Visit www.scilinks.org and enter the SciLinks code for more information about the topic listed.

opic: What Causes Diseases?
SciLinks code: HSM1653

Topic: Pathogens
SciLinks code: HSM1118

Topic: Body Defenses
SciLinks code: HSM0181

Topic: Allergies
SciLinks code: HSM0048

Topic: Cancer and HIV
SciLinks code: HSM0208

6

Overview

Tell students that this chapter will help them learn about diseases, including how infectious diseases are spread, how the body defends against diseases, and what problems can arise with the immune system.

Assessing Prior Knowledge

Students should be familiar with the following topics:

- bacteria and viruses
- body systems

Identifying Misconceptions

As students learn the material in this chapter, some of them may be confused about how diseases are spread. For example, warnings to bundle up during cold weather may lead students to think that cold weather causes infectious disease. Help them understand that pathogens cause infectious disease. Furthermore, some students may think that all microorganisms are harmful. Emphasize that although some microorganisms can cause disease, life as we know it would be impossible without microbes.

6

Body Defenses and Disease

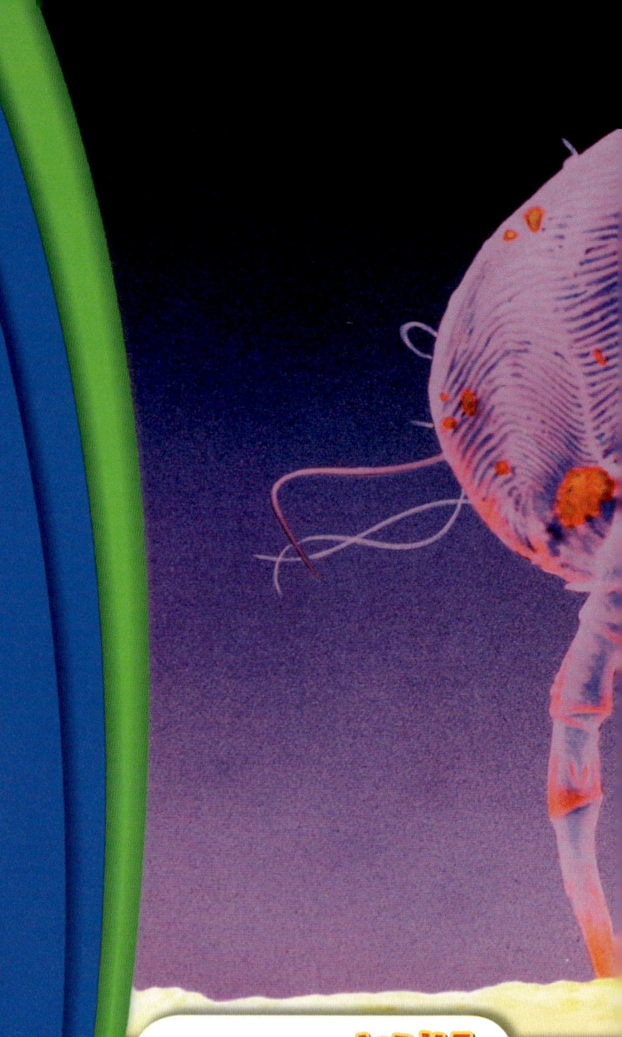

About the PHOTO

No, this photo is not from a sci-fi movie. It is not an alien insect soldier. This is, in fact, a greatly enlarged image of a house dust mite that is tinier than the dot of an *i*. Huge numbers of these creatures live in carpets, beds, and sofas in every home. Dust mites often cause problems for people who have asthma or allergies. The body's immune system fights diseases and alien factors, such as dust mites, that cause allergies.

PRE-READING ACTIVITY

FOLDNOTES

Tri-Fold Before you read the chapter, create the FoldNote entitled "Tri-Fold" described in the **Study Skills** section of the Appendix. Write what you know about the body's defenses in the column labeled "Know." Then, write what you want to know in the column labeled "Want." As you read the chapter, write what you learn about the body's defenses in the column labeled "Learn."

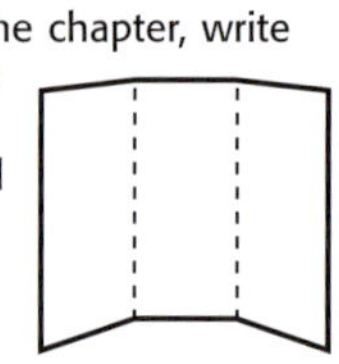

Standards Correlations

National Science Education Standards

The following codes indicate the National Science Education Standards that correlate to this chapter. The full text of the standards is at the front of the book.

Chapter Opener
SAI 1, 2; LS 1f

Section 1 Disease
SAI 1, 2; ST 2; SPSP 1, 5; HNS 1, 3; LS 1f

Section 2 Your Body's Defenses
UCP 1, 2, 3, 4, 5; SAI 1; HNS 3; LS 1b, 1d, 1e, 1f, 3a, 3b; *LabBook:* UCP2; SAI 1

Chapter Lab
UCP 1; SAI 1

Chapter Review
LS 1b, 1d, 1e, 1f, 3a, 3b

Science in Action
HNS 1; LS 1f

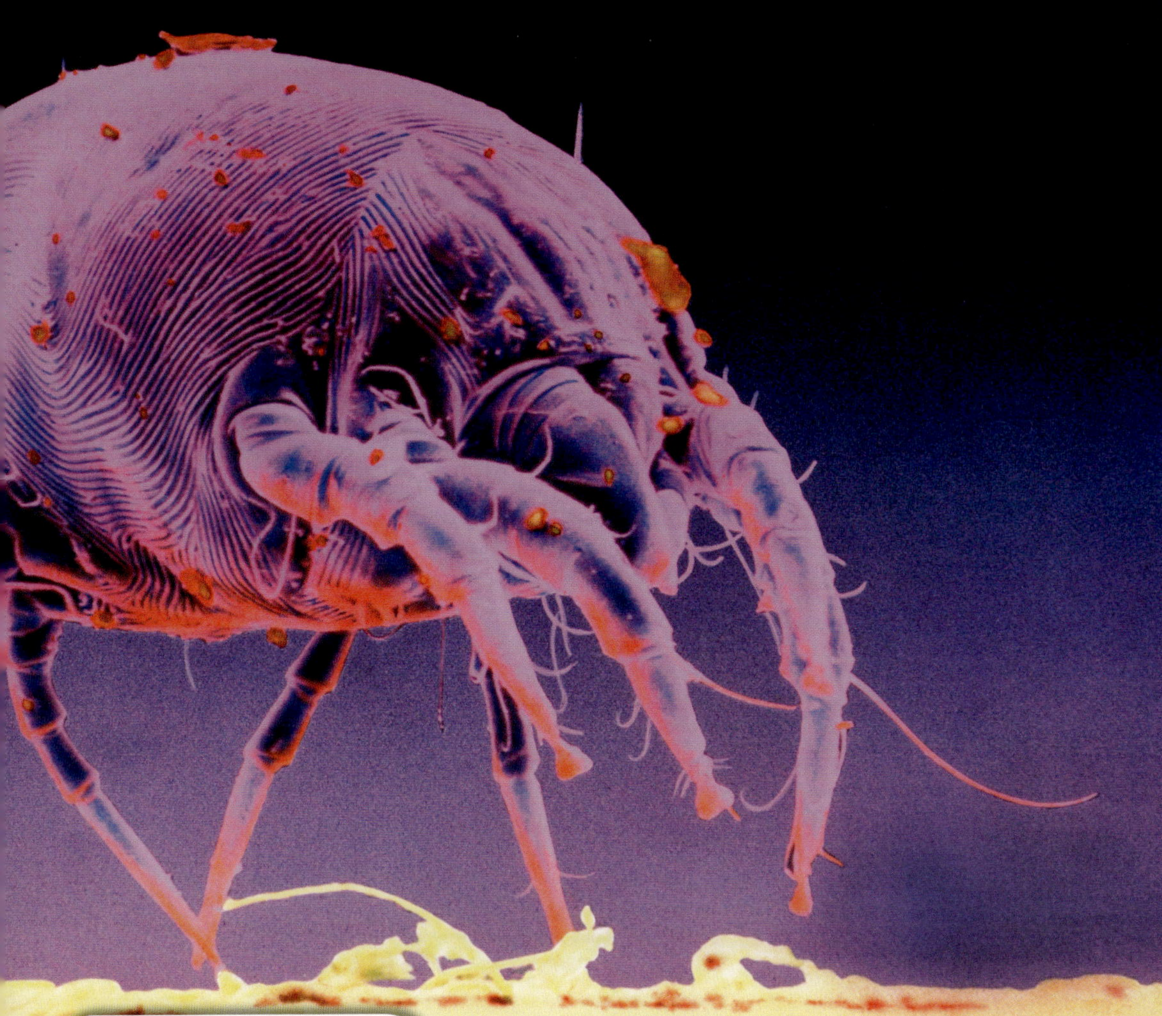

START-UP ACTIVITY

Invisible Invaders

In this activity, you will see tiny organisms grow.

Procedure

1. Obtain **two Petri dishes containing nutrient agar.** Label them "Washed" and "Unwashed."
2. Rub **two marbles** between the palms of your hands. Observe the appearance of the marbles.
3. Roll one marble in the Petri dish labeled "Unwashed."
4. Put on a pair of **disposable gloves.** Wash the other marble with **soap** and **warm water** for 4 min.
5. Roll the washed marble in the Petri dish labeled "Washed."
6. Secure the lids of the Petri dishes with **transparent tape.** Place the dishes in a warm, dark place. Do not open the Petri dishes after they are sealed.
7. Record changes in the Petri dishes for 1 week.

Analysis

1. How did the washed and unwashed marbles compare? How did the Petri dishes differ after several days?
2. Why is it important to wash your hands before eating?

START-UP ACTIVITY

MATERIALS

FOR EACH GROUP
- agar, nutrient
- gloves, protective
- marbles (2)
- Petri dishes (2)
- soap
- tape, transparent
- water, warm

Safety Caution: Remind students to review all safety cautions and icons before beginning this lab activity. Tell students not to open the Petri dishes once they are sealed. Treat all growth in the Petri dishes as pathogenic, and dispose of the dishes as you would any other biohazard.

Teacher's Notes: Keep the lids on the Petri dishes except when rolling the marble on the agar. Doing so will help keep outside contamination to a minimum. In step 6, it might be helpful to use an incubator set at 37°C.

Answers

1. Descriptions may vary. The Petri dish labeled "Unwashed" should have the most bacterial growth.
2. Sample answer: It is important to wash my hands to help decrease the number of microorganisms I put in my mouth.

Body Defenses and Disease — CHAPTER STARTER

Imagine . . .

A doctor arrives at your house with a cow. Your mom isn't surprised at all, and she invites the doctor and the cow right into your living room. The doctor explains that his cow has a disease that he plans to give to you. Your father is thrilled with the idea. This really scares you. Has everyone gone mad? The grown-ups seem to think it's okay to give you a cow disease!

Don't worry, the adults aren't crazy. Around 700 C.E., the Chinese did a similar thing. They realized a dangerous disease called smallpox could be prevented by exposing healthy people to the powdered scabs of infected people. It wasn't until the 1700s that Europeans learned of this technique. In 1796, a British physician named Edward Jenner observed that people who'd had a similar but less serious disease called cowpox did not get smallpox. Jenner started exposing patients to fluid from cowpox blisters. He called his treatment a *vaccine,* from the Latin word *vacca,* meaning "cow."

Read on to find out what causes diseases such as smallpox and how your body responds to them.

Chapter Starter Transparency
Use this transparency to help students begin thinking about the way that vaccines help defend the body.

CHAPTER RESOURCES

Technology

Transparencies READING SKILLS
- Chapter Starter Transparency

Student Edition CD-ROM

Guided Reading Audio CD
- English or Spanish

Classroom Videos
- Brain Food Video Quiz

Workbooks

Science Puzzlers, Twisters & Teasers
- Body Defenses and Disease GENERAL

SECTION 1

Focus

Overview

This section introduces the difference between noninfectious diseases and infectious diseases. Students will learn how they can come into contact with pathogens and how cleanliness, pasteurization, vaccines, and antibiotics can help reduce the spread of pathogens.

Bellringer

Writing Ask students to list as many diseases as they can. After students have completed their lists, you may want to make a master list on the board or on a transparency. The list might include physical illnesses as well as mental illnesses. Explain that in this chapter, students will learn how pathogens cause illnesses and how illnesses can be prevented.

SECTION 1

Disease

READING WARM-UP

Objectives

- Explain the difference between infectious diseases and noninfectious diseases.
- Identify five ways that you might come into contact with a pathogen.
- Discuss four methods that have helped reduce the spread of disease.

Terms to Learn

noninfectious disease
infectious disease
pathogen
immunity

READING STRATEGY

Paired Summarizing Read this section silently. In pairs, take turns summarizing the material. Stop to discuss ideas that seem confusing.

You've probably heard it before: "Cover your mouth when you sneeze!" "Wash your hands!" "Don't put that in your mouth!"

What is all the fuss about? When people say these things to you, they are concerned about the spread of disease.

Causes of Disease

When you have a *disease,* your normal body functions are disrupted. Some diseases, such as most cancers and heart disease, are not spread from one person to another. They are called **noninfectious diseases.**

Noninfectious diseases can be caused by a variety of factors. For example, a genetic disorder causes the disease hemophilia (HEE moh FIL ee uh), in which a person's blood does not clot properly. Smoking, lack of physical activity, and a high-fat diet can greatly increase a person's chances of getting certain noninfectious diseases. Avoiding harmful habits may help you avoid noninfectious diseases.

A disease that can be passed from one living thing to another is an **infectious disease.** Infectious diseases are caused by agents called **pathogens.** Viruses and some bacteria, fungi, protists, and worms may all cause diseases. **Figure 1** shows some enlarged images of common pathogens.

noninfectious disease a disease that cannot spread from one individual to another

infectious disease a disease that is caused by a pathogen and that can be spread from one individual to another

pathogen a virus, microorganism, or other organism that causes disease

Figure 1 Pathogens

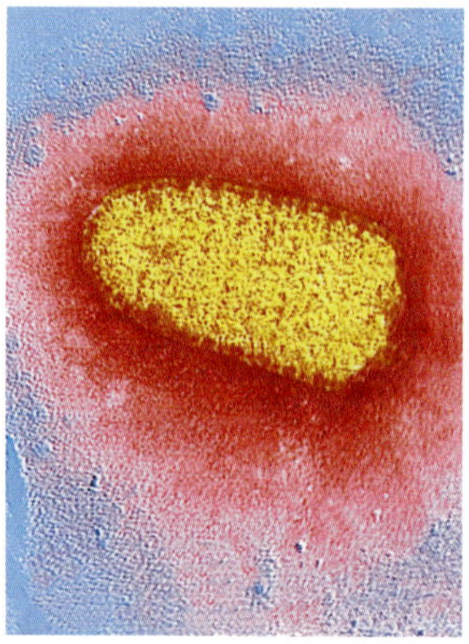

▲ This virus causes rabies.

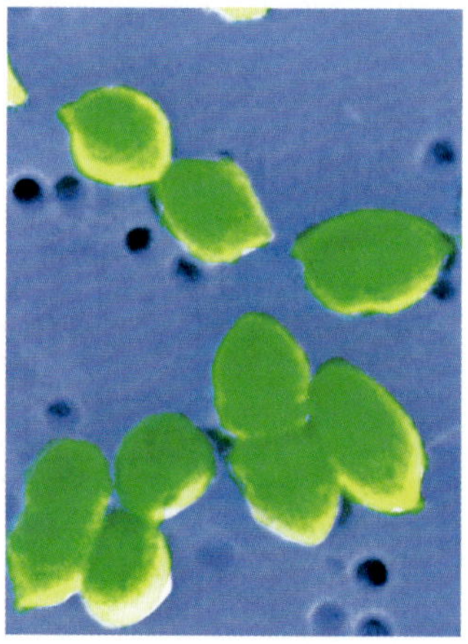

▲ *Streptococcus* bacteria can cause strep throat.

CHAPTER RESOURCES

Chapter Resource File

- Lesson Plan
- Directed Reading A BASIC
- Directed Reading B SPECIAL NEEDS

Technology

Transparencies

- Bellringer

MISCONCEPTION ALERT

Helpful Bacteria Students may think that all bacteria are pathogens. Point out that very few bacteria are pathogenic. Bacteria serve many useful—and, in some cases, essential—functions. Bacteria fix nitrogen and assist in digestion. They are used to make foods such as yogurt and cheese, to clean up oil spills, and to recycle wastes. Also, they can be genetically engineered to produce insulin and other human proteins that are needed to treat disease.

Pathways to Pathogens

There are many ways pathogens can be passed from one person to another. Being aware of them can help you stay healthy.

Air

Some pathogens travel through the air. For example, a single sneeze, such as the one shown in **Figure 2**, releases thousands of tiny droplets of moisture that can carry pathogens.

Figure 2 *A sneeze can force thousands of pathogen-carrying droplets out of your body at up to 160 km/h.*

Contaminated Objects

You may already know that if you drink from a glass that an infected person has just used, you could become infected with a pathogen. A person who is sick may leave bacteria or viruses on many other objects, too. For example, contaminated doorknobs, keyboards, combs, and towels can pass pathogens.

Person to Person

Some pathogens are spread by direct person-to-person contact. You can become infected with some illnesses by kissing, shaking hands, or touching the sores of an infected person.

Animals

Some pathogens are carried by animals. For example, humans can get a fungus called *ringworm* from handling an infected dog or cat. Also, ticks may carry bacteria that cause Lyme disease or Rocky Mountain spotted fever.

Food and Water

Drinking water in the United States is generally safe. But water lines can break, or treatment plants can become flooded. These problems may allow microorganisms to enter the public water supply. Bacteria growing in foods and beverages can cause illness, too. For example, meat, fish, and eggs that are not cooked enough can still contain dangerous bacteria or parasites. Even leaving food out at room temperature can give bacteria such as salmonella the chance to grow and produce toxins in the food. Refrigerating foods can slow the growth of many of these pathogens. Because bacteria grow in food, washing all used cooking surfaces and tools is also important.

✓ Reading Check Why must you cook meat and eggs thoroughly? (*See the Appendix for answers to Reading Checks.*)

CONNECTION TO Social Studies

Disease and History Many diseases have shaped history. For example, yellow fever, which is caused by a virus that is spread by mosquitoes, was one of the obstacles in building the Panama Canal. Only after people learned how to prevent the spread of the yellow fever virus could the canal be completed.

Use information from Internet and library research to create a poster describing how one infectious disease affected history.

ACTIVITY

Motivate

Discussion — GENERAL

Diseases Invite students to describe a disease. Encourage students to describe symptoms of the disease, how the disease is caught, and how the disease is treated or cured. Lead a discussion about how diseases are caused and treated by asking the following questions:

- Was the disease you described infectious or noninfectious? (Answers may vary.)
- What kind of pathogen caused the disease? (Answers will depend on the disease that the student describes. Allow students to research the cause of the disease if time permits. Most high school and college biology textbooks have a chart of common diseases and their pathogenic causes.)

LS Auditory

Answer to Reading Check

Cooking kills dangerous bacteria or parasites living in meat, fish, and eggs.

Teach

ACTIVITY — BASIC

Using Microscopes Provide students with a compound light microscope and prepared slides of different kinds of bacteria, including rod-shaped bacteria. Have them observe the specimens, compare the different kinds of bacteria, and make sketches in their **science journal.** English Language Learners

LS Visual

INCLUSION Strategies

- *Attention Deficit Disorder*
- *Developmentally Delayed*

Use this activity to explain the difference between infectious and noninfectious diseases.

1. Give half the students bright blue or green chalk, and ask them to make large colored dots on each fingertip.
2. Give the other half of the students black or red markers, and have them make a large dot on each fingertip.
3. Discuss that the chalk dots represent infectious diseases and the marker dots represent noninfectious diseases.
4. Ask students to shake hands with everyone near them.
5. Have students look at their hands to see that the "infectious diseases" have "spread" and that the "noninfectious" diseases have not.

LS Visual

Close

Answer to School-to-Home Activity

Most products that require pasteurization also require refrigeration because once the products are opened, they are vulnerable to getting bacteria in them. Refrigeration helps slow the growth of the bacteria.

Reteaching — BASIC

Reviewing Pathways Have students make a chart titled "How to Stay Healthy." In one column, they can list the ways that pathogens are encountered. In a second column, students can list ways to prevent becoming infected. **LS Logical**

Quiz — GENERAL

Ask students whether the following statements are true or false.

1. All diseases are caused by pathogens. (false)
2. You can become infected with some pathogens by shaking hands with an infected person. (true)
3. Pasteurization uses very cold temperatures to kill bacteria. (false)

Alternative Assessment — GENERAL

Writing **Pamphlet** Have students develop a disease-prevention pamphlet. Encourage students to include illustrations. **LS Verbal**

SCHOOL to HOME

Label Check

At home or in a local store, find a product that has been pasteurized. In your **science journal,** write down other safety information you find on the label, including the product's refrigeration needs. Why do you think most products that require pasteurization also require refrigeration?

ACTIVITY

immunity the ability to resist or to recover from an infectious disease

Putting Pathogens in Their Place

Until the twentieth century, surgery patients often died of bacterial infections. But doctors learned that simple cleanliness could help prevent the spread of some diseases. Today, hospitals and clinics use a variety of technologies to prevent the spread of pathogens. For example, ultraviolet radiation, boiling water, and chemicals are used to kill pathogens in health facilities.

Pasteurization

During the mid-1800s, Louis Pasteur, a French scientist, discovered that microorganisms caused wine to spoil. The uninvited microorganisms were bacteria. Pasteur devised a method of using heat to kill most of the bacteria in the wine. This method is called *pasteurization* (PAS tuhr i ZAY shuhn), and it is still used today. The milk that the girl in **Figure 3** is drinking has been pasteurized.

Vaccines and Immunity

In the late 1700s, no one knew what a pathogen was. During this time, Edward Jenner studied a disease called *smallpox*. He observed that people who had been infected with cowpox seemed to have protection against smallpox. These people had a resistance to the disease. The ability to resist or recover from an infectious disease is called **immunity.** Jenner's work led to the first modern vaccine. A *vaccine* is a substance that helps your body develop immunity to a disease.

Today, vaccines are used all over the world to prevent many serious diseases. Modern vaccines contain pathogens that are killed or specially treated so that they can't make you very sick. The vaccine is enough like the pathogen to allow your body to develop a defense against the disease.

Figure 3 *Today, pasteurization is used to kill pathogens in many different types of food, including dairy products, shellfish, and juices.*

CONNECTION to Physical Science — GENERAL

Pasteurization Pasteurization is the process of killing harmful bacteria by using thermal energy. Milk is pasteurized in one of two ways: by heating it to 63°C for 30 min or by heating it to 72°C for 16 s. The organisms that can survive pasteurization will eventually spoil the milk, but they are generally not harmful to people. Boiling the milk would kill even more of the bacteria —not just the harmful ones—but would also change the milk. Use the teaching transparency titled "Thermal Energy in Water" to illustrate how increasing the temperature of a substance increases the substance's thermal energy. **LS Visual**

Antibiotics

Have you ever had strep throat? If so, you have had a bacterial infection. Bacterial infections can be a serious threat to your health. Fortunately, doctors can usually treat these kinds of infections with antibiotics. An *antibiotic* is a substance that can kill bacteria or slow the growth of bacteria. Antibiotics may also be used to treat infections caused by other microorganisms, such as fungi. You may take an antibiotic when you are sick. Always take antibiotics according to your doctor's instructions to ensure that all the pathogens are killed.

Viruses, such as those that cause colds, are not affected by antibiotics. Antibiotics can kill only living things, and viruses are not alive. The only way to destroy viruses in your body is to locate and kill the cells they have invaded.

Reading Check **Frank caught a bad cold just before the opening night of a school play. He visited his doctor and asked her to prescribe antibiotics for his cold. The doctor politely refused and advised Frank to stay home and get plenty of rest. Why do you think the doctor refused to give Frank antibiotics?**

MATH PRACTICE

Epidemic!

You catch a cold and return to your school while sick. Your friends don't have immunity to your cold. On the first day, you expose five friends to your cold. The next day, each of those friends passes the virus to five more people. If this pattern continues for 5 more days, how many people will be exposed to the virus?

SECTION Review

Summary

- Noninfectious diseases cannot be spread from one person to another.
- Infectious diseases are caused by pathogens that are passed from one living thing to another.
- Pathogens can travel through the air or can be spread by contact with other people, contaminated objects, animals, food, or water.
- Cleanliness, pasteurization, vaccines, and antibiotics help control the spread of pathogens.

Using Key Terms

1. In your own words, write a definition for each of the following terms: *infectious disease, noninfectious disease,* and *immunity.*

Understanding Key Ideas

2. Vaccines contain
   a. treated pathogens.
   b. heat.
   c. antibiotics.
   d. pasteurization.
3. List five ways that you might come into contact with a pathogen.
4. Name four ways to help keep safe from pathogens.

Math Skills

5. If 10 people with the virus each expose 25 more people to the virus, how many people will be exposed to the virus?

Critical Thinking

6. **Identifying Relationships** Why might the risk of infectious disease be high in a community that has no water treatment facility?
7. **Analyzing Methods** Explain what might happen if a doctor did not wear gloves when treating patients.
8. **Applying Concepts** Why do vaccines for diseases in animals help prevent some illnesses in people?

SCILINKS. NSTA Developed and maintained by the National Science Teachers Association

For a variety of links related to this chapter, go to www.scilinks.org

Topic: Pathogens; What Causes Diseases?
SciLinks code: HSM1118; HSM1653

Answers to Math Practice

1st day: five people exposed
2nd day: 25 (5^2) people exposed
7th day: 78,125 (5^7) people exposed

Answers to Section Review

1. Sample answer: An infectious disease is an illness caused by a pathogen. A noninfectious disease is an illness that is not caused by a pathogen. Immunity is the ability to resist or recover from an infections disease.
2. a
3. Answers may vary but may include being sneezed on by an infected person, touching an infected person, having sexual contact with an infected person, eating infected food, or breathing in the pathogen.
4. Sample answer: keeping clean, pasteurization, vaccines, and antibiotics
5. $10 \times 25 = 250$
6. Sample answer: Without water treatment, a community's water supply can easily become infected with pathogens. These pathogens can be ingested or can contaminate objects washed in the water.
7. Sample answer: A doctor who does not wear gloves might spread an infectious disease from one patient to another.
8. Sample answer: Some animal diseases, such as rabies, can also infect people. Preventing the spread of these diseases in animals makes the diseases less likely to be spread to humans.

Answer to Reading Check

Frank's doctor did not prescribe antibiotics because Frank had a cold. Colds are caused by viruses. Antibiotics cannot stop viruses.

CHAPTER RESOURCES

Chapter Resource File

- Section Quiz GENERAL
- Section Review GENERAL
- Vocabulary and Section Summary GENERAL
- Reinforcement Worksheet BASIC

Technology

Transparencies

- LINK TO PHYSICAL SCIENCE Thermal Energy in Water

SECTION 2

Focus

Overview

In this section, students will learn how the skin keeps pathogens out of the body and how the immune system works. Students will also learn about allergies and other immune system problems.

Bellringer

Have students make a list of the ways that pathogens might enter the body. (Examples include through the mouth, ears, nose, and cuts in the skin. Pathogens can travel in the water, in the air, and in food.)

Motivate

ACTIVITY — GENERAL

Reactions to Illness Ask students to think of a time when they were ill. Have students list the ways in which their body reacted to the illness. (Answers might include having a fever, chills, a runny nose, a sore throat, a rash, and throbbing pain.) Encourage students to share their lists with a partner. Then, have students skim this lesson and try to link their body's reactions with the reactions of the immune system to pathogenic invasions.
LS Interpersonal

SECTION 2

Your Body's Defenses

Bacteria and viruses can be in the air, in the water, and on all the surfaces around you.

READING WARM-UP

Objectives

- Describe how your body keeps out pathogens.
- Explain how the immune system fights infections.
- Describe four challenges to the immune system.

Terms to Learn

immune system	memory B cell
macrophage	allergy
T cell	autoimmune disease
B cell	cancer
antibody	

READING STRATEGY

Reading Organizer As you read this section, make a flowchart of the steps of how your body responds to a virus.

Your body must constantly protect itself against pathogens that are trying to invade it. But how does your body do that? Luckily, your body has its own built-in defense system.

First Lines of Defense

For a pathogen to harm you, it must attack a part of your body. Usually, though, very few of the pathogens around you make it past your first lines of defense.

Many organisms that try to enter your eyes or mouth are destroyed by special enzymes. Pathogens that enter your nose are washed down the back of your throat by mucus. The mucus carries the pathogens to your stomach, where most are quickly digested.

Your skin is made of many layers of flat cells. The outermost layers are dead. As a result, many pathogens that land on your skin have difficulty finding a live cell to infect. As **Figure 1** shows, the dead skin cells are constantly dropping off your body as new skin cells grow from beneath. As the dead skin cells flake off, they carry away viruses, bacteria, and other microorganisms. In addition, glands secrete oil onto your skin's surface. The oil contains chemicals that kill many pathogens.

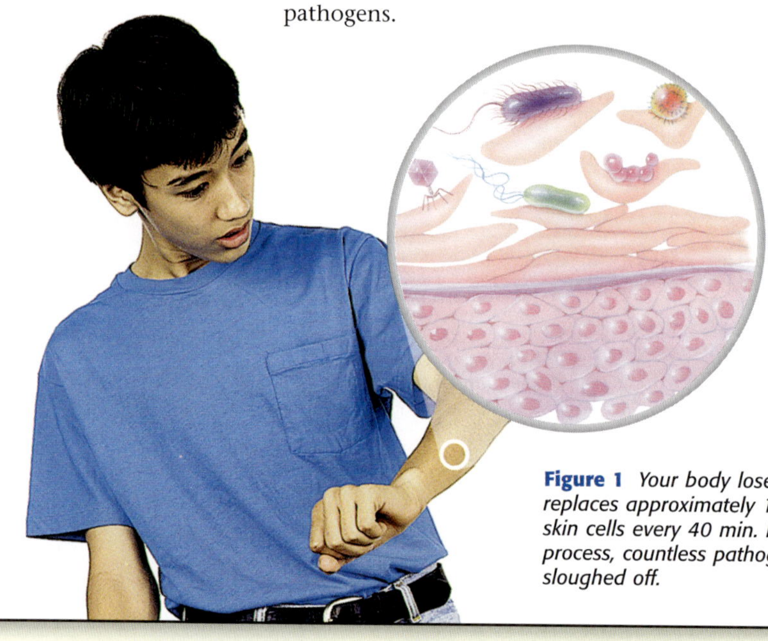

Figure 1 *Your body loses and replaces approximately 1 million skin cells every 40 min. In the process, countless pathogens are sloughed off.*

CHAPTER RESOURCES

Chapter Resource File

- Lesson Plan
- Directed Reading A BASIC
- Directed Reading B SPECIAL NEEDS

Technology

Transparencies

- Bellringer
- An Antibody's Shape Fits an Antigen

WEIRD SCIENCE

Earwax is one of the body's defenses against foreign invaders. Wax collects dirt, bacteria, fungi, and other foreign matter that could cause an ear infection.

Failure of First Lines

Sometimes, skin is cut or punctured and pathogens can enter the body. The body acts quickly to keep out as many pathogens as possible. Blood flow to the injured area increases. Cell parts in the blood called *platelets* help seal the open wound so that no more pathogens can enter.

The increased blood flow also brings cells that belong to the **immune system,** the body system that fights pathogens. The immune system is not localized in any one place in your body. It is not controlled by any one organ, such as the brain. Instead, it is a team of individual cells, tissues, and organs that work together to keep you safe from invading pathogens.

Cells of the Immune System

The immune system consists mainly of three kinds of cells. One kind is the macrophage (MAK roh FAYJ). **Macrophages** engulf and digest many microorganisms or viruses that enter your body. If only a few microorganisms or viruses have entered a wound, the macrophages can easily stop them.

The other two main kinds of immune-system cells are T cells and B cells. **T cells** coordinate the immune system and attack many infected cells. **B cells** are immune-system cells that make antibodies. **Antibodies** are proteins that attach to specific antigens. *Antigens* are substances that stimulate an immune response. Your body is capable of making billions of different antibodies. Each antibody usually attaches to only one kind of antigen, as illustrated in **Figure 2.**

Reading Check How do macrophages help fight disease? *(See the Appendix for answers to Reading Checks.)*

immune system the cells and tissues that recognize and attack foreign substances in the body

macrophage an immune system cell that engulfs pathogens and other materials

T cell an immune system cell that coordinates the immune system and attacks many infected cells

B cell a white blood cell that makes antibodies

antibody a protein made by B cells that binds to a specific antigen

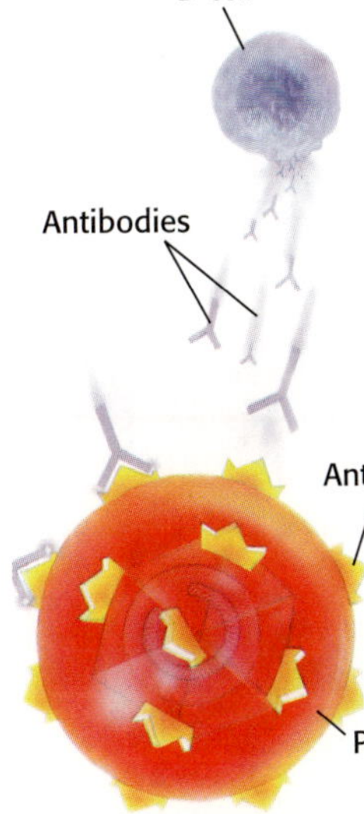

Figure 2 *An antibody's shape is very specialized. It matches an antigen like a key fits a lock.*

Quick Lab

Only Skin Deep

1. Cut an **apple** in half.
2. Place **plastic wrap** over each half. The plastic wrap will act as skin.
3. Use **scissors** to cut the plastic wrap on one of the apple halves, and then use an **eyedropper** to drip **food coloring** on each apple half. The food coloring represents pathogens coming into contact with your body.
4. What happened to each apple half?
5. How is the plastic wrap similar to skin?
6. How is the plastic wrap different from skin?

Cultural Awareness

Advanced

Disease and Conquest In the 16th century, Hernando Cortes and the Spanish conquistadors overwhelmed and conquered the flourishing cultures of what is now Mexico. One reason the Europeans were able to overtake the Native Americans so quickly is that the Europeans incidentally brought with them diseases such as smallpox. These diseases were new to the populations in the Americas, and they devastated entire nations. In the span of just two generations, an estimated 12,000,000 to 25,000,000 Native Americans died as a result of European diseases. Have interested students research and report on the role that disease played when the Europeans settled in North and South America. **LS Verbal**

Teach

Quick Lab

MATERIALS

For Each Student

- apple (cut in half)
- eyedropper
- food coloring
- plastic wrap
- scissors

Safety Caution: Tell students to cut in a direction away from their face and hands. Students should not ingest the apple or the food coloring after the activity. Food coloring can stain hands and clothes.

Teacher's Note: You may want to provide gloves and smocks for this activity. Dark food coloring, such as blue, green, and red, works best in this activity. Yellow does not show up as well as the darker colors. Instruct students to limit themselves to three drops of food coloring per apple half.

Answers

4. The apple half with the uncut wrap was not affected by the food coloring. The apple half with the cut wrap was dyed by the food coloring that seeped through the slits.
5. Sample answer: The plastic wrap is similar to skin because it keeps foreign substances out and protects the soft tissue underneath it.
6. Sample answer: The plastic wrap is different from skin because it cannot heal the wound or swell to help close the wound.

Answer to Reading Check

Macrophages engulf, or eat, any microorganisms or viruses that enter your body.

Teach, *continued*

ACTIVITY — BASIC

Follow the Path Have students trace the proper sequence of the pictures across these two pages. Point out that each picture has its own caption that begins with a bold heading. Then, have students outline the information on these two pages by listing the bold headings in order on a sheet of paper. Have pairs of students compare their outlines. English Language Learners
LS Visual

Using the Figure — GENERAL

Identifying Cells As students look at the pictures on these pages, point out that the pictures are numbered. After students have studied the sequence of events portrayed in the pictures, have them point to each kind of white blood cell shown in the pictures on these two pages as you call out the names of the cells: *helper T cell, killer T cell, B cell,* and *macrophage.* Then, ask students to describe the role of each kind of white blood cell. English Language Learners
LS Visual

INTERNET ACTIVITY

For another activity related to this chapter, go to **go.hrw.com** and type in the keyword **HL5BD6W.**

Responding to a Virus

If virus particles enter your body, some of the particles may pass into body cells and begin to replicate. Other virus particles will be engulfed and broken up by macrophages. This is just the beginning of the immune response. The process your immune system uses to fight an invading virus is summarized in the figure below.

Reading Check What are two things that can happen to virus particles when they enter the body?

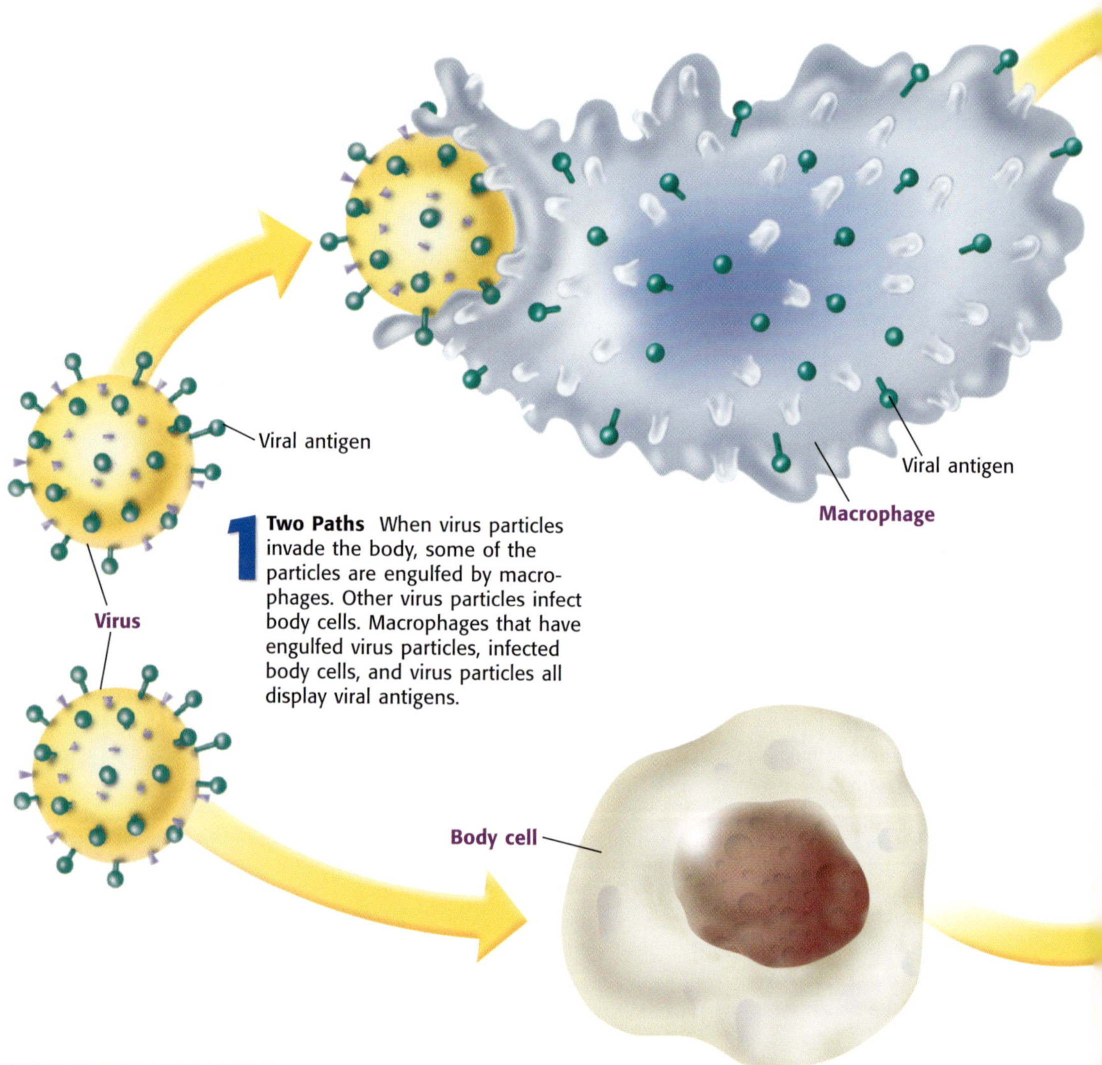

1 Two Paths When virus particles invade the body, some of the particles are engulfed by macrophages. Other virus particles infect body cells. Macrophages that have engulfed virus particles, infected body cells, and virus particles all display viral antigens.

CHAPTER RESOURCES

Technology

Transparencies
- Immune Response: A
- Immune Response: B

Answer to Reading Check

If a virus particle enters the body, it may pass into body cells and begin to replicate. Or it may be engulfed and broken up by macrophages.

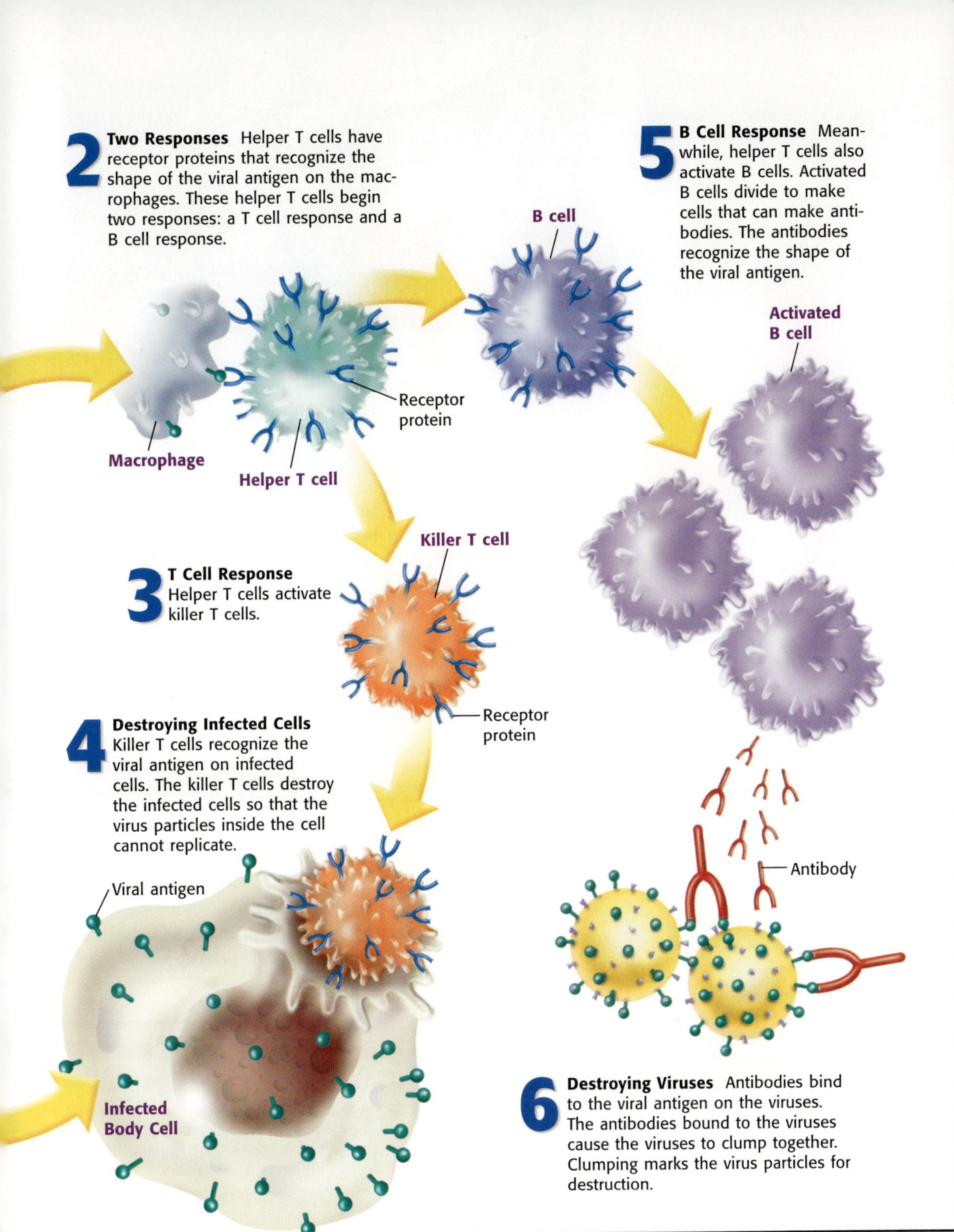

Research — GENERAL

Writing **Immune Deficiency** Have students research the immune problems of "bubble" children, such as David Vetter (the "Bubble Boy"). These children suffer from severe combined immune deficiency (SCID). (Students should find that children afflicted with this disease lack T cells and B cells and have a very weak immune system. Childhood illnesses that most children can shrug off can kill a child who has SCID. People who have SCID must live in a germ-free environment, such as in a sterile plastic bubble or cubicle in which the air is cleaned.) **LS Verbal**

Group Activity — ADVANCED

Immunity Skit Have students perform a skit that demonstrates the immune system's response to an invasion of viruses. Assign a role to each student, and have students study the text to learn about their role. Allow students time to prepare their skit and make any props that they need. For example, the students playing B cells may wish to make antibodies, and the students playing viruses may wish to construct simple body cells to infect. **LS Kinesthetic/Interpersonal**

Weird Science

In some cases, a pregnant woman may form antibodies against the blood of the baby she is carrying. The situation is likely to occur when the mother's blood is Rh-negative, meaning it lacks the Rh antigen, and the baby's blood is Rh-positive, meaning it carries the Rh antigen. The mother's immune system recognizes the Rh antigen as foreign and begins to attack the baby's blood cells.

Is That a Fact!

In 1918, a strain of flu viruses called the *Spanish flu* killed at least 20 million people. That number is more than the number of people killed in combat in World War I.

Teach, *continued*

CONNECTION ACTIVITY
Real World — GENERAL

Treating Allergies An allergy is usually treated in two ways once the allergen is identified. An allergist may try to desensitize a patient to the allergen by administering small, periodic doses of the allergen. In addition, an allergist may prescribe drugs to minimize the body's reactions to the allergen. Have interested students contact an allergist and conduct an interview about his or her work.
LS Interpersonal/Verbal

Research — GENERAL

Writing **Immune Disorders** Have students find an article or news report about one of the immune disorders discussed in this lesson. Have students work in small groups to prepare a written or oral report that summarizes one of the articles and links the article it to the concepts presented in this chapter. LS Inerpersonal/Verbal

Answer to Connection to Chemistry

As the egg white cooks, the proteins change shape. The change in shape causes the proteins to form a solid mass.

Figure 3 *You may not feel well when you have a fever. But a fever is one way that your body fights infections.*

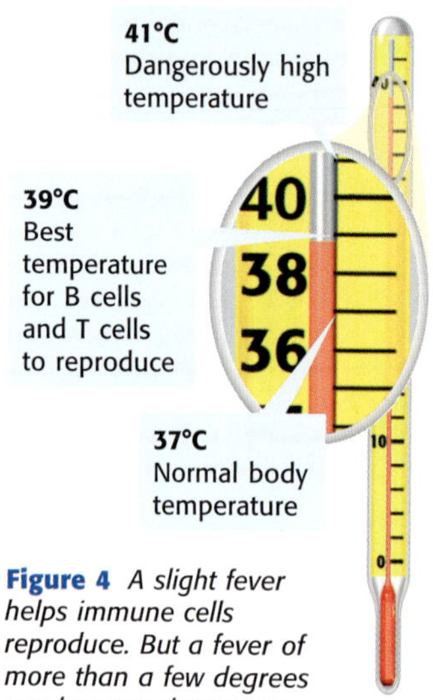

Figure 4 *A slight fever helps immune cells reproduce. But a fever of more than a few degrees can become dangerous.*

Fevers

The man in **Figure 3** is sick and has a fever. What is a fever? When macrophages activate the helper T cells, they send a chemical signal that tells your brain to turn up the thermostat. In a few minutes, your body's temperature can rise several degrees. A moderate fever of one or two degrees actually helps you get well faster because it slows the growth of some pathogens. As shown in **Figure 4,** a fever also helps B cells and T cells multiply faster.

Memory Cells

Your immune system can respond to a second encounter faster than it can respond the first time. B cells must have had previous contact with a pathogen before they can make the correct antibodies. During the first encounter with a new pathogen, specialized B cells make antibodies that are effective against that particular invader. This process takes about 2 weeks, which is far too long to prevent an infection. Therefore, the first time you are infected, you usually get sick.

A few of the B cells become memory B cells. **Memory B cells** are cells in your immune system that "remember" how to make an antibody for a particular pathogen. If the pathogen shows up again, the memory B cells produce B cells that make enough antibodies in just 3 or 4 days to protect you.

memory B cell a B cell that responds to an antigen more strongly when the body is reinfected with an antigen than it does during its first encounter with the antigen

CONNECTION TO Chemistry

Bent out of Shape When you have a fever, the heat of the fever changes the shape of viral or bacterial proteins, slowing or preventing the reproduction of the pathogen. With an adult present, observe how an egg white changes as it cooks. What do you think happens to the protein in the egg white as it cooks?

ACTIVITY

INCLUSION *Strategies*

- ***Learning Disabled***
- ***Attention Deficit Disorder***

Many students have trouble differentiating between allergies and autoimmune diseases. Help students by repeating that autoimmune diseases happen when the body mistakenly attacks its own tissues, and that allergies happen when the immune system mistakenly attacks harmless substances that enter the body. Then, use the following drill:

Which of the following are allergic reactions and which are autoimmune responses?

- sneezing at pollen (allergic reaction)
- the immune system attacking a nerve cell (autoimmune response)
- the immune system attacking the body cells that make insulin (autoimmune response)
- the immune system attacking the proteins in peanut butter (allergic reaction)

LS Verbal

Challenges to the Immune System

The immune system is a very effective body-defense system, but it is not perfect. The immune system is unable to deal with some diseases. There are also conditions in which the immune system does not work properly.

allergy a reaction to a harmless or common substance by the body's immune system

autoimmune disease a disease in which the immune system attacks the organism's own cells

Allergies

Sometimes, the immune system overreacts to antigens that are not dangerous to the body. This inappropriate reaction is called an **allergy.** Allergies may be caused by many things, including certain foods and medicines. Many people have allergic reactions to pollen, shown in **Figure 5.** Symptoms of allergic reactions range from a runny nose and itchy eyes to more serious conditions, such as asthma.

Doctors are not sure why the immune system overreacts in some people. Scientists think allergies might be useful because the mucus draining from your nose carries away pollen, dust, and microorganisms.

Figure 5 *Pollen is one substance that can cause allergic reactions.*

Autoimmune Diseases

A disease in which the immune system attacks the body's own cells is called an **autoimmune disease.** In an autoimmune disease, immune-system cells mistake body cells for pathogens. One autoimmune disease is rheumatoid arthritis (ROO muh TOYD ahr THRIET IS), in which the immune system attacks the joints. A common location for rheumatoid arthritis is the joints of the hands, as shown in **Figure 6.** Other autoimmune diseases include type 1 diabetes, multiple sclerosis, and lupus.

Reading Check Name four autoimmune diseases.

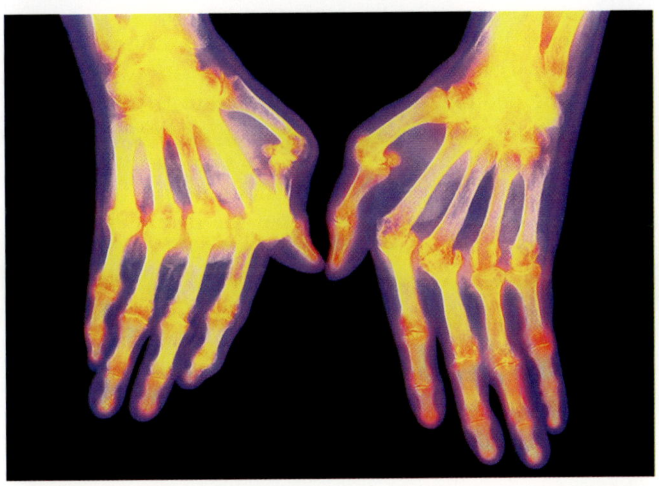

Figure 6 *In rheumatoid arthritis, immune-system cells cause joint-tissue swelling, which can lead to joint deformities.*

Discussion — GENERAL

Allergies Ask students to tell you about any allergies that they know about and the reactions that people experience when exposed to the particular allergens. Write their responses on the board or an overhead transparency, and use this list, especially the body responses, as a way of introducing how the immune system can make peoples's lives miserable—or at least uncomfortable—when it does not function properly.
LS Auditory

Homework — GENERAL

Writing **Anaphylactic Shock** Ask students to research and report on the topic of anaphylactic shock. (Students should find that anaphylactic shock is a serious allergic reaction that can result in death. A small amount of an allergen can produce anaphylactic shock. People who are prone to such an allergic reaction may carry with them a syringe containing the hormone epinephrine to combat the onset of anaphylactic shock.) **LS Verbal**

Is That a Fact!

Peanuts are dangerous, even potentially deadly, to people who are allergic to them. A person nearly died from eating a tuna sandwich because the knife used to cut the sandwich had just been used to make a peanut butter sandwich.

Answer to Reading Check

rheumatoid arthritis, diabetes, multiple sclerosis, and lupus

Close

Reteaching — Basic

Reviewing HIV You can use the figures from this section that illustrate an immune response to help students understand how HIV hurts the immune system. HIV infects and destroys helper T cells. Have students cover and hide the helper T cell illustrated in the figure. Then, ask students how destroying helper T cells would affect the immune response. (Students should be able to see that both the B cell and T cell responses to infection would be stopped if the helper T cells were eliminated.) **LS Visual/Auditory**

Quiz — General

Ask students whether the following statements are true or false.

1. Antibodies are specific to certain pathogens. (true)
2. An allergy is caused by the immune system attacking the cells of the body. (false)
3. People with AIDS have too many killer T cells. These cells begin to attack the body's cells, and AIDS is the result. (false)

Alternative Assessment — Advanced

Writing Have students choose one of the disorders discussed in this section and prepare a report or a presentation about it. Encourage students to discuss the immune response and the cells of the immune system. **LS Verbal**

Figure 7 **Immune Cells Fighting Cancer**

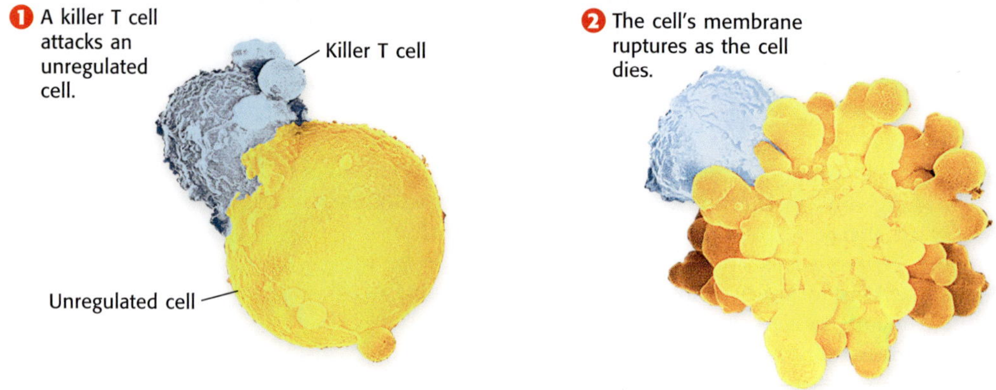

Cancer

Healthy cells divide at a carefully regulated rate. Occasionally, a cell doesn't respond to the body's regulation and begins dividing at an uncontrolled rate. As can be seen in **Figure 7,** killer T cells destroy this type of cell. Sometimes, the immune system cannot control the division of these cells. **Cancer** is the condition in which cells divide at an uncontrolled rate.

cancer a disease in which the cells begin dividing at an uncontrolled rate and become invasive

Many cancers will invade nearby tissues. They can also enter the cardiovascular system or lymphatic system. Cancers can then be transported to other places in the body. Cancers disrupt the normal activities of the organs they have invaded, sometimes leading to death. Today, though, there are many treatments for cancer. Surgery, radiation, and certain drugs can be used to remove or kill cancer cells or slow their division.

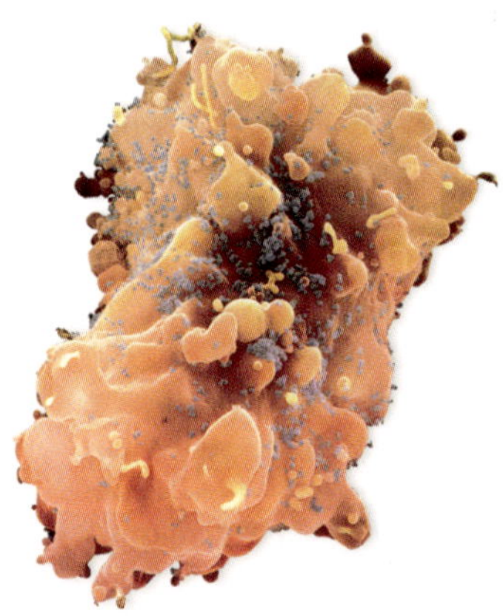

Figure 8 *The blue particles on this helper T cell are human immunodeficiency viruses. They replicated inside the T cell.*

AIDS

The human immunodeficiency virus (HIV) causes acquired immune deficiency syndrome (AIDS). Most viruses infect cells in the nose, mouth, lungs, or intestines, but HIV is different. HIV infects the immune system itself, using helper T cells as factories to produce more viruses. You can see HIV particles in **Figure 8.** The helper T cells are destroyed in the process. Remember that the helper T cells put the B cells and killer T cells to work.

People with AIDS have very few helper T cells, so nothing activates the B cells and killer T cells. Therefore, the immune system cannot attack HIV or any other pathogen. People with AIDS don't usually die of AIDS itself. They die of other diseases that they are unable to fight off.

Reading Check What virus causes AIDS?

Activity — General

Writing **Concept Mapping** Have students define the new terms presented in this chapter and then use the terms in a concept map. Have students use the chapter title as the first concept in their map. **LS Visual/Logical**

Answer to Reading Check

HIV causes AIDS.

SECTION Review

Summary

- Macrophages engulf pathogens, display antigens on their surface, and activate helper T cells. The helper T cells put the killer T cells and B cells to work.
- Killer T cells kill infected cells. B cells make antibodies.
- Fever helps speed immune-cell growth and slow pathogen growth.
- Memory B cells remember how to make an antibody for a pathogen that the body has previously fought.
- An allergy is the overreaction of the immune system to a harmless antigen.
- Autoimmune diseases are responses in which the immune system attacks healthy tissue.
- Cancer cells are cells that undergo uncontrolled division.
- AIDS is a disease that results when the human immunodeficiency virus kills helper T cells.

Using Key Terms

For each pair of terms, explain how the meanings of the terms differ.

1. *B cell* and *T cell*
2. *autoimmune disease* and *allergy*

Understanding Key Ideas

3. Your body's first line of defense against pathogens includes
   - a. skin.
   - b. macrophages.
   - c. T cells.
   - d. B cells.
4. List three ways your body defends itself against pathogens.
5. Name three different cells in the immune system, and describe how they respond to pathogens.
6. Describe four challenges to the immune system.
7. What characterizes a cancer cell?

Critical Thinking

8. **Identifying Relationships** Can your body make antibodies for pathogens that you have never been in contact with? Why or why not?
9. **Applying Concepts** If you had chickenpox at age 7, what might prevent you from getting chickenpox again at age 8?

Interpreting Graphics

10. Look at the graph below. Over time, people with AIDS become very sick and are unable to fight off infection. Use the information in the graph below to explain why this occurs.

T Cell Count of a Person with AIDS

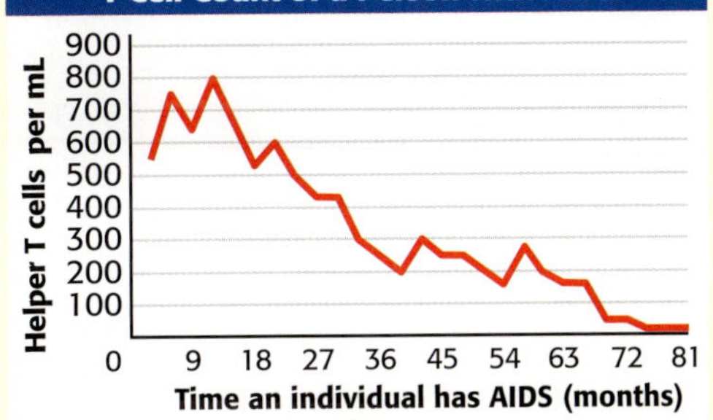

SCILINKS. NSTA Developed and maintained by the National Science Teachers Association

For a variety of links related to this chapter, go to www.scilinks.org

Topic: Body Defenses; Allergies
SciLinks code: HSM0181; HSM0048

CHAPTER RESOURCES

Chapter Resource File

- Section Quiz GENERAL
- Section Review GENERAL
- Vocabulary and Section Summary GENERAL
- SciLinks Activity GENERAL
- Datasheet for Quick Lab

Answers to Section Review

1. Sample answer: B cells pathogens by making antibodies. T cells fight pathogens by coordinating responses and killing infected cells directly.
2. Sample answer: In an autoimmune disease, the immune system attacks the body's tissues. In an allergy, the immune system overreacts to antigens that are not part of the body.
3. a
4. Sample answer: Skin keeps most pathogens out. Chemical defenses are in your eyes, stomach, and mouth. If you have a cut in your skin, blood platelets help close the wound so that more microorganisms cannot enter. Microorganisms that do enter encounter immune system cells.
5. Sample answer: a macrophage engulfs pathogens and stick antigens on their outer membranes, helper T cells activate killer T cells and B cells, and killer T cells kill any body cell infected with pathogens.
6. allergy, the body has an inappropriate reaction to a harmless antigen; autoimmune disease: the immune system reacts to body tissues; cancer: cells of the body divide uncontrollably; AIDS: HIV reduces the number of helper T cells, so the body cannot mount a defense against pathogens
7. A cell becomes cancerous when it starts dividing at an uncontrolled rate.
8. Sample answer: No, the body makes antibodies against specific pathogens. If a pathogen has never been met, the immune system would not have had any reason to make an antibody against it.
9. Sample answer: memory B cells
10. Sample answer: People who have AIDS become less able to fight infections because their helper T cell count decreases over time.

Skills Practice Lab

Passing the Cold

Teacher's Notes

Time Required

One 45-minute class period

Lab Ratings

EASY → HARD

Teacher Prep 2

Student Set-Up 2

Concept Level 2

Clean Up 2

MATERIALS

Prepare a phenolphthalein indicator solution ahead of time. Dilute the indicator solution in water. Add 10 mL of the indicator to 40 mL of water. This solution is enough for one student. Mix a 1.5 % NaOH solution. Mix 15 g NaOH with 1 L of water. All but one student will receive 50 mL of this solution.

Safety Caution: Remind students that although they are working with a very low concentration of an alkaline solution, they should work safely with all materials. All spills should be cleaned up immediately. Skin exposed to solutions should be washed immediately with plenty of running water. Remind students that they must never mix unknown solutions without teacher supervision and approval.

Using Scientific Methods

Skills Practice Lab

Passing the Cold

OBJECTIVES

Investigate how diseases spread.

Analyze data about how diseases spread.

MATERIALS

- beaker or a cup, 200 mL
- eyedropper
- gloves, protective
- solution, unknown, 50 mL

SAFETY

There are more than 100 viruses that cause the symptoms of the common cold. Any of the viruses can be passed from person to person—through the air or through direct contact. In this activity, you will track the progress of an outbreak in your class.

Ask a Question

1. With other members of your group, form a question about the spread of disease. For example "How are cold viruses passed from person to person?" or "How can the progress of an outbreak be modeled?"

Form a Hypothesis

2. Form a hypothesis based on the question you asked.

Test the Hypothesis

3. Obtain an empty cup or beaker, an eyedropper, and 50 mL of one of the solutions from your teacher. Only one student will have the "cold virus" solution. You will see a change in your solution when you have become "infected."
4. Your teacher will divide the class into two equal groups. If there is an extra student, that person will record data on the board. Otherwise, the teacher will act as the recorder.
5. The two groups should form straight lines, facing each other.
6. Each time your teacher says the word *mix,* fill your eyedropper with your solution, and place 10 drops of your solution in the beaker of the person in the line opposite you without touching your eyedropper to the other liquid.
7. Gently stir the liquid in your cup with your eyedropper. Do not put your eyedropper in anyone else's solution.
8. If your solution changes color, raise your hand so that the recorder can record the number of students who have been "infected."
9. Your teacher will instruct one line to move one person to the right. Then, the person at the end of the line without a partner should go to the other end of the line.

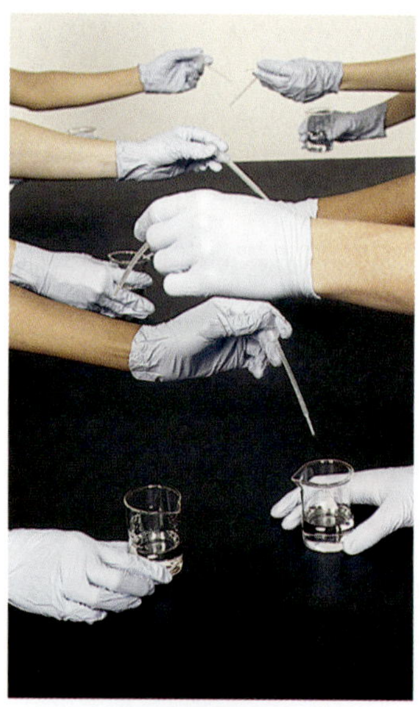

Edith McAlanis
Socorro Middle School
El Paso, Texas

CHAPTER RESOURCES

Chapter Resource File

- Datasheet for Chapter Lab
- Lab Notes and Answers

Technology

Classroom Videos
- Lab Video

LabBook
- Antibodies to the Rescue

Trial	Number of infected people	Total number of people	Percentage of infected people
1			
2			
3			
4			
5			
6			
7			
8			
9			
10			

Results of Experiment

10. Repeat steps 5–9 nine more times for a total of 10 trials.
11. Return to your desk, and create a data table in your notebook similar to the table above. The column with the title "Total number of people" will remain the same in every row. Enter the data from the board into your data table.
12. Find the percentage of infected people for the last column by dividing the number of infected people by the total number of people and multiplying by 100 in each line.

Analyze the Results

1. **Describing Events** Did you become infected? If so, during which trial did you become infected?
2. **Examining Data** Did everyone eventually become infected? If so, how many trials were necessary to infect everyone?

Draw Conclusions

3. **Interpreting Information** Explain at least one reason why this simulation may underestimate the number of people who might have been infected in real life.
4. **Applying Conclusions** Use your results to make a line graph showing the change in the infection percentage per trial.

Applying Your Data

Do research in the library or on the Internet to find out some of the factors that contribute to the spread of a cold virus. What is the best and easiest way to reduce your chances of catching a cold? Explain your answer.

Preparation Notes

Phenolphthalein is a base indicator and will turn pink in the presence of NaOH. One student (or two, if your class is large) will be given 50 mL of the indicator. This student will represent the original "infected" individual. (It is more fun if no one knows who the original infected student is at first.) All other students will be given the NaOH solution.

Prepare a results chart similar to the student table on the board. You will need to record results while the students are performing the experiment.

When you switch the students between trials, the student on the end will need to move to the other end of the line so that all students will have a new partner for each trial. In case of an odd number of students, you will need to participate, or you may use one student volunteer to record results on the board.

Analyze the Results

1.–2. If there are 10 students (two rows of 5 students facing each other), it will take six trials for everyone to become infected.

Draw Conclusions

3. In real life, colds are spread by more than one means. Coughing, sneezing, and touching with unwashed hands are ways to spread a cold. More than one person in a classroom may have a cold.

4. The line graphs should accurately reflect student data.

Applying Your Data

Colds are usually spread by close contact with a person who is infected. Colds may be more prevalent in the winter because people tend to stay indoors, where they are in closer contact with each other. The cold virus is carried on the microdroplets in a cough or sneeze and by unwashed hands. The best and easiest way to reduce the chances of catching a cold is to avoid crowded places, wash one's hands thoroughly and frequently when people in close proximity have colds, and avoid touching one's face near one's eyes. Eating a healthy diet and getting plenty of rest and exercise will help a person stay in good general health.

CHAPTER RESOURCES

Workbooks

Long-Term Projects & Research Ideas
- A Chuckle a Day Keeps the Doctor Away ADVANCED

Chapter Review

Assignment Guide	
Section	**Questions**
1	1, 7–9, 15, 19
2	2–6, 10–14, 16–18, 20–21

ANSWERS

Using Key Terms

1. infectious disease
2. pathogen
3. T cells
4. antibody
5. allergy
6. Cancer

Understanding Key Ideas

7. b
8. d
9. c
10. d
11. d
12. d
13. Sample answer: When a macrophage engulfs a pathogen, it places pieces of the pathogen called *antigens* on its outer membrane. The antigens attract helper T cells.
14. Helper T cells activate B cells and killer T cells.
15. from animals and from the air

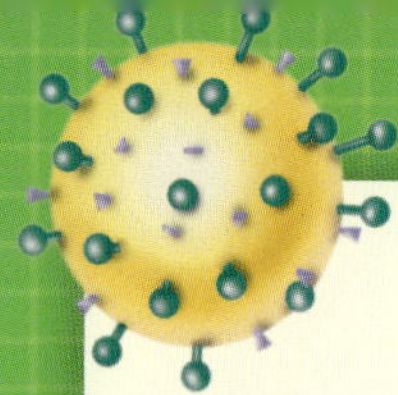

Chapter Review

USING KEY TERMS

Complete each of the following sentences by choosing the correct term from the word bank.

antibody	cancer
infectious disease	B cell
noninfectious disease	T cell
pathogen	allergy

1. A(n) ____ is caused by a pathogen.
2. Antibiotics can be used to kill a(n) ____.
3. Macrophages attract helper ____.
4. A(n) ____ binds to an antigen.
5. An immune-system overreaction to a harmless substance is a(n) ____.
6. ____ is the unregulated growth of cells.

UNDERSTANDING KEY IDEAS

Multiple Choice

7. Pathogens are
   - **a.** all viruses and microorganisms.
   - **b.** viruses and microorganisms that cause disease.
   - **c.** noninfectious organisms.
   - **d.** all bacteria that live in water.
8. Which of the following is an infectious disease?
   - **a.** allergies
   - **b.** rheumatoid arthritis
   - **c.** asthma
   - **d.** a common cold
9. The skin keeps pathogens out by
   - **a.** staying warm enough to kill pathogens.
   - **b.** releasing killer T cells onto the surface.
   - **c.** shedding dead cells and secreting oils.
   - **d.** All of the above

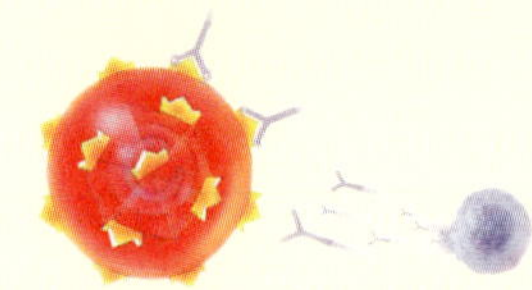

10. Memory B cells
    - **a.** kill pathogens.
    - **b.** activate killer T cells.
    - **c.** activate killer B cells.
    - **d.** produce B cells that make antibodies.
11. A fever
    - **a.** slows pathogen growth.
    - **b.** helps B cells multiply faster.
    - **c.** helps T cells multiply faster.
    - **d.** All of the above
12. Macrophages
    - **a.** make antibodies.
    - **b.** release helper T cells.
    - **c.** live in the gut.
    - **d.** engulf pathogens.

Short Answer

13. Explain how macrophages start an immune response.
14. Describe the role of helper T cells in responding to an infection.
15. Name two ways that you come into contact with pathogens.

CRITICAL THINKING

16 **Concept Mapping** Use the following terms to create a concept map: *macrophages, helper T cells, B cells, antibodies, antigens, killer T cells,* and *memory B cells.*

17 **Identifying Relationships** Why does the disappearance of helper T cells in AIDS patients damage the immune system?

18 **Predicting Consequences** Many people take fever-reducing drugs as soon as their temperature exceeds 37°C. Why might it not be a good idea to reduce a fever immediately with drugs?

19 **Evaluating Data** The risk of dying from a whooping cough vaccine is about one in 1 million. In contrast, the risk of dying from whooping cough is about one in 500. Discuss the pros and cons of this vaccination.

INTERPRETING GRAPHICS

The graph below compares the concentration of antibodies in the blood the first time you are exposed to a pathogen with the concentration of antibodies the next time you are exposed to the pathogen. Use the graph below to answer the questions that follow.

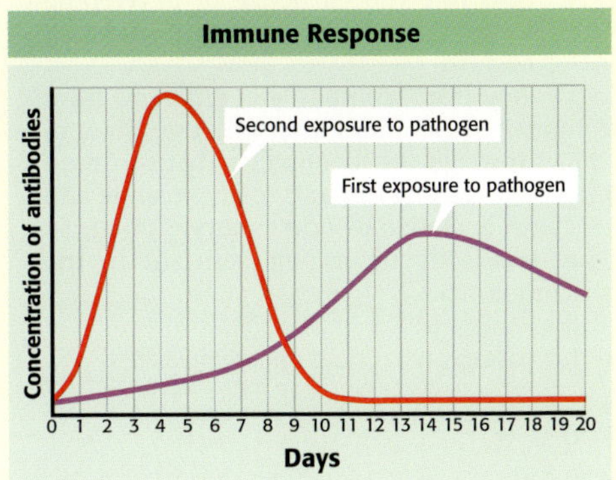

20 Are there more antibodies present during the first week of the first exposure or the first week of the second exposure? Why do you think this is so?

21 What is the difference in recovery time between the first exposure and second exposure? Why?

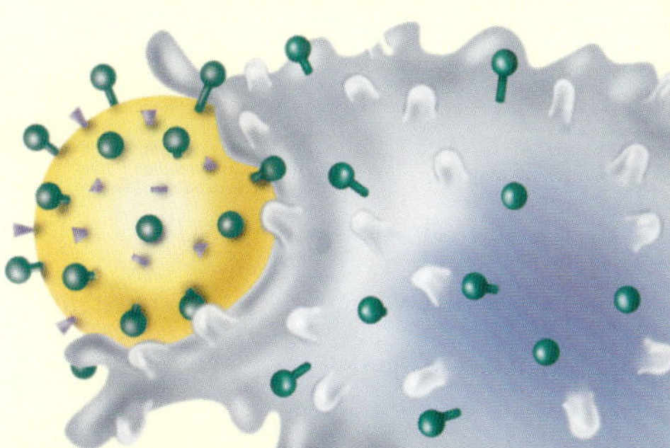

CHAPTER RESOURCES

Chapter Resource File

- Chapter Review GENERAL
- Chapter Test A GENERAL
- Chapter Test B ADVANCED
- Chapter Test C SPECIAL NEEDS
- Vocabulary Activity GENERAL

Workbooks

Study Guide

- Assessment resources are also available in Spanish.

Critical Thinking

16. 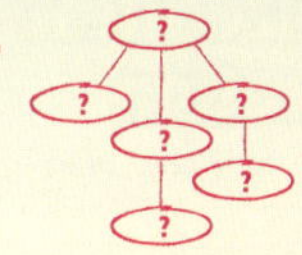
An answer to this exercise can be found at the end of this book.

17. Sample answer: Helper T cells are needed to activate B cells and killer T cells. Without B cells, the immune system cannot produce antibodies. Without killer T cells, the immune system cannot destroy infected cells.

18. Sample answer: Fevers can help T cells and B cells multiply faster. Fevers can also slow the growth of some pathogens. High fevers can be dangerous, though. If a fever exceeds 40.6°C, fever-reducing drugs can help bring the body's temperature down to a safer temperature.

19. Sample answer: The vaccine gives you an immunity to whooping cough, which is a dangerous disease. However, you may never come into contact with whooping cough and therefore may not need the vaccine. If you receive the vaccine anyway, you will have a small risk of dying from the vaccine. If no one is vaccinated, however, the risk of contracting the disease rises.

Interpreting Graphics

20. Sample answer: There are more antibodies present in the second exposure because memory B cells recognize the pathogen and immediately start to produce antibody-producing B cells.

21. Sample answer: Recovery time is much shorter for the second exposure because memory B cells allow the body to produce more antibodies in less time than the body produced during the first exposure.

Standardized Test Preparation

READING

Read each of the passages below. Then, answer the questions that follow each passage.

Passage 1 Bacteria are becoming resistant to many human-made antibiotics, which means that the drugs no longer affect the bacteria. Scientists now face the challenge of developing new antibiotics that can overcome the resistant strains of bacteria.

Antibiotics from animals are different from some human-made antibiotics. These antibiotics bore holes through the membranes that surround bacterial cells, causing the cells to disintegrate and die. Bacterial membranes don't mutate often, so they are less likely to become resistant to the animal antibiotics.

1. In this passage, what does *mutate* mean?
- **A** to change
- **B** to grow
- **C** to form
- **D** to degrade

2. Based on the passage, which of the following statements is a fact?
- **F** Bacterial membranes are on the inside of the bacterial cell.
- **G** Bacterial membranes are on the outside of the bacterial cell.
- **H** All strains of bacteria mutate.
- **I** Bacterial membranes never change.

3. Based on the passage, which of the following sentences is false?
- **A** Antibiotics from animals are different from human-made antibiotics.
- **B** Antibiotics from animals bore holes in bacterial membranes.
- **C** Bacterial membranes don't change very often.
- **D** Bacteria rarely develop resistance to human-made antibiotics.

Passage 2 Drinking water in the United States is generally safe, but water lines can break, or treatment plants can become flooded, allowing microorganisms to enter the public water supply. Bacteria growing in foods and beverages can cause illness, too. Refrigerating foods can slow the growth of many of these pathogens, but meat, fish, and eggs that are not cooked enough can still contain dangerous bacteria or parasites. Leaving food out at room temperature can give bacteria such as *salmonella* time to grow and produce toxins in the food. For these reasons, it is important to wash all used cooking tools.

1. Which of the following statements can you infer from this passage?
- **A** Treatment plants help keep drinking water safe.
- **B** Treatment plants never become flooded.
- **C** Eliminating treatment plants would help keep water safe.
- **D** New treatment plants are better than old ones.

2. Which of the following statements can you infer from the passage?
- **F** Bacteria that live in food produce more toxins than molds produce.
- **G** Cooking food thoroughly kills bacteria living in the food.
- **H** Some bacteria are helpful to humans.
- **I** Illnesses caused by bacteria living in food are seldom serious.

3. According to this passage, what do pathogens cause?
- **A** disease
- **B** flooding
- **C** water-line breaks
- **D** water supplies

Standardized Test Preparation

Teacher's Note

To provide practice under more realistic testing conditions, give students 20 minutes to answer all of the questions in this Standardized Test Preparation.

MISCONCEPTION ALERT

Answers to the standardized test preparation can help you identify student misconceptions and misunderstandings.

READING

Passage 1

1. A
2. G
3. D

TEST DOCTOR

Question 2: Students may mistakenly select answer H, but the passage does not state that all bacteria mutate. Tell students to be wary of answers that use the words *always* or *never*. Answers framed in absolute terms may be overstating ideas from the passage. Although the word *outside* is not used explicitly in this passage, the passage does use the word *surround*. To surround a cell, the membrane would have to be located outside the cell.

Passage 2

1. A
2. G
3. A

TEST DOCTOR

Question 1: Students may mistakenly select answer D, which is a probable answer. But the passage does not distinguish between the effectiveness of new and old plants. Although the passage does not state explicitly that water treatment plants help keep drinking water safe, the first sentence does imply that microorganisms can endanger public health when treatment plants stop working.

INTERPRETING GRAPHICS

The graph below shows the reported number of people living with HIV/AIDS. Use the graph to answer the questions that follow.

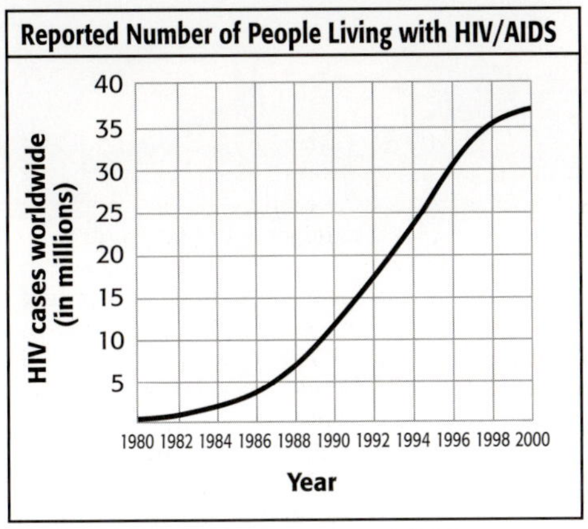

Source: Joint United Nations Progam on HIV/AIDS

1. When did the number of people living with HIV/AIDS reach 5 million?

A 1985
B 1986
C 1987
D 1988

2. When did the number of people living with HIV/AIDS reach 30 million?

F 1996
G 1997
H 1998
I 1999

3. When was the rate of increase of people with HIV/AIDS the **greatest**?

A from 1980 to 1982
B from 1984 to 1986
C from 1988 to 1990
D from 1998 to 2000

4. What percentage of the people who are infected with HIV do not yet have AIDS?

F 10%
G 24%
H 75%
I There is not enough information to determine the answer.

5. If the virus continued to spread as the graph indicates, in the year 2002, about how many people would be infected with HIV?

A 30 million
B 35 million
C 39 million
D 60 million

6. Which part of the graph indicates the rate of infection?

F *x*-axis
G *y*-axis
H slope of the line being graphed
I number of years in the sample

MATH

Read each question below, and choose the best answer.

1. Suppose you have 50,000 flu viruses on your fingers and you rub your eyes. Only 20,000 viruses enter your eyes, 10,000 dissolve in chemicals, and 10,000 are washed down into your nose. Of those, you sneeze out 2,000. How many viruses are left to wash down the back of your throat and possibly start an infection?

A 50,000
B 10,000
C 8,000
D 5,000

2. In which of the following lists are the numbers in order from smallest to greatest?

F 0.027, 0.072, 0.270, 0.720
G 0.270, 0.072, 0.720, 0.270
H 0.072, 0.027, 0.270, 0.720
I 0.720, 0.270, 0.072, 0.027

Standardized Test Preparation

INTERPRETING GRAPHICS

1. C
2. F
3. C
4. I
5. C
6. H

TEST DOCTOR

Question 3: Students may struggle to determine when the rate of increase was the greatest. Help them understand that the rate of increase is the same as the number of cases divided by the time in years. Thus, the rate of increase is also the slope of the line. Use a ruler or other straight object to show students how to determine where the line on the graph is steepest.

Question 5: Students may mistakenly choose answer D simply because it is the highest number, but in this case, the highest number is incorrect. Students may struggle to extrapolate information from the graph. Show them how to continue the line at its latest rate to estimate infection rates for years later than the rates shown on the graph.

MATH

1. C
2. F

TEST DOCTOR

Question 1: Students may mistakenly choose answer B because the question states that 10,000 flu viruses are washed into the nose. But show students how to track word problems by using a small chart or table. By keeping track of the parts of the problem, students can isolate the information needed to answer the question.

CHAPTER RESOURCES

Chapter Resource File

- Standardized Test Preparation GENERAL

State Resources

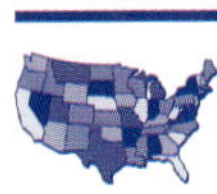

For specific resources for your state, visit **go.hrw.com** and type in the keyword **HSMSTR.**

Science in Action

Weird Science

Background

Modern scientific study of antibiotics began in the 1800s. Louis Pasteur discovered that bacteria spread infectious diseases, and Robert Koch developed methods for isolating and growing different kinds of bacteria. A breakthrough in treating bacterial diseases came in the early 1900s, when Alexander Fleming discovered penicillin, an antibiotic formed from mold. Streptomycin, a fungal antibiotic, was discovered by Selman A. Waksman in 1943. Doctors now use antibiotics to treat diseases such as strep throat, bacterial meningitis, and tuberculosis.

Scientific Discoveries

Background

The number of people who are allergic to peanuts has increased significantly in recent years, and scientists are not sure why. There has been speculation about the connection between infant creams containing peanut oil and the development of peanut allergies, but nothing has been proven yet. A child is more likely to have a peanut allergy if one of his or her parents is also allergic. Most people who have a peanut allergy also have asthma.

Science in Action

Weird Science

Frogs in the Medicine Cabinet?

Frog skin, mouse intestines, cow lungs, and shark stomachs are all being tested to make more effective medicines to combat harmful bacteria. In 1896, a biologist named Michael Zasloff was studying African clawed frogs. He noticed that cuts in the frogs' skin healed quickly and never became infected. Zasloff decided to investigate further. He found that when a frog was cut, its skin released a liquid antibiotic that killed invading bacteria. Furthermore, sand sharks, moths, pigs, mice, and cows also contain chemicals that kill bacteria and other microorganisms. These useful antibiotics are even found in the small intestines of humans!

Social Studies ACTIVITY

Many medicines were discovered in plants or animals by people living near those plants or animals. Research the origin of one or two common medicines discovered this way. Make a poster showing a world map and the location of the medicines that you researched.

Scientific Discoveries

Medicine for Peanut Allergies

Scientists estimate that 1.5 million people in the United States suffer from peanut allergies. Every year 50 to 100 people in the United States die from an allergic reaction to peanuts. Peanuts and peanut oil are used to make many foods. People who have a peanut allergy sometimes mistakenly eat these foods and suffer severe reactions. A new drug has been discovered to help people control severe reactions. The drug is called TNX-901. The drug is actually an antibody that binds to the antibodies that the body makes during the allergic reaction to the peanuts. By binding these antibodies, the drug controls the allergic response.

Math ACTIVITY

During the testing of the new drug, 84 people were given four injections over the course of 4 months. One-fourth of the people participating received injections of a control that had no medicine in it. The rest of the people participating received different doses of the drug. How many people received the control? How many people received medicine? How many shots containing medicine were administered during the 4-month test?

Answer to Social Studies Activity

Answers may vary but may discuss the origins of the drugs aspirin and digitalis.

Answer to Math Activity

Twenty-one people received a control, and 63 people received medicine. Two hundred fifty-six shots containing medicine were administered.

Careers

Terrel Shepherd III

Nurse Terrel Shepherd III is a registered nurse (RN) at Texas Children's Hospital in Houston, Texas. RNs have many responsibilities. These responsibilities include giving patients their medications, assessing patients' health, and establishing intravenous access. Nurses also serve as a go-between for the patient and the doctor. Although most nurses work in hospitals or clinics, some nurses work for corporations. Pediatric nurses such as Shepherd work specifically with infants, children, and adolescents. The field of nursing offers a wide variety of job opportunities including home-care nurses, traveling nurses, and flight nurses. The hospital alone has many areas of expertise for nurses, including geriatrics (working with the elderly), intensive care, administration, and surgery. Traditionally, nursing has been considered to be a woman's career. However, since nursing began as a profession, men and women have practiced nursing. A career in nursing is possible for anyone who does well in science, enjoys people, and wants to make a difference in people's lives.

Language Arts Activity

WRITING SKILL Create a brochure that persuades people to consider a career in nursing. Describe nursing as a career, the benefits of becoming a nurse, and the education needed to be a nurse. Illustrate the brochure with pictures of nurses from the Internet or from magazines.

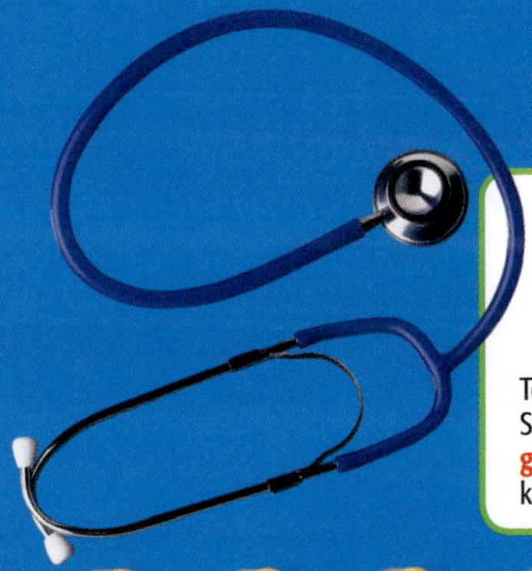

go.hrw.com

To learn more about these Science in Action topics, visit go.hrw.com and type in the keyword HL5BD6F.

Current Science

Check out Current Science® articles related to this chapter by visiting go.hrw.com. Just type in the keyword HL5CS27.

Careers Activity — General

Interested students can research any of the following:

1. How did the American Red Cross originate, and what are its responsibilities?
2. What are the responsibilities of nurses who work on emergency helicopter response teams?
3. How does a nurse practitioner differ from other RNs?
4. How has the field of nursing changed during the last 50 years?

Answer to Language Arts Activity

Student brochures should be written in language that persuades people to try nursing. If students struggle to be persuasive, ask them to think about the kind of language that persuades them to try new opportunities. The language should be honest and compelling.

7 Staying Healthy
Chapter Planning Guide

Compression guide:
To shorten instruction because of time limitations, omit Section 3.

OBJECTIVES	LABS, DEMONSTRATIONS, AND ACTIVITIES	TECHNOLOGY RESOURCES
PACING • 90 min pp. 152–159 **Chapter Opener**	SE **Start-up Activity,** p. 153 GENERAL	OSP **Parent Letter** ■ GENERAL CD **Student Edition on CD-ROM** CD **Guided Reading Audio CD** ■ TR **Chapter Starter Transparency*** VID **Brain Food Video Quiz**
Section 1 Good Nutrition • Identify the six groups of nutrients and explain their importance to good health. • Describe the Food Guide Pyramid. • Understand how to read Nutrition Facts labels. • Explain the dangers of various nutritional disorders.	TE **Activity** Nutrient Table, p. 155 GENERAL TE **Connection Activity** Math, p. 155 ADVANCED SE **Quick Lab** Brown Bag Test, p. 157 ◆ GENERAL CRF **Datasheet for Quick Lab*** TE **Activity** Food Guide Pyramid, p. 157 BASIC SE **Inquiry Lab** To Diet or Not to Diet, p. 187 GENERAL CRF **Datasheet for LabBook*** LB **Labs You Can Eat** Snack Attack* ◆ GENERAL LB **Long-Term Projects & Research Ideas** Breakfast, Lunch, and Dinner of Champions* ADVANCED	CRF **Lesson Plans*** TR **Bellringer Transparency*** TR *LINK TO PHYSICAL SCIENCE* Covalent Bonds in a Water Molecule* TR The Food Guide Pyramid* TR Nutrition Facts Label* CRF **SciLinks Activity*** GENERAL
PACING • 45 min pp. 160–165 **Section 2 Risks of Alcohol and Other Drugs** • Describe the difference between psychological and physical dependence. • Explain the hazards of tobacco, alcohol, and illegal drugs. • Distinguish between positive and negative uses of drugs.	TE **Activity** Poster Project, p. 160 GENERAL TE **Connection Activity** Real World, p. 163 ADVANCED SE **School-to-Home Activity** Good Reasons, p. 164 GENERAL	CRF **Lesson Plans*** TR **Bellringer Transparency***
PACING • 90 min pp. 166–171 **Section 3 Healthy Habits** • Describe three important aspects of good hygiene. • Explain why exercise and sleep are important to good health. • Describe methods of handling stress. • List three ways to stay safe at home, on the road, and outdoors. • Plan what you would do in the case of an accident.	TE **Activity** Don't Kick These Habits!, p. 166 GENERAL TE **Connection Activity** Real World, p. 168 GENERAL TE **Group Activity** Poster Project, p. 169 GENERAL SE **Skills Practice Lab** Keep It Clean, p. 172 ◆ GENERAL CRF **Datasheet for Chapter Lab*** LB **Inquiry Labs** Consumer Challenge* GENERAL LB **Calculator-Based Labs** Counting Calories* ADVANCED	CRF **Lesson Plans*** TR **Bellringer Transparency*** SE **Internet Activity,** p. 168 GENERAL VID **Lab Videos for Life Science**

PACING • 90 min

CHAPTER REVIEW, ASSESSMENT, AND STANDARDIZED TEST PREPARATION

CRF **Vocabulary Activity*** GENERAL
SE **Chapter Review,** pp. 174–175 GENERAL
CRF **Chapter Review*** ■ GENERAL
CRF **Chapter Tests A*** ■ GENERAL, **B*** ADVANCED, **C*** SPECIAL NEEDS
SE **Standardized Test Preparation,** pp. 176–177 GENERAL
CRF **Standardized Test Preparation*** GENERAL
CRF **Performance-Based Assessment*** GENERAL
OSP **Test Generator** GENERAL
CRF **Test Item Listing*** GENERAL

Online and Technology Resources

Visit **go.hrw.com** for a variety of free resources related to this textbook. Enter the keyword **HL5BD7.**

Holt Online Learning
Students can access interactive problem-solving help and active visual concept development with the *Holt Science and Technology* Online Edition available at **www.hrw.com.**

Guided Reading Audio CD
A direct reading of each chapter using instructional visuals as guideposts. For auditory learners, reluctant readers, and Spanish-speaking students. Available in English and Spanish.

KEY

SE Student Edition	**CRF** Chapter Resource File	**SS** Science Skills Worksheets	* Also on One-Stop Planner
TE Teacher Edition	**OSP** One-Stop Planner	**MS** Math Skills for Science Worksheets	◆ Requires advance prep
	LB Lab Bank	**CD** CD or CD-ROM	■ Also available in Spanish
	TR Transparencies	**VID** Classroom Video/DVD	

SKILLS DEVELOPMENT RESOURCES	SECTION REVIEW AND ASSESSMENT	STANDARDS CORRELATIONS
SE **Pre-Reading Activity,** p. 152 GENERAL OSP **Science Puzzlers, Twisters & Teasers** GENERAL		National Science Education Standards SAI 1, 2; SPSP 1, 4
CRF **Directed Reading A*** ■ BASIC, **B*** SPECIAL NEEDS CRF **Vocabulary and Section Summary*** ■ GENERAL SE **Reading Strategy** Reading Organizer, p. 154 GENERAL SE **Connection to Oceanography** Nutritious Seaweed, p. 156 GENERAL SE **Math Practice** What Percentage?, p. 158 GENERAL CRF **Critical Thinking** A Daily Routine* ADVANCED CRF **Reinforcement Worksheet** To Eat or Not to Eat…* BASIC	SE **Reading Checks,** pp. 155, 157, 158 GENERAL TE **Homework,** p. 156 GENERAL TE **Reteaching,** p. 158 BASIC TE **Quiz,** p. 158 GENERAL TE **Alternative Assessment,** p. 158 GENERAL SE **Section Review,*** p. 159 ■ GENERAL CRF **Section Quiz*** ■ GENERAL	UCP 1, 4; SAI 1; SPSP 1, 4; *LabBook:* SAI 1, 2; SPSP 1, 4
CRF **Directed Reading A*** ■ BASIC, **B*** SPECIAL NEEDS CRF **Vocabulary and Section Summary*** ■ GENERAL SE **Reading Strategy** Reading Organizer, p. 160 GENERAL TE **Inclusion Strategies,** p. 162	SE **Reading Checks,** pp. 161, 163, 164 GENERAL TE **Homework,** p. 161 GENERAL TE **Homework,** p. 162 GENERAL TE **Reteaching,** p. 164 BASIC TE **Quiz,** p. 164 GENERAL TE **Alternative Assessment,** p. 164 GENERAL TE **Homework,** p. 164 GENERAL SE **Section Review,*** p. 165 ■ GENERAL CRF **Section Quiz*** ■ GENERAL	SAI 1; SPSP 1, 4, 5
CRF **Directed Reading A*** ■ BASIC, **B*** SPECIAL NEEDS CRF **Vocabulary and Section Summary*** ■ GENERAL SE **Reading Strategy** Prediction Guide, p. 166 GENERAL SE **Connection to Language Arts** Dreamy Poetry, p. 167 GENERAL TE **Reading Strategy** Prediction Guide, p. 167 GENERAL TE **Inclusion Strategies,** p. 170	SE **Reading Checks,** pp. 167, 169, 171 GENERAL TE **Homework,** p. 168 GENERAL TE **Homework,** p. 169 GENERAL TE **Reteaching,** p. 170 BASIC TE **Quiz,** p. 170 GENERAL TE **Alternative Assessment,** p. 170 ADVANCED SE **Section Review,*** p. 171 ■ GENERAL CRF **Section Quiz*** ■ GENERAL	UCP 3; SAI 1; SPSP 1, 4, 5; *Chapter Lab:* UCP 2; SAI 1, 2; ST 2, SPSP 1, 3, 4, 5

One-Stop Planner® CD-ROM

This convenient CD-ROM includes:
- **Lab Materials QuickList Software**
- **Holt Calendar Planner**
- **Customizable Lesson Plans**
- **Printable Worksheets**
- **ExamView® Test Generator**

CNN Student News

cnnstudentnews.com

Find the latest news, lesson plans, and activities related to important scientific events.

SciLinks NSTA

www.scilinks.org

Maintained by the **National Science Teachers Association.** See Chapter Enrichment pages for a complete list of topics.

Current Science®

Check out ***Current Science*** articles and activities by visiting the HRW Web site at **go.hrw.com.** Just type in the keyword **HL5CS28T.**

Classroom Videos

- **Lab Videos** demonstrate the chapter lab.
- **Brain Food Video Quizzes** help students review the chapter material.

Visual Resources

CHAPTER STARTER TRANSPARENCY

BELLRINGER TRANSPARENCIES

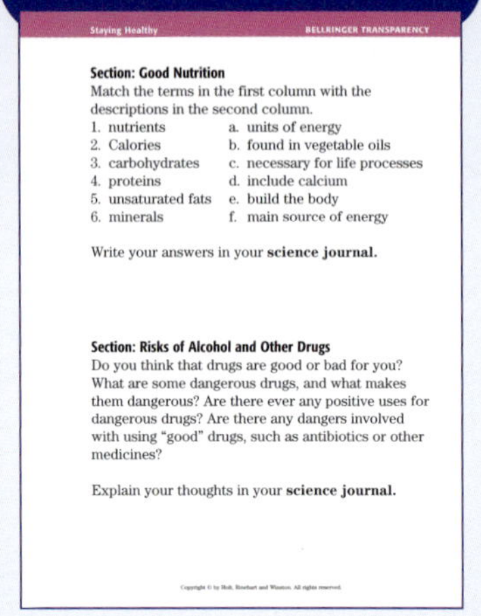
Staying Healthy BELLRINGER TRANSPARENCY

Section: Good Nutrition
Match the terms in the first column with the descriptions in the second column.

1. nutrients a. units of energy
2. Calories b. found in vegetable oils
3. carbohydrates c. necessary for life processes
4. proteins d. include calcium
5. unsaturated fats e. build the body
6. minerals f. main source of energy

Write your answers in your **science journal.**

Section: Risks of Alcohol and Other Drugs
Do you think that drugs are good or bad for you? What are some dangerous drugs, and what makes them dangerous? Are there ever any positive uses for dangerous drugs? Are there ever any dangers involved with using "good" drugs, such as antibiotics or other medicines?

Explain your thoughts in your **science journal.**

TEACHING TRANSPARENCIES

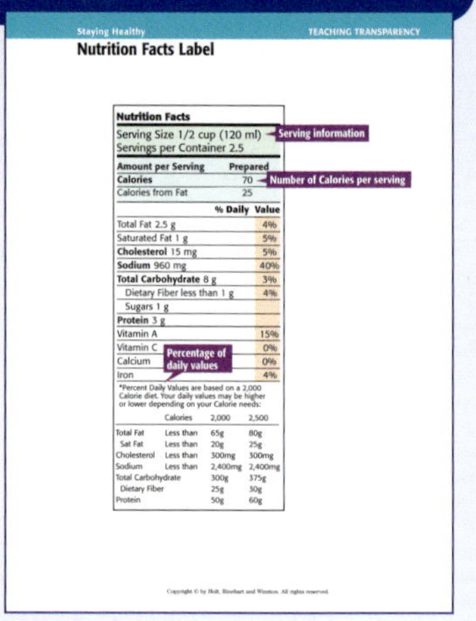

TEACHING TRANSPARENCIES

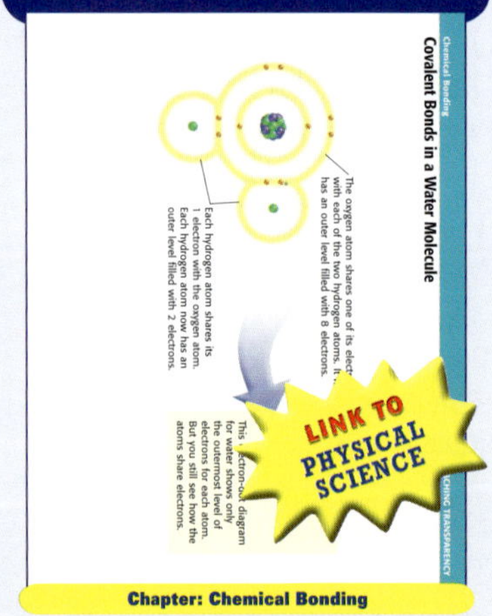

CONCEPT MAPPING TRANSPARENCY

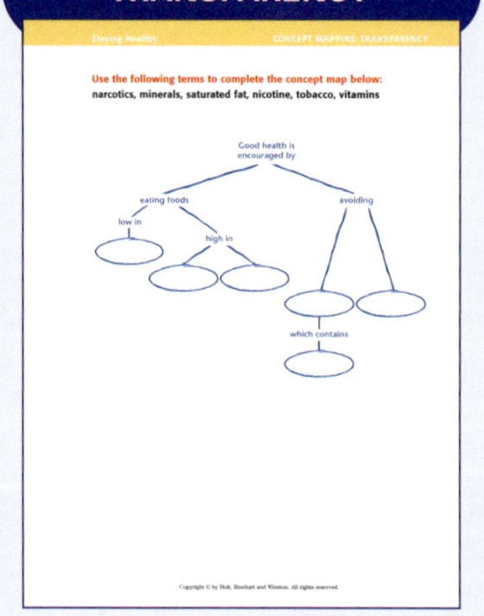

Planning Resources

LESSON PLANS

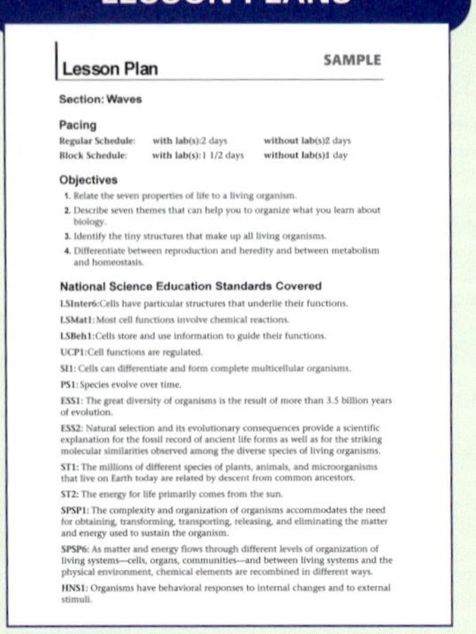

PARENT LETTER

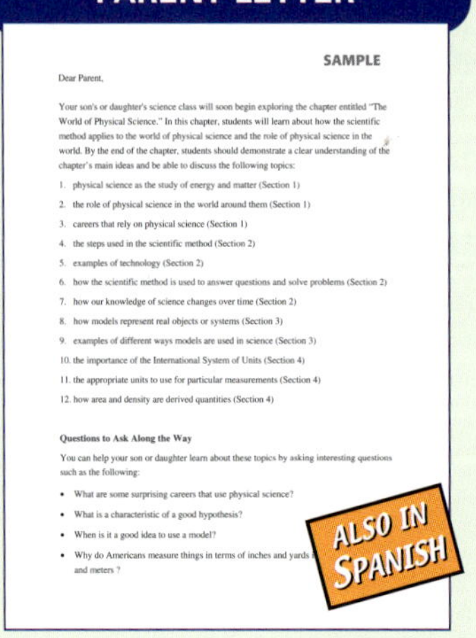

TEST ITEM LISTING

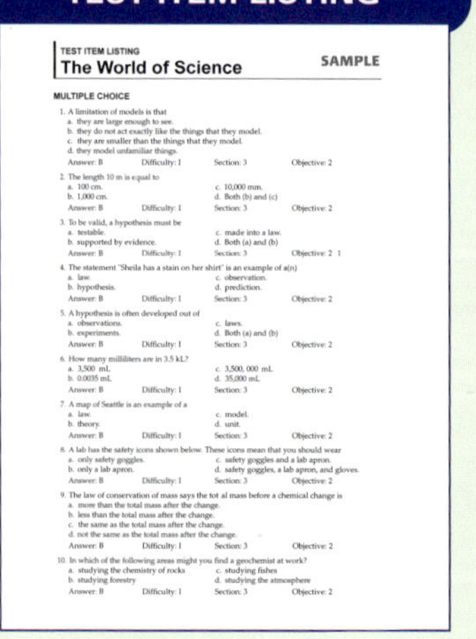

One-Stop Planner® CD-ROM

This CD-ROM includes all of the resources shown here and the following time-saving tools:

- *Lab Materials QuickList Software*
- *Customizable lesson plans*
- *Holt Calendar Planner*
- *The powerful ExamView® Test Generator*

For a preview of available worksheets covering math and science skills, see pages T12–T19. All of these resources are also on the One-Stop Planner®.

Meeting Individual Needs

DIRECTED READING A
Directed Reading A — SAMPLE
BASIC · ALSO IN SPANISH

VOCABULARY ACTIVITY
Vocabulary Activity — SAMPLE
GENERAL

REINFORCEMENT
Reinforcement — SAMPLE
BASIC

SCILINKS ACTIVITY
SciLinks Activity — SAMPLE
GENERAL

DIRECTED READING B
Directed Reading B — SAMPLE
SPECIAL NEEDS

VOCABULARY AND SECTION SUMMARY
Vocabulary & Notes — SAMPLE
GENERAL · ALSO IN SPANISH

CRITICAL THINKING
Critical Thinking — SAMPLE
ADVANCED

SCIENCE PUZZLERS, TWISTERS & TEASERS
GENERAL

Labs and Activities

LONG-TERM PROJECTS & RESEARCH IDEAS
ADVANCED

LABS YOU CAN EAT
GENERAL

INQUIRY LABS
GENERAL

CALCULATOR-BASED LABS
ADVANCED

DATASHEETS FOR QUICK LABS
Reaction to Stress — SAMPLE

DATASHEETS FOR CHAPTER LABS
Using Scientific Methods — SAMPLE

DATASHEETS FOR LABBOOK
Does It All Add Up? — SAMPLE

Review and Assessments

SECTION QUIZ
Section Quiz — SAMPLE
GENERAL · ALSO IN SPANISH

CHAPTER REVIEW
Chapter Review — SAMPLE
GENERAL · ALSO IN SPANISH

CHAPTER TEST B
Chapter Test B — SAMPLE
ADVANCED

STANDARDIZED TEST PREPARATION
Standardized Test Preparation — SAMPLE
GENERAL

SECTION REVIEW
Section Review — SAMPLE
GENERAL · ALSO IN SPANISH

CHAPTER TEST A
Chapter Test A — SAMPLE
GENERAL · ALSO IN SPANISH

CHAPTER TEST C
Chapter Test C — SAMPLE
SPECIAL NEEDS

PERFORMANCE-BASED ASSESSMENT
Performance-Based Assessment — SAMPLE
GENERAL

7 Chapter Enrichment

This Chapter Enrichment provides relevant and interesting information to expand and enhance your presentation of the chapter material.

Section 1

Good Nutrition

Nutrient Needs in Adolescence

- Puberty is a period of rapid growth. Changes brought on by puberty affect every organ of the body. Adolescents need extra nutrients to meet their bodies' needs during the growth spurt that accompanies puberty.

- The onset of menstruation in girls and the change in lean body mass in boys increase an adolescent's need for iron. During adolescence, the Recommended Daily Allowance (RDA) for iron is 12–15 mg.

- The increase in skeletal mass that occurs during adolescence boosts the body's calcium needs, which increases the RDA for calcium to 1,200–1,500 mg. About 75 to 80% of the skeleton is built during adolescence.

- Rapid growth during adolescence increases the body's caloric needs. Energy needs vary depending on growth rate, body composition, and activity level. In general, boys have a larger proportion of lean body mass to fat and require more Calories than girls do.

Fiber

- Soluble fiber, found in fruit, beans, peas, and other legumes, can reduce the risk of heart disease by lowering cholesterol levels in the blood. Legume-based fiber also aids in the regulation of blood glucose levels.

- Although fiber contains no vitamins or minerals, it is essential to good health. Insoluble fiber is found in whole-grain foods. Along with fluids, insoluble fiber helps the colon remove waste (fecal matter) from the bowels. A lack of insoluble fiber in a person's diet can increase the risk for constipation, colon and bowel cancer, and diverticulosis, a disorder in which pouches form in the bowel wall.

Is That a Fact!

◆ Just before astronauts Neil Armstrong (1930–) and Edwin Aldrin Jr. (1930–1999) embarked on the first moonwalk on July 20, 1969, each astronaut ate four bacon squares, three sugar cookies, and peaches. They drank pineapple-grapefruit juice and coffee.

Adolescence and Nutrition

- The physical, emotional, and social changes of adolescence have a great impact on teen nutrition. Adolescence is typically a period of increased autonomy, so teens have increased opportunity to make food choices. But rather than making food choices based on long-term health, teens tend to be influenced by social pressures to reach cultural ideals of thinness, to gain the acceptance of peers, and to assert their independence from their parents. The characteristic adolescent preoccupation with body image extends to nutrition. If sound nutritional principles are taught and reinforced during adolescence, long-term health can benefit.

For background information about teaching strategies and issues, refer to the ***Professional Reference for Teachers.***

Section 2

Risks of Alcohol and Other Drugs

Drug Abuse

- According to yearly surveys conducted by the University of Michigan's Institute for Social Research (funded by the National Institute on Drug Abuse), drug abuse among 8th, 10th, and 12th grade students has decreased over the last few years. That doesn't mean that drug abuse is no longer a problem. The following are some statistics from the 2002 survey:

  - About 15% of 8th grade students surveyed indicated that they had tried inhalants in their lifetime.
  - About 19% of 8th grade students surveyed indicated that they had tried marijuana in their lifetime.
  - About 31% of 8th grade students surveyed indicated that they had tried cigarettes. More than 2% of 8th grade students indicated that they smoked more than half a pack of cigarettes a day.
  - About 47% of 8th grade students surveyed indicated that they had tried alcohol in their lifetime.

Is That a Fact!

- In the 1800s, scientists learned how to isolate drug compounds from plants. Morphine, cocaine, and heroin became readily available. These drugs were welcomed as safe, powerful pain relievers because, at first, no one knew of their addictive properties. By the early 1900s, there was an epidemic of drug abuse in the United States.

Section 3

Healthy Habits

Safety

- In the United States, accidental injuries are the leading cause of death for children and adolescents between the ages of 1 and 19.

- In 2002, motor-vehicle accidents claimed the lives of almost 43,000 Americans. More than 1,600 victims were between the ages of 8 and 15. Many of these deaths could have been prevented—59% of all motor-vehicle fatalities in 2002 were unrestrained passengers.

Excuse Me?

- When sounds enter the ear, they cause the eardrum to vibrate. Tiny hair cells in the inner ear respond to these vibrations by sending nerve impulses to the brain. Ordinarily, these hair cells slowly die as a person ages. When exposed to loud noises, however, more hair cells die than usual. Fewer impulses are sent to the brain. Once damage occurs, it is irreversible.
- Many teens are at risk for preventable hearing loss. Experts attribute this trend to increased exposure to damaging levels of noise, such as loud music. Sources of loud noise include CD players, concerts, car stereos, video games, and television. Hearing loss can be prevented by turning down the volume on audio devices and by wearing earplugs when loud noise cannot be controlled.

SCILINKS NSTA Developed and maintained by the National Science Teachers Association

SciLinks is maintained by the National Science Teachers Association to provide you and your students with interesting, up-to-date links that will enrich your classroom presentation of the chapter.

Visit www.scilinks.org and enter the SciLinks code for more information about the topic listed.

Topic: Food Pyramids
SciLinks code: HSM0598

Topic: Drug and Alcohol Abuse
SciLinks code: HSM0428

Topic: Nutritional Disorders
SciLinks code: HSM1057

Topic: Safety
SciLinks code: HSM1339

Overview

Tell students that this chapter will help them learn about good nutrition, the risks of drugs, and how they can take care of themselves to stay healthy.

Assessing Prior Knowledge

Students should be familiar with the following topics:

- body systems
- disease

Identifying Misconceptions

As students learn the material in this chapter, some of them may express misconceptions about drugs. Some students may not recognize that drugs, such as caffeine (found in many sodas, teas, and coffee) or herbal additives (including chamomile, ginseng, and echinacea), are found in some of the foods they eat. You may also want to point out that many of the products that claim to make people think more clearly, control weight, and increase energy contain drugs. Some students may have misconceptions about alcohol, including the belief that alcohol is not a drug, that beer and wine are safer than liquor, and that alcohol is not addictive.

7

Staying Healthy

About the PHOTO

What do you see in this photo? Sure, you can see five students facing the camera, but what else does the picture tell you? The bright eyes, happy smiles, and shiny hair show radiant health. Having a clear mind and a long, active life depend on having a healthy body. Keeping your body healthy depends on eating well; avoiding drugs, cigarettes, and alcohol; and staying safe.

PRE-READING ACTIVITY

FOLDNOTES

Booklet Before you read the chapter, create the FoldNote entitled "Booklet" described in the **Study Skills** section of the Appendix. Label each page of the booklet with a main idea from the chapter. As you read the chapter, write what you learn about each main idea on the appropriate page of the booklet.

Standards Correlations

National Science Education Standards

The following codes indicate the National Science Education Standards that correlate to this chapter. The full text of the standards is at the front of the book.

Chapter Opener
SAI 1, 2; SPSP 1, 4

Section 1 Good Nutrition
UCP 1, 4; SAI 1; SPSP 1, 4; *LabBook:* SAI 1, 2; SPSP 1, 4

Section 2 Risks of Alcohol and Other Drugs
SAI 1; SPSP 1, 4, 5

Section 3 Taking Care of Your Body
UCP 3; SAI 1; SPSP 1, 4, 5

Chapter Lab
UCP 2; SAI 1, 2; ST 2; SPSP 1, 3, 4, 5

Chapter Review
UCP 1, 3; SAI 1; SPSP 1, 3, 4, 5

Science in Action
SPSP 1, 4, 5

START-UP ACTIVITY

Conduct a Survey

How healthy are the habits of your classmates? Find out for yourself.

Procedure

1. Copy and answer yes or no to each of the five questions at right. Do not put your name on the survey.

Analysis

1. As a class, record the data from the completed surveys in a chart. For each question, calculate the percentage of your class that answered yes.
2. What good and bad habits do your classmates have?

1. Do you exercise at least three times a week?
2. Do you wear a seat belt every time you ride in a car?
3. Do you eat five or more servings of fruits and vegetables every day?
4. Do you use sunscreen to protect your skin when you are outdoors?
5. Do you eat a lot of high-fat foods?

START-UP ACTIVITY

MATERIALS

For Each Group

- paper
- pen or pencil
- questionnaire

Teacher's Notes: To ensure privacy, you may want to have students complete their questionnaires at home. Many students will be uncomfortable sharing personal information.

Answers

1. Answers may vary.
2. Answers may vary. Students should recognize that the first four items on the questionnaire are considered good habits, and the fifth item is considered a bad habit.

Chapter Starter Transparency
Use this transparency to help students begin thinking about health and good nutrition.

CHAPTER RESOURCES

Technology

- **Transparencies** READING SKILLS
  - Chapter Starter Transparency
- **Student Edition on CD-ROM**
- **Guided Reading Audio CD**
  - English or Spanish
- **Classroom Videos**
  - Brain Food Video Quiz

Workbooks

- **Science Puzzlers, Twisters & Teasers**
  - Staying Healthy GENERAL

SECTION 1

Focus

Overview

In this section, students will learn about the six essential nutrients and why these nutrients are important for good health. Dietary guidelines and the food pyramid are used to illustrate the principles of sound nutrition, and students will learn how to use these tools. Finally, students will learn about nutritional disorders.

Bellringer

Post the following lists on the board or an overhead projector:

Column A

1. nutrients
2. Calories
3. carbohydrates
4. proteins
5. unsaturated fats
6. minerals

Column B

a. units of energy
b. found in vegetable oils
c. necessary for life processes
d. include calcium
e. build the body
f. main source of energy

Challenge students to match the terms in column A with the descriptions in column B.
(1. c; 2. a; 3. f; 4. e; 5. b; 6. d)

SECTION 1

Good Nutrition

READING WARM-UP

Objectives

- Identify the six groups of nutrients, and explain their importance to good health.
- Describe the Food Guide Pyramid.
- Understand how to read Nutrition Facts labels.
- Explain the dangers of various nutritional disorders.

Terms to Learn

nutrient	mineral
carbohydrate	vitamin
protein	malnutrition
fat	

READING STRATEGY

Reading Organizer As you read this section, create an outline of the section. Use the headings from the section in your outline.

Does the saying "You are what you eat" mean that you are pizza? No, but substances in pizza help build your body.

Protein in the cheese may become part of your hair. Carbohydrates in the crust can give you energy for your next race.

Nutrients

Are you more likely to have potato chips or broccoli for a snack? If you eat many foods that are high in fat, such as potato chips, your food choices probably are not as healthy as they could be. Broccoli is a healthier food than potato chips. But eating only broccoli, as the person in **Figure 1** is doing, does not give you a balanced diet.

To stay healthy, you need to take in **nutrients,** or substances that provide the materials needed for life processes. Nutrients are grouped into six classes: *carbohydrates, proteins, fats, water, vitamins,* and *minerals*. Carbohydrates, proteins, and fats provide energy for the body in units called *Calories* (Cal).

nutrient a substance in food that provides energy or helps form body tissues and that is necessary for life and growth

Carbohydrates

Carbohydrates are your body's main source of energy. A **carbohydrate** is a chemical composed of simple sugars. There are two types of carbohydrates: simple and complex. *Simple carbohydrates* are sugars. They are easily digested and give you quick energy. *Complex carbohydrates* are made up of many sugar molecules linked together. They are digested slowly and give you long-lasting energy. Some complex carbohydrates are good sources of fiber. Fiber is a part of a healthy diet and is found in whole-grain foods, such as brown rice and whole-wheat bread. Many fruits and vegetables also contain fiber.

carbohydrate a class of energy-giving nutrients that includes sugars, starches, and fiber

Figure 1 *Eating only one food, even a healthy food, will not give you all the substances your body needs.*

CHAPTER RESOURCES

Chapter Resource File

- Lesson Plan
- Directed Reading A BASIC
- Directed Reading B SPECIAL NEEDS

Technology

Transparencies

- Bellringer
- LINK TO PHYSICAL SCIENCE Covalent Bonds in a Water Molecule

CONNECTION to Physical Science GENERAL

Water Molecules Water is a very simple molecule that is made up of one oxygen atom and two hydrogen atoms. Water is vital to all living organisms. Remind students that water makes up nearly 70% of the human body. Illustrate the water molecule using the teaching transparency titled "Covalent Bonds in a Water Molecule."
LS Visual

Protein

Proteins are found in body fluids, muscle, bone, and skin. **Proteins** are nutrients used to build and repair your body. Your body makes the proteins it needs, but it must have the necessary building blocks, called *amino acids*. Your digestive system breaks down protein into individual amino acids that are then used to make new proteins. Some foods, such as poultry, fish, milk, and eggs, provide all of the amino acids your body needs. Foods that contain all of these essential amino acids are called *complete proteins*. *Incomplete proteins* contain only some of the essential amino acids. Most plant foods contain incomplete protein, but eating a variety of plant foods will provide all of the amino acids your body needs.

Reading Check What is an incomplete protein? (*See the Appendix for answers to Reading Checks.*)

Figure 2 *This sample meal provides many of the nutrients a growing teenager needs.*

Fats

Another class of nutrients that is important to a healthy meal, such as the meal shown in **Figure 2,** is fat. **Fats** are energy-storage nutrients. Fats are needed to store and transport vitamins, produce hormones, keep skin healthy, and provide insulation. Fats also provide more energy than either proteins or carbohydrates. There are two types of fats: saturated and unsaturated. *Saturated fats* are found in meat, dairy products, coconut oil, and palm oil. Saturated fats raise blood cholesterol levels. Although *cholesterol* is a fat-like substance found naturally in the body, high levels can increase the risk of heart disease. *Unsaturated fats* and foods high in fiber may help reduce blood cholesterol levels. Your body cannot make unsaturated fats. They must come from vegetable oils and fish in your diet. The body needs both kinds of fats.

protein a molecule that is made up of amino acids and that is needed to build and repair body structures and to regulate processes in the body

fat an energy-storage nutrient that helps the body store some vitamins

Water

You cannot survive for more than a few days without water. Your body is about 70% water. Water is in every cell of your body. The main functions of water are to transport substances, regulate body temperature, and provide lubrication. Some scientists think you should drink at least eight glasses of water a day. When you exercise you need more water, as shown in **Figure 3.** You also get water from other liquids you drink and the foods you eat. Fresh fruits and vegetables, juices, soups, and milk are good sources of water.

Figure 3 *When you exercise, you need to drink more water.*

Motivate

ACTIVITY GENERAL

Nutrient Table Ask students to create a table with the following headings across the top: "Body fuel," "Body builders," and "Energy storage." On the side of the table, have them list the following information categories: "Type of nutrient," and "Food sources." Have students complete the table. (Students should classify carbohydrates under "Body fuel," and list rice, whole-wheat bread, fruits, and vegetables as food sources. They should classify protein under "Body builders" and list poultry, fish, milk, and eggs as food sources. Finally, students should classify fats under "Energy storage" and list meat, dairy products, vegetable oils, and fish as food sources.) **LS Logical/Visual**

Teach

CONNECTION ACTIVITY
Math ADVANCED

Calories Ask students the following question: "If there are 3,312 Cal in 1 lb of butter, and a single serving of butter is 14.2 g, how many Calories are in a single serving of butter?" (You may want to tell students that 1 kg is equal to 2.2 lbs and that there are 1,000 g in 1 kg.)
(3,312 Cal/lb × 2.2 lb/kg = 7,286 Cal/kg;
7,286 Cal/kg ÷ 1,000 g/kg = 7.3 Cal/g;
7.3 Cal/g × 14.2 g = 104 Cal)
LS Logical

Answer to Reading Check

An incomplete protein does not contain all of the essential amino acids.

Is That a Fact!

Frozen vegetables are often just as nutritious as fresh vegetables are.

Teach, *continued*

CONNECTION to Real World — ADVANCED

Writing **Diets** Have students examine a popular diet trend. Then, have students evaluate the diet in terms of its ability to meet the body's nutrient and energy needs. If a diet does not meet the body's needs, encourage students to identify changes to the diet that will make the diet healthful. Have students write reports about their findings. LS **Verbal**

Homework — GENERAL

PORTFOLIO **Essential Vitamins** Have students research the physical effects of vitamin deficiencies for at least three essential vitamins. Ask students to create informative brochures about their findings. LS **Verbal**

BRAIN FOOD

Vitamins Point out that many vitamin manufacturers promote their products by claiming that their vitamins are "natural" or "organic," rather than synthetic (manufactured). These vitamins are almost always more expensive than their synthetic counterparts. Point out that most experts agree that organic and synthetic vitamins are identical in both structure and function.

Table 1 Some Essential Vitamins

Vitamin	What it does	Where you get it
A	keeps skin and eyes healthy; builds strong bones and teeth	yellow and orange fruits and vegetables, leafy greens, meats, and milk
B (various forms)	helps body use carbohydrates; helps blood, nerves, and heart function	meats, whole grains, beans, peas, nuts, and seafood
C	strengthens tissues; helps the body absorb iron, fight disease	citrus fruits, leafy greens, broccoli, peppers, and cabbage
D	builds strong bones and teeth; helps the body use calcium and phosphorus	sunlight, enriched milk, eggs, and fish
E	protects red blood cells from destruction; keeps skin healthy	oils, fats, eggs, whole grains, wheat germ, liver, and leafy greens
K	assists with blood clotting	leafy greens, tomatoes, and potatoes

Minerals

If you eat a balanced diet, you should get all of the vitamins and minerals you need. **Minerals** are elements that are essential for good health. You need six minerals in large amounts: calcium, chloride, magnesium, phosphorus, potassium, and sodium. There are at least 12 minerals that are required in very small amounts. These include fluorine, iodine, iron, and zinc. Calcium is necessary for strong bones and teeth. Magnesium and sodium help the body use proteins. Potassium is needed to regulate your heartbeat and produce muscle movement, and iron is necessary for red blood cell production.

mineral a class of nutrients that are chemical elements that are needed for certain body processes

vitamin a class of nutrients that contain carbon and that are needed in small amounts to maintain health and allow growth

Vitamins

Vitamins are another class of nutrients. **Vitamins** are compounds that control many body functions. Only vitamin D can be made by the body, so you have to get most vitamins from food. **Table 1** provides information about six essential vitamins.

CONNECTION TO Oceanography

Nutritious Seaweed Kelp, a type of seaweed, is a good source of iodine. This nutritious food is grown on special farms off the coasts of China and Japan. What other nutritious foods come from the sea?

WEIRD SCIENCE

You are what you eat, and sometimes, you can even look like it! Eating a lot of carrots can give skin a yellowish color, a harmless condition. Tomatoes can give skin a reddish color. In 1960, a doctor examined a patient whose skin was orange. It turned out that the patient was eating a lot of carrots and tomatoes. The two colors mixed, and *voilà*, the patient had orange skin!

Answer to Connection to Oceanography

Sample answer: Other nutritious seafoods include fish and shellfish.

Eating for Good Health

Now you have learned which nutrients you need for good health. But how can you be sure to get all the important nutrients in the right amounts? To begin, keep in mind that most teenage girls need about 2,200 Cal per day, and most boys need about 2,800 Cal. Because different foods contain different nutrients, *where* you get your Calories is as important as *how many* you get. The Food Guide Pyramid, shown in **Figure 4,** can help you make good food choices.

Reading Check Using the Food Guide Pyramid below, design a healthy lunch that includes one food from each food group.

Figure 4 The Food Guide Pyramid

The U.S. Department of Agriculture and the Department of Health and Human Services developed the Food Guide Pyramid to help Americans make healthy food choices. The Food Guide Pyramid divides foods into six groups. It shows how many servings you need daily from each group and gives examples of foods for each. This pyramid also provides sample serving sizes for each group. Within each group, the food choices are up to you.

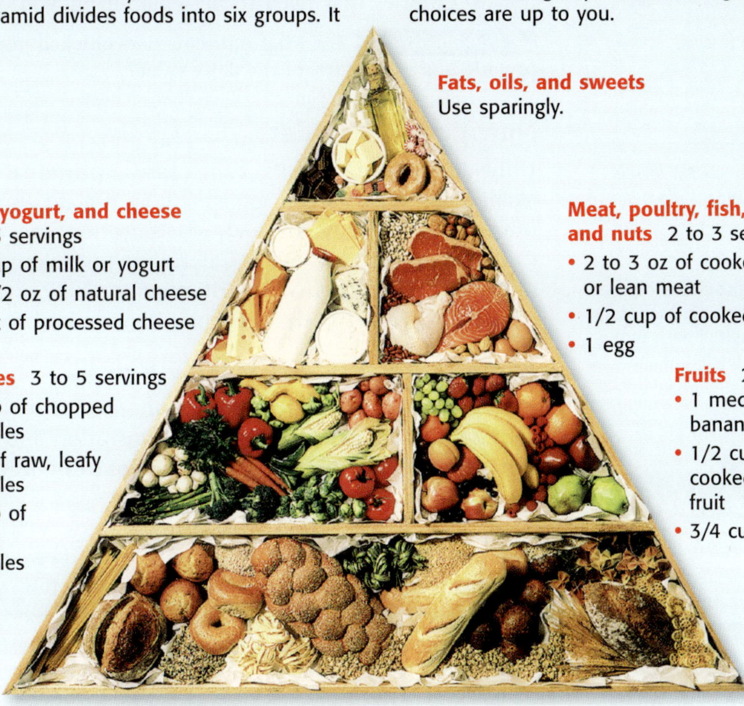

Fats, oils, and sweets
Use sparingly.

Milk, yogurt, and cheese 2 to 3 servings
- 1 cup of milk or yogurt
- 1 1/2 oz of natural cheese
- 2 oz of processed cheese

Meat, poultry, fish, beans, eggs, and nuts 2 to 3 servings
- 2 to 3 oz of cooked poultry, fish, or lean meat
- 1/2 cup of cooked dried beans
- 1 egg

Vegetables 3 to 5 servings
- 1/2 cup of chopped vegetables
- 1 cup of raw, leafy vegetables
- 3/4 cup of cooked vegetables

Fruits 2 to 4 servings
- 1 medium apple, banana, or orange
- 1/2 cup of chopped, cooked, or canned fruit
- 3/4 cup of fruit juice

Bread, cereal, rice, and pasta 6 to 11 servings
- 1 slice of bread
- 1 oz of ready-to-eat cereal
- 1/2 cup of rice or pasta
- 1/2 cup of cooked cereal

Quick Lab

Brown Bag Test

1. Cut a **brown paper bag** into squares.
2. Gather a variety of **foods.** Place a piece of each food on a different square and leave overnight.
3. What do you see when you remove the food?
4. Which food had the most oil? Which food had the least?

Quick Lab

MATERIALS

For Each Group
- bag, paper, brown
- foods (5–10 different kinds)

Answers

3. Students should see oil spots after the food is removed. Foods which lack oils will likely not leave spots.
4. Answers may vary. The food with the highest fat content will likely leave the largest amount of oil behind.

Activity — Basic

Food Guide Pyramid Have students keep track of the food that they eat during a single day. Then, have students determine how many servings from each of the food groups they consumed. Have students determine the percentage of recommended servings these numbers represent. For example, two glasses of milk and a grilled cheese sandwich represent 100% of the recommended daily dairy servings. One apple is 25–50% of the recommended daily fruit servings. Encourage students to identify how they could change their diet to make it more healthful. **LS Logical**

Debate — Advanced

Changing the Pyramid Many nutrition experts think that the Food Guide Pyramid should be updated to reflect new discoveries about healthy eating habits. Ask students to research the proposed revisions to the Food Guide Pyramid. Then, have students debate whether the Food Guide Pyramid should be changed. **LS Verbal/Logical**

Answer to Reading Check

Sample answer: a peanut butter sandwich, a glass of milk, and fresh fruit and vegetable slices

CHAPTER RESOURCES

Technology

Transparencies
- The Food Guide Pyramid

Close

Reteaching — Basic

Memory Game Ask students to create a memory game about the six essential nutrients. Give students 12 index cards. On six of the cards, students should list each nutrient. On the other six cards, they should list the role each nutrient plays in the body. Ask students to turn over the cards, scramble them, and turn over cards individually to match each with the appropriate nutrient or function. LS **Kinesthetic/Verbal**

Quiz — General

1. What are the six essential nutrients? (carbohydrates, proteins, fats, water, vitamins, and minerals)
2. What are the effects of obesity? (Sample answer: Obesity increases the risk of high blood pressure, heart disease, and diabetes. Some obese people may suffer from malnutrition.)

Alternative Assessment — General

Concept Mapping Have students create a concept map using the new terms from the section and any additional words that are necessary. Have students begin their concept map with the phrase "Good nutrition." LS **Logical**

Nutrition Facts

Nutrition Facts	
Serving Size 1/2 cup (120 ml) ← Serving information	
Servings per Container 2.5	
Amount per Serving	**Prepared**
Calories	70 ← Number of Calories per serving
Calories from Fat	25
	% Daily Value
Total Fat 2.5 g	4%
Saturated Fat 1 g	5%
Cholesterol 15 mg	5%
Sodium 960 mg	40%
Total Carbohydrate 8 g	3%
Dietary Fiber less than 1 g	4%
Sugars 1 g	
Protein 3 g	
Vitamin A	15%
Vitamin C	0%
Calcium	0%
Iron	4%

Percentage of daily values

*Percent Daily Values are based on a 2,000 Calorie diet. Your daily values may be higher or lower depending on your Calorie needs:

	Calories	2,000	2,500
Total Fat	Less than	65g	80g
Sat Fat	Less than	20g	25g
Cholesterol	Less than	300mg	300mg
Sodium	Less than	2,400mg	2,400mg
Total Carbohydrate		300g	375g
Dietary Fiber		25g	30g
Protein		50g	60g

Figure 5 *Nutrition Facts labels provide a lot of information.*

malnutrition a disorder of nutrition that results when a person does not consume enough of each of the nutrients that are needed by the human body

MATH PRACTICE

What Percentage?

Use the Nutrition Facts label above to answer the following question. The recommended daily value of fat is 72 g for teenage girls and 90 g for teenage boys. What percentage of the daily recommended fat value is provided in one cup of soup?

Reading Food Labels

Packaged foods must have Nutrition Facts labels. **Figure 5** shows a Nutrition Facts label for chicken noodle soup. Nutrition Facts labels show what amount of each nutrient is in one serving of the food. You can tell whether a food is high or low in a nutrient by looking at its daily value. Reading food labels can help you make healthy eating choices. The percentage of daily values shown is based on a diet that consists of 2,000 Cal per day. Most teenagers need more than 2,000 Cal per day. The number of Calories needed depends on factors such as height, weight, age, and level of activity. Playing sports and exercising use up Calories that need to be replaced for you to grow.

Reading Check **For what nutrients does chicken noodle soup provide more than 10% of the daily value?**

Nutritional Disorders

Unhealthy eating habits can cause nutritional disorders. **Malnutrition** occurs when someone does not eat enough of the nutrients needed by the body. Malnutrition can result from eating too few or too many Calories or not taking in enough of the right nutrients. Malnutrition affects how one looks and how quickly one's body can repair damage and fight illness.

Anorexia Nervosa and Bulimia Nervosa

Anorexia nervosa (AN uh REKS ee uh nuhr VOH suh) is an eating disorder characterized by self-starvation and an intense fear of gaining weight. Anorexia nervosa can lead to severe malnutrition.

Bulimia nervosa (boo LEE mee uh nuhr VOH suh) is a disorder characterized by binge eating followed by induced vomiting. Sometimes, people suffering from bulimia nervosa use laxatives or diuretics to rid their bodies of food and water. Bulimia nervosa can damage teeth and the digestive system and can lead to kidney and heart failure.

Both anorexia and bulimia can cause weak bones, low blood pressure, and heart problems. These eating disorders can be fatal if not treated. If you are worried that you or someone you know may have an eating disorder, talk to an adult.

Math Practice

3.5% for girls (2.5 g ÷ 72 g × 100 = 3.5%)
2.8% for boys (2.5 g ÷ 90 g × 100 = 2.8%)

Answer to Reading Check

One serving of chicken noodle soup provides more than 10% of the daily recommended allowance of vitamin A and sodium.

Obesity

Eating too much food that is high in fat and low in other nutrients, such as junk food and fast food, can lead to malnutrition. *Obesity* (oh BEE suh tee) is having an extremely high percentage of body fat. People suffering from obesity may not be eating a variety of foods that provide them with the correct balance of essential nutrients. Eating too many foods from the top of the Food Guide Pyramid and having an inactive lifestyle can contribute to obesity.

Obesity increases the risk of high blood pressure, heart disease, and diabetes. Eating a more balanced diet and exercising regularly can help reduce obesity. Obesity may also be caused by other factors. Scientists are studying the links between obesity and heredity.

SECTION Review

Summary

- A healthy diet has a balance of carbohydrates, proteins, fats, water, vitamins, and minerals.
- The Food Guide Pyramid is a good guide for healthy eating.
- Nutrition Facts labels provide information needed to plan a healthy diet.
- Anorexia nervosa and bulimia nervosa cause malnutrition and damage to many body systems.
- Obesity can lead to heart disease and diabetes.

Using Key Terms

1. In your own words, write a definition for each of the following terms: *nutrient, mineral,* and *vitamin.*

Understanding Key Ideas

2. Malnutrition can be caused by
   a. obesity.
   b. bulimia nervosa.
   c. anorexia nervosa.
   d. All of the above
3. What information is found on a Nutrition Facts label?
4. Give an example of a carbohydrate, a protein, and a fat.
5. If vitamins and minerals do not supply energy, why are they important to a healthy diet?
6. How do anorexia nervosa and bulimia nervosa differ?
7. How can someone who is obese suffer from malnutrition?

Math Skills

8. If you eat 2,500 Cal per day and 20% are from fat, 30% are from protein, and 50% are from carbohydrates, how many Calories of each nutrient do you eat?

Critical Thinking

9. **Applying Concepts** Name some of the nutrients that can be found in a glass of milk.
10. **Identifying Relationships** Explain how eating a variety of foods can help ensure good nutrition.
11. **Predicting Consequences** How would your growth be affected if your diet consistently lacked important nutrients?
12. **Applying Concepts** Explain how you can use the Nutrition Facts label to choose food that is high in calcium.

SCILINKS. NSTA Developed and maintained by the National Science Teachers Association

For a variety of links related to this chapter, go to www.scilinks.org

Topic: Food Pyramids; Nutritional Disorders
SciLinks code: HSM0598; HSM1057

Answers to Section Review

1. Sample answer: A nutrient is a substance that your body needs to keep working. A mineral is nutrient that your body needs for certain body processes. A vitamin is a nutrient that controls many body functions and must be obtained from food.
2. d
3. Sample answer: A Nutrition Facts label shows how much of each nutrient is in a single serving of a packaged food.
4. Sample answer: Bread is a carbohydrate, oil is a fat, and fish contains protein.
5. Sample answer: Vitamins and minerals help control body functions, such as heartbeat and muscle movement, and they help build body parts, such as bones and blood cells.
6. Sample answer: Anorexia nervosa is an eating disorder in which a person starves himself or herself. Bulimia nervosa is an eating disorder in which a person eats and then induces vomiting.
7. Sample answer: A person who is obese can suffer from malnutrition because his or her diet may be high in fat, and he or she does not get enough of the other essential nutrients.
8. 500 Cal from fat ($2,500 \text{ Cal} \times 0.2 = 500 \text{ Cal}$), 750 Cal from protein ($2,500 \text{ Cal} \times 0.3 = 750 \text{ Cal}$), 1,250 Cal from carbohydrates ($2,500 \text{ Cal} \times 0.5 = 1,250 \text{ Cal}$)
9. Accept all reasonable answers. All six types of nutrients can be found in milk.
10. Sample answer: Because different foods contain different nutrients, eating a variety of foods can help people get a variety of nutrients. Many of these nutrients are needed for good nutrition.
11. Sample answer: If your diet lacked Calories, your body would not have the energy to grow. If your diet lacked minerals, such as calcium, your body would not have the building blocks needed to remain healthy and grow.
12. Sample answer: The Nutrition Facts label lists the percentage of daily value for calcium found in a single serving of food. Reading the labels of foods helps people make sure that they eat foods high in calcium.

CHAPTER RESOURCES

Chapter Resource File

- Section Quiz GENERAL
- Section Review GENERAL
- Vocabulary and Section Summary GENERAL
- Reinforcement Worksheet BASIC
- Critical Thinking ADVANCED
- SciLinks Activity GENERAL
- Datasheet for Quick Lab

Technology

Transparencies
- Nutrition Facts Label

SECTION 2

Focus

Overview

In this section, students will learn that drugs can have positive and negative effects on the body. Students will learn about different types of drugs. Finally, they will learn about the causes, effects, and treatments of drug abuse and addiction.

Bellringer

Ask students, "Are drugs good or bad for you?" (Explain that a drug is not always inherently good or bad. Any chemical substance that causes a physical or emotional change in a person is a drug.)

Motivate

ACTIVITY — GENERAL

Poster Project Have students collect magazine advertisements for prescription drugs, over-the-counter drugs, alcohol, and tobacco. Ask students to make a poster that identifies the drug in each advertisement and the type of person that would find the advertisement appealing.
LS Visual

SECTION 2

Risks of Alcohol and Other Drugs

READING WARM-UP

Objectives

- Describe the difference between psychological and physical dependence.
- Explain the hazards of tobacco, alcohol, and illegal drugs.
- Distinguish between the positive and negative uses of drugs.

Terms to Learn

drug	alcoholism
addiction	narcotic
nicotine	

READING STRATEGY

Reading Organizer As you read this section, make a table comparing the positive and negative uses of drugs.

You see them in movies and on television and read about them in magazines. But what are drugs?

You are exposed to information, and misinformation, about drugs every day. So, how can you make the best decisions?

What Is a Drug?

Any chemical substance that causes a physical or psychological change is called a **drug.** Drugs come in many forms, as shown in **Figure 1.** Some drugs enter the body through the skin. Other drugs are swallowed, inhaled, or injected. Drugs are classified by their effects. *Analgesics* (AN'l JEE ziks) relieve pain. *Antibiotics* (AN tie bie AHT iks) fight bacterial infections, and *antihistamines* (AN tie HIS tuh MEENZ) control cold and allergy symptoms. *Stimulants* speed up the central nervous system, and *depressants* slow it down. When used correctly, legal drugs can help your body heal. When used illegally or improperly, however, drugs can do great harm.

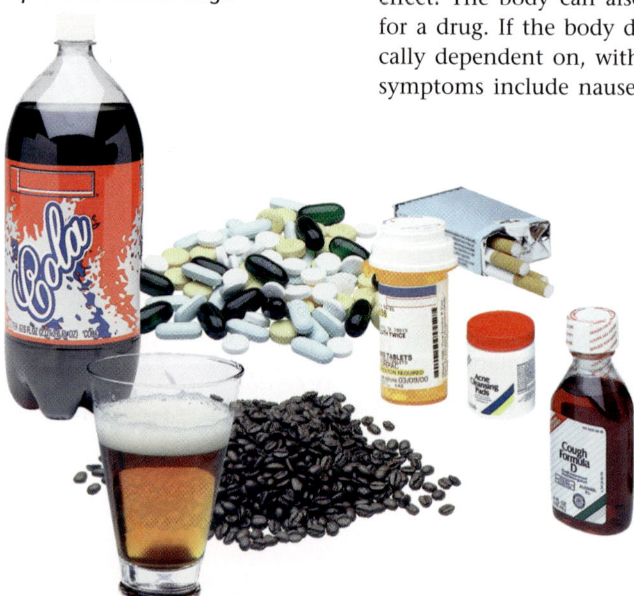

Figure 1 *All of these products contain drugs.*

Dependence and Addiction

The body can develop *tolerance* to a drug. Tolerance means that larger and larger doses of the drug are needed to get the same effect. The body can also form a *physical dependence* or need for a drug. If the body doesn't receive a drug that it is physically dependent on, withdrawal symptoms occur. Withdrawal symptoms include nausea, vomiting, pain, and tremors.

Addiction is the loss of control of drug-taking behavior. Once addicted, a person finds it very hard to stop taking a drug. Sometimes, the need for a drug is not due only to physical dependence. Some people also form *psychological dependence* on a drug, which means that they feel powerful cravings for the drug.

CHAPTER RESOURCES

Chapter Resource File

- Lesson Plan
- Directed Reading A BASIC
- Directed Reading B SPECIAL NEEDS

Technology

Transparencies

- Bellringer

WEIRD SCIENCE

Humans may not be the only organisms that use drugs. Chimpanzees have been observed swallowing the leaves of a plant that is known to be effective against infections and parasites. Muriqui monkeys of Brazil may be able to control their fertility with certain plants, and howler monkeys may even select the sex of their offspring by eating different plants.

Types of Drugs

There are many kinds of drugs. Some drugs are made from plants, and some are made in a lab. You can buy some drugs at the grocery store, while others can be prescribed only by a doctor. Some drugs are illegal to buy, sell, or possess.

drug any substance that causes a change in a person's physical or psychological state

addiction a dependence on a substance, such as alcohol or another drug

Herbal Medicines

Information about herbal medicines has been handed down for centuries, and some herbs contain chemicals with important healing properties. The tea in **Figure 2** contains chamomile and is made from a plant. Chamomile has chemicals in it that can help you sleep. However, herbs are drugs and should be used carefully. The Federal Drug Administration does not regulate herbal medicines or teas and cannot guarantee their safety.

Figure 2 *Some herbs can be purchased in health-food stores. Medicinal herbs should always be used with care.*

Over-the-Counter and Prescription Drugs

Over-the-counter drugs can be bought without a prescription. A prescription is written by a doctor and describes the drug, directions for use, and the amount of the drug to be taken.

Many over-the-counter and prescription drugs are powerful healing agents. However, some drugs also produce unwanted side effects. *Side effects* are uncomfortable symptoms, such as nausea, headaches, drowsiness, or more serious problems.

Whether purchased with or without a prescription, all drugs must be used with care. Information on proper use can be found on the label. **Figure 3** shows some general drug safety tips.

Reading Check **What is the difference between an over-the-counter drug and a prescription drug?** (*See the Appendix for answers to Reading Checks.*)

Figure 3 Drug Safety Tips

- *Never take another person's prescription medicine.*
- *Read the label before each use. Always follow the instructions on the label and those provided by your doctor or pharmacist.*
- *Do not take more or less medication than prescribed.*
- *Consult a doctor if you have any side effects.*
- *Throw away leftover and out-of-date medicines.*

Teach

Using the Figure—ADVANCED

Writing **Drug Safety Tips** Refer students to **Figure 3.** Encourage students to research how following each of the drug safety tips listed in the figure can keep them safe. Ask students to write an informative magazine article about their findings. Encourage students to use the tips as headings in their articles. (Students should discover that taking someone else's prescription medicine is risky because they don't know how their bodies are going to react to the medicine. They may have an allergic reaction or the medicine may interfere with their own medications. Students should recognize that reading the label ensures that they take the medicine correctly. Students should discover that taking less medicine than prescribed can make the medicine ineffective. If they take more medicine than prescribed, they may be injured or overdose. Students should recognize that side effects may indicate that they are allergic to the medicine, that the medicine is not effective, or that they might need another medicine to meet their needs. Students should recognize that leftover medicine might be discovered and taken by younger children and that out-of-date medicines likely are no longer effective.) **LS Verbal**

Homework — GENERAL

Pharmacy Encourage students to visit a pharmacy and make a list of 25 over-the-counter drugs. Challenge students to include drugs that have different modes of administration. Have students prepare a table that categorizes the drugs by their active ingredients and by their uses. **LS Verbal**

Cultural Awareness — GENERAL

Aspirin The Chickasaw Indians relieved pain and fever by chewing on the bark of the willow tree. Today many people relieve their minor ills with a chemically related substance—aspirin. Derived from a chemical that is also found in willow bark, aspirin is the most widely used drug in the world. Aspirin relieves pain and fever and prevents blood clotting.

Answer to Reading Check

Over-the-counter drugs can be bought without a prescription. Prescription drugs can be bought only with a prescription from a doctor or other medical professional.

Teach, continued

INCLUSION Strategies

- **Behavior Control Issues**
- **Hearing Impaired**
- **Learning Disabled**

Students often benefit from small-group work. Have students work in groups of four. Have them create information webs with large circles in the center and small circles surrounding the large circles. Tell students to write each of the following phrases inside a large circle: "Problems caused by cigarettes" and "Smokeless tobacco hazards." Tell students to add six small circles to the cigarette web and three small circles to the smokeless tobacco web and to write a risk in each of the small circles. LS Visual

Discussion — GENERAL

Advertising Alcohol Tell students that beer, liquor, and wine companies in the United States spend billions of dollars annually on advertising and promotion. Display for students a selection of advertisements for beer, liquor, or wine taken from magazines. Encourage students to identify advertising tactics that might attract youth. (Students might note that the ads often show groups of young, beautiful, and active people in outdoor settings. These ads imply that people cannot have fun without alcohol.) LS Verbal

Figure 4 Effects of Smoking

Healthy lung tissue of a nonsmoker

Damaged lung tissue of a smoker

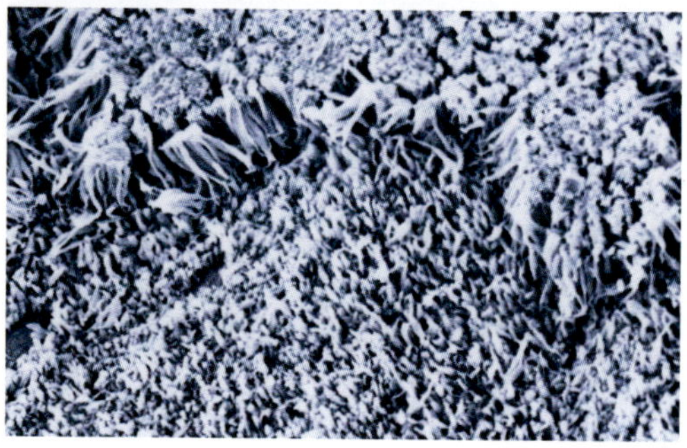

nicotine a toxic, addictive chemical that is found in tobacco and that is one of the major contributors to the harmful effects of smoking

alcoholism a disorder in which a person repeatedly drinks alcoholic beverages in an amount that interferes with the person's health and activities

Tobacco

Cigarettes are addictive, and smoking has serious health effects. **Nicotine** (NIK uh TEEN) is a chemical in tobacco that increases heart rate and blood pressure and is extremely addictive. Smokers experience a decrease in physical endurance. **Figure 4** shows the effects of smoking on the cilia of your lungs. Cilia clean the air you breathe and prevent debris from entering your lungs. Smoking increases the chances of lung cancer, and it has been linked to other cancers, emphysema, chronic bronchitis, and heart disease. Experts estimate that there are more than 430,000 deaths related to smoking each year in the United States. Secondhand smoke also poses significant health risks.

Like cigarettes, smokeless, or chewing, tobacco is addictive and can cause health problems. Nicotine is absorbed through the lining of the mouth. Smokeless tobacco increases the risk of several cancers, including mouth and throat cancer. It also causes gum disease and yellowing of the teeth.

Figure 5 *This car was in an accident involving a drunk driver.*

Alcohol

It is illegal in most of the United States for people under the age of 21 to use alcohol. Alcohol slows down the central nervous system and can cause memory loss. Excessive use of alcohol can damage the liver, pancreas, brain, nerves, and cardiovascular system. In very large quantities, alcohol can cause death. Alcohol is a factor in more than half of all suicides, murders, and accidental deaths. **Figure 5** shows the results of one alcohol-related accident. Alcohol also affects decision making and can lead you to take unhealthy risks.

People can suffer from **alcoholism,** which means that they are physically and psychologically dependent on alcohol. Alcoholism is considered a disease, and genetic factors are thought to influence the development of alcoholism in some people.

Is That a Fact!

About 90% of first-time cigarette smokers will become addicted to cigarettes. Fewer than half of these people are able to quit smoking, which puts their health at great risk. About 90% of people who get lung cancer are smokers or former smokers. Cigarette smokers are also at greater risk of getting cancers of the larynx, lip, esophagus, bladder, pancreas, and kidneys.

Homework — GENERAL

PORTFOLIO **Avoiding Alcohol** Tell students that counteradvertising is an effective technique for deterring alcohol abuse. Have students work in small groups to create a magazine advertisement against alcohol abuse. Consider hanging the advertisements in the classroom or around the school. LS Verbal

Figure 6 *Smoking marijuana can make your health and dreams go up in smoke.*

Marijuana

Marijuana is an illegal drug that comes from the Indian hemp plant. Marijuana affects different people in different ways. It may increase anxiety or cause feelings of paranoia. Marijuana slows reaction time, impairs thinking, and causes a loss of coordination. Regular use of marijuana can affect many areas of your life, as described in **Figure 6.**

narcotic a drug that is derived from opium and that relieves pain and induces sleep

Cocaine

Cocaine and its more purified form, crack, are made from the coca plant. Both drugs are illegal and highly addictive. Users can become addicted to them in a very short time. Cocaine can produce feelings of intense excitement followed by anxiety and depression. Both drugs increase heart rate and blood pressure and can cause heart attacks, even among first-time users.

Reading Check What are two dangers to users of cocaine?

Narcotics and Designer Drugs

Drugs made from the opium plant are called **narcotics.** Some narcotics are used to treat severe pain. Narcotics are illegal unless prescribed by a doctor. Some narcotics are never legal. For example, heroin is one of the most addictive narcotics and is always illegal. Heroin is usually injected, and users often share needles. Therefore, heroin users have a high risk of becoming infected with diseases such as hepatitis and AIDS. Heroin users can also die of an overdose of the drug.

Other illegal drugs include inhalants, barbiturates (bahr BICH uhr itz), amphetamines (am FET uh MEENZ), and *designer drugs*. Designer drugs are made by making small changes to existing drugs. Ecstasy, or "X," is a designer drug that causes feelings of well-being. Over time, the drug causes lesions (LEE zhuhnz), or holes, in a user's brain, as shown in **Figure 7.** Ecstasy users are also more likely to develop depression.

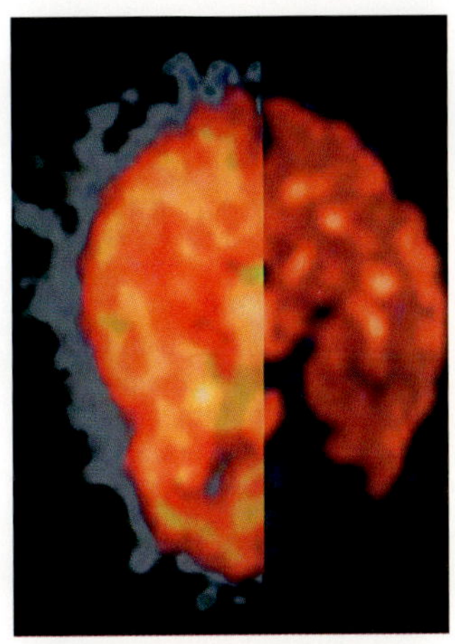

Figure 7 *The brain scan on the left shows a healthy brain. The scan on the right is from a teenager who has regularly used Ecstasy.*

CONNECTION ACTIVITY
Real World — ADVANCED

Alcohol and Drugs Current studies suggest that young people are making their first choices about alcohol and drugs between the ages of 12 and 14. However, only a generation ago, those choices were made between the ages of 16 and 18. Ask interested students to research the incidence of drinking and the use of drugs in their age group. Have students create a table demonstrating their findings. (A 2002 study funded by the National Institute on Drug Abuse revealed the following percentages for the 8th graders surveyed: 31.4% had tried cigarettes, 11.2% had tried smokeless tobacco, 47% had tried alcohol, 19.2% had tried marijuana, 3.6% had tried cocaine, 2.5% had tried crack cocaine, and 4.3% had tried MDMA, or ecstacy.)
LS Logical/Visual

CONNECTION to
Language Arts — BASIC

Writing **Letter** Have students work in small groups to write a letter to a friend that discourages drug abuse. Ask students to describe the hazards of drug abuse in their letters.
LS Verbal

MISCONCEPTION ALERT

Beer One of the most common alcohol-related myths is that beer is not as intoxicating as liquor. In fact, a serving of beer has the same effects as a serving of liquor. Alcohol from any source is addictive. It can cause liver damage, memory loss, heart damage, stomach problems, and sexual impotence. It impairs judgment, contributes to traffic deaths, and plays a role in many suicides and homicides.

Answer to Reading Check

First-time use of cocaine can cause a heart attack or can cause a person to become addicted.

Close

Reteaching — Basic

Reviewing Drugs Ask students to briefly describe the drugs discussed in the section. Have students identify effects that these drugs can have on the body. LS Verbal

Quiz — General

1. What is drug abuse? (Sample answer: Drug abuse is the improper or illegal use of drugs.)
2. What is nicotine? (Nicotine is the drug in tobacco that increases heart rate and blood pressure and is extremely addictive.)
3. What are designer drugs? Give an example. (Sample answer: Designer drugs are drugs made by making small changes to existing drugs. Ecstasy is a designer drug.)

Alternative Assessment — General

Making Tables Have students prepare tables describing alcohol, tobacco, marijuana, cocaine, LSD, and heroin. Have students define each and describe the effects of the drug on the body. In addition, have students indicate whether the drug is legal and whether it is addictive. LS Logical/Verbal

Figure 8 *Drug abuse can leave you depressed and feeling alone.*

Figure 9 Drug Myths

Myth ***"It's only alcohol, not drugs."***

Reality Alcohol is a mood-altering and mind-altering drug. It affects the central nervous system and is addictive.

Myth ***"I won't get hooked on one or two cigarettes a day."***

Reality Addiction is not related to the amount of a drug used. Some people become addicted after using a drug once or twice.

Myth ***"I can quit any time I want."***

Reality Addicts may quit and return to drug usage many times. Their inability to stay drug-free shows how powerful the addiction is.

Hallucinogens

Hallucinogens (huh LOO si nuh juhnz) distort the senses and cause mood changes. Users have hallucinations, which means that they see and hear things that are not real. LSD and PCP are powerful, illegal hallucinogens. Sniffing glue or solvents can also cause hallucinations and serious brain damage.

Drug Abuse

A drug user takes a drug to prevent or improve a medical condition. The drug user obtains the drug legally and uses the drug properly. A drug abuser does not take a drug to relieve a medical condition. An abuser may take drugs for the temporary good feelings they produce, to escape from problems, or to belong to a group. The drug is often obtained illegally, and it is often taken without knowledge of the drug's dangers.

Reading Check **What is the difference between drug use and drug abuse?**

How Drug Abuse Starts

Nicotine, alcohol, and marijuana are sometimes called *gateway drugs* because they are often the first drugs a person abuses. The abuse of other, more dangerous drugs may follow the abuse of gateway drugs. Peer pressure is often the reason that young people begin to use drugs. Teenagers may drink, smoke, or try marijuana to make friends or avoid being teased. Because drug abusers often stand out, it can sometimes be hard to see that many teenagers do not abuse drugs.

Many teenagers begin using illegal drugs to feel part of a group, but drug abuse has many serious consequences. Drug abuse can lead to problems with friends, family, school, and handling money. These problems often lead to depression and social isolation, as shown in **Figure 8.**

Many people who start using drugs do not recognize the dangers. Misinformation about drugs is everywhere. Several common drug myths are discussed in **Figure 9.**

School to Home

Good Reasons

Writing Skill Discuss with your parent the possible effects of drug abuse on your family. Then, write yourself a letter giving reasons why you should stay drug-free. Put your letter in a safe place. If you ever find yourself thinking about using drugs, take out your letter and read it.

Activity

Homework — General

Drug Report Have students write a short paper about one of the following drugs: alcohol, marijuana, cocaine, LSD, PCP, ecstasy, or heroin. Ask students to describe the type of drug, common names for the drug, what the drug is made from, the effects of the drug, the penalty for possession of the drug, and what help is available to those who are addicted to the drug. LS Verbal

Answer to Reading Check

Drug use is the proper use of a legal drug. Drug abuse is either the use of an illegal drug or the improper use of a legal drug.

Getting Off Drugs

People who abuse drugs undergo emotional and physical changes. Teenagers who had few problems often begin to have problems with school, family, and money when they start to use drugs.

The first step to quitting drugs is to admit to abusing drugs and to decide to stop. It is important for the addicted person to get the proper medical treatment. There are drug treatment centers, like the one shown in **Figure 10,** available to help. Getting off drugs can be extremely difficult. Withdrawal symptoms are often painful, and powerful cravings for a drug can continue long after a person quits. But people who stop abusing drugs lead happier and healthier lives.

Figure 10 *Drug treatment centers help people get off drugs and back on track to healthier, happier lives.*

SECTION Review

Summary

- Physical dependence causes withdrawal symptoms when a person stops using a drug. Psychological dependence causes powerful cravings.
- There are many types of drugs, including over-the-counter, prescription, and herbal medicines.
- Tobacco contains the highly addictive chemical nicotine.
- Abuse of alcohol can lead to alcoholism.
- Illegal drugs include marijuana, cocaine, hallucinogens, designer drugs, and many narcotics.
- Getting off drugs requires proper medical treatment.

Using Key Terms

1. In your own words, write a definition for the terms *drug, addiction,* and *narcotic.*

Understanding Key Ideas

2. Which of the following products does NOT contain a drug?
   - **a.** cola
   - **b.** sports drink
   - **c.** herbal tea
   - **d.** cough syrup
3. Describe the difference between physical and psychological dependence.
4. What is the difference between drug use and drug abuse?
5. How does addiction occur, and what are two consequences of drug addiction?
6. Name two different kinds of illegal drugs, and give examples of each.

Math Skills

7. If 2,200 people between the ages of 16 and 20 die every year in alcohol-related car crashes, how many die every day?

Critical Thinking

8. **Analyzing Relationships** How are nicotine, alcohol, heroin, and cocaine similar? How are they different?
9. **Analyzing Ideas** What are two ways that a person who abuses drugs can get in trouble with the law?
10. **Predicting Consequences** How can drug abuse damage family relationships?
11. **Making Inferences** Driving a car while under the influence of drugs can put others in danger. Describe another situation in which one person's drug abuse could put other people in danger.

SCILINKS. NSTA Developed and maintained by the National Science Teachers Association

For a variety of links related to this chapter, go to www.scilinks.org

Topic: Drug and Alcohol Abuse
SciLinks code: HSM0428

Answers to Section Review

1. Sample answer: A drug is any substance that changes a person's physical or emotional state. Addiction happens when someone becomes dependent on a drug. A narcotic is a drug that is derived from opium and that is used to treat pain.
2. b
3. Sample answer: When someone forms a physical dependence, he or she suffers from withdrawal symptoms if he or she stops using a drug. Someone who has a psychological dependence on a drug feels powerful cravings for the drug.
4. Sample answer: Drug use is the legal and proper use of a drug to prevent or improve a medical condition. Drug abuse is the improper and illegal use of drugs, often for the feelings the drugs produce, to escape from problems, and to fit in.
5. Sample answer: Addiction is the loss of control of drug-taking behavior. Addiction can cause problems at school, at home, and with money.
6. Sample answer: Narcotics include opium and heroin. LSD and PCP are hallucinogens.
7. 6 people/day (2,200 people ÷ 365 days = 6 people/day)
8. Sample answer: Nicotine, alcohol, heroin, and cocaine are all addictive. Nicotine and alcohol are legal for many people, but cocaine and heroin are illegal.
9. Sample answers: A person who abuses drugs can get in trouble with the law because many drugs are illegal. Also drug abuse can cause money troubles, which might lead someone to steal. Some people may drink and drive, which is illegal even if a person is old enough to legally drink alcohol.
10. Sample answer: Drug abuse can cause stress in a family because drug abuse causes failing grades and depression.
11. Sample answer: People who abuse drugs often make bad decisions. For example, a drug abuser may try to go swimming in dangerous water. A person who tried to help a drug abuser who makes a bad decision could be hurt.

CHAPTER RESOURCES

Chapter Resource File

- Section Quiz GENERAL
- Section Review GENERAL
- Vocabulary and Section Summary GENERAL

SECTION 3

Focus

Overview

In this section, students will learn how hygiene, exercise, rest, and stress management can enhance their well-being. Students will learn how stress affects the body. Finally, students will learn how injuries can be prevented.

Bellringer

Tell students the following information: "In the 1800s and earlier, people who sustained simple cuts on their skin often ended up with serious infections. That rarely happens now." Then, ask students to explain why infections are less common now than they once were. (Sample answer: People now understand the importance of washing wounds with soap to keep them clean. People also have antiseptics and antibiotics today.)

Motivate

ACTIVITY — GENERAL

Don't Kick These Habits! Ask students to list some of their daily hygienic habits. (Sample answers: washing hands, bathing, brushing teeth, washing dishes, and using clean utensils to eat) **LS Verbal**

SECTION 3

Healthy Habits

Do you like playing sports or acting in plays? How does your health affect your favorite activities?

Whatever you do, the better your health is, the better you can perform. Keeping yourself healthy is a daily responsibility.

READING WARM-UP

Objectives

- Describe three important aspects of good hygiene.
- Explain why exercise and sleep are important to good health.
- Describe methods of handling stress.
- List three ways to stay safe at home, on the road, and outdoors.
- Plan what you would do in the case of an accident.

Terms to Learn

hygiene
aerobic exercise
stress

READING STRATEGY

Prediction Guide Before reading this section, write the title of each heading in this section. Next, under each heading, write what you think you will learn.

Taking Care of Your Body

The science of preserving and protecting your health is known as **hygiene.** It sounds simple, but washing your hands is the best way to prevent the spread of disease and infection. You should always wash your hands after using the bathroom and before and after handling food. Taking care of your skin, hair, and teeth is important for good hygiene. Good hygiene includes regularly using sunscreen, shampooing your hair, and brushing and flossing your teeth daily.

hygiene the science of health and ways to preserve health

Good Posture

Posture is also important to health. Good posture helps you look and feel your best. Bad posture strains your muscles and ligaments and makes breathing difficult. To have good posture, imagine a vertical line passing through your ear, shoulder, hip, knee, and ankle when you stand, as shown in **Figure 1.** When working at a desk, you should maintain good posture by pulling your chair forward and planting your feet firmly on the floor.

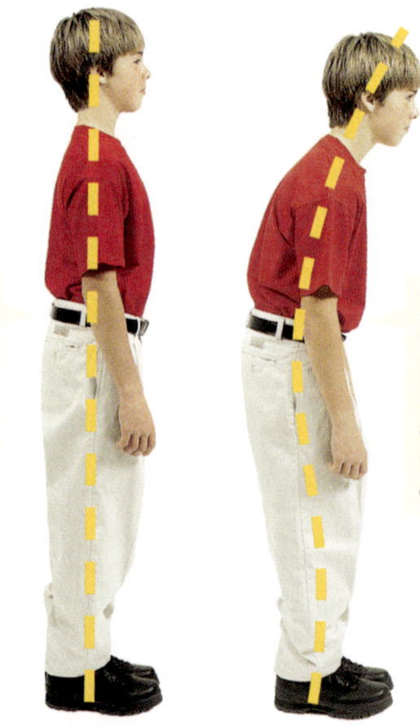

When you have good posture, your ear, shoulder, hip, knee, and ankle are in a straight line.

Bad posture strains your muscles and ligaments and can make breathing difficult.

Figure 1 *A slumped posture strains your lower back.*

CHAPTER RESOURCES

Chapter Resource File

- Lesson Plan
- Directed Reading A BASIC
- Directed Reading B SPECIAL NEEDS

Technology

Transparencies
- Bellringer

CONNECTION to Earth Science — ADVANCED

Fossil Hygiene Tell students that they may use fossils as part of their daily hygienic habits. Many toothpastes contain diatomaceous earth, which consists of fossilized diatoms—protist producers with silica shells. The silica in diatomaceous earth polishes the surface of teeth. Ask interested students to research how diatomaceous earth formed and other uses for it. Have students make a poster about their findings. **LS Visual**

Exercise

Aerobic exercise at least three times a week is essential to good health. **Aerobic exercise** is vigorous, constant exercise of the whole body for 20 minutes or more. Walking, running, swimming, and biking are all examples of aerobic exercise. **Figure 2** shows another popular aerobic exercise—basketball.

Aerobic exercise increases the heart rate. As a result, more oxygen is taken in and distributed throughout the body. Over time, aerobic exercise strengthens the heart, lungs, and bones. It burns Calories, helps your body conserve nutrients, and aids digestion. It also gives you more energy and stamina. Aerobic exercise protects your physical and mental health.

Figure 2 *Aerobic exercise can be fun if you choose an activity you enjoy.*

Reading Check What are two benefits of regular exercise? *(See the Appendix for answers to Reading Checks.)*

Sleep

Believe it or not, teenagers actually need more sleep than younger children. Do you ever fall asleep in class, like the girl in **Figure 3,** or feel tired in the middle of the afternoon? If so, you may not be getting enough sleep. Scientists say that teenagers need about 9.5 hours of sleep each night.

At night, the body goes through several cycles of progressively deeper sleep, with periods of lighter sleep in between. If you do not sleep long enough, you will not enter the deepest, most restful period of sleep.

aerobic exercise physical exercise intended to increase the activity of the heart and lungs to promote the body's use of oxygen

Figure 3 *If you fall asleep easily during the day, you are probably not getting enough sleep.*

CONNECTION TO Language Arts

Dreamy Poetry

You are not wrong, who deem
That my days have been a dream;
Yet if hope has flown away
In a night, or in a day,
In a vision, or in none,
Is it therefore the less gone?
All that we see or seem
Is but a dream within a dream.

(Edgar Allan Poe, "A Dream Within a Dream")

What do you think Poe means by "a dream within a dream?" Why do you think there are many poems written about dreams or sleep?

Teach

READING STRATEGY — GENERAL

Prediction Guide Before students read this page, have them answer the following question:

Which of the following pairs of exercises lists aerobic exercises?

a. walking and running
b. swimming and stretching
c. biking and yoga
d. running and weight lifting

LS Logical

CONNECTION to Language Arts — BASIC

Writing **Adages** Recite for students the adage, "An apple a day keeps the doctor away." Tell students to write adages that encourage washing hands, brushing teeth, and good posture. LS Verbal

Research — ADVANCED

Writing **Bone Density** Encourage students to use library or Internet resources to investigate how exercise affects bone density. Have students identify which exercises have the greatest effect on bone density and why. Have students prepare brief reports and share their findings with the class. LS Verbal

Answer to Reading Check

Aerobic exercise strengthens the heart, lungs, and bones and reduces stress. Regular exercise also burns Calories and can give you more energy.

WEIRD SCIENCE

Regular, intense exercise decreases body fat. For female athletes who participate in intense exercise frequently, this loss of body fat can restrict the release of estrogen. When this occurs in prepubescent athletes, it delays the onset of puberty. Menstruation is delayed, and the period of bone growth is sometimes extended.

Answer to Connection to Language Arts

Sample answer: "A dream within a dream" could refer to the sense that everything people feel is based on events that have already happened. There are many poems written about dreams because dreams are often strange and can cause strong emotions.

Teach, *continued*

Discussion — GENERAL

Happy Stress Point out to students that although many people think that the term *stress* refers only to the emotional and physical response to negative events, many positive situations also cause stress. Encourage students to identify "happy" events that can create stress. (Sample answers: a new job, the addition of a sibling, and moving to a new house) Point out that stress, whatever its source, produces a physiological reaction that is designed to prepare the body for difficult situations. **LS Logical/Verbal**

CONNECTION ACTIVITY
Real World — GENERAL

Writing **Sleep Disorders** Scientists who study people's sleep patterns have learned that about 33% of the general population suffers from a sleep disorder. Ask interested students to research some of these sleep disorders. Students should identify the disorder, characteristics of the disorder, symptoms of the disorder, and ways in which the disorder is treated. Ask students to write a newspaper article about their findings. **LS Verbal**

Figure 4 *Can you identify all of the things in this picture that could cause stress?*

stress a physical or mental response to pressure

INTERNET ACTIVITY

For another activity related to this chapter, go to **go.hrw.com** and type in the keyword **HL5BD7W.**

Coping with Stress

You have a big soccer game tomorrow. Are you excited and ready for action? You got a low grade on your English paper. Are you upset or angry? The game and the test are causing you stress. **Stress** is the physical and mental response to pressure.

Some stress is a normal part of life. Stress stimulates your body to prepare for difficult or dangerous situations. However, sometimes you may have no outlet for the stress, and it builds up. Many things are causing stress for the girl shown in **Figure 4.** Excess stress is harmful to your health and can decrease your ability to carry out your daily activities.

You may not even realize you are stressed until your body reacts. Perhaps you get a headache, have an upset stomach, or lie awake at night. You might feel tired all the time or begin an old nervous habit, such as nail-biting. You may become irritable or resentful. All of these things can be signs of too much stress.

Dealing with Stress

Different people are stressed by different things. Once you identify the source of the stress, you can find ways to deal with it. If you cannot remove the cause of stress, here are some ideas for handling stress.

- Share your problems. Talk things over with someone you trust, such as a parent, friend, teacher, or school counselor.
- Make a list of all the things you would like to get done, and rank the things in order of importance. Do the most important things first.
- Exercise regularly, and get enough sleep.
- Pet a friendly animal.
- Spend some quiet time alone, or practice deep breathing or other relaxation techniques.

Homework — GENERAL

Stress Management Ask students, "What are the three things in your life that cause you the most stress?" Have students list the sources of stress in their life and come up with ways of eliminating, reducing, or coping with this stress. Keep in mind that some students are uncomfortable sharing personal information. **LS Verbal**

Injury Prevention

Have you ever fallen off your bike or sprained your ankle? Accidents happen, and they can cause injury and even death. It is impossible to prevent all accidents, but you can decrease your risk by using your common sense and following basic safety rules.

Safety Outdoors

Always dress appropriately for the weather and for the activity. Never hike or camp alone. Tell someone where you are going and when you expect to return. If you do not bring water from home, be sure to purify any water you drink in the wilderness.

Learn how to swim. It could save your life! Never swim alone, and do not dive into shallow water or water of unknown depth. When in a boat, wear a life jacket. If a storm threatens, get out of the water and seek shelter.

Reading Check Name three safety tips for the outdoors.

Safety at Home

Many accidents can be avoided. **Figure 5** shows tips for safety around the house.

Figure 5 Home Safety Tips

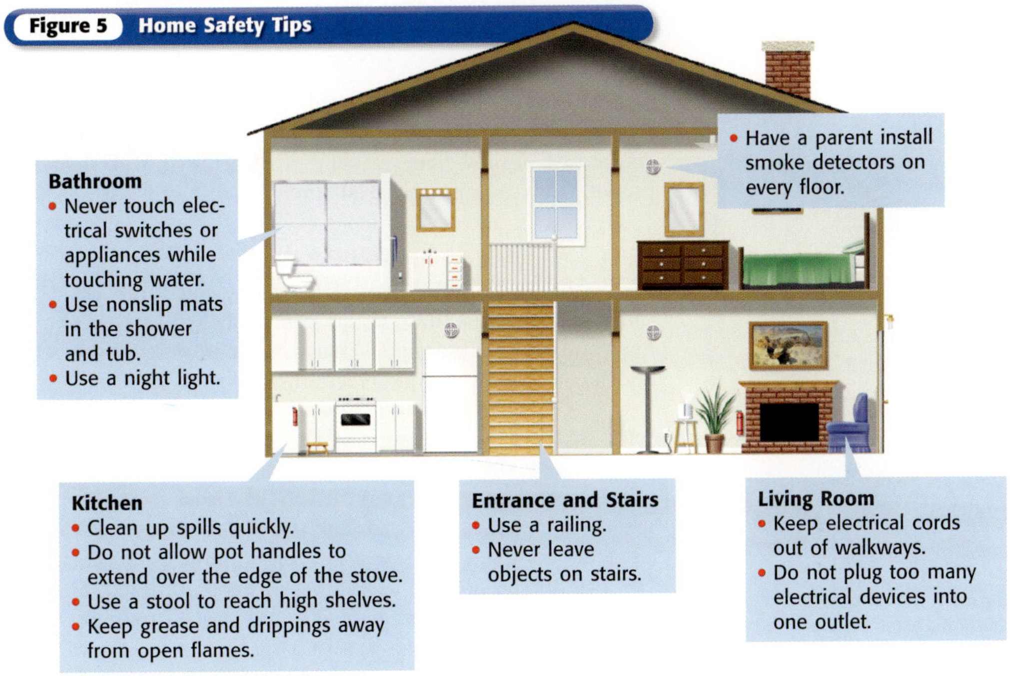

Discussion — GENERAL

Fatal Injuries Tell students that injuries cause more deaths among children between the ages of 1 and 19 than any major diseases. Point out that bicycle injuries alone cause the death of more than 500 children and teens in the United States each year. Most of these deaths involve head trauma. Remind students that many accidents can be prevented. Experts estimate that the simple act of wearing a bicycle helmet can reduce the risk of head injury by 85%, yet many children don't wear helmets. Ask students why they think people don't wear helmets. (Sample answers: Perhaps some people cannot afford helmets. Some people think that helmets are "funny-looking" or "uncool." Other people may think helmets are uncomfortable.) Ask students if there are other activities in which a helmet protects participants. (Sample answers: skating, snowboarding, football, and hockey) Ask students to brainstorm ways to get more children and teens to wear helmets during physical activity. **LS Verbal**

Group Activity — GENERAL

Poster Project Have students work in groups of four, and provide each group with poster board and markers. Have students create posters designed to educate the public about injury prevention. After students present their posters to the class, display the posters in the classroom. **LS Visual**

Homework — GENERAL

Writing **Temperature Injuries** Have students research and write a report about the physical symptoms and emergency treatments for hypothermia and heatstroke. **LS Verbal**

Answer to Reading Check

Sample answers: Never hike or camp alone, dress for the weather, learn how to swim, wear a life jacket, and never drink unpurified water.

Close

Reteaching — Basic

Taking Care Write the following headings on the board: "Healthy habits," "Coping with stress," and "Preventing injuries." Ask students to write examples of each under the headings. LS Logical/Verbal

Quiz — General

1. List three ways to deal with stress. (Sample answers: Share your problems. Make a list. Exercise regularly. Get enough sleep. Pet a friendly animal. Spend some quiet time alone. Practice relaxation techniques.)
2. What do bicycle helmets and seat belts have in common? (Sample answer: They both decrease the risk of injury from accidents.)
3. When you call for emergency help, what should you tell the operator? (the location of the emergency, the type of accident, the number of people injured, and the types of injuries)

Alternative Assessment — Advanced

Writing **Analyzing Habits** Have students write an analysis of their hygienic habits, exercise regimens, sleep patterns, and injury risks. Tell students to explain whether their activities and habits are encouraging good health and to describe what changes they might need to make to their routines. LS Verbal

Figure 6 *It is always important to use the appropriate safety equipment.*

Safety on the Road

In the car, always wear a seat belt, even if you are traveling only a short distance. Never ride in a car with someone who has been drinking. Safety equipment and common sense are your best defense against injury. When riding a bicycle, always wear a helmet like those shown in **Figure 6.** Ride with traffic, and obey all traffic rules. Be sure to signal when stopping or turning.

Safety in Class

Accidents can happen in school, especially in a lab class or during woodworking class. To avoid hurting yourself and others, always follow your teacher's instructions, and wear the proper safety equipment at all times.

When Accidents Happen

No matter how well you practice safety measures, accidents can still happen. What should you do if a friend chokes on food and cannot breathe? What if a friend is stung by a bee and has a violent allergic reaction?

Figure 7 *When calling 911, stay calm and listen carefully to what the dispatcher tells you.*

Call for Help

Once you've checked for other dangers, call for medical help immediately, as the person shown in **Figure 7** is doing. In most communities, you can dial 911. Speak slowly and clearly. Give the complete address and a description of the location. Describe the accident, the number of people injured, and the types of injuries. Ask what to do, and listen carefully to the instructions. Let the other person hang up first to be sure there are no more questions or instructions for you.

INCLUSION Strategies

• Developmentally Delayed • Visually Impaired • Learning Disabled

Students often benefit from verbal activities. Give students a chance to participate by having the class role-play 911 calls. Tell students that the calls should include the following parts:

- slow, clear speech
- description of emergency
- complete address or description of location
- type of injury
- number of people injured
- what has been done to help the victim
- request for instructions
- careful listening

LS Kinesthetic/Verbal

Learn First Aid

If you want to learn more about what to do in an emergency, you can take a first-aid or CPR course, such as the one shown in **Figure 8.** *CPR* can revive a person who is not breathing and has no heartbeat. If you are over 12 years old, you can become certified in both CPR and first aid. Some baby-sitting classes also provide information on first aid. The American Red Cross, community organizations, and local hospitals offer these classes. However, you should not attempt any lifesaving procedure unless you have been trained.

Reading Check What is CPR, and how can you learn it?

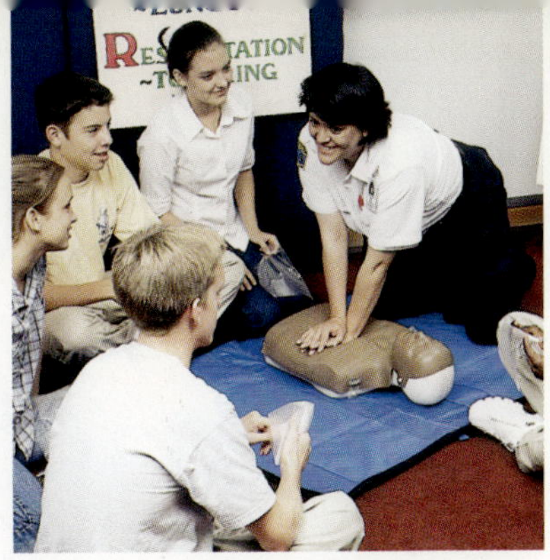

Figure 8 *These teenagers are taking a CPR course to prepare themselves for emergency situations.*

SECTION Review

Summary

- Good hygiene includes taking care of your skin, hair, and teeth.
- Good posture is important to health.
- Exercise keeps your heart, lungs, and bones healthy.
- Teenagers need more than 9 hours of sleep to stay rested and healthy.
- Coping with stress is an important part of staying physically and emotionally healthy.
- It is important to be aware of the possible hazards around your home, outdoors, and at school. Using the appropriate safety equipment can also help keep you safe.

Using Key Terms

Complete each of the following sentences by choosing the correct term from the word bank.

hygiene, aerobic exercise, sleep, stress

1. The science of protecting your health is called ___.
2. ___ strengthens your heart, lungs, and bones.
3. ___ is the physical and mental response to pressure.

Understanding Key Ideas

4. Which of the following is important for good health?
   a. irregular exercise
   b. getting your hair cut
   c. taking care of your teeth
   d. getting plenty of sun
5. List two things you should do when calling for help in a medical emergency.
6. List three ways to stay safe when you are outside, and three ways to stay safe at home.
7. How do seat belts and safety equipment protect you?

Math Skills

8. It is estimated that only 65% of adults wear their seat belts. If there are 10,000 people driving in your area right now, how many of them are wearing their seat belts?

Critical Thinking

9. **Applying Concepts** What situations cause you stress? What can you do to help relieve the stress you are feeling?
10. **Making Inferences** According to the newspaper, the temperature outside is 61°F right now. Later, it will be 90°F outside. If you and your friends want to play soccer in the park, what should you wear? What should you bring with you?

SCILINKS. NSTA Developed and maintained by the National Science Teachers Association

For a variety of links related to this chapter, go to www.scilinks.org

Topic: Safety
SciLinks code: HSM1339

Answers to Section Review

1. hygiene
2. Aerobic exercise
3. Stress
4. c
5. Sample answers: Speak slowly and clearly, give the address and a description of the location, describe the accident and the number of people injured, describe the injuries, ask what to do, listen carefully, and let the other person hang up first.
6. Sample answer: When outdoors, always dress for the weather and activity, tell someone where you are going and when you will be back, and purify all of the water that you drink in the wilderness. When at home, clean up spills quickly, never touch electrical switches while touching water, and use a night light.
7. Sample answer: A seat belt can protect you from being thrown from a car in case of an accident. Other safety equipment protects your body (such as your head, knees, and elbows) from injury during an accident.
8. 6,500 people ($10,000 \text{ people} \times 0.65 = 6,500 \text{ people}$)
9. Sample answer: Tests and quizzes at school cause stress. I can exercise regularly, spend some quiet time alone, and take deep breaths.
10. Sample answer: I should dress in layers so that when the temperature gets warmer, I can take off my jacket or sweater. I should also bring water and snacks with me.

Answer to Reading Check

CPR is a way to revive someone whose heart has stopped beating. CPR classes are available in many places in the community.

CHAPTER RESOURCES

Chapter Resource File

- Section Quiz GENERAL
- Section Review GENERAL
- Vocabulary and Section Summary GENERAL

Skills Practice Lab

Keep It Clean

Teacher's Notes

Time Required

One 45-minute class period

Lab Ratings

EASY → HARD

Teacher Prep 3

Student Set-Up 1

Concept Level 2

Clean Up 2

MATERIALS

You can reduce the number of materials needed by having students work in groups and having each group select a volunteer.

Safety Caution

Remind students to review all safety cautions and icons before beginning this lab activity. Check for known allergies to antibacterial soaps or other soaps before beginning the activity. If possible, use plastic Petri dishes.

Skills Practice Lab

Keep It Clean

OBJECTIVES

Investigate how well antibacterial soap works.

Practice counting bacterial colonies.

MATERIALS

- incubator
- pencil, wax
- Petri dishes, nutrient agar–filled, sterile (3)
- scrub brush, new
- soap, liquid antibacterial
- stopwatch
- tape, transparent

SAFETY

One of the best ways to prevent the spread of bacterial and viral infections is to frequently wash your hands with soap and water. Many companies advertise that their soap ingredients can destroy bacteria normally found on the body. In this activity, you will investigate how effective antibacterial soaps are at killing bacteria.

Procedure

1. Keeping the agar plates closed at all times, use the wax pencil to label the bottoms of three agar plates. Label one plate "Control," one plate "No soap," and one plate "Soap."
2. Without washing your hands, carefully press several surfaces of your hands on the agar plate marked "Control." Have your partner immediately put the cover back on the plate. After you touch the agar, do not touch anything with either hand.
3. Hold your right hand under running water for 2 min. Ask your partner to scrub all surfaces of your right hand with the scrub brush throughout these 2 min. Be sure that he or she scrubs under your fingernails. After scrubbing, your partner should turn off the water and open the plate marked "No soap." Touching only the agar, carefully press on the "No soap" plate with the same surfaces of your right hand that you used to press on the "Control" plate.

Elizabeth Rustad
Crane Junior High School
Yuma, Arizona

CHAPTER RESOURCES

Chapter Resource File

- Datasheet for Chapter Lab
- Lab Notes and Answers

Technology

Classroom Videos

- Lab Video

LabBook

- To Diet or Not to Diet

4. Repeat step 3, but use your left hand instead of your right. This time, ask your partner to scrub your left hand with liquid antibacterial soap and the scrub brush. Use the plate marked "Soap" instead of the plate marked "No soap."
5. Secure the lid of each plate to its bottom half with transparent tape. Place the plates upside down in the incubator. Incubate all three plates overnight at 37°C.
6. Remove the plates from the incubator, and turn them right side up. Check each plate for the presence of bacterial colonies, and count the number of colonies present on each splate. Record this information. **Caution:** Do not remove the lids on any of the plates.

Analyze the Results

1. **Examining Data** Compare the bacterial growth on the plates. Which plate contained the most growth? Which contained the least?

Draw Conclusions

2. **Drawing Conclusions** Does water alone effectively kill bacteria? Explain.

Applying Your Data

Repeat this experiment, but scrub with regular, not antibacterial, liquid soap. Describe how the results of the two experiments differ.

Analyze the Results

1. Students should observe that the control dish shows the most bacterial growth. Ideally, the dish inoculated after scrubbing hands with antibacterial soap should show the least bacterial growth.

Draw Conclusions

2. Because of chlorine and other agents used to purify drinking water, water alone kills many bacteria. Washing with antibacterial soap for a short time may be only slightly better than washing with water alone.

Applying Your Data

On average, hands must be scrubbed with antibacterial soap for a minimum of 4 minutes to see a significant reduction or absence of bacterial growth in nutrient agar. Ask students if they know how long doctors scrub their hands before surgery. (4–5 min)

CHAPTER RESOURCES

Workbooks

Labs You Can Eat
- Snack Attack GENERAL

Inquiry Labs
- Consumer Challenge GENERAL

Long-Term Projects & Research Ideas
- Breakfast, Lunch, and Dinner of Champions ADVANCED

Calculator-Based Labs
- Counting Calories ADVANCED

Chapter Review

Assignment Guide	
Section	**Questions**
1	1–3, 14–15, 17–18
2	4–7, 9, 11–13
3	8, 10, 16, 19–21

ANSWERS

Using Key Terms

1. nutrients
2. food pyramid
3. malnutrition
4. addiction
5. drug

Understanding Key Ideas

6. d
7. b
8. c
9. d
10. c
11. a

Chapter Review

USING KEY TERMS

Complete each of the following sentences by choosing the correct term from the word bank.

nutrients	Food Guide Pyramid
addiction	
malnutrition	drug

1. Carbohydrates, proteins, fats, vitamins, minerals, and water are the six categories of ___.
2. The ___ divides foods into six groups and gives a recommended number of servings for each group.
3. Both bulimia nervosa and anorexia nervosa cause ___.
4. A physical or psychological dependence on a drug can lead to ___.
5. A(n) ___ is any substance that causes a change in a person's physical or psychological state.

UNDERSTANDING KEY IDEAS

Multiple Choice

6. Which of the following statements about drugs is true?
   - **a.** A child cannot become addicted to drugs.
   - **b.** Smoking just one or two cigarettes is safe for anyone.
   - **c.** Alcohol is not a drug.
   - **d.** Withdrawal symptoms may be painful.
7. What does alcohol do to the central nervous system (CNS)?
   - **a.** It speeds the CNS up.
   - **b.** It slows the CNS down.
   - **c.** It keeps the CNS regulated.
   - **d.** It has no effect on the CNS.
8. To keep your teeth healthy,
   - **a.** brush your teeth as hard as you can.
   - **b.** use a toothbrush until it is worn out.
   - **c.** brush at least twice a day.
   - **d.** floss at least once a week.
9. According to the Food Guide Pyramid, what foods should you eat most?
   - **a.** meats
   - **b.** milk, yogurt, and cheese
   - **c.** fruits and vegetables
   - **d.** bread, cereal, rice, and pasta
10. Which of the following can help you deal with stress?
    - **a.** ignoring your homework
    - **b.** drinking a caffeinated drink
    - **c.** talking to a friend
    - **d.** watching television
11. Tobacco use increases the risk of
    - **a.** lung cancer.
    - **b.** car accidents.
    - **c.** liver damage.
    - **d.** depression.

Short Answer

12 Are all narcotics illegal? Explain.

13 What are three dangers of tobacco and alcohol use?

14 What are the three types of nutrients that provide energy in Calories, and what is the main function of each type in the body?

15 Name two conditions that can lead to malnutrition.

16 Explain why you should always wear safety equipment when you ride your bicycle.

CRITICAL THINKING

17 **Concept Mapping** Use the following terms to create a concept map: *carbohydrates, water, proteins, nutrients, fats, vitamins, minerals, saturated fats,* and *unsaturated fats.*

18 **Applying Concepts** You have recently become a vegetarian, and you worry that you are not getting enough protein. Name two foods that you could eat to get more protein.

19 **Analyzing Ideas** Your two-year-old cousin will be staying with your family. Name three things that you can do to make sure that the house is safe for a young child.

INTERPRETING GRAPHICS

Look at the photos below. The people in the photos are not practicing safe habits. List the unsafe habits shown in these photos. For each unsafe habit, tell what the corresponding safe habit is.

20

21

12. Sample answer: No, not all narcotics are illegal. Some narcotics are prescribed by doctors for the treatment of pain.

13. Sample answer: Alcohol use is a factor in car accidents, and it can cause liver damage. Tobacco use can damage the lungs.

14. Sample answer: Carbohydrates are the body's main source of energy. Fats are energy-storage nutrients. Proteins are used for building and repairing the body.

15. Sample answers: anorexia nervosa, bulimia nervosa, and obesity

16. Sample answer: Safety equipment, such as a helmet, can prevent a head injury if I fall off my bicycle or if I'm involved in an accident.

Critical Thinking

17. 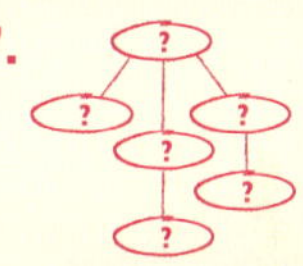

An answer to this exercise can be found at the end of this book.

18. Sample answers: nuts, beans, eggs, and dairy foods

19. Sample answer: Keep objects off the stairs, use a night light, and keep electrical cords out of walkways.

Interpreting Graphics

20. Sample answer: The girl is riding without her seat belt. She needs to fasten her seat belt.

21. Sample answer: A pot handle is hanging over the edge of the stove. The woman should use a pot holder to turn the handle of the pot so that it faces inward.

CHAPTER RESOURCES

Chapter Resource File

- Chapter Review GENERAL
- Chapter Test A GENERAL
- Chapter Test B ADVANCED
- Chapter Test C SPECIAL NEEDS
- Vocabulary Activity GENERAL

Workbooks

Study Guide

- Assessment resources are also available in Spanish.

Teacher's Note

To provide practice under more realistic testing conditions, give students 20 minutes to answer all of the questions in this Standardized Test Preparation.

MISCONCEPTION ALERT

Answers to the standardized test preparation can help you identify student misconceptions and misunderstandings.

READING

Passage 1

1. C
2. G
3. B

TEST DOCTOR

Question 3: The passage states that holes in the air sacks of the lungs do not heal. Students may infer that cigarette smoking does not cause all cases of emphysema and bronchitis, but cigarette smoking does cause 80% of cases. Students may think that chronic bronchitis will go away if someone stopped smoking, but the passage states that a chronic disease is a disease that does not go away.

Standardized Test Preparation

READING

Read each of the passages below. Then, answer the questions that follow each passage.

Passage 1 A chronic disease is a disease that, once developed, is always present and will not go away. Chronic bronchitis is a disease that causes the airways in the lungs to become swollen. This irritation causes a lot of mucus to form in the lungs. As a result, a person who has chronic bronchitis coughs a lot. Another chronic condition is emphysema. Emphysema destroys the tiny air sacs and the walls in the lungs. The holes in the air sacs cannot heal. Eventually, the lung tissue dies, and the lungs can no longer work. Cigarette smoking causes more than 80% of all cases of chronic bronchitis and emphysema.

1. In the passage, what does the word *chronic* mean?
A disappearing
B temporary
C always present
D mucus filled

2. According to the passage, what disease destroys the tiny air sacs and walls of the lungs?
F chronic bronchitis
G emphysema
H chronic cough
I cigarette smoking

3. Which of the following is a true statement according to the passage?
A Holes in the air sacs of lungs heal very quickly.
B Cigarette smoking causes more than 80% of all cases of chronic bronchitis and emphysema.
C Cigarette smoking does not cause chronic bronchitis or emphysema.
D Chronic bronchitis will go away after a person stops smoking cigarettes.

Passage 2 Each body reacts differently to alcohol. Several factors affect how a body reacts to alcohol. A person who has several drinks in a short time is likely to be affected more than a person who has a single drink in the same amount of time. Food in a drinker's stomach can also slow alcohol absorption into the blood. Finally, the way that women absorb and process alcohol differs from the way that men do. If a man and a woman drink the same amount of alcohol, the woman's blood alcohol content (BAC) will be higher than the man's. As BAC increases, mental and physical abilities decline. Muscle coordination, which is especially important for walking and driving, decreases. Vision becomes blurred. Speech and memory are impaired. A high BAC can cause a person to pass out or even die.

1. According to the passage, what does *BAC* stand for?
A blood alcohol content
B blood alcohol contaminant
C blurred alcohol capacity
D blood alcoholic coordination

2. According to the passage, which of the following factors can affect BAC?
F time of day
G food in the stomach
H age
I physical activity

3. Which of the following is a fact according to the passage?
A Alcohol does not affect mood or mental abilities.
B Men absorb alcohol in the same way that women do.
C Alcohol decreases muscle coordination.
D Everybody reacts to alcohol in the same way.

Passage 2

1. A
2. G
3. C

TEST DOCTOR

Question 2: Each body reacts differently to alcohol. However, time of day, age, and physical activity are not mentioned as factors which affect how a person reacts to alcohol. The factors mentioned in the passage include how many drinks a person has over a period of time, whether a person has food in his or her stomach, and gender.

INTERPRETING GRAPHICS

The figure below shows a sample prescription drug label. Use this figure to answer the questions that follow.

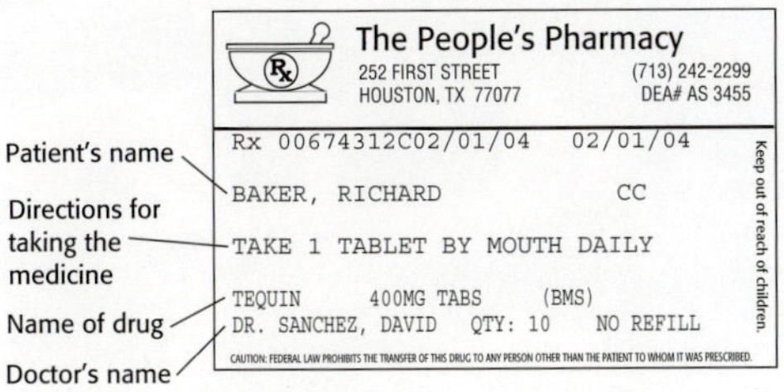

1. According to the label, what is the patient's name?

A Richard Baker
B Baker Richard
C David Sanchez
D James Beard

2. According to the label, how often should the medication be taken?

F once a day
G twice a day
H three times a day
I once a week

3. According to the label, how many refills remain on the prescription?

A 0
B 1
C 2
D 3

4. If this patient follows the directions exactly, how long will he need to take this medicine?

F 1 day
G 5 days
H 10 days
I There is not enough data to determine the answer.

MATH

Read each question below, and choose the best answer.

1. Which of the following ratios is equal to 2/4?

A 1/2
B 17/18
C 5/2
D 7/2

2. If 1 gal = 3.79 L, how many liters are in 3 gal?

F 3.79 L
G 7.58 L
H 11.37 L
I 15.16 L

3. Approximately how many liters are in 5 gal?

A 5 L
B 10 L
C 20 L
D 30 L

4. Ada has just built a car for a Pinewood Derby. She wants to find the average speed of her new car. During her first test run, she goes 5 mi/h. During her second run, she goes 4 mi/h, and in her third run, she goes 6 mi/h. What is her average speed?

F 4 mi/h
G 5 mi/h
H 6 mi/h
I 7 mi/h

5. Which of the following numbers is largest?

A 1×10^2
B 1×10^5
C 3×10^5
D 5×10^4

6. On Saturday, Mae won a goldfish at the school carnival. On the way home, Mae and her mother bought a fishbowl for $10.25, a container of fish food for $3.75, and a plastic coral for $8.15. How much money did Mae and her mother spend?

F $11.90
G $18.40
H $22.15
I $30.30

INTERPRETING GRAPHICS

1. A
2. F
3. A
4. H

TEST DOCTOR

Question 1: Baker Richard is the reversed form of the patient's name. It starts with the patient's last name. David Sanchez is the doctor who prescribed the drug, and James Beard is not a name that appears on the label.

MATH

1. A
2. H
3. C
4. G
5. C
6. H

TEST DOCTOR

Question 1: 2/4 can be simplified to 1/2, or one-half of the whole. 17/18 is almost equal to the whole and 5/2 and 7/2 are greater than the whole. Students struggling with fractions may benefit from working with visual aids, such as wooden blocks.

Question 5: 1×10^2 is the smallest number in the answer choices. Some students may base their answer on the power of ten alone and answer 1×10^5, not realizing that 3×10^5 is a larger number. Some students may answer 5×10^4 because 5 is larger than 1 and 3, which are used in the other answer choices. However, a higher power of ten is used in two of the other answer choices.

CHAPTER RESOURCES

Chapter Resource File

- Standardized Test Preparation GENERAL

State Resources

For specific resources for your state, visit **go.hrw.com** and type in the keyword **HSMSTR.**

Science in Action

Science, Technology, and Society

Teaching Strategy-ADVANCED

Explain that a protein consists of a long chain of amino acids. Assign each student one of the 20 common amino acids to investigate. Tell students to find out what foods the amino acid can be found in and what function it serves in the body. Then, on a large index card, have students write information about the amino acid and make illustrations of food sources for the amino acid. Use the index cards to make a poster for the class.

Scientific Discoveries

Background

The female athlete triad is most likely to begin with an attempt to lose weight, which leads to disordered eating. Disordered eating ranges from moderate restriction of food intake to eating disorders, such as anorexia nervosa and bulimia nervosa. Signs that a girl may have disordered eating include a preoccupation with food and weight, a constant expression of being too fat, eating alone, use of laxatives, and trips to the bathroom immediately after eating.

Science in Action

Bones can become severely weakened by the female athlete triad.

Science, Technology, and Society

Meatless Munching

Recent studies suggest that a vegetarian diet may reduce the risk of heart disease, adult-onset diabetes, and some forms of cancer. However, a vegetarian diet takes careful planning. Vegetarians must ensure that they get the proper balance of protein and vitamins in their diet. New foods that can help vegetarians remain healthy are being developed constantly. Meat substitutes are now made from soybeans, textured vegetable protein, and tofu. One new food, which is shown above, is made of a fungus that is a relative of mushrooms and truffles.

Social Studies ACTIVITY

WRITING SKILL Research a culture that has a mostly vegetarian diet, such as Hindu or Buddhist. What kinds of food do the people eat? Why don't they eat animals? Write a short report on your findings.

Scientific Discoveries

Female Athlete Triad

Getting enough exercise is an important part of staying healthy. But in 1992, doctors learned that too much exercise can be harmful for women. When a girl or woman exercises too much, three things can happen. She may lose too much weight. She may stop having her period. And her bones may become very weak. These three symptoms form the female athlete triad. To prevent this condition, female athletes need to take in enough Calories. Women who exercise heavily and try to lose weight may have a reduction in estrogen. Estrogen is the hormone that helps regulate the menstrual cycle. Low levels of estrogen and inadequate nutrition can cause bones to become weak and brittle. The photo above shows bone that has been weakened greatly.

Math ACTIVITY

Some scientists recommend that teenagers get 1,200 to 1,500 mg of calcium every day. A cup of milk has 300 mg of calcium, and a serving of yogurt has 400 mg of calcium. Calculate two combinations of milk and yogurt that would give you the recommended 1,500 mg of calcium.

Answer to Social Studies Activity

Answers may vary. Students should identify some foods eaten by the people in the cultures that they research. Students should also identify reasons such as religious beliefs, cultural traditions, and the availability of meat as reasons why people in these cultures have vegetarian diets.

Answer to Math Activity

Sample answer: 1 serving of yogurt and 4 cups of milk (4×300 mg = 1200 mg calcium; 1200 mg + 400 mg = 1600 mg calcium), or 3 cups of milk and 2 servings of yogurt (3×300 mg = 900 mg calcium; 2×400 mg = 800 mg calcium; 900 mg + 800 mg = 1700 mg calcium)

Careers

Russell Selger

Guidance Counselor Guidance counselors help students think about their future by helping them discover their interests. After focusing their interests, a guidance counselor helps students plan a good academic schedule. A guidance counselor might talk to you about taking an art or computer science class that may help you discover a hidden talent. Many skills are vital to being a good guidance counselor. The job requires empathy, which is the ability to understand and sympathize with another person's feelings. Counselors also need patience, good listening skills, and a love of helping young people. Russell Selger, a guidance counselor at Timberlane Middle School, has a great respect for middle school students. "The kids are just alive. They want to learn. There's something about the spark that they have, and it's so much fun to guide them through all of this stuff," he explains.

Language Arts ACTIVITY

WRITING SKILL Visit the guidance counselor's office at your school. What services does your guidance counselor offer? Conduct an interview with a guidance counselor. Ask why he or she became a counselor. Write an article for the school paper about your findings.

go.hrw.com

To learn more about these Science in Action topics, visit go.hrw.com and type in the keyword HL5BD7F.

Current Science

Check out Current Science® articles related to this chapter by visiting go.hrw.com. Just type in the keyword HL5CS28.

Careers

Discussion — GENERAL

Ask students the following questions: "What do you think guidance counselors talk to students about?" (Many students will likely say grades and future career options. Help students understand that guidance counselors can also help them with other problems, such as stress management, conflict management, and mediation.) "What skills does a good guidance counselor need?" (Sample answers: good listening skills, patience, and empathy)

Answer to Language Arts Activity

Answers may vary.

Inquiry Lab

Muscles at Work

Teacher's Notes

Time Required
One 45-minute class period

Lab Ratings
EASY → HARD

Teacher Prep 1 flask
Student Set-Up 1 flask
Concept Level 2 flasks
Clean Up 1 flask

Safety Caution
A digital thermometer that measures temperature from the ear is recommended.

Because of the vigorous nature of the exercise, you may want to ask for volunteers to do the exercising. Also, you should be aware of any health concerns your students have.

Using Scientific Methods

Inquiry Lab

Muscles at Work

Have you ever exercised outside on a cold fall day wearing only a thin warm-up suit or shorts? How did you stay warm? The answer is that your muscle cells contracted, and when contraction takes place, some energy is used to do work, and the rest is converted to thermal energy. This process helps your body maintain a constant temperature in cold conditions. In this activity, you will learn how the release of energy can cause a change in your body temperature.

MATERIALS

- clock (or watch) with a second hand
- thermometer, small, hand held
- other materials as approved by your teacher

Ask a Question

1. Write a question that you can test about how activity affects body temperature.

Form a Hypothesis

2. Form a group of four students. In your group, discuss several exercises that can produce a change in body temperature. Write a hypothesis that could answer the question you asked.

Test the Hypothesis

3. Develop an experimental procedure that includes the steps necessary to test your hypothesis. Be sure to get your teacher's approval before you begin.
4. Assign tasks to individuals in the group, such as note taking, data recording, and timing. What observations and data will you be recording? Design your data tables accordingly.
5. Perform your experiment as planned by your group. Be sure to record all observations in your data tables.

Analyze the Results

1. How did you determine if muscle contractions cause the release of thermal energy? Was your hypothesis supported by your data? Explain your results in a written report. Describe how you could improve your experimental method.

Applying Your Data

Why do humans shiver in the cold? Do all animals shiver? Find out why shivering is one of the first signs that your body is becoming too cold.

Analyze the Results

1. All answers will depend on the students' observations and their own hypotheses.

Applying Your Data

Sample answer: In a process known as *shivering thermogenesis,* muscle tone is gradually increased. Shivering increases the workload of the muscles and elevates oxygen and energy consumption. The heat that is produced warms the deep vessels. Shivering can elevate body temperature effectively. It can increase the rate of heat generation by as much as 400%. Endothermic animals have the capacity to shiver. Shivering is an automatic response of the body to cold.

CHAPTER RESOURCES

Chapter Resource File

- Datasheet for LabBook
- Lab Notes and Answers

CLASSROOM TESTED & APPROVED

Kathy LaRoe
East Valley Middle School
East Helena, Montana

Model-Making Lab

Build a Lung

When you breathe, you actually pull air into your lungs because your diaphragm muscle causes your chest to expand. You can see this is true by placing your hands on your ribs and inhaling slowly. Did you feel your chest expand?

In this activity, you will build a model of a lung by using some common materials. You will see how the diaphragm muscle works to inflate your lungs. Refer to the diagrams at right as you construct your model.

MATERIALS

- bag, trash, small plastic
- balloon, small
- bottle, top half, 2 L
- clay, golf-ball-sized piece
- rubber bands (2)
- ruler, metric
- straw, plastic
- tape, transparent

Procedure

1. Attach the balloon to the end of the straw with a rubber band. Make a hole through the clay, and insert the other end of the straw through the hole. Be sure at least 8 cm of the straw extends beyond the clay. Squeeze the ball of clay gently to seal the clay around the straw.
2. Insert the balloon end of the straw into the neck of the bottle. Use the ball of clay to seal the straw and balloon into the bottle.
3. Turn the bottle gently on its side. Place the trash bag over the cut end of the bottle. Expand a rubber band around the bottom of the bottle to secure the bag. You may wish to reinforce the seal with tape. Before the plastic is completely sealed, gather the excess material of the bag into your hand, and press toward the inside of the bottle slightly. (You may need to tie a knot about halfway up from the bottom of the bag to take up excess material.) Use tape to finish sealing the bag to the bottle with the bag in this position. The excess air will be pushed out of the bottle.

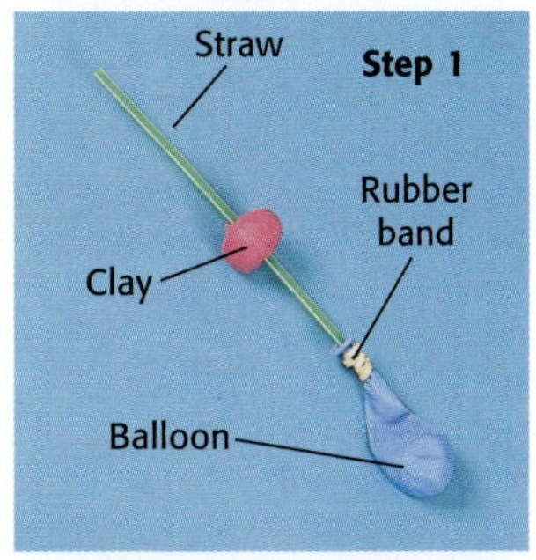

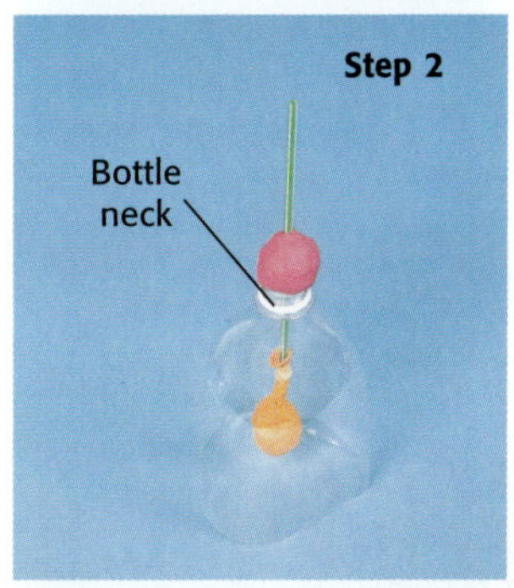

Analyze the Results

1. What can you do with your model to make the "lung" inflate?
2. What do the balloon, the plastic wrap, and the straw represent in your model?
3. Using your model, demonstrate to the class how air enters the lung and how air exits the lung.

Applying Your Data

Do some research to find out what an "iron lung" is and why it was used in the past. Research and write a report about what is used today to help people who have difficulty breathing.

LabBook

Model-Making Lab

Build a Lung

Teacher's Notes

Time Required

One 45-minute class period

Lab Ratings

EASY → HARD

Teacher Prep 2
Student Set-Up 2
Concept Level 1
Clean Up 2

MATERIALS

You may want to build a model first to use as a reference for students. If so, you may want to substitute a bag smaller than the one that students use to model the diaphragm.

Analyze the Results

1. The balloon will inflate when the plastic bag is pulled down.
2. The balloon represents a lung, the plastic wrap represents a diaphragm, and the straw represents a trachea. (Students may also note that the bottle represents the part of the body in which the lungs are located.)
3. Air enters the lungs when the diaphragm moves down and creates more space inside the chest cavity. Air is forced out of the lungs when the diaphragm moves up. Students should demonstrate this process by moving the plastic bag up and down.

CHAPTER RESOURCES

Chapter Resource File

- Datasheet for LabBook
- Lab Notes and Answers

Applying Your Data

From the late 1920s to the 1950s, iron lungs were used to treat respiratory paralysis due to poliomyelitis. The patient was encased within an airtight chamber from the neck down. A large set of leather bellows mounted in a separate pumping unit expanded, causing pressure changes inside the chamber. This, in turn, caused the chest of the patient to expand, drawing fresh air into the lungs through the mouth. Now, several models of portable ventilators allow patients much more freedom to move about.

Yvonne Brannum
Hine Junior High School
Washington, D.C.

LabBook

Skills Practice Lab

Enzymes in Action

Teacher's Notes

Time Required

One 45-minute class period

Lab Ratings

EASY → HARD

Teacher Prep 2

Student Set-Up 2

Concept Level 3

Clean Up 2

MATERIALS

The materials listed on the student page are enough for 1 student or a group of 2–4 students. If you do not have enough mortar and pestles or small plates, you may have students use any small container to mash the liver with a fork. Beef liver is readily obtained at the grocery store.

Safety Caution

Remind students to review all safety cautions and icons before beginning this lab activity. Use a dilute solution of hydrogen peroxide. Hydrogen peroxide can be harmful to skin and clothing. Caution students to be careful not to spill or splatter the solution while pouring. If hydrogen peroxide comes into contact with skin, wash the skin immediately with plenty of running water.

Skills Practice Lab

Enzymes in Action

You know how important enzymes are in the process of digestion. This lab will help you see enzymes at work. Hydrogen peroxide is continuously produced by your cells. If it is not quickly broken down, hydrogen peroxide will kill your cells. Luckily, your cells contain an enzyme that converts hydrogen peroxide into two nonpoisonous substances. This enzyme is also present in the cells of beef liver. In this lab, you will observe the action of this enzyme on hydrogen peroxide.

MATERIALS

- beef liver, 1 cm cubes (3)
- gloves, protective
- graduated cylinder, 10 mL
- hydrogen peroxide, fresh (4 mL)
- mortar and pestle (or fork and watch glass)
- plate, small
- spatula
- test tube (3)
- test-tube rack
- tweezers
- water

SAFETY

Procedure

1. Draw a data table similar to the one below. Be sure to leave enough space to write your observations.

Data Table

Size and condition of liver	Experimental liquid	Observations
1 cm cube beef liver	2 mL water	
1 cm cube beef liver	2 mL hydrogen peroxide	DO NOT WRITE IN BOOK
1 cm cube beef liver (mashed)	2 mL hydrogen peroxide	

James Chin
Frank A. Day Middle School
Newtonville, Massachusetts

CHAPTER RESOURCES

Chapter Resource File

- Datasheet for LabBook
- Lab Notes and Answers

2. Get three equal-sized pieces of beef liver from your teacher, and use your forceps to place them on your plate.
3. Pour 2 mL of water into a test tube labeled "Water and liver."
4. Using the tweezers, carefully place one piece of liver in the test tube. Record your observations in your data table.
5. Pour 2 mL of hydrogen peroxide into a second test tube labeled "Liver and hydrogen peroxide."
Caution: Do not splash hydrogen peroxide on your skin. If you do get hydrogen peroxide on your skin, rinse the affected area with running water immediately, and tell your teacher.
6. Using the tweezers, carefully place one piece of liver in the test tube. Record your observations of the second test tube in your data table.
7. Pour another 2 mL of hydrogen peroxide into a third test tube labeled "Ground liver and hydrogen peroxide."
8. Using a mortar and pestle (or fork and watch glass), carefully grind the third piece of liver.
9. Using the spatula, scrape the ground liver into the third test tube. Record your observations of the third test tube in your data table.

Analyze the Results

1. What was the purpose of putting the first piece of liver in water? Why was this a necessary step?
2. Describe the difference you observed between the liver and the ground liver when each was placed in the hydrogen peroxide. How can you account for this difference?

Applying Your Data

Do plant cells contain enzymes that break down hydrogen peroxide? Try this experiment using potato cubes instead of liver to find out.

Analyze the Results

1. Sample answer: The piece of liver in water is the control. A control is necessary so that differences in reactions to other substances can be observed.
2. Students should observe that the cube of liver produced a quick, foaming reaction to the hydrogen peroxide as the solution came into contact with the cells and destroyed them. The mashed liver should also produce a very quick reaction; however, students should observe that the effervescence subsides faster with the mashed liver than with the cube. The reason is that the mashed liver has a greater surface area than the cube, exposing more liver enzymes to the hydrogen peroxide. There should be no reaction in water.

Applying Your Data

Students will discover that plant cells do contain enzymes that will break down hydrogen peroxide. Have them experiment with different types of plants and vegetables so they can observe the different rates of reaction.

LabBook

Skills Practice Lab

My, How You've Grown!

Teacher's Notes

Time Required

One 45-minute class period

Lab Ratings

EASY → HARD

Teacher Prep: 1 of 4

Student Set-Up: 1 of 4

Concept Level: 2 of 4

Clean Up: 1 of 4

Procedure

2. Students' graphs should look like those below.

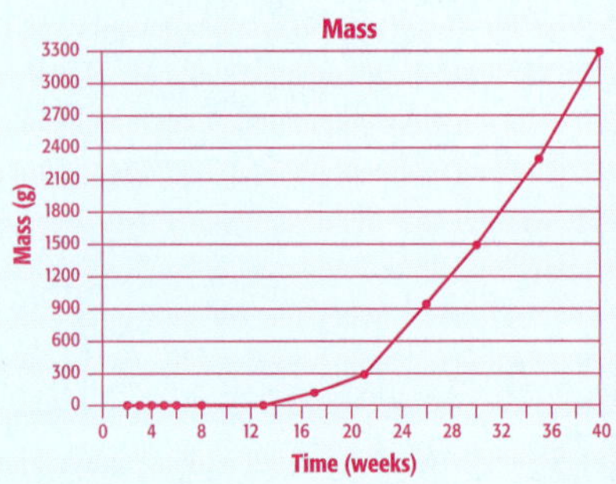

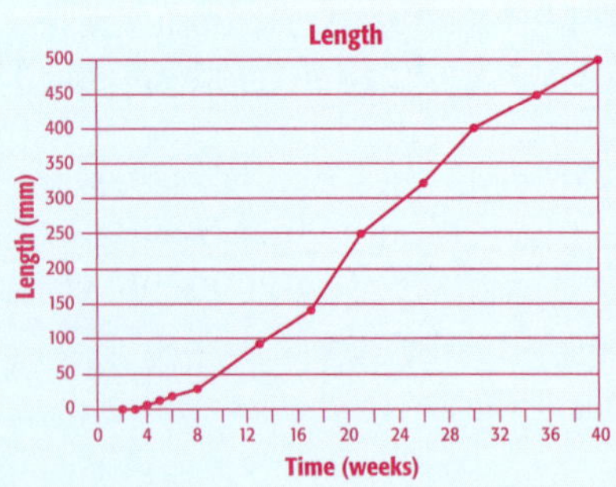

Skills Practice Lab

My, How You've Grown!

In humans, the process of development that takes place between fertilization and birth lasts about 266 days. In 4 weeks, the new individual grows from a single fertilized cell to an embryo whose heart is beating and pumping blood. All of the organ systems and body parts are completely formed by the end of the seventh month. During the last 2 months before birth, the baby grows, and its organ systems mature. At birth, the average mass of a baby is about 33,000 times as much as that of an embryo at 2 weeks of development! In this activity, you will discover just how fast a fetus grows.

MATERIALS

- paper, graph
- pencils, colored

Procedure

1. Using graph paper, make two graphs—one entitled "Length" and one entitled "Mass." On the length graph, use intervals of 25 mm on the *y*-axis. Extend the *y*-axis to 500 mm. On the mass graph, use intervals of 100 g on the *y*-axis. Extend this *y*-axis to 3,300 g. Use 2-week intervals for time on the *x*-axes for both graphs. Both *x*-axes should extend to 40 weeks.
2. Examine the data table at right. Plot the data in the table on your graphs. Use a colored pencil to draw the curved line that joins the points on each graph.

Increase of Mass and Length of Average Human Fetus

Time (weeks)	Mass (g)	Length (mm)
2	0.1	1.5
3	0.3	2.3
4	0.5	5.0
5	0.6	10.0
6	0.8	15.0
8	1.0	30.0
13	15.0	90.0
17	115.0	140.0
21	300.0	250.0
26	950.0	320.0
30	1,500.0	400.0
35	2,300.0	450.0
40	3,300.0	500.0

Analyze the Results

1. Describe the change in mass of a developing fetus. How can you explain this change?
2. Describe the change in length of a developing fetus. How does the change in mass compare to the change in length?

Applying Your Data

Using the information in your graphs, estimate how tall a child would be at age 3 if he or she continued to grow at the same average rate that a fetus grows.

LabBook

Analyze the Results

1. The change in mass of a developing fetus is steadily increasing, approximately tripling each month of the first and second trimesters. This period is one of rapid cell division.
2. Fetal length steadily increases, doubling and even tripling each month in the first two trimesters. In the third trimester, the rate of lengthening slows. Mass continues to increase, even in the last trimester.

Applying Your Data

The child would be 2.45 m (8.04 ft) tall!

CHAPTER RESOURCES

Chapter Resource File

- Datasheet for LabBook
- Lab Notes and Answers

Randy Christian
Stovall Junior High School
Houston, Texas

Model-Making Lab

Antibodies to the Rescue

Some cells of the immune system, called *B cells,* make antibodies that attack and kill invading viruses and microorganisms. These antibodies help make you immune to disease. Have you ever had chickenpox? If you have, your body has built up antibodies that can recognize that particular virus. Antibodies will attach themselves to the virus, tagging it for destruction. If you are exposed to the same disease again, the antibodies remember that virus. They will attack the virus even quicker and in greater number than they did the first time. This is the reason that you will probably never have chickenpox more than once.

In this activity, you will construct simple models of viruses and their antibodies. You will see how antibodies are specific for a particular virus.

MATERIALS

- craft materials, such as buttons, fabric scraps, pipe cleaners, and recycled materials
- paper, colored
- scissors
- tape (or glue)

Procedure

1. Draw the virus patterns shown on this page on a separate piece of paper, or design your own virus models from the craft supplies. Remember to design different receptors on each of your virus models.
2. Write a few sentences describing how your viruses are different.
3. Cut out the viruses, and attach them to a piece of colored paper with tape or glue.

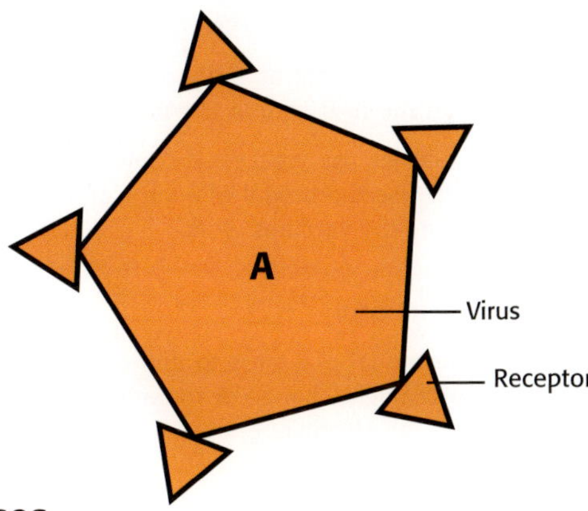

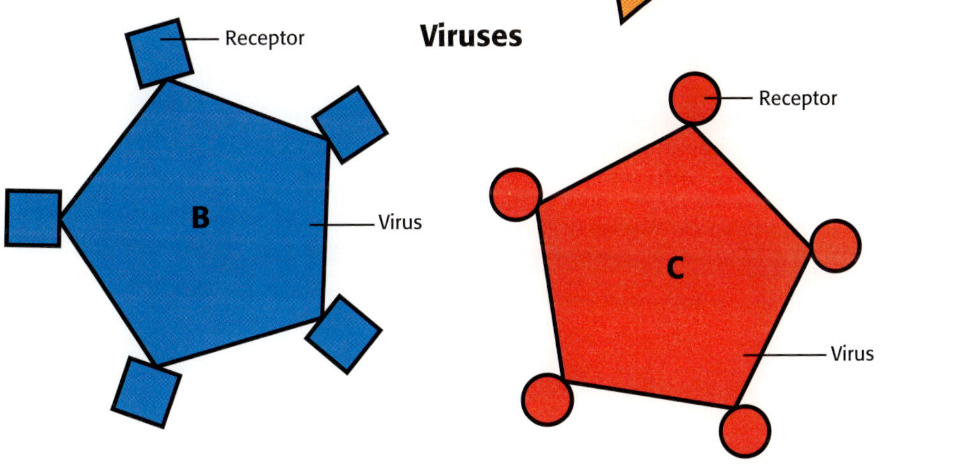

Model-Making Lab

Antibodies to the Rescue

Teacher's Notes

Time Required

One 45-minute class period

Lab Ratings

EASY → HARD

Teacher Prep
Student Set-Up
Concept Level
Clean Up

MATERIALS

You will be able to expand this activity from very simple to a grand art project! Have students bring craft supplies from home. You may want to include poster paper in the materials list.

Preparation Notes

Encourage students to be imaginative in this exercise. Have them come up with interesting shapes for the viruses and the antibodies that fit them. Remind students that the main lesson of this exercise is that the antibodies can fit only a specific pathogen and in only one specific way, just as a key fits only one lock.

CHAPTER RESOURCES

Chapter Resource File

- Datasheet for LabBook
- Lab Notes and Answers

Martha Kisiah
Fairview Middle School
Tallahassee, Florida

Analyze the Results

1. Sample answer: Antibodies recognize and bind to specific pathogens because antibodies are shaped to match the specific three-dimensional shape of the antigen.
2. Sample answer: Antibodies bind to specific pathogens and either inactivate the pathogen or trigger its destruction by macrophages.
3. Sample answer: Vaccines produce immunity because they contain antigens that stimulate an immune response in which memory cells are produced. Students can use their models to demonstrate or give examples.

Draw Conclusions

4. Sample answer: The antibodies that are specific to a pathogen will be ineffective when the pathogen changes its receptors. Students can show this mutation by slightly changing their model so that the virus and the antibody no longer fit as a lock and key do.

4. Select the antibodies drawn below, or design your own antibodies that will exactly fit on the receptors on your virus models. Draw or create each antibody enough times to attach one to each receptor site on the virus.

Antibodies

5. Cut out the antibodies you have drawn. Arrange the antibodies so that they bind to the virus at the appropriate receptor. Attach them to the virus with tape or glue.

Analyze the Results

1. Explain how an antibody "recognizes" a particular virus.
2. After the attachment of antibodies to the receptors, what would be the next step in the immune response?
3. Many vaccines use weakened copies of the virus to protect the body. Use the model of a virus and its specific antibody to explain how vaccines work.

Draw Conclusions

4. Use your model of a virus to demonstrate to the class how a receptor might change or mutate so that a vaccine would no longer be effective.

Applying Your Data

Research in the library or on the Internet to find information about the discovery of the Salk vaccine for polio. Include information on how polio affects people today.

Research in the library or on the Internet to find information and write a report about filoviruses. What do they look like? What diseases do they cause? Why are they especially dangerous? Is there an effective vaccine against any filovirus? Explain.

Applying Your Data

Dr. Jonas Edward Salk (1914–1995) was trying to develop an anti-influenza vaccine. This work led him and his colleagues to develop an inactivated vaccine against polio in 1952. Wide-scale testing in 1954 was successful, and the vaccine was distributed nationally. The vaccine greatly reduced the incidence of the polio. Partly because of the Salk vaccine, the number of cases of polio have decreased year after year—the Western Hemisphere was declared polio-free in 1994, and only 4,000 polio cases were reported worldwide in 1996. The World Health Organization is planning to end routine polio vaccination around the year 2005.

Filoviruses belong to the family Filoviridae, one of several groups of viruses that can cause hemorrhagic fever, such as Ebola, in animals and humans. When magnified many thousands of times by an electron microscope, filoviruses resemble long filaments or threads. Because filoviruses can be extremely hazardous, laboratory studies of these viruses must be conducted in special maximum-containment facilities. The reservoir and natural history of filoviruses remain unknown. Filoviruses have the highest fatality rates (as high as 90% for epidemics of hemorrhagic fever caused by Ebola-Zaire virus). No vaccine exists to protect against filovirus infection, and no specific treatment is available for diseases caused by these viruses.

Skills Practice Lab

To Diet or Not to Diet

There are six main classes of foods that we need in order to keep our bodies functioning properly: water, vitamins, minerals, carbohydrates, fats, and proteins. In this activity you will investigate the importance of a well-balanced diet in maintaining a healthy body. Then you will create a poster or picture that illustrates the importance of one of the three energy-producing nutrients—carbohydrates, fats, and proteins.

MATERIALS

- crayons (or markers), assorted colors
- diet books
- menus, fast-food (optional)
- nutrition reference books
- paper, white unlined

Procedure

1. Draw a table like the one below. Research in the library, on nutrition labels, in nutrition or diet books, or on the Internet to find the information you need to fill out the chart.

Nutrition Data Table

	Fats	Carbohydrates	Proteins
Found in which foods			
Functions in the body	DO NOT WRITE IN BOOK		
Consequences of deficiency			

2. Choose one of the foods you have learned about in your research, and create a poster or picture that describes its importance in a well-balanced diet.

Analyze the Results

1. Based on what you have learned in this lab, how might you change your eating habits to have a well-balanced diet? Does the nutritional value of foods concern you? Why or why not? Write down your answers, and explain your reasoning.

Communicating Your Data

Write a paragraph explaining why water is a nutrient. Analyze a typical fast-food meal, and determine its overall nutritional value.

LabBook

Skills Practice Lab

To Diet or Not to Diet

Teacher's Notes

Time Required

Two 45-minute class periods

Lab Ratings

EASY → HARD

Teacher Prep: 2
Student Set-Up: 1
Concept Level: 2
Clean Up: 2

MATERIALS

A copy of a chart that fast-food restaurants post in their dining rooms outlining the nutritional information of the food they serve would be very helpful in analyzing a fast-food meal.

Analyze the Results

1. Answers may vary. Students may say they might try to eat fewer fatty foods. The nutritional value of foods is a subject that may be new to many students.

Communicating Your Data

Sample answer: Without water, most of the reactions that maintain life could not take place. Water carries the other essential nutrients to all parts of the body and is the medium in which waste products are dissolved and carried away from body tissues.

Answers will vary depending on the type of meal a student chooses. Fast-food restaurants usually post Nutrition Facts in public view. Have students obtain this information and evaluate a fast-food meal based on those Nutrition Facts and daily recommended amounts. Examining the Nutrition Facts labels on food packaging is helpful before beginning this activity.

CHAPTER RESOURCES

Chapter Resource File

- Datasheet for LabBook
- Lab Notes and Answers

Classroom Tested & Approved

Ivora Washington
Hyattsville Middle School
Hyattsville, Maryland

Contents

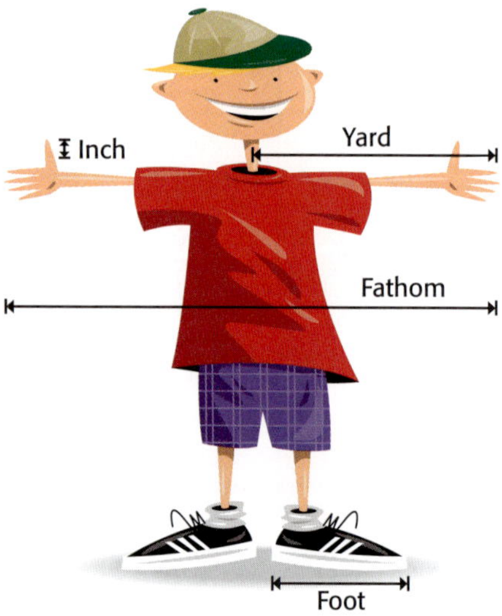

✓Reading Check Answers

Chapter 1 Body Organization and Structure

Section 1

Page 5: The stomach works with other organs, such as the small and large intestines, to digest food.

Page 6: Sample answer: The cardiovascular system includes the heart and blood vessels. These organs are also part of the circulatory system, which includes blood. Together, these systems deliver the materials cells need to survive.

Section 2

Page 9: Sample answer: As people grow, most of the cartilage that they start out with is replaced with bone.

Page 10: Sample answer: Joints are held together by ligaments. Cartilage cushions the area in a joint where bones meet.

Section 3

Page 13: Sample answer: One muscle, the flexor, bends part of the body. Another muscle, the extensor, straightens part of the body.

Page 15: Sample answer: Anabolic steroids can damage the heart, liver, and kidneys. They can also cause high blood pressure. Anabolic steroids can cause bones to stop growing.

Section 4

Page 17: The dermis is the layer of skin that lies beneath the epidermis. It is composed of a protein called *collagen,* while the epidermis contains keratin.

Page 18: Sample answer: A nail grows from living cells in the nail root at the base of the nail. As new cells form, the nail grows longer.

Chapter 2 Circulation and Respiration

Section 1

Page 30: The four main parts of the cardiovascular system are the heart and the arteries, capillaries, and veins.

Page 32: Arteries have thick, stretchy walls and carry blood away from the heart. Capillaries are tiny blood vessels that allow the exchange of oxygen, carbon dioxide, and nutrients between cells and blood. Veins are blood vessels that carry blood back to the heart.

Page 34: Atherosclerosis is dangerous because it can cause the buildup of material inside an artery. When the artery becomes blocked, blood can't flow and can't reach the cells. In some cases, a person can have a heart attack from a blocked artery.

Section 2

Page 36: plasma, red blood cells, white blood cells, and platelets

Page 37: White blood cells identify and attack pathogens that may make you sick.

Page 38: Systolic pressure is the pressure inside arteries when the ventricles contract. Diastolic pressure is the pressure inside the arteries when the ventricles are relaxed.

Page 39: The red blood cells of a person who has type O blood have no A or B antigens. The A or B antibodies in another person's blood will not react to the type O cells. It is safe for anyone to receive type O blood.

Section 3

Page 40: The lymphatic system is a secondary circulatory system in the body. The lymphatic system collects fluid and particles from between the cells and returns them to the cardiovascular system.

Page 42: The white pulp of the spleen is part of the lymphatic system. It helps fight infections by storing and producing lymphocytes. The red pulp of the spleen removes unwanted material, such as defective red blood cells, from the circulatory system.

Section 4

Page 45: nose, pharynx, larynx, trachea, bronchi, bronchioles, alveoli

Page 46: Cellular respiration is the process inside a cell in which oxygen is used to release energy stored in molecules of glucose. During the process, carbon dioxide (CO_2) and water are released.

Chapter 3 The Digestive and Urinary Systems

Section 1

Page 59: Enzymes cut proteins into amino acids that the body can use.

Page 61: Chyme is a soupy mixture of partially digested food in the stomach.

Page 63: Bile breaks large fat droplets into very small droplets. This process allows more fat molecules to be exposed to digestive enzymes.

Page 64: Fiber keeps the stool soft and keeps material moving through the large intestine.

Section 2

Page 67: Nephrons are microscopic filters inside the kidneys.

Page 68: Diuretics are chemicals that cause the kidneys to make more urine.

Chapter 4 Communication and Control

Section 1

Page 80: The CNS is the brain and the spinal cord. The PNS is all of the parts of the nervous system except the brain and the spinal cord.

Page 81: A neuron is a cell that has a cell body and a nucleus. A neuron also has dendrites that receive signals from other neurons and axons that send signals to other neurons.

Page 82: A nerve is a collection of nerve fibers, or axons, bundled together with blood vessels through which impulses travel between the central nervous system and other parts of the body.

Page 83: The PNS connects your CNS to the rest of your body, controls voluntary movements, and keeps your body's functions in balance.

Page 84: A voluntary action is an action over which you have conscious control. Voluntary activities include throwing a ball, playing a video game, talking to your friends, taking a bite of food, and raising your hand to answer a question in class. An involuntary action is an action that happens automatically. It is an action or process over which you do not have conscious control.

Page 85: The medulla is important because it controls your heart rate, blood pressure, and ordinary breathing.

Page 86: When someone touches your skin, an impulse that travels along a sensory neuron to your spinal cord and then to your brain is created. The response travels back from your brain to your spinal cord and then along a motor neuron to a muscle.

Section 2

Page 88: Skin can detect pressure, temperature, pain, and vibration.

Page 89: Reflexes are important because they can protect you from injury.

Page 90: Light strikes cells on the retina and triggers impulses in those cells. The impulses are carried to the brain, which interprets the impulses as images that you "see."

Page 91: In bright light, your iris contracts and reduces the amount of light entering the eye.

Page 92: Neurons in the cochlea convert waves into electrical impulses that the brain interprets as sound.

Section 3

Page 95: Sample answer: The thyroid gland increases the rate at which the body uses energy. The thymus gland regulates the immune system, which helps your body fight disease.

Page 96: Insulin helps regulate the amount of glucose in the blood.

Chapter 5 Reproduction and Development

Section 1

Page 109: Sexual reproduction is reproduction in which the sex cells (egg and sperm) of two parents unite to form a new individual.

Page 110: External fertilization happens when the sex cells unite outside of the female's body. Internal fertilization happens when the sex cells unite inside the female's body.

Page 111: All mammals reproduce sexually and nurture their young with milk.

Section 2

Page 112: testes, epididymis, vas deferens, urethra, penis

Page 114: Twins happen about 30 times in every 1,000 births.

Section 3

Page 116: Fertilization happens when the nucleus of a sperm unites with the nucleus of an egg. Implantation happens after the fertilized egg travels down the fallopian tube to the uterus and embeds itself in the wall of the uterus.

Page 117: The placenta is important because it provides the embryo with oxygen and nutrients from the mother's blood. Wastes from the embryo also travel to the placenta, where they are carried to the mother so that she can excrete them.

Page 118: The embryo is now called a *fetus.* The fetus's face begins to look more human, and the fetus can swallow, grows rapidly (triples in size), and begins to make movements that the mother can feel.

Page 120: A person's reproductive system becomes mature.

Chapter 6 Body Defenses and Disease

Section 1

Page 135: Cooking kills dangerous bacteria or parasites living in meat, fish, and eggs.

Page 137: Frank's doctor did not prescribe antibiotics because Frank had a cold. Colds are caused by viruses. Antibiotics can't stop viruses.

Section 2

Page 139: Macrophages engulf, or eat, any microorganisms or viruses that enter your body.

Page 140: If a virus particle enters the body, it may pass into body cells and begin to replicate. Or it may be engulfed and broken up by macrophages.

Page 143: rheumatoid arthritis, diabetes, multiple sclerosis, and lupus

Page 144: HIV causes AIDS.

Appendix

Chapter 7 Staying Healthy

Section 1

Page 157: An incomplete protein does not contain all of the essential amino acids.

Page 159: Sample answer: a peanut butter sandwich, a glass of milk, and fresh fruit and vegetable slices

Page 160: One serving of chicken noodle soup provides more than 10% of the daily recommended allowance of vitamin A and sodium.

Section 2

Page 163: Over-the-counter drugs can be bought without a prescription. Prescription drugs can be bought only with a prescription from a doctor or other medical professional.

Page 165: First-time use of cocaine can cause a heart attack or can cause a person to become addicted.

Page 166: Drug use is the proper use of a legal drug. Drug abuse is either the use of an illegal drug or the improper use of a legal drug.

Section 3

Page 169: Aerobic exercise strengthens the heart, lungs, and bones and reduces stress. Regular exercise also burns Calories and can give you more energy.

Page 171: Sample answers: Never hike or camp alone, dress for the weather, learn how to swim, wear a life jacket, and never drink unpurified water.

Page 173: CPR is a way to revive someone whose heart has stopped beating. CPR classes are available in many places in the community.

Study Skills

FoldNote Instructions

Have you ever tried to study for a test or quiz but didn't know where to start? Or have you read a chapter and found that you can remember only a few ideas? Well, FoldNotes are a fun and exciting way to help you learn and remember the ideas you encounter as you learn science!

FoldNotes are tools that you can use to organize concepts. By focusing on a few main concepts, FoldNotes help you learn and remember how the concepts fit together. They can help you see the "big picture." Below you will find instructions for building 10 different FoldNotes.

Pyramid

1. Place a sheet of paper in front of you. Fold the lower left-hand corner of the paper diagonally to the opposite edge of the paper.
2. Cut off the tab of paper created by the fold (at the top).
3. Open the paper so that it is a square. Fold the lower right-hand corner of the paper diagonally to the opposite corner to form a triangle.
4. Open the paper. The creases of the two folds will have created an X.
5. Using scissors, cut along one of the creases. Start from any corner, and stop at the center point to create two flaps. Use tape or glue to attach one of the flaps on top of the other flap.

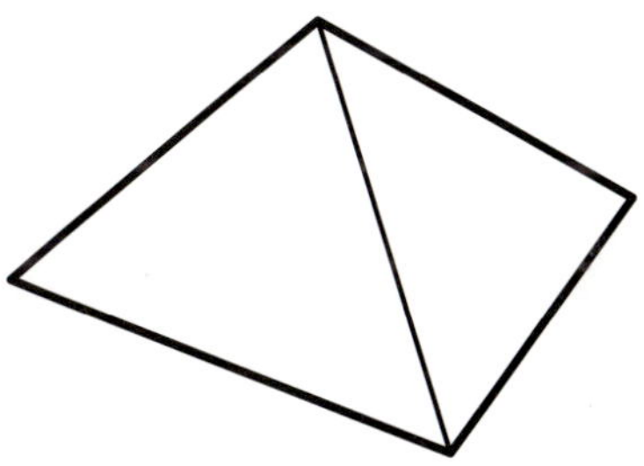

Double Door

1. Fold a sheet of paper in half from the top to the bottom. Then, unfold the paper.
2. Fold the top and bottom edges of the paper to the crease.

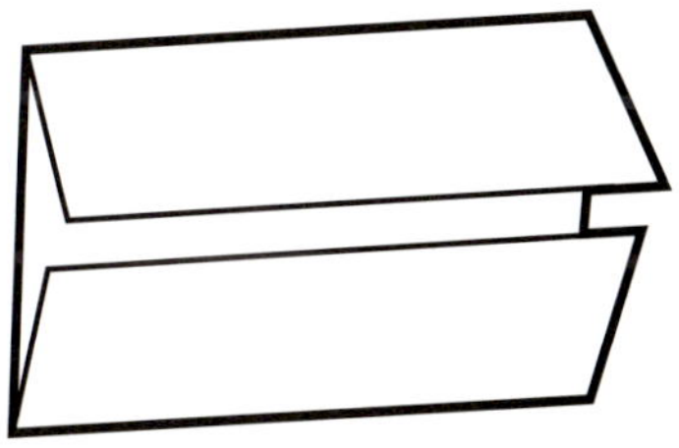

Appendix

Booklet

1. Fold a sheet of paper in half from left to right. Then, unfold the paper.
2. Fold the sheet of paper in half again from the top to the bottom. Then, unfold the paper.
3. Refold the sheet of paper in half from left to right.
4. Fold the top and bottom edges to the center crease.
5. Completely unfold the paper.
6. Refold the paper from top to bottom.
7. Using scissors, cut a slit along the center crease of the sheet from the folded edge to the creases made in step 4. Do not cut the entire sheet in half.
8. Fold the sheet of paper in half from left to right. While holding the bottom and top edges of the paper, push the bottom and top edges together so that the center collapses at the center slit. Fold the four flaps to form a four-page book.

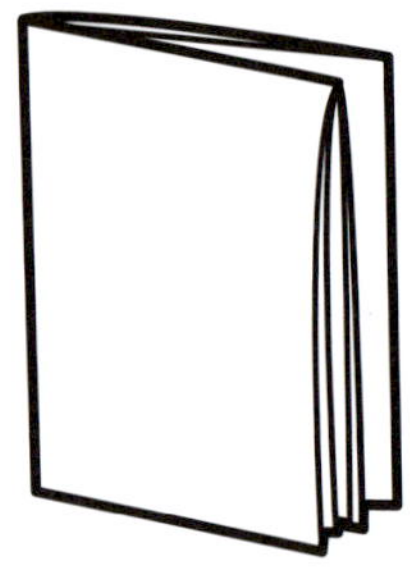

Layered Book

1. Lay one sheet of paper on top of another sheet. Slide the top sheet up so that 2 cm of the bottom sheet is showing.
2. Hold the two sheets together, fold down the top of the two sheets so that you see four 2 cm tabs along the bottom.
3. Using a stapler, staple the top of the FoldNote.

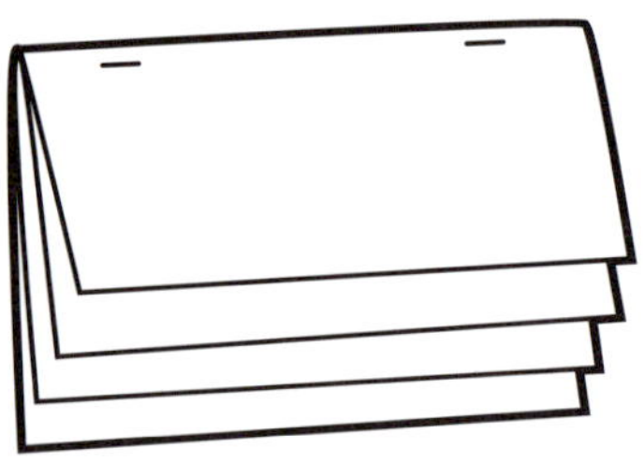

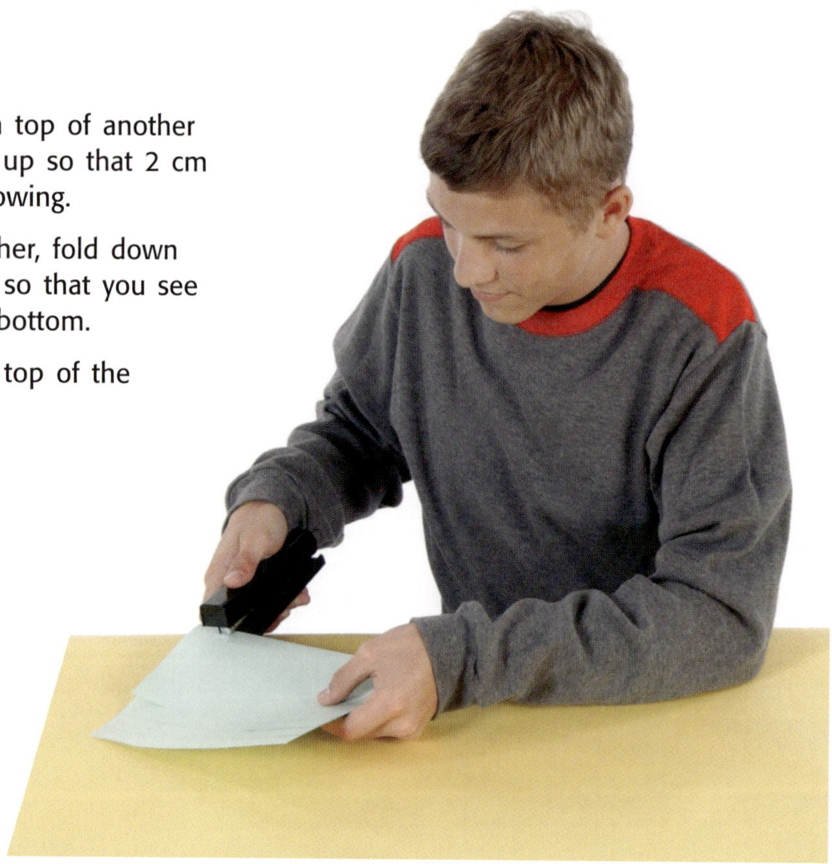

Key-Term Fold

1. Fold a sheet of lined notebook paper in half from left to right.
2. Using scissors, cut along every third line from the right edge of the paper to the center fold to make tabs.

Four-Corner Fold

1. Fold a sheet of paper in half from left to right. Then, unfold the paper.
2. Fold each side of the paper to the crease in the center of the paper.
3. Fold the paper in half from the top to the bottom. Then, unfold the paper.
4. Using scissors, cut the top flap creases made in step 3 to form four flaps.

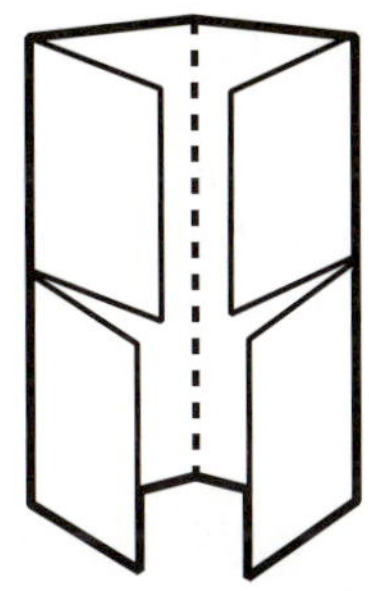

Three-Panel Flip Chart

1. Fold a piece of paper in half from the top to the bottom.
2. Fold the paper in thirds from side to side. Then, unfold the paper so that you can see the three sections.
3. From the top of the paper, cut along each of the vertical fold lines to the fold in the middle of the paper. You will now have three flaps.

Table Fold

1. Fold a piece of paper in half from the top to the bottom. Then, fold the paper in half again.
2. Fold the paper in thirds from side to side.
3. Unfold the paper completely. Carefully trace the fold lines by using a pen or pencil.

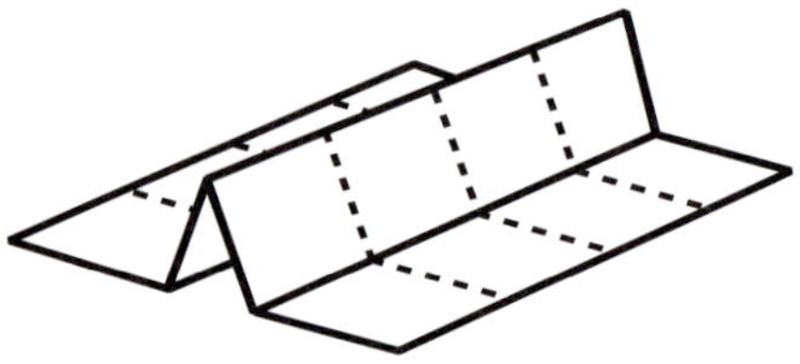

Two-Panel Flip Chart

1. Fold a piece of paper in half from the top to the bottom.
2. Fold the paper in half from side to side. Then, unfold the paper so that you can see the two sections.
3. From the top of the paper, cut along the vertical fold line to the fold in the middle of the paper. You will now have two flaps.

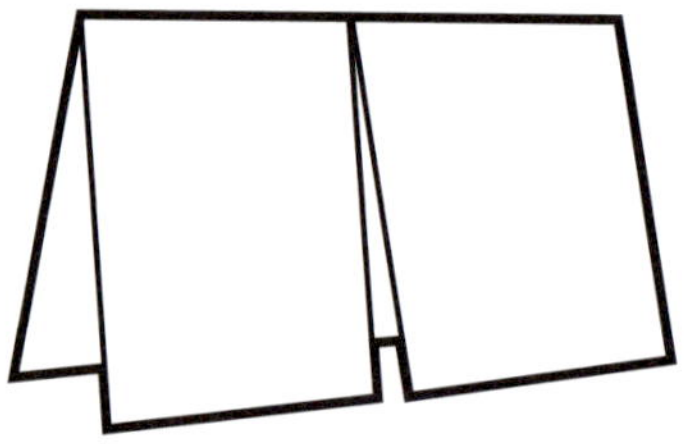

Tri-Fold

1. Fold a piece a paper in thirds from the top to the bottom.
2. Unfold the paper so that you can see the three sections. Then, turn the paper sideways so that the three sections form vertical columns.
3. Trace the fold lines by using a pen or pencil. Label the columns "Know," "Want," and "Learn."

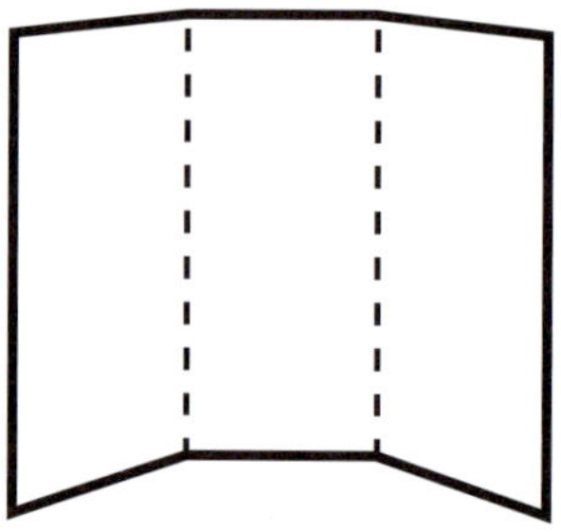

Graphic Organizer Instructions

Have you ever wished that you could "draw out" the many concepts you learn in your science class? Sometimes, being able to *see* how concepts are related really helps you remember what you've learned. Graphic Organizers do just that! They give you a way to draw or map out concepts.

All you need to make a Graphic Organizer is a piece of paper and a pencil. Below you will find instructions for four different Graphic Organizers designed to help you organize the concepts you'll learn in this book.

Spider Map

1. Draw a diagram like the one shown. In the circle, write the main topic.
2. From the circle, draw legs to represent different categories of the main topic. You can have as many categories as you want.
3. From the category legs, draw horizontal lines. As you read the chapter, write details about each category on the horizontal lines.

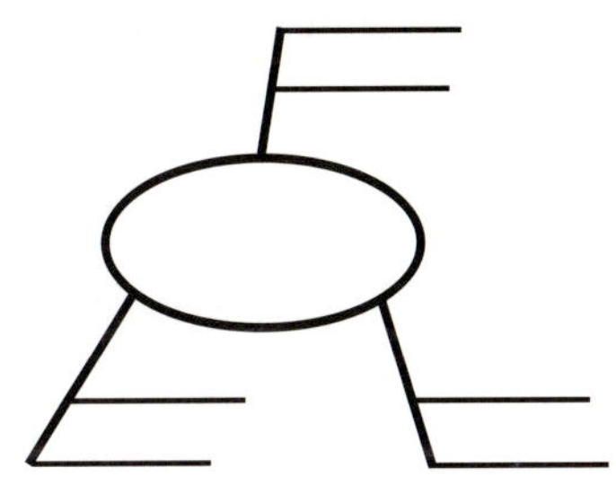

Comparison Table

1. Draw a chart like the one shown. Your chart can have as many columns and rows as you want.
2. In the top row, write the topics that you want to compare.
3. In the left column, write characteristics of the topics that you want to compare. As you read the chapter, fill in the characteristics for each topic in the appropriate boxes.

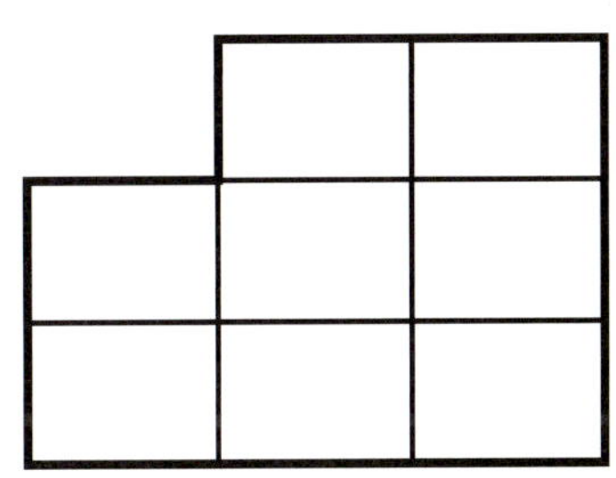

Chain-of-Events-Chart

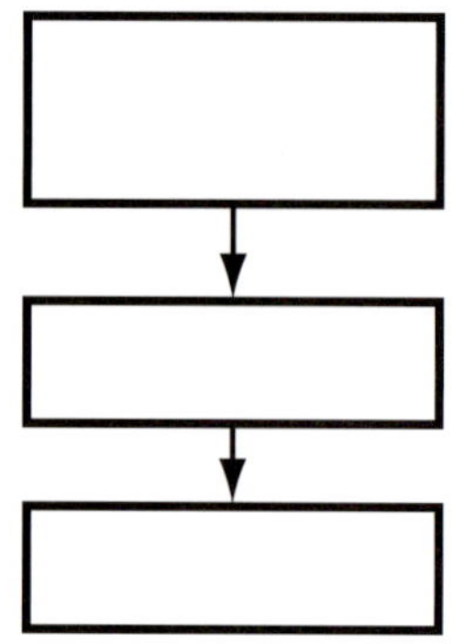

1. Draw a box. In the box, write the first step of a process or the first event of a timeline.
2. Under the box, draw another box, and use an arrow to connect the two boxes. In the second box, write the next step of the process or the next event in the timeline.
3. Continue adding boxes until the process or timeline is finished.

Concept Map

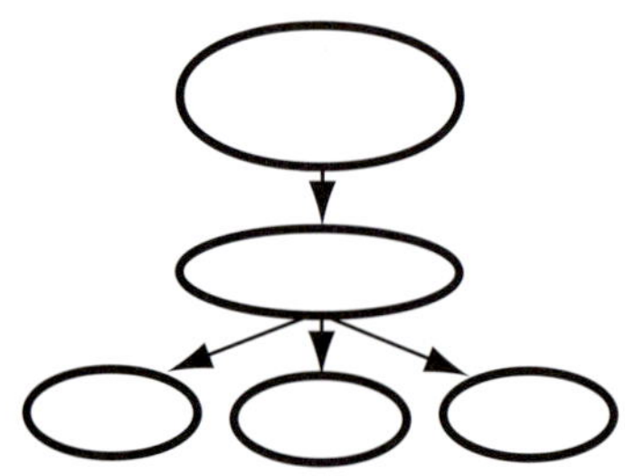

1. Draw a circle in the center of a piece of paper. Write the main idea of the chapter in the center of the circle.
2. From the circle, draw other circles. In those circles, write characteristics of the main idea. Draw arrows from the center circle to the circles that contain the characteristics.
3. From each circle that contains a characteristic, draw other circles. In those circles, write specific details about the characteristic. Draw arrows from each circle that contains a characteristic to the circles that contain specific details. You may draw as many circles as you want.

Appendix

SI Measurement

The International System of Units, or SI, is the standard system of measurement used by many scientists. Using the same standards of measurement makes it easier for scientists to communicate with one another.

SI works by combining prefixes and base units. Each base unit can be used with different prefixes to define smaller and larger quantities. The table below lists common SI prefixes.

SI Prefixes

Prefix	Symbol	Factor	Example
kilo-	k	1,000	kilogram, 1 kg = 1,000 g
hecto-	h	100	hectoliter, 1 hL = 100 L
deka-	da	10	dekameter, 1 dam = 10 m
		1	meter, liter, gram
deci-	d	0.1	decigram, 1 dg = 0.1 g
centi-	c	0.01	centimeter, 1 cm = 0.01 m
milli-	m	0.001	milliliter, 1 mL = 0.001 L
micro-	μ	0.000 001	micrometer, 1 μm = 0.000 001 m

SI Conversion Table

SI units	From SI to English	From English to SI
Length		
kilometer (km) = 1,000 m	1 km = 0.621 mi	1 mi = 1.609 km
meter (m) = 100 cm	1 m = 3.281 ft	1 ft = 0.305 m
centimeter (cm) = 0.01 m	1 cm = 0.394 in.	1 in. = 2.540 cm
millimeter (mm) = 0.001 m	1 mm = 0.039 in.	
micrometer (μm) = 0.000 001 m		
nanometer (nm) = 0.000 000 001 m		
Area		
square kilometer (km^2) = 100 hectares	1 km^2 = 0.386 mi^2	1 mi^2 = 2.590 km^2
hectare (ha) = 10,000 m^2	1 ha = 2.471 acres	1 acre = 0.405 ha
square meter (m^2) = 10,000 cm^2	1 m^2 = 10.764 ft^2	1 ft^2 = 0.093 m^2
square centimeter (cm^2) = 100 mm^2	1 cm^2 = 0.155 $in.^2$	1 $in.^2$ = 6.452 cm^2
Volume		
liter (L) = 1,000 mL = 1 dm^3	1 L = 1.057 fl qt	1 fl qt = 0.946 L
milliliter (mL) = 0.001 L = 1 cm^3	1 mL = 0.034 fl oz	1 fl oz = 29.574 mL
microliter (μL) = 0.000 001 L		
Mass		
kilogram (kg) = 1,000 g	1 kg = 2.205 lb	1 lb = 0.454 kg
gram (g) = 1,000 mg	1 g = 0.035 oz	1 oz = 28.350 g
milligram (mg) = 0.001 g		
microgram (μg) = 0.000 001 g		

Appendix

Measuring Skills

Using a Graduated Cylinder

When using a graduated cylinder to measure volume, keep the following procedures in mind:

1. Place the cylinder on a flat, level surface before measuring liquid.
2. Move your head so that your eye is level with the surface of the liquid.
3. Read the mark closest to the liquid level. On glass graduated cylinders, read the mark closest to the center of the curve in the liquid's surface.

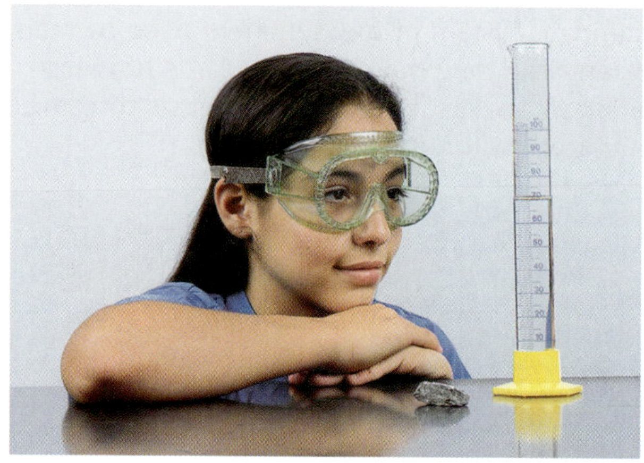

Using a Meterstick or Metric Ruler

When using a meterstick or metric ruler to measure length, keep the following procedures in mind:

1. Place the ruler firmly against the object that you are measuring.
2. Align one edge of the object exactly with the 0 end of the ruler.
3. Look at the other edge of the object to see which of the marks on the ruler is closest to that edge. (Note: Each small slash between the centimeters represents a millimeter, which is one-tenth of a centimeter.)

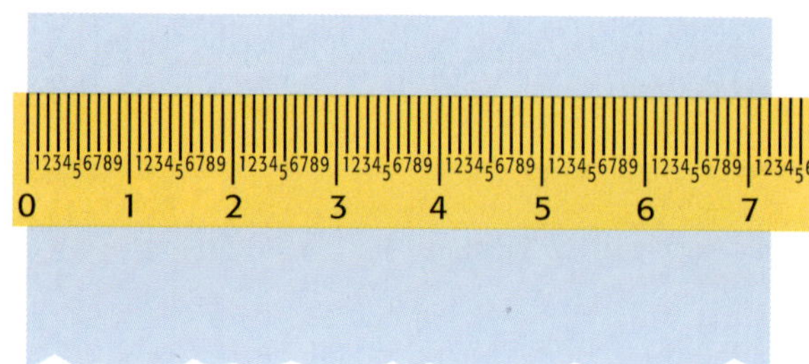

Using a Triple-Beam Balance

When using a triple-beam balance to measure mass, keep the following procedures in mind:

1. Make sure the balance is on a level surface.
2. Place all of the countermasses at 0. Adjust the balancing knob until the pointer rests at 0.
3. Place the object you wish to measure on the pan. **Caution:** Do not place hot objects or chemicals directly on the balance pan.
4. Move the largest countermass along the beam to the right until it is at the last notch that does not tip the balance. Follow the same procedure with the next-largest countermass. Then, move the smallest countermass until the pointer rests at 0.
5. Add the readings from the three beams together to determine the mass of the object.
6. When determining the mass of crystals or powders, first find the mass of a piece of filter paper. Then, add the crystals or powder to the paper, and remeasure. The actual mass of the crystals or powder is the total mass minus the mass of the paper. When finding the mass of liquids, first find the mass of the empty container. Then, find the combined mass of the liquid and container. The mass of the liquid is the total mass minus the mass of the container.

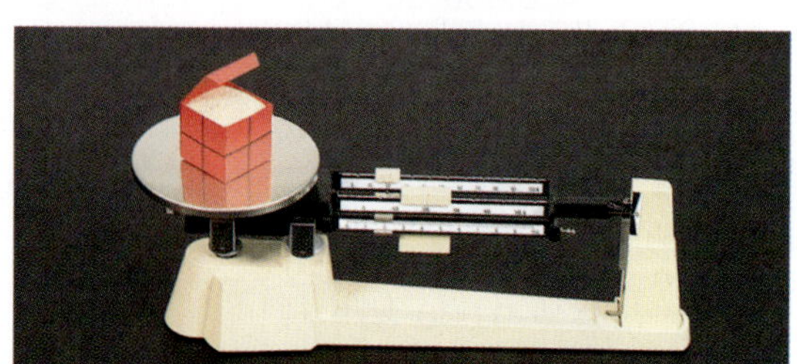

Scientific Methods

The ways in which scientists answer questions and solve problems are called **scientific methods.** The same steps are often used by scientists as they look for answers. However, there is more than one way to use these steps. Scientists may use all of the steps or just some of the steps during an investigation. They may even repeat some of the steps. The goal of using scientific methods is to come up with reliable answers and solutions.

Six Steps of Scientific Methods

1 Ask a Question

Good questions come from careful **observations.** You make observations by using your senses to gather information. Sometimes, you may use instruments, such as microscopes and telescopes, to extend the range of your senses. As you observe the natural world, you will discover that you have many more questions than answers. These questions drive investigations.

Questions beginning with *what, why, how,* and *when* are important in focusing an investigation. Here is an example of a question that could lead to an investigation.

Question: How does acid rain affect plant growth?

2 Form a Hypothesis

After you ask a question, you need to form a **hypothesis.** A hypothesis is a clear statement of what you expect the answer to your question to be. Your hypothesis will represent your best "educated guess" based on what you have observed and what you already know. A good hypothesis is testable. Otherwise, the investigation can go no further. Here is a hypothesis based on the question, "How does acid rain affect plant growth?"

Hypothesis: Acid rain slows plant growth.

The hypothesis can lead to predictions. A prediction is what you think the outcome of your experiment or data collection will be. Predictions are usually stated in an if-then format. Here is a sample prediction for the hypothesis that acid rain slows plant growth.

Prediction: If a plant is watered with only acid rain (which has a pH of 4), then the plant will grow at half its normal rate.

3 Test the Hypothesis

After you have formed a hypothesis and made a prediction, your hypothesis should be tested. One way to test a hypothesis is with a controlled experiment. A **controlled experiment** tests only one factor at a time. In an experiment to test the effect of acid rain on plant growth, the **control group** would be watered with normal rain water. The **experimental group** would be watered with acid rain. All of the plants should receive the same amount of sunlight and water each day. The air temperature should be the same for all groups. However, the acidity of the water will be a variable. In fact, any factor that is different from one group to another is a **variable.** If your hypothesis is correct, then the acidity of the water and plant growth are *dependant variables.* The amount a plant grows is dependent on the acidity of the water. However, the amount of water each plant receives and the amount of sunlight each plant receives are *independent variables.* Either of these factors could change without affecting the other factor.

Sometimes, the nature of an investigation makes a controlled experiment impossible. For example, the Earth's core is surrounded by thousands of meters of rock. Under such circumstances, a hypothesis may be tested by making detailed observations.

4 Analyze the Results

After you have completed your experiments, made your observations, and collected your data, you must analyze all the information you have gathered. Tables and graphs are often used in this step to organize the data.

5 Draw Conclusions

After analyzing your data, you can determine if your results support your hypothesis. If your hypothesis is supported, you (or others) might want to repeat the observations or experiments to verify your results. If your hypothesis is not supported by the data, you may have to check your procedure for errors. You may even have to reject your hypothesis and make a new one. If you cannot draw a conclusion from your results, you may have to try the investigation again or carry out further observations or experiments.

6 Communicate Results

After any scientific investigation, you should report your results. By preparing a written or oral report, you let others know what you have learned. They may repeat your investigation to see if they get the same results. Your report may even lead to another question and then to another investigation.

Scientific Methods in Action

Scientific methods contain loops in which several steps may be repeated over and over again. In some cases, certain steps are unnecessary. Thus, there is not a "straight line" of steps. For example, sometimes scientists find that testing one hypothesis raises new questions and new hypotheses to be tested. And sometimes, testing the hypothesis leads directly to a conclusion. Furthermore, the steps in scientific methods are not always used in the same order. Follow the steps in the diagram, and see how many different directions scientific methods can take you.

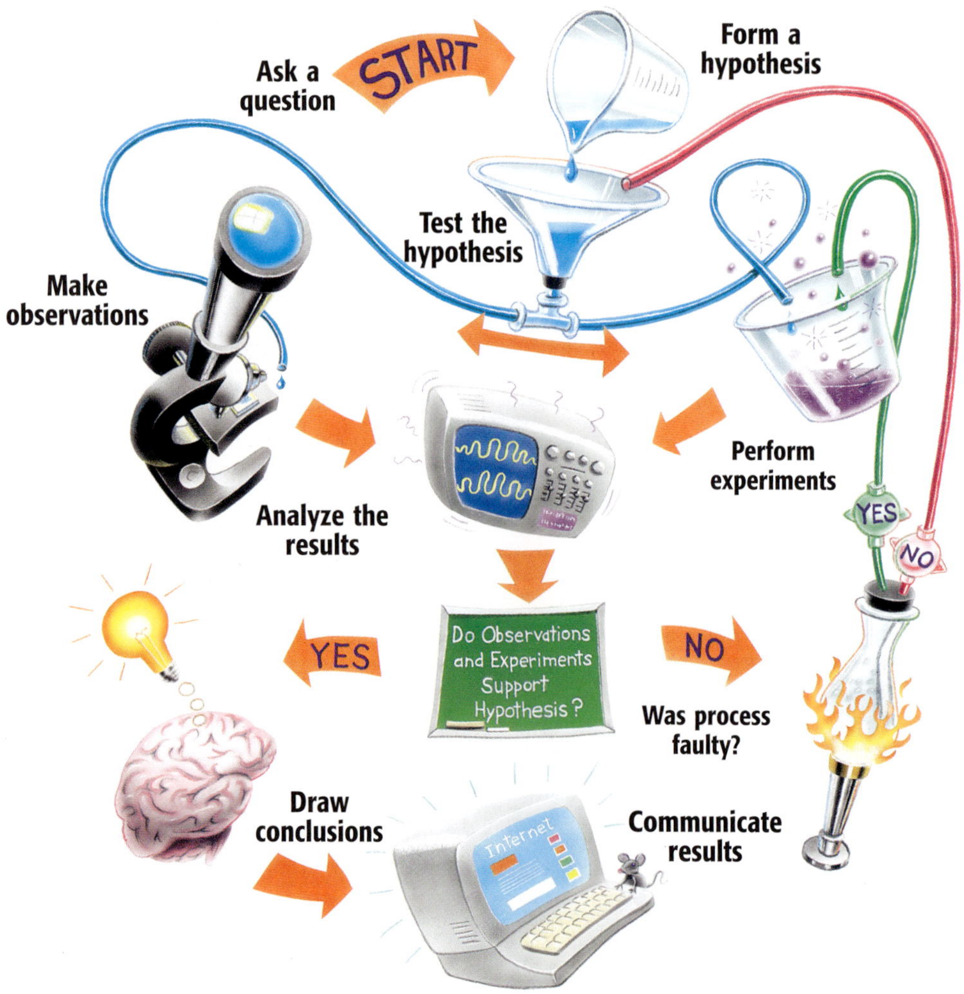

Making Charts and Graphs

Pie Charts

A pie chart shows how each group of data relates to all of the data. Each part of the circle forming the chart represents a category of the data. The entire circle represents all of the data. For example, a biologist studying a hardwood forest in Wisconsin found that there were five different types of trees. The data table at right summarizes the biologist's findings.

Wisconsin Hardwood Trees	
Type of tree	**Number found**
Oak	600
Maple	750
Beech	300
Birch	1,200
Hickory	150
Total	3,000

How to Make a Pie Chart

1. To make a pie chart of these data, first find the percentage of each type of tree. Divide the number of trees of each type by the total number of trees, and multiply by 100.

$$\frac{600 \text{ oak}}{3{,}000 \text{ trees}} \times 100 = 20\%$$

$$\frac{750 \text{ maple}}{3{,}000 \text{ trees}} \times 100 = 25\%$$

$$\frac{300 \text{ beech}}{3{,}000 \text{ trees}} \times 100 = 10\%$$

$$\frac{1{,}200 \text{ birch}}{3{,}000 \text{ trees}} \times 100 = 40\%$$

$$\frac{150 \text{ hickory}}{3{,}000 \text{ trees}} \times 100 = 5\%$$

2. Now, determine the size of the wedges that make up the pie chart. Multiply each percentage by 360°. Remember that a circle contains 360°.

$20\% \times 360° = 72°$ $25\% \times 360° = 90°$

$10\% \times 360° = 36°$ $40\% \times 360° = 144°$

$5\% \times 360° = 18°$

3. Check that the sum of the percentages is 100 and the sum of the degrees is 360.

$20\% + 25\% + 10\% + 40\% + 5\% = 100\%$

$72° + 90° + 36° + 144° + 18° = 360°$

4. Use a compass to draw a circle and mark the center of the circle.
5. Then, use a protractor to draw angles of 72°, 90°, 36°, 144°, and 18° in the circle.
6. Finally, label each part of the chart, and choose an appropriate title.

A Community of Wisconsin Hardwood Trees

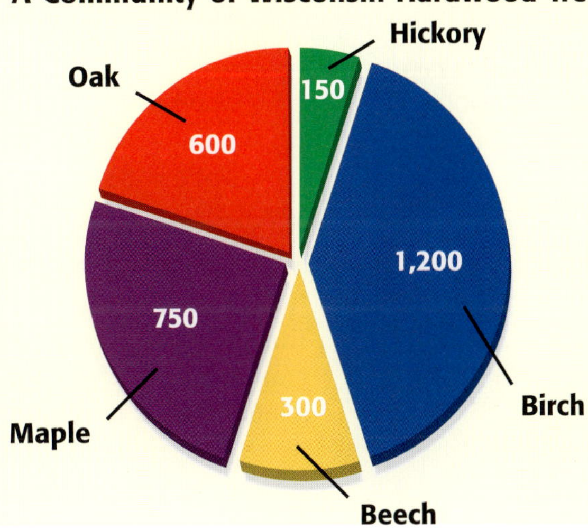

Line Graphs

Line graphs are most often used to demonstrate continuous change. For example, Mr. Smith's students analyzed the population records for their hometown, Appleton, between 1900 and 2000. Examine the data at right.

Because the year and the population change, they are the *variables*. The population is determined by, or dependent on, the year. Therefore, the population is called the **dependent variable,** and the year is called the **independent variable.** Each set of data is called a **data pair.** To prepare a line graph, you must first organize data pairs into a table like the one at right.

Population of Appleton, 1900–2000

Year	Population
1900	1,800
1920	2,500
1940	3,200
1960	3,900
1980	4,600
2000	5,300

How to Make a Line Graph

1. Place the independent variable along the horizontal (x) axis. Place the dependent variable along the vertical (y) axis.
2. Label the x-axis "Year" and the y-axis "Population." Look at your largest and smallest values for the population. For the y-axis, determine a scale that will provide enough space to show these values. You must use the same scale for the entire length of the axis. Next, find an appropriate scale for the x-axis.
3. Choose reasonable starting points for each axis.
4. Plot the data pairs as accurately as possible.
5. Choose a title that accurately represents the data.

Population of Appleton, 1900–2000

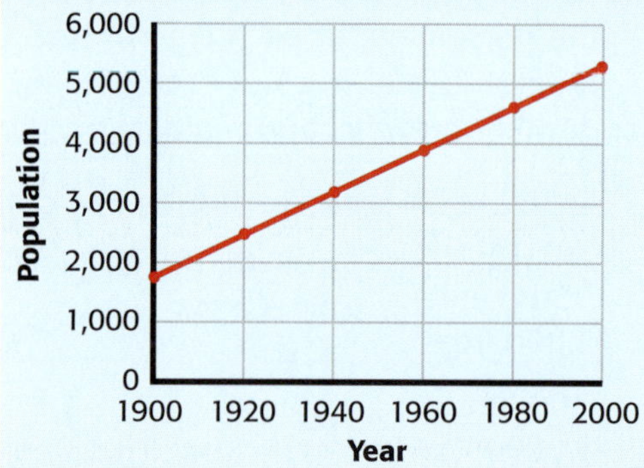

How to Determine Slope

Slope is the ratio of the change in the y-value to the change in the x-value, or "rise over run."

1. Choose two points on the line graph. For example, the population of Appleton in 2000 was 5,300 people. Therefore, you can define point a as (2000, 5,300). In 1900, the population was 1,800 people. You can define point b as (1900, 1,800).
2. Find the change in the y-value. (y at point a) − (y at point b) = 5,300 people − 1,800 people = 3,500 people
3. Find the change in the x-value. (x at point a) − (x at point b) = 2000 − 1900 = 100 years
4. Calculate the slope of the graph by dividing the change in y by the change in x.

$$slope = \frac{change\ in\ y}{change\ in\ x}$$

$$slope = \frac{3{,}500\ \text{people}}{100\ \text{years}}$$

$$slope = 35\ \text{people per year}$$

In this example, the population in Appleton increased by a fixed amount each year. The graph of these data is a straight line. Therefore, the relationship is **linear.** When the graph of a set of data is not a straight line, the relationship is **nonlinear.**

Using Algebra to Determine Slope

The equation in step 4 may also be arranged to be

$$y = kx$$

where y represents the change in the y-value, k represents the slope, and x represents the change in the x-value.

$$slope = \frac{change\ in\ y}{change\ in\ x}$$

$$k = \frac{y}{x}$$

$$k \times x = \frac{y \times x}{x}$$

$$kx = y$$

Bar Graphs

Bar graphs are used to demonstrate change that is not continuous. These graphs can be used to indicate trends when the data cover a long period of time. A meteorologist gathered the precipitation data shown here for Hartford, Connecticut, for April 1–15, 1996, and used a bar graph to represent the data.

Precipitation in Hartford, Connecticut April 1–15, 1996

Date	Precipitation (cm)	Date	Precipitation (cm)
April 1	0.5	April 9	0.25
April 2	1.25	April 10	0.0
April 3	0.0	April 11	1.0
April 4	0.0	April 12	0.0
April 5	0.0	April 13	0.25
April 6	0.0	April 14	0.0
April 7	0.0	April 15	6.50
April 8	1.75		

How to Make a Bar Graph

1. Use an appropriate scale and a reasonable starting point for each axis.
2. Label the axes, and plot the data.
3. Choose a title that accurately represents the data.

Precipitation in Hartford, Connecticut, April 1–15, 1996

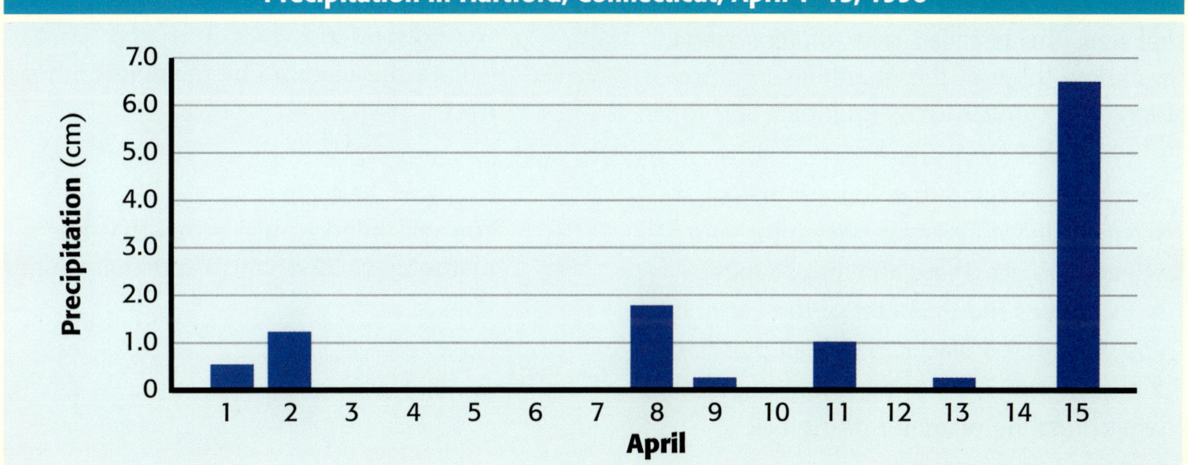

Math Refresher

Science requires an understanding of many math concepts. The following pages will help you review some important math skills.

Averages

An **average,** or **mean,** simplifies a set of numbers into a single number that *approximates* the value of the set.

Example: Find the average of the following set of numbers: 5, 4, 7, and 8.

Step 1: Find the sum.

$$5 + 4 + 7 + 8 = 24$$

Step 2: Divide the sum by the number of numbers in your set. Because there are four numbers in this example, divide the sum by 4.

$$\frac{24}{4} = 6$$

The average, or mean, is **6.**

Ratios

A **ratio** is a comparison between numbers, and it is usually written as a fraction.

Example: Find the ratio of thermometers to students if you have 36 thermometers and 48 students in your class.

Step 1: Make the ratio.

$$\frac{36 \text{ thermometers}}{48 \text{ students}}$$

Step 2: Reduce the fraction to its simplest form.

$$\frac{36}{48} = \frac{36 \div 12}{48 \div 12} = \frac{3}{4}$$

The ratio of thermometers to students is **3 to 4,** or $\frac{3}{4}$. The ratio may also be written in the form 3:4.

Proportions

A **proportion** is an equation that states that two ratios are equal.

$$\frac{3}{1} = \frac{12}{4}$$

To solve a proportion, first multiply across the equal sign. This is called *cross-multiplication*. If you know three of the quantities in a proportion, you can use cross-multiplication to find the fourth.

Example: Imagine that you are making a scale model of the solar system for your science project. The diameter of Jupiter is 11.2 times the diameter of the Earth. If you are using a plastic-foam ball that has a diameter of 2 cm to represent the Earth, what must the diameter of the ball representing Jupiter be?

$$\frac{11.2}{1} = \frac{x}{2 \text{ cm}}$$

Step 1: Cross-multiply.

$$\frac{11.2}{1} \times \frac{x}{2}$$

$$11.2 \times 2 = x \times 1$$

Step 2: Multiply.

$$22.4 = x \times 1$$

Step 3: Isolate the variable by dividing both sides by 1.

$$x = \frac{22.4}{1}$$

$$x = 22.4 \text{ cm}$$

You will need to use a ball that has a diameter of **22.4** cm to represent Jupiter.

Percentages

A **percentage** is a ratio of a given number to 100.

Example: What is 85% of 40?

Step 1: Rewrite the percentage by moving the decimal point two places to the left.

0.85

Step 2: Multiply the decimal by the number that you are calculating the percentage of.

$0.85 \times 40 = 34$

85% of 40 is **34.**

Decimals

To **add** or **subtract decimals,** line up the digits vertically so that the decimal points line up. Then, add or subtract the columns from right to left. Carry or borrow numbers as necessary.

Example: Add the following numbers: 3.1415 and 2.96.

Step 1: Line up the digits vertically so that the decimal points line up.

$$\begin{array}{r} 3.1415 \\ +\ 2.96 \\ \hline \end{array}$$

Step 2: Add the columns from right to left, and carry when necessary.

$$\begin{array}{r} {\scriptstyle 1\ 1} \\ 3.1415 \\ +\ 2.96 \\ \hline 6.1015 \end{array}$$

The sum is **6.1015.**

Fractions

Numbers tell you how many; **fractions** tell you *how much of a whole*.

Example: Your class has 24 plants. Your teacher instructs you to put 5 plants in a shady spot. What fraction of the plants in your class will you put in a shady spot?

Step 1: In the denominator, write the total number of parts in the whole.

$$\frac{?}{24}$$

Step 2: In the numerator, write the number of parts of the whole that are being considered.

$$\frac{5}{24}$$

So, $\mathbf{\frac{5}{24}}$ of the plants will be in the shade.

Reducing Fractions

It is usually best to express a fraction in its simplest form. Expressing a fraction in its simplest form is called *reducing* a fraction.

Example: Reduce the fraction $\frac{30}{45}$ to its simplest form.

Step 1: Find the largest whole number that will divide evenly into both the numerator and denominator. This number is called the *greatest common factor* (GCF).

Factors of the numerator 30:

1, 2, 3, 5, 6, 10, **15,** 30

Factors of the denominator 45:

1, 3, 5, 9, **15,** 45

Step 2: Divide both the numerator and the denominator by the GCF, which in this case is 15.

$$\frac{30}{45} = \frac{30 \div 15}{45 \div 15} = \frac{2}{3}$$

Thus, $\frac{30}{45}$ reduced to its simplest form is $\mathbf{\frac{2}{3}}$.

Appendix

Adding and Subtracting Fractions

To **add** or **subtract fractions** that have the **same denominator,** simply add or subtract the numerators.

Examples:

$$\frac{3}{5} + \frac{1}{5} = ? \text{ and } \frac{3}{4} - \frac{1}{4} = ?$$

Step 1: Add or subtract the numerators.

$$\frac{3}{5} + \frac{1}{5} = \frac{4}{\ } \text{ and } \frac{3}{4} - \frac{1}{4} = \frac{2}{\ }$$

Step 2: Write the sum or difference over the denominator.

$$\frac{3}{5} + \frac{1}{5} = \frac{4}{5} \text{ and } \frac{3}{4} - \frac{1}{4} = \frac{2}{4}$$

Step 3: If necessary, reduce the fraction to its simplest form.

$\mathbf{\frac{4}{5}}$ cannot be reduced, and $\frac{2}{4} = \mathbf{\frac{1}{2}}$.

To **add** or **subtract fractions** that have **different denominators,** first find the least common denominator (LCD).

Examples:

$$\frac{1}{2} + \frac{1}{6} = ? \text{ and } \frac{3}{4} - \frac{2}{3} = ?$$

Step 1: Write the equivalent fractions that have a common denominator.

$$\frac{3}{6} + \frac{1}{6} = ? \text{ and } \frac{9}{12} - \frac{8}{12} = ?$$

Step 2: Add or subtract the fractions.

$$\frac{3}{6} + \frac{1}{6} = \frac{4}{6} \text{ and } \frac{9}{12} - \frac{8}{12} = \frac{1}{12}$$

Step 3: If necessary, reduce the fraction to its simplest form.

The fraction $\frac{4}{6} = \mathbf{\frac{2}{3}}$, and $\mathbf{\frac{1}{12}}$ cannot be reduced.

Multiplying Fractions

To **multiply fractions,** multiply the numerators and the denominators together, and then reduce the fraction to its simplest form.

Example:

$$\frac{5}{9} \times \frac{7}{10} = ?$$

Step 1: Multiply the numerators and denominators.

$$\frac{5}{9} \times \frac{7}{10} = \frac{5 \times 7}{9 \times 10} = \frac{35}{90}$$

Step 2: Reduce the fraction.

$$\frac{35}{90} = \frac{35 \div 5}{90 \div 5} = \mathbf{\frac{7}{18}}$$

Dividing Fractions

To **divide fractions,** first rewrite the divisor (the number you divide by) upside down. This number is called the *reciprocal* of the divisor. Then multiply and reduce if necessary.

Example:

$$\frac{5}{8} \div \frac{3}{2} = ?$$

Step 1: Rewrite the divisor as its reciprocal.

$$\frac{3}{2} \rightarrow \frac{2}{3}$$

Step 2: Multiply the fractions.

$$\frac{5}{8} \times \frac{2}{3} = \frac{5 \times 2}{8 \times 3} = \frac{10}{24}$$

Step 3: Reduce the fraction.

$$\frac{10}{24} = \frac{10 \div 2}{24 \div 2} = \mathbf{\frac{5}{12}}$$

Appendix

Scientific Notation

Scientific notation is a short way of representing very large and very small numbers without writing all of the place-holding zeros.

Example: Write 653,000,000 in scientific notation.

Step 1: Write the number without the place-holding zeros.

653

Step 2: Place the decimal point after the first digit.

6.53

Step 3: Find the exponent by counting the number of places that you moved the decimal point.

6.53000000

The decimal point was moved eight places to the left. Therefore, the exponent of 10 is positive 8. If you had moved the decimal point to the right, the exponent would be negative.

Step 4: Write the number in scientific notation.

$\mathbf{6.53 \times 10^8}$

Area

Area is the number of square units needed to cover the surface of an object.

Formulas:

area of a square = *side* × *side*

area of a rectangle = *length* × *width*

area of a triangle = $\frac{1}{2}$ × *base* × *height*

Examples: Find the areas.

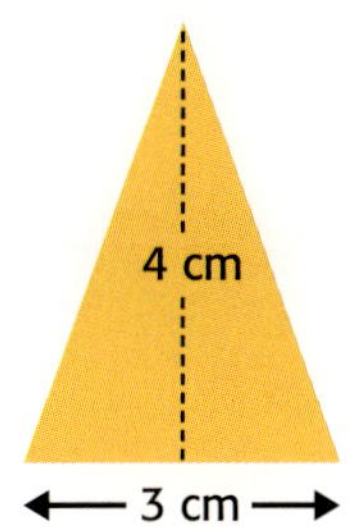

Triangle

area = $\frac{1}{2}$ × *base* × *height*

area = $\frac{1}{2}$ × 3 cm × 4 cm

area = **6 cm²**

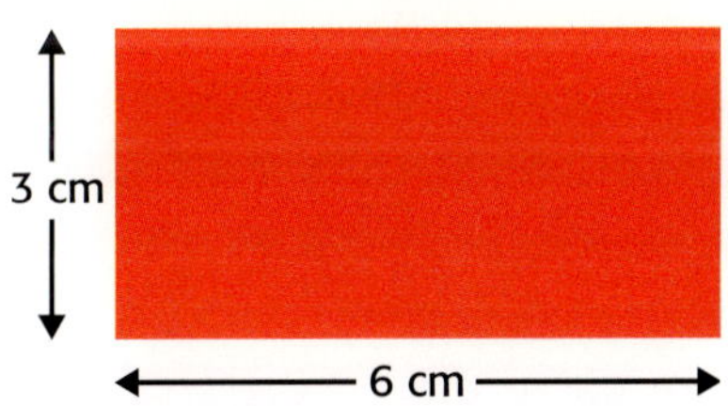

Rectangle

area = *length* × *width*

area = 6 cm × 3 cm

area = **18 cm²**

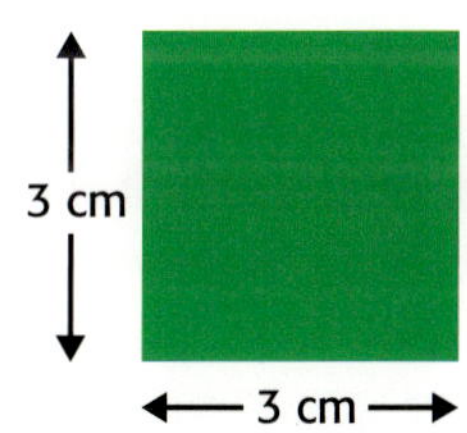

Square

area = *side* × *side*

area = 3 cm × 3 cm

area = **9 cm²**

Volume

Volume is the amount of space that something occupies.

Formulas:

volume of a cube = *side* × *side* × *side*

volume of a prism = *area of base* × *height*

Examples:

Find the volume of the solids.

Cube

volume = *side* × *side* × *side*

volume = 4 cm × 4 cm × 4 cm

volume = **64 cm³**

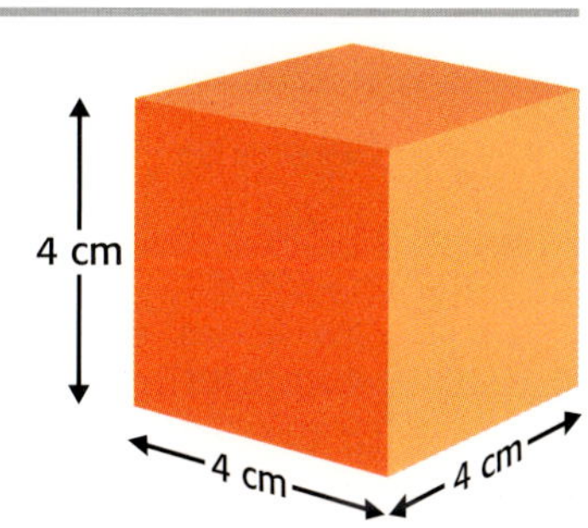

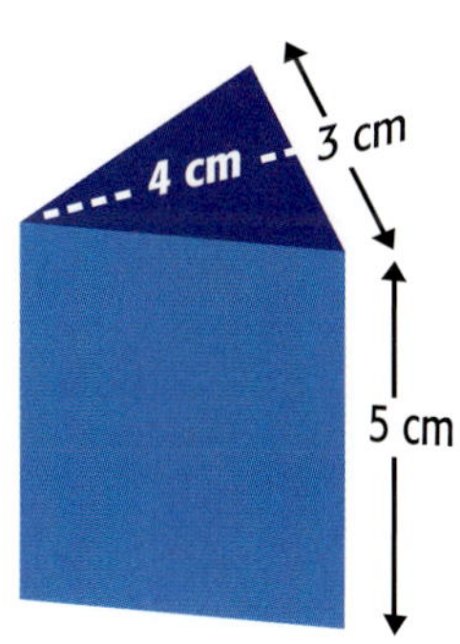

Prism

volume = *area of base* × *height*

volume = (*area of triangle*) × *height*

volume = ($\frac{1}{2}$ × 3 cm × 4 cm) × 5 cm

volume = 6 cm² × 5 cm

volume = **30 cm³**

Temperature Scales

Temperature can be expressed by using three different scales: Fahrenheit, Celsius, and Kelvin. The SI unit for temperature is the kelvin (K).

Although 0 K is much colder than 0°C, a change of 1 K is equal to a change of 1°C.

Three Temperature Scales

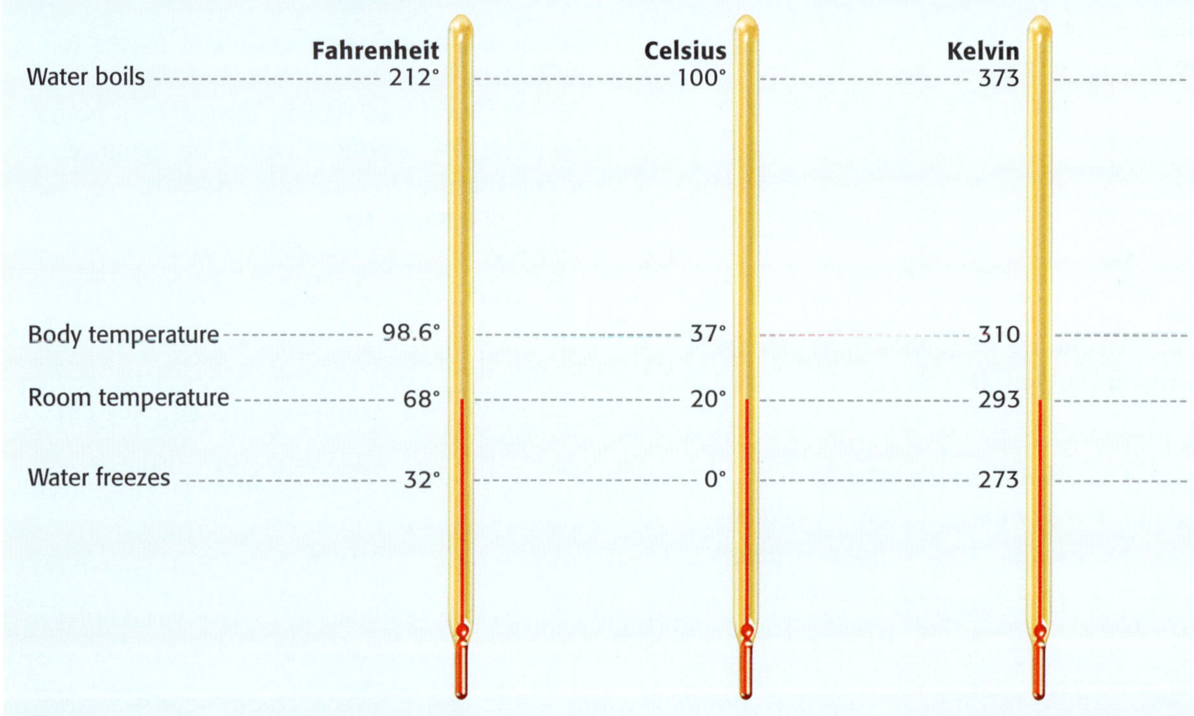

Temperature Conversions Table

To convert	Use this equation:	Example
Celsius to Fahrenheit °C → °F	$°F = \left(\frac{9}{5} \times °C\right) + 32$	Convert 45°C to °F. $°F = \left(\frac{9}{5} \times 45°C\right) + 32 = 113°F$
Fahrenheit to Celsius °F → °C	$°C = \frac{5}{9} \times (°F - 32)$	Convert 68°F to °C. $°C = \frac{5}{9} \times (68°F - 32) = 20°C$
Celsius to Kelvin °C → K	$K = °C + 273$	Convert 45°C to K. $K = 45°C + 273 = 318\ K$
Kelvin to Celsius K → °C	$°C = K - 273$	Convert 32 K to °C. $°C = 32K - 273 = -241°C$

Using the Microscope

Parts of the Compound Light Microscope

- The **ocular lens** magnifies the image 10×.
- The **low-power objective** magnifies the image 10×.
- The **high-power objective** magnifies the image either 40× or 43×.
- The **revolving nosepiece** holds the objectives and can be turned to change from one magnification to the other.
- The **body tube** maintains the correct distance between the ocular lens and objectives.
- The **coarse-adjustment knob** moves the body tube up and down to allow focusing of the image.
- The **fine-adjustment knob** moves the body tube slightly to bring the image into sharper focus.
- The **stage** supports a slide.
- **Stage clips** hold the slide in place for viewing.
- The **diaphragm** controls the amount of light coming through the stage.
- The light source provides a **light** for viewing the slide.
- The **arm** supports the body tube.
- The **base** supports the microscope.

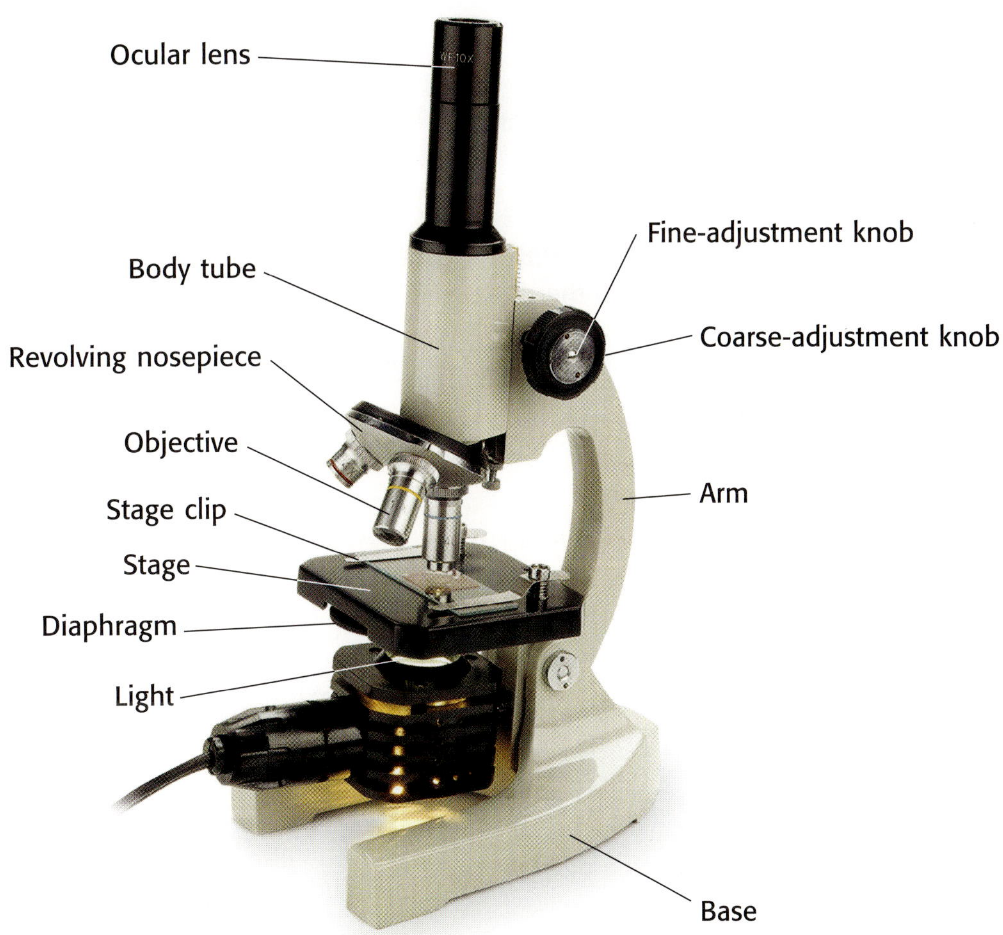

Proper Use of the Compound Light Microscope

1. Use both hands to carry the microscope to your lab table. Place one hand beneath the base, and use the other hand to hold the arm of the microscope. Hold the microscope close to your body while carrying it to your lab table.
2. Place the microscope on the lab table at least 5 cm from the edge of the table.
3. Check to see what type of light source is used by your microscope. If the microscope has a lamp, plug it in and make sure that the cord is out of the way. If the microscope has a mirror, adjust the mirror to reflect light through the hole in the stage. **Caution:** If your microscope has a mirror, do not use direct sunlight as a light source. Direct sunlight can damage your eyes.
4. Always begin work with the low-power objective in line with the body tube. Adjust the revolving nosepiece.
5. Place a prepared slide over the hole in the stage. Secure the slide with the stage clips.
6. Look through the ocular lens. Move the diaphragm to adjust the amount of light coming through the stage.
7. Look at the stage from eye level. Slowly turn the coarse adjustment to lower the objective until the objective almost touches the slide. Do not allow the objective to touch the slide.
8. Look through the ocular lens. Turn the coarse adjustment to raise the low-power objective until the image is in focus. Always focus by raising the objective away from the slide. Never focus the objective downward. Use the fine adjustment to sharpen the focus. Keep both eyes open while viewing a slide.
9. Make sure that the image is exactly in the center of your field of vision. Then, switch to the high-power objective. Focus the image by using only the fine adjustment. Never use the coarse adjustment at high power.
10. When you are finished using the microscope, remove the slide. Clean the ocular lens and objectives with lens paper. Return the microscope to its storage area. Remember to use both hands when carrying the microscope.

Making a Wet Mount

1. Use lens paper to clean a glass slide and a coverslip.
2. Place the specimen that you wish to observe in the center of the slide.
3. Using a medicine dropper, place one drop of water on the specimen.
4. Hold the coverslip at the edge of the water and at a 45° angle to the slide. Make sure that the water runs along the edge of the coverslip.
5. Lower the coverslip slowly to avoid trapping air bubbles.
6. Water might evaporate from the slide as you work. Add more water to keep the specimen fresh. Place the tip of the medicine dropper next to the edge of the coverslip. Add a drop of water. (You can also use this method to add stain or solutions to a wet mount.) Remove excess water from the slide by using the corner of a paper towel as a blotter. Do not lift the coverslip to add or remove water.

Glossary

A

addiction a dependence on a substance, such as alcohol or another drug (161)

aerobic exercise physical exercise intended to increase the activity of the heart and lungs to promote the body's use of oxygen (167)

alcoholism a disorder in which a person repeatedly drinks alcoholic beverages in an amount that interferes with the person's health and activities (162)

allergy a reaction to a harmless or common substance by the body's immune system (141)

alveoli (al VEE uh LIE) any of the tiny air sacs of the lungs where oxygen and carbon dioxide are exchanged (45)

antibody a protein made by B cells that binds to a specific antigen (137)

artery a blood vessel that carries blood away from the heart to the body's organs (32)

asexual reproduction reproduction that does not involve the union of sex cells and in which one parent produces offspring that are genetically identical to the parent (108)

autoimmune disease a disease in which the immune system attacks the organism's own cells (141)

B

B cell a white blood cell that makes antibodies (137)

blood the fluid that carries gases, nutrients, and wastes through the body and that is made up of platelets, white blood cells, red blood cells, and plasma (36)

blood pressure the force that blood exerts on the walls of the arteries (38)

brain the mass of nerve tissue that is the main control center of the nervous system (84)

bronchus (BRAHNG kuhs) one of the two tubes that connect the lungs with the trachea (45)

C

cancer a tumor in which the cells begin dividing at an uncontrolled rate and become invasive (142)

capillary a tiny blood vessel that allows an exchange between blood and cells in tissue (32)

carbohydrate a class of energy-giving nutrients that includes sugars, starches, and fiber; contains carbon, hydrogen, and oxygen (154)

cardiovascular system a collection of organs that transport blood throughout the body (30)

central nervous system the brain and the spinal cord; its main function is to control the flow of information in the body (80)

cochlea (KAHK lee uh) a coiled tube that is found in the inner ear and that is essential to hearing (92)

D

dermis the layer of skin below the epidermis (17)

digestive system the organs that break down food so that it can be used by the body (58)

drug any substance that causes a change in a person's physical or psychological state (160)

E

egg a sex cell produced by a female (109)

embryo (EM bree OH) a plant or an animal in an early stage of development; a developing human, from fertilization through the first 8 weeks of development (the 10th week of pregnancy) (116)

endocrine system a collection of glands and groups of cells that secrete hormones that regulate growth, development, and homeostasis; includes the pituitary, thyroid, parathyroid, and adrenal glands, the hypothalamus, the pineal body, and the gonads (94)

epidermis (EP uh DUHR mis) the surface layer of cells on a plant or animal (17)

esophagus (i SAHF uh guhs) a long, straight tube that connects the pharynx to the stomach (60)

external fertilization the union of sex cells outside the bodies of the parents (110)

F

fat an energy-storage nutrient that helps the body store some vitamins (155)

feedback mechanism a cycle of events in which information from one step controls or affects a previous step (89)

fetus (FEET uhs) a developing human from seven or eight weeks after fertilization until birth (118)

G

gallbladder a sac-shaped organ that stores bile produced by the liver (63)

gland a group of cells that make special chemicals for the body (94)

H

homeostasis (HOH mee OH STAY sis) the maintenance of a constant internal state in a changing environment (4)

hormone a substance that is made in one cell or tissue and that causes a change in another cell or tissue in a different part of the body (94)

hygiene the science of health and ways to preserve health (166)

I

immune system the cells and tissues that recognize and attack foreign substances in the body (137)

immunity the ability to resist or recover from an infectious disease (134)

infectious disease a disease that is caused by a pathogen and that can be spread from one individual to another (132)

integumentary system (in TEG yoo MEN tuhr ee SIS tuhm) the organ system that forms a protective covering on the outside of the body (16, 88)

internal fertilization fertilization of an egg by sperm that occurs inside the body of a female (110)

J

joint a place where two or more bones meet (10)

K

kidney one of the pair of organs that filter water and wastes from the blood and that excrete products as urine (67)

L

large intestine the wider and shorter portion of the intestine that removes water from mostly digested food and that turns the waste into semisolid feces, or stool (64)

larynx (LAR ingks) the area of the throat that contains the vocal cords and produces vocal sounds (45)

liver the largest organ in the body; it makes bile, stores and filters blood, and stores excess sugars as glycogen (63)

lymph the fluid that is collected by the lymphatic vessels and nodes (40)

lymphatic system (lim FAT ik SIS tuhm) a collection of organs whose primary function is to collect extracellular fluid and return it to the blood; the organs in this system include the lymph nodes and the lymphatic vessels (40)

lymph nodes oval masses of lymphatic tissue found in the lymphatic vessels that filter lymph (41)

M

macrophage (MAK roh FAYJ) an immune system cell that engulfs pathogens and other materials (137)

malnutrition a disorder of nutrition that results when a person does not consume enough of each of the nutrients that are needed by the human body (158)

memory B cell a B cell that responds to an antigen more strongly when the body is reinfected with an antigen than it does during its first encounter with the antigen (140)

mineral a class of nutrients that are chemical elements that are needed for certain body processes (156)

muscular system the organ system whose primary function is movement and flexibility (12)

N

narcotic a drug that is derived from opium and that relieves pain and induces sleep; examples include heroine, morphine, and codeine (163)

nephron the unit in the kidney that filters blood (67)

nerve a collection of nerve fibers through which impulses travel between the central nervous system and other parts of the body (82)

neuron (NOO RAHN) a nerve cell that is specialized to receive and conduct electrical impulses (81)

nicotine (NIK uh TEEN) a toxic, addictive chemical that is found in tobacco and that is one of the major contributors to the harmful effects of smoking (162)

noninfectious disease a disease that cannot spread from one individual to another (132)

nutrient a substance in food that provides energy or helps form body tissues and that is necessary for life and growth (154)

O

organ a collection of tissues that carry out a specialized function of the body (5)

ovary in flowering plants, the lower part of a pistil that produces eggs in ovules (320); in the female reproductive system of animals, an organ that produces eggs (113)

P

pancreas the organ that lies behind the stomach and that makes digestive enzymes and hormones that regulate sugar levels (62)

pathogen a virus, microorganism, or other organism that causes disease (132)

penis the male organ that transfers sperm to a female and that carries urine out of the body (112)

peripheral nervous system (puh RIF uhr uhl NUHR vuhs SIS tuhm) all of the parts of the nervous system except for the brain and the spinal cord (80)

pharynx (FAR ingks) in flatworms, the muscular tube that leads from the mouth to the gastrovascular cavity; in animals with a digestive tract, the passage from the mouth to the larynx and esophagus (45)

placenta (pluh SEN tuh) the structure that attaches a developing fetus to the uterus and that enables the exchange of nutrients, wastes, and gases between the mother and the fetus (116)

protein a molecule that is made up of amino acids and that is needed to build and repair body structures and to regulate processes in the body (155)

pulmonary circulation (PUL muh NER ee SUHR kyoo LAY shuhn) the flow of blood from the heart to the lungs and back to the heart through the pulmonary arteries, capillaries, and veins (33)

R

reflex an involuntary and almost immediate movement in response to a stimulus (89)

respiration in biology, the exchange of oxygen and carbon dioxide between living cells and their environment; includes breathing and cellular respiration (44)

respiratory system a collection of organs whose primary function is to take in oxygen and expel carbon dioxide; the organs of this system include the lungs, the throat, and the passageways that lead to the lungs (44)

retina the light-sensitive inner layer of the eye, which receives images formed by the lens and transmits them through the optic nerve to the brain (90)

S

sexual reproduction reproduction in which the sex cells from two parents unite to produce offspring that share traits from both parents (109)

skeletal system the organ system whose primary function is to support and protect the body and to allow the body to move (8)

small intestine the organ between the stomach and the large intestine where most of the breakdown of food happens and most of the nutrients from food are absorbed (62)

sperm the male sex cell (109)

spleen the largest lymphatic organ in the body; serves as a blood reservoir, disintegrates old red blood cells, and produces lymphocytes and plasmids (42)

stomach the saclike, digestive organ between the esophagus and the small intestine that breaks down food by the action of muscles, enzymes, and acids (61)

stress a physical or mental response to pressure (168)

systemic circulation (sis TEM ik SUHR kyoo LAY shuhn) the flow of blood from the heart to all parts of the body and back to the heart (33)

T

T cell an immune system cell that coordinates the immune system and attacks many infected cells (137)

testes the primary male reproductive organs, which produce sperm cells and testosterone (singular, *testis*) (112)

thymus the main gland of the lymphatic system; it releases mature T lymphocytes (41)

tissue a group of similar cells that perform a common function (4)

tonsils small, rounded masses of lymphatic tissue located in the pharynx and in the passage from the mouth to the pharynx (43)

trachea (TRAY kee uh) in insects, myriapods, and spiders, one of a network of air tubes; in vertebrates, the tube that connects the larynx to the lungs (45)

U

umbilical cord (uhm BIL i kuhl KAWRD) the structure that connects an embryo and then the fetus to the placenta and through which blood vessels pass (117)

urinary system the organs that make, store, and eliminate urine (66)

uterus in female mammals, the hollow, muscular organ in which a fertilized egg is embedded and in which the embryo and fetus develop (113)

V

vagina the female reproductive organ that connects the outside of the body to the uterus (113)

vein in biology, a vessel that carries blood to the heart (32)

vitamin a class of nutrients that contain carbon and that are needed in small amounts to maintain health and allow growth (156)

Glossary

Spanish Glossary

A

addiction/adicción una dependencia de substancia, tal como el alcohol u otra droga (161)

aerobic exercise/ejercicio aeróbico ejercicio físico cuyo objetivo es aumentar la actividad del corazón y los pulmones para hacer que el cuerpo use más oxígeno (167)

alcoholism/alcoholismo un trastorno en el cual una persona consume bebidas alcohólicas repetidamente en una cantidad tal que interfiere con su salud y sus actividades (162)

allergy/alergia una reacción del sistema inmunológico del cuerpo a una substancia inofensiva o común (141)

alveoli/alveolo cualquiera de las diminutas bolsas de aire de los pulmones, en donde ocurre el intercambio de oxígeno y dióxido de carbono (45)

antibody/anticuerpo una proteína producida por las células B que se une a un antígeno específico (137)

artery/arteria un vaso sanguíneo que transporta sangre del corazón a los órganos del cuerpo (32)

asexual reproduction/reproducción asexual reproducción que no involucra la unión de células sexuales, en la que un solo progenitor produce descendencia que es genéticamente igual al progenitor (108)

autoimmune disease/enfermedad autoinmune una enfermedad en la que el sistema inmunológico ataca las células del propio organismo (141)

B

B cell/célula B un glóbulo blanco de la sangre que fabrica anticuerpos (137)

blood/sangre el líquido que lleva gases, nutrientes y desechos por el cuerpo y que está formado por plaquetas, glóbulos blancos, glóbulos rojos y plasma (36)

blood pressure/presión sanguínea la fuerza que la sangre ejerce en las paredes de las arterias (38)

brain/encéfalo la masa de tejido nervioso que es el centro principal de control del sistema nervioso (84)

bronchus/bronquio uno de los dos tubos que conectan los pulmones con la tráquea (45)

C

cancer/cáncer un tumor en el cual las células comienzan a dividirse a una tasa incontrolable y se vuelven invasivas (142)

capillary/capilar diminuto vaso sanguíneo que permite el intercambio entre la sangre y las células de los tejidos (32)

carbohydrate/carbohidrato una clase de nutrientes que proporcionan energía; incluye los azúcares, los almidones y las fibras; contiene carbono, hidrógeno y oxígeno (154)

cardiovascular system/aparato cardiovascular un conjunto de órganos que transportan la sangre a través del cuerpo (30)

central nervous system/sistema nervioso central el cerebro y la médula espinal; su principal función es controlar el flujo de información en el cuerpo (80)

cochlea/cóclea un tubo enrollado que se encuentra en el oído interno y es esencial para poder oír (92)

D

dermis/dermis la capa de piel que está debajo de la epidermis (17)

digestive system/aparato digestivo los órganos que descomponen la comida de modo que el cuerpo la pueda usar (58)

drug/droga cualquier substancia que produce un cambio en el estado físico o psicológico de una persona (160)

E

egg/óvulo una célula sexual producida por una hembra (109)

embryo/embrión una planta o un animal en una de las primeras etapas de su desarrollo; un ser humano desde la fecundación hasta las primeras 8 semanas de desarrollo (décima semana del embarazo) (116)

endocrine system/sistema endocrino un conjunto de glándulas y grupos de células que secretan hormonas que regulan el crecimiento, el desarrollo y la homeostasis; incluye las glándulas pituitaria, tiroides, paratiroides y suprarrenal, el hipotálamo, el cuerpo pineal y las gónadas (94)

epidermis/epidermis la superficie externa de las células de una planta o animal (17)

esophagus/esófago un conducto largo y recto que conecta la faringe con el estómago (60)

external fertilization/fecundación externa la unión de células sexuales fuera del cuerpo de los progenitores (110)

F

fat/grasa un nutriente que almacena energía y ayuda al cuerpo a almacenar algunas vitaminas (155)

feedback mechanism/mecanismo de retroalimentación un ciclo de sucesos en el que la información de una etapa controla o afecta a una etapa anterior (89)

fetus/feto un ser humano en desarrollo de las semanas siete a ocho después de la fecundación hasta el nacimiento (118)

G

gallbladder/vesícula biliar un órgano que tiene la forma de una bolsa y que almacena la bilis producida por el hígado (63)

gland/glándula un grupo de células que elaboran ciertas substancias químicas para el cuerpo (94)

H

homeostasis/homeostasis la capacidad de mantener un estado interno constante en un ambiente en cambio (4)

hormone/hormona una substancia que es producida en una célula o tejido, la cual causa un cambio en otra célula o tejido ubicado en una parte diferente del cuerpo (94)

hygiene/higiene la ciencia de la salud y las formas de preservar la salud (166)

I

immune system/sistema inmunológico las células y tejidos que reconocen y atacan substancias extrañas en el cuerpo (137)

immunity/inmunidad la capacidad de resistir una enfermedad infecciosa o recuperarse de ella (134)

infectious disease/enfermedad infecciosa una enfermedad que es causada por un patógeno y que puede transmitirse de un individuo a otro (132)

integumentary system/sistema integumentario el sistema de órganos que forma una cubierta de protección en la parte exterior del cuerpo (16, 88)

internal fertilization/fecundación interna fecundación de un óvulo por un espermatozoide, la cual ocurre dentro del cuerpo de la hembra (110)

J

joint/articulación un lugar donde se unen dos o más huesos (10)

K

kidney/riñón uno de los dos órganos que filtran el agua y los desechos de la sangre y excretan productos en fomra de orina (67)

L

large intestine/intestino grueso la porción más ancha y más corta del intestino, que elimina el agua de los alimentos casi totalmente digeridos y convierte los desechos en heces semisólidas o excremento (64)

larynx/laringe el área de la garganta que contiene las cuerdas vocales y que produce sonidos vocales (45)

liver/hígado el órgano más grande del cuerpo; produce bilis, almacena y filtra la sangre, y almacena el exceso de azúcares en forma de glucógeno (63)

lymph/linfa el fluido que es recolectado por los vasos y nodos linfáticos (40)

lymphatic system/sistema linfático un conjunto de órganos cuya función principal es recolectar el fluido extracelular y regresarlo a la sangre; los órganos de este sistema incluyen los nodos linfáticos y los vasos linfáticos (40)

lymph nodes/nodos linfáticos masas ovaladas de tejido linfático que se encuentran en los vasos linfáticos y filtran la linfa (41)

M

macrophage/macrófago una célula del sistema inmunológico que envuelve a los patógenos y otros materiales (137)

malnutrition/desnutrición un trastorno de nutrición que resulta cuando una persona no consume una cantidad suficiente de cada nutriente que el cuerpo humano necesita (158)

memory B cell/célula B de memoria una célula B que responde con mayor eficacia a un antígeno cuando el cuerpo vuelve a infectarse con él que cuando lo encuentra por primera vez (140)

mineral/mineral una clase de nutrientes que son elementos químicos necesarios para ciertos procesos del cuerpo (156)

muscular system/sistema muscular el sistema de órganos cuya función principal es permitir el movimiento y la flexibilidad (12)

N

narcotic/narcótico una droga que proviene del opio, la cual alivia el dolor e induce el sueño; entre los ejemplos se encuentran la heroína, morfina y codeína (163)

nephron/nefrona la unidad del riñón que filtra la sangre (67)

nerve/nervio un conjunto de fibras nerviosas a través de las cuales se desplazan los impulsos entre el sistema nervioso central y otras partes del cuerpo (82)

neuron/neurona una célula nerviosa que está especializada en recibir y transmitir impulsos eléctricos (81)

nicotine/nicotina una substancia química tóxica y adictiva que se encuentra en el tabaco y que es una de las principales causas de los efectos dañinos de fumar (162)

noninfectious disease/enfermedad no infecciosa una enfermedad que no se contagia de una persona a otra (132)

nutrient/nutriente una substancia de los alimentos que proporciona energía o ayuda a formar tejidos corporales y que es necesaria para la vida y el crecimiento (154)

O

organ/órgano un conjunto de tejidos que desempeñan una función especializada en el cuerpo (5)

ovary/ovario en las plantas con flores, la parte inferior del pistilo que produce óvulos (320); en el aparato reproductor femenino de los animales, un órgano que produce óvulos (113)

P

pancreas/páncreas el órgano que se encuentra detrás del estómago y que produce las enzimas digestivas y las hormonas que regulan los niveles de azúcar (62)

pathogen/patógeno un virus, microorganismo u otra substancia que causa enfermedades (132)

penis/pene el órgano masculino que transfiere espermatozoides a una hembra y que lleva la orina hacia el exterior del cuerpo (112)

peripheral nervous system/sistema nervioso periférico todas las partes del sistema nervioso, excepto el encéfalo y la médula espinal (80)

pharynx/faringe en los gusanos planos, el tubo muscular que va de la boca a la cavidad gastrovascular; en los animales que tienen tracto digestivo, el conducto que va de la boca a la laringe y al esófago (45)

placenta/placenta la estructura que une al feto en desarrollo con el útero y que permite el intercambio de nutrientes, desechos y gases entre la madre y el feto (116)

protein/proteína una molécula formada por aminoácidos que es necesaria para construir y reparar estructuras corporales y para regular procesos del cuerpo (155)

pulmonary circulation/circulación pulmonar el flujo de sangre del corazón a los pulmones y de vuelta al corazón a través de las arterias, los capilares y las venas pulmonares (33)

R

reflex/reflejo un movimiento involuntario y prácticamente inmediato en respuesta a un estímulo (89)

respiration/respiración en biología, el intercambio de oxígeno y dióxido de carbono entre células vivas y su ambiente; incluye la respiración y la respiración celular (44)

respiratory system/aparato respiratorio un conjunto de órganos cuya función principal es tomar oxígeno y expulsar dióxido de carbono; los órganos de este aparato incluyen a los pulmones, la garganta y las vías que llevan a los pulmones (44)

retina/retina la capa interna del ojo, sensible a la luz, que recibe imágenes formadas por el lente ocular y las transmite al cerebro por medio del nervio óptico (90)

S

sexual reproduction/reproducción sexual reproducción en la que se unen las células sexuales de los dos progenitores para producir descendencia que comparte caracteres de ambos progenitores (109)

skeletal system/sistema esquelético el sistema de órganos cuya función principal es sostener y proteger el cuerpo y permitir que se mueva (8)

small intestine/intestino delgado el órgano que se encuentra entre el estómago y el intestino grueso en el cual se produce la mayor parte de la descomposición de los alimentos y se absorben la mayoría de los nutrientes (62)

sperm/espermatozoide la célula sexual masculina (109)

spleen/bazo el órgano linfático más grande del cuerpo; funciona como depósito para la sangre, desintegra los glóbulos rojos viejos y produce linfocitos y plásmidos (42)

stomach/estómago el órgano digestivo con forma de bolsa ubicado entre el esófago y el intestino delgado, que descompone los alimentos por la acción de músculos, enzimas y ácidos (61)

stress/estrés una respuesta física o mental a la presión (168)

systemic circulation/circulación sistémica el flujo de sangre del corazón a todas las partes del cuerpo y de vuelta al corazón (33)

T

T cell/célula T una célula del sistema inmunológico que coordina el sistema inmunológico y ataca a muchas células infectadas (137)

testes/testículos los principales órganos reproductores masculinos, los cuales producen espermatozoides y testosterona (112)

thymus/timo la glándula principal del sistema linfático; libera linfocitos T maduros (41)

tissue/tejido un grupo de células similares que llevan a cabo una función común (4)

tonsils/amígdalas masas pequeñas y redondas de tejido linfático, ubicadas en la faringe y en el paso de la boca a la faringe (43)

trachea/tráquea en los insectos, miriápodos y arañas, uno de los conductos de una red de conductos de aire; en los vertebrados, el conducto que une la laringe con los pulmones (45)

U

umbilical cord/cordón umbilical la estructura que une al embrión y después al feto con la placenta, a través de la cual pasan vasos sanguíneos (117)

urinary system/sistema urinario los órganos que producen, almacenan y eliminan la orina (66)

uterus/útero en los mamíferos hembras, el órgano hueco y muscular en el que se incrusta el óvulo fecundado y en el que se desarrollan el embrión y el feto (113)

V

vagina/vagina el órgano reproductivo femenino que conecta la parte exterior del cuerpo con el útero (113)

vein/vena en biología, un vaso que lleva sangre al corazón (32)

vitamin/vitamina una clase de nutrientes que contiene carbono y que es necesaria en pequeñas cantidades para mantener la salud y permitir el crecimiento (156)

Index

Boldface page numbers refer to illustrative material, such as figures, tables, margin elements, photographs, and illustrations.

Index

E

F

G

H

I

J

K

L

M

N

O

P

Q

R

S

T

U

V

W

Y

Z

Credits

Abbreviations used: (t) top, (c) center, (b) bottom, (l) left, (r) right, (bkgd) background

PHOTOGRAPHY

Front Cover Mehau Kulyk/Photo Researchers, Inc.

Skills Practice Lab Teens Sam Dudgeon/HRW

Connection to Astrology Corbis Images; **Connection to Biology** David M. Phillips/Visuals Unlimited; **Connection to Chemistry** Digital Image copyright © 2005 PhotoDisc; **Connection to Environment** Digital Image copyright © 2005 PhotoDisc; **Connection to Geology** Letraset Phototone; **Connection to Language Arts** Digital Image copyright © 2005 PhotoDisc; **Connection to Meteorology** Digital Image copyright © 2005 PhotoDisc; **Connection to Oceanography** © ICONOTEC; **Connection to Physics** Digital Image copyright © 2005 PhotoDisc

Table of Contents iv (tl), Sam Dudgeon/HRW; iv–v (t), © Nih / Science Source/Photo Researchers, Inc.; v (bl), © Rob Van Petten/Getty Images/The Image Bank; x (bl), Sam Dudgeon/HRW; xi (tl), John Langford/HRW; xi (b), Sam Dudgeon/HRW; xii (tl), Victoria Smith/HRW; xii (bl), Stephanie Morris/HRW; xii (br), Sam Dudgeon/HRW; xiii (tl), Patti Murray/Animals, Animals; xiii (tr), Jana Birchum/HRW; xiii (b), Peter Van Steen/HRW

Chapter One 2–3 AFP/CORBIS; 4–5 (b–bkgd), © David Madison/Getty Images/Stone; 6 Sam Dudgeon/HRW; 8 Sam Dudgeon/HRW; 10 (c), HRW Photo by Sergio Purtell/FOCA; 10 (r), HRW Photo by Sergio Purtell/FOCA; 10 (l), SP/FOCA/HRW Photo; 11 Scott Camazine/Photo Researchers, Inc.; 12 (bkgd), © Bob Torrez/Getty Images/Stone; 12 (bl–inset), Dr. E.R. Degginger; 12 (r–inset), Manfred Kage/Peter Arnold, Inc. ; 12 (tl–inset), © G.W. Willis/Biological Photo Service; 14 (r), Sam Dudgeon/HRW; 14 (l), Chris Hamilton; 16 Sam Dudgeon/HRW; 18 (bkgd), Peter Van Steen/HRW; 18 (l), Dr. Robert Becker/Custom Medical Stock Photo; 18 (r), Peter Van Steen/HRW; 21 Sam Dudgeon/HRW; 22 Sam Dudgeon/HRW; 23 (t), Sam Dudgeon/HRW; 23 (b), Peter Van Steen/HRW; 26 (l), © Dan McCoy/Rainbow; 26 (r), Reuters/David Gray/NewsCom; 27 (t), Photo courtesy of Dr. Zahra Beheshti; 27 (b), Creatas/PictureQuest

Chapter Two 28–29 © Nih/Science Source/Photo Researchers, Inc.; 32 (l), O. Meckes/Nicole Ottawa/Photo Researchers; 32 (r), O. Meckes/Nicole Ottawa/Photo Researchers; 34 © John Bavosi/Photo Researchers, Inc.; 36 Susumu Nishinaga/Science Photo Library/Photo Researchers, Inc.; 37 (b), Don Fawcett/Photo Researchers; 39 © Getty Images/The Image Bank; 42 © Collection CNRI/Phototake Inc./Alamy Photos; 47 (l), Matt Meadows/Peter Arnold, Inc.; 47 (r), Matt Meadows/Peter Arnold, Inc.; 48 (b), Sam Dudgeon/HRW; 54 (l), Richard T. Nowitz/Phototake; 54 (r), © Paul A. Souders/CORBIS; 55 Courtesy of the Boggy Creek Gang Camp

Chapter Three 56–57 © ISM/Phototake; 65 (t), Victoria Smith/HRW; 68 Getty Images/The Image Bank; 69 Stephen J. Krasemann/DRK Photo; 70 (b), Sam Dudgeon/HRW; 76 (l), J.H. Robinson/Photo Researchers; 76 (r), REUTERS/David Gray/NewsCom; 77 (t), Peter Van Steen/HRW

Chapter Four 78–79 Omikron/Photo Researchers, Inc.; 85 (t), Sam Dudgeon/HRW; 89 Sam Dudgeon/HRW; 96 (c), Sam Dudgeon/HRW; 97 Will & Deni McIntyre/Photo Researchers; 98 (r), Sam Dudgeon/HRW; 98 (l), Sam Dudgeon/HRW; 99 Sam Dudgeon/HRW; 101 (t), Sam Dudgeon/HRW; 104 (l), Victoria Smith/HRW; 104 (r), Mike Derer/AP/Wide World Photos; 105 (t), Photo courtesy of Dr. Bertha Madras; 105 (b), SPL/Photo Researchers, Inc.

Chapter Five 106–107 Lennart Nilsson/Albert Bonniers Forlag AB, A Child Is Born, Dell Publishing Company; 108 (r), Visuals Unlimited/Cabisco; 108 (l), Innerspace Visions; 110 (b), Photo Researchers; 110 (t), Digital Image copyright © 2005 PhotoDisc Green; 111 © Charles Phillip/CORBIS; 114 Chip Henderson; 119 (tl), Petit Format/Nestle/Science Source/Photo Researchers, Inc.; 119 (cl), Photo Lennart Nilsson/Albert Bonniers Forlag AB, A Child Is Born, Dell Publishing Company; 119 (cr), Photo Lennart Nilsson/Albert Bonniers Forlag AB, A Child Is Born, Dell Publishing Company; 119 (br), Keith/Custom Medical Stock Photo; 119 (tr), David M. Phillips/Photo Researchers, Inc.; 120 (l), Peter Van Steen/HRW; 120 (cl), Peter Van Steen/HRW; 120 (c), Peter Van Steen/HRW; 120 (cr), Peter Van Steen/HRW; 120 (r), Peter Van Steen/HRW; 121 © Mark Harmel/Getty Images/FPG International; 123 Digital Image copyright © 2005 PhotoDisc; 124 Peter Van Steen/HRW; 125 Photo Lennart Nilsson/Albert Bonniers Forlag AB, A Child Is Born, Dell Publishing Company; 128 (r), Jim Tunell/Zuma Press/NewsCom; 128 (l), © Michael Clancy; 129 (l), ZEPHYR/Science Photo Library/Photo Researchers, Inc.; 129 (r), Salem Community College

Chapter Six 130–131 (t), © K. Kjeldsen/Photo Researchers, Inc.; 132 (br), CNRI/Science Photo Library/Photo Researchers; 132 (bl), Tektoff–RM/CNRI/Science Photo Library/Photo Researchers; 133 (t), Kent Wood/Photo Researchers; 134 (b), Peter Van Steen/HRW ; 136 (b), Peter Van Steen/HRW; 140 (t), John Langford/HRW Photo; 141 (b), Clinical Radiology Dept., Salisbury District Hospital/Science Photo Library/Photo Researchers; 141 (t), SuperStock, 142 (b), Photo Lennart Nilsson/Albert Bonniers Forlag AB; 142 (tl), Dr. A. Liepins/Science Photo Library/Photo Researchers; 142 (tr), Dr. A. Liepins/Science Photo Library/Photo Researchers; 144 Sam Dudgeon/HRW; 147 (t), Peter Van Steen/HRW ; 150 (l), E. R. Degginger/Bruce Coleman; 150 (r), Chris Rogers/Index Stock Imagery, Inc.; 151 (t), Peter Van Steen/HRW; 151 (b), Corbis

Chapter Seven 152–153 © Arthur Tilley/Getty Images/Taxi; 154 Peter Van Steen/HRW; 155 (b), Peter Van Steen/HRW; 155 (t), Sam Dudgeon/HRW; 156 (c), Image Copyright ©2004 PhotoDisc, Inc./HRW; 156 (t), Image Copyright ©2004 PhotoDisc, Inc./HRW; 156 (bl), CORBIS Images/HRW; 156 (br), CORBIS Images/HRW; 157 © John Kelly/Getty Images/Stone; 158 John Burwell/FoodPix; 159 Peter Van Steen/HRW; 160 Peter Van Steen/HRW; 161 (b), Peter Van Steen/HRW; 161 (t), ©1999 Steven Foster; 162 (tl), E. Dirksen/Photo Researchers; 162 (b), Spencer Grant/Photo Researchers, Inc.; 162 (tr), Dr. Andrew P. Evans/Indiana University; 164 Jeff Greenberg/PhotoEdit; 165 Mike Siluk/The Image Works; 166 Sam Dudgeon/HRW; 167 (t), © Rob Van Petten/Getty Images/The Image Bank; 167 (b), Peter Van Steen/HRW; 168 Sam Dudgeon/HRW; 170 (b), Peter Van Steen/HRW; 170 (t), © Mug Shots/CORBIS; 171 Peter Van Steen/HRW; 172 Digital Image copyright © 2005 PhotoDisc; 174 (t), © John Kelly/Getty Images/Stone; 174 (b), Peter Van Steen/HRW; 175 (t), Peter Van Steen/HRW; 175 (b), Peter Van Steen/HRW; 178 (l), Brian Hagiwara/FoodPix; 179 (r), Courtesy Russell Selger; 179 (l), © Eyebyte/Alamy Photos

Lab Book/Appendix "LabBook Header", "L", Corbis Images; "a", Letraset Phototone; "b", and "B", HRW; "o", and "k", images ©2006 PhotoDisc/HRW; 183 Peter Van Steen/HRW; 184 (t), Sam Dudgeon/HRW; 184 (b), Sam Dudgeon/HRW; 194 Sam Dudgeon/HRW; 195 Sam Dudgeon/HRW; 200 (t), Peter Van Steen/HRW; 200 (b), Sam Dudgeon/HRW; 211 CENCO

TEACHER EDITION CREDITS

1E (r), Sam Dudgeon/HRW Photo; 1F (tr), Dr. Robert Becker/Custom Medical Stock Photo; 27E (r), SUSUMU NISHINAGA/SCIENCE PHOTO LIBRARY/Photo Researchers, Inc.; 27E (l), John Bavosi /Photo Researchers, Inc.; 105E (tl), Innerspace Visions; 105E (bl), Charles Phillip/CORBIS; 105F (r), Photo Lennart Nilsson/Albert Bonniers Forlag AB, A Child Is Born, Dell Publishing Company; 129E (l), Tektoff-RM/CNRI/Science Photo Library/Photo Researchers ; 129E (r), Kent Wood/Photo Researchers, Inc.; 129F (l), Peter Van Steen/HRW Photo; 129F (r), SuperStock; 151E (l), Peter Van Steen/HRW Photo; 151E (r), John Kelly/Stone /Getty Images; 151F (r), Spencer Grant/Photo Researchers, Inc.; 151F (l), Peter Van Steen/HRW Photo.

Answers to Concept Mapping Questions

The following pages contain sample answers to all of the concept mapping questions that appear in the Chapter Reviews. Because there is more than one way to do a concept map, your students' answers may vary.

CHAPTER 1 Body Organization and Structure

17.

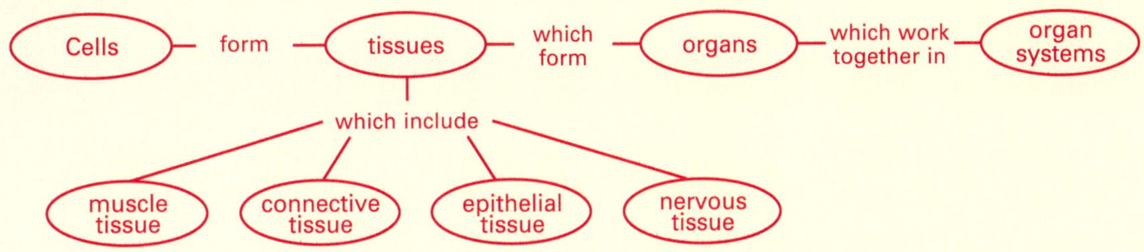

CHAPTER 2 Circulation and Respiration

18.

CHAPTER 3 The Digestive and Urinary Systems

19.

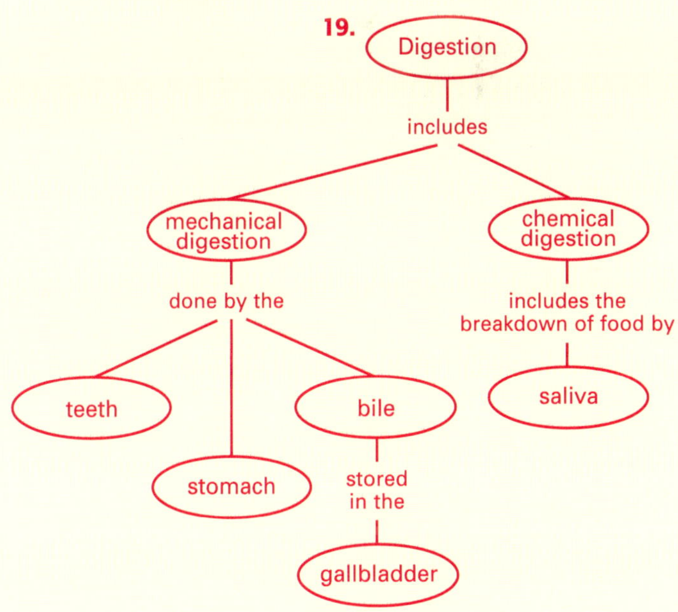

CHAPTER 4 Communication and Control

18.

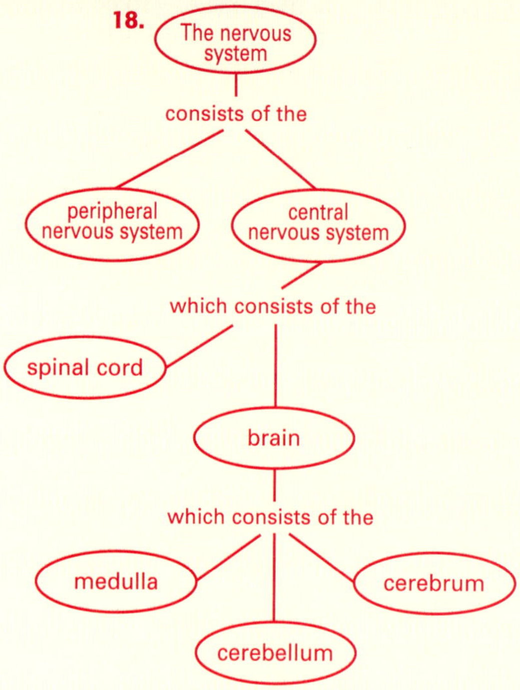

CHAPTER 5 Reproduction and Development

18.

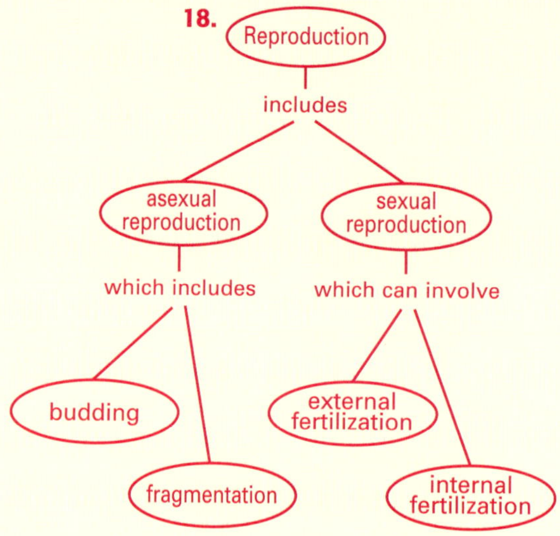

CHAPTER 6 Body Defenses and Disease

16.

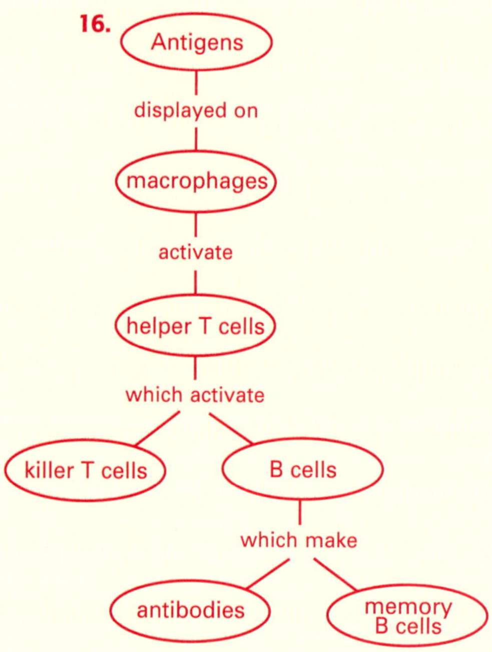

CHAPTER 7 Staying Healthy

17.

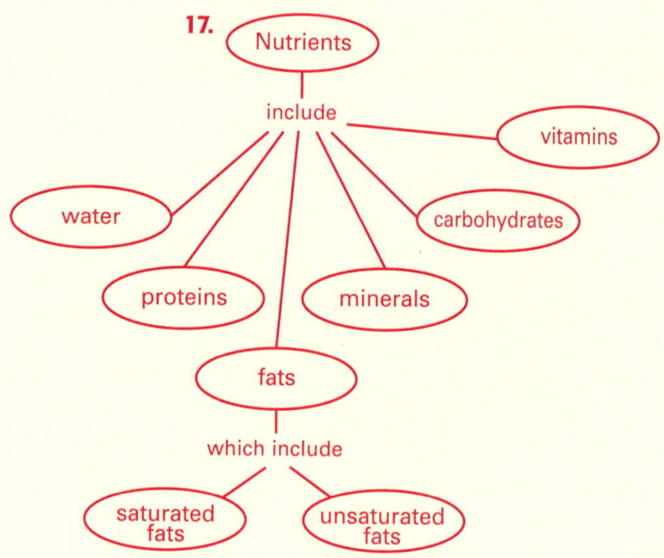